T0419374

STOCHASTIC PROCESSES IN NON-ARCHIMEDEAN BANACH SPACES, MANIFOLDS AND TOPOLOGICAL GROUPS

STOCHASTIC PROCESSES IN NON-ARCHIMEDEAN BANACH SPACES, MANIFOLDS AND TOPOLOGICAL GROUPS

SERGEY V. LUDKOVSKY

Nova Science Publishers, Inc.
New York

Copyright © 2010 by Nova Science Publishers, Inc.

All rights reserved. No part of this book may be reproduced, stored in a retrieval system or transmitted in any form or by any means: electronic, electrostatic, magnetic, tape, mechanical photocopying, recording or otherwise without the written permission of the Publisher.

For permission to use material from this book please contact us:
Telephone 631-231-7269; Fax 631-231-8175
Web Site: http://www.novapublishers.com

NOTICE TO THE READER

The Publisher has taken reasonable care in the preparation of this book, but makes no expressed or implied warranty of any kind and assumes no responsibility for any errors or omissions. No liability is assumed for incidental or consequential damages in connection with or arising out of information contained in this book. The Publisher shall not be liable for any special, consequential, or exemplary damages resulting, in whole or in part, from the readers' use of, or reliance upon, this material. Any parts of this book based on government reports are so indicated and copyright is claimed for those parts to the extent applicable to compilations of such works.

Independent verification should be sought for any data, advice or recommendations contained in this book. In addition, no responsibility is assumed by the publisher for any injury and/or damage to persons or property arising from any methods, products, instructions, ideas or otherwise contained in this publication.

This publication is designed to provide accurate and authoritative information with regard to the subject matter covered herein. It is sold with the clear understanding that the Publisher is not engaged in rendering legal or any other professional services. If legal or any other expert assistance is required, the services of a competent person should be sought. FROM A DECLARATION OF PARTICIPANTS JOINTLY ADOPTED BY A COMMITTEE OF THE AMERICAN BAR ASSOCIATION AND A COMMITTEE OF PUBLISHERS.

Additional color graphics may be available in the e-book version of this book.

LIBRARY OF CONGRESS CATALOGING-IN-PUBLICATION DATA

L|dkovsky, Sergey V.
Stochastic processes in non-Archimedean Banach spaces, manifolds, and topological groups / [edited by] S.V. Ludkovsky.
p. cm.
Includes index.
ISBN 978-1-61668-787-8 (hardcover)
1. Stochastic processes. 2. Group theory. 3. Banach spaces. 4. Random walks (Mathematics) I. Title.
QA274.L886 2009
519.2'3--dc22

2010025686

Published by Nova Science Publishers, Inc. † New York

Contents

Preface

This book presents the new technique for description of stochastic functions in non-archimedean spaces. Stochastic processes and functions with values in topological vector spaces over a non-archimedean normed field $\mathbf{K}$ can serve for descriptions of various processes.

The utilization of topological vector spaces over the non-archimedean normed fields for stochastic functions is natural not only in non-archimedean quantum mechanics, but also for various systems with finite or countable set of states or more generally with a totally disconnected set of states. Since a graph can be embedded into the field $\mathbf{K}$ using decompositions of numbers in $\mathbf{K}$ into series, this opens possibilities for studying processes in graphs with the help of this new technique. For example, biological or economic processes usually have some finite or countable set of states of the system such that they fit this condition. The metric in the graph inherited from $\mathbf{K}^n$ is non-archimedean.

If the initial system is considered in the discrete topology, then the inherited topology is not stronger than the initial one. The non-archimedean metric or uniformity is frequently essential for studying stochastic flows of an information, that is in the information theory. In cryptology generators of random functions or numbers are frequently constructed in the non-archimedean normed spaces.

It is well-known that fields with non-archimedean norm such as the field of p-adic numbers were first introduced by K. Hensel in the 19-th century [36]. Then it was proved by A. Ostrowski [90] that on the field of rational numbers each multiplicative norm is either the usual norm as in the real field $\mathbf{R}$ or is equivalent to a non-archimedean norm $|x| = p^{-k}$, where $x = np^k/m \in \mathbf{Q}$, $n, m, k \in \mathbf{Z}$ are non-zero integers, $p \geq 2$ is a prime number, n and m and p are mutually pairwise prime numbers. It is well known, that each locally compact infinite field with a non trivial non-archimedean norm is either a finite algebraic extension of the field $\mathbf{Q_p}$ of p-adic numbers or is isomorphic to the field $\mathbf{F}_{p^k}(\theta)$ of power series of the variable θ with expansion coefficients in the finite field $\mathbf{F}_{p^k}$ of p^k elements, where $p \geq 2$ is a prime number, $k \in \mathbf{N}$ is a natural number [99, 111]. Non locally compact fields are also wide spread [17, 99, 103].

Key words and phrases: infinitely divisible distribution, non-archimedean field, zero characteristic, positive characteristic, random process, linear space, manifold, topological group, Markov process, spectral function, stochastic differential equation, transition probability, Brownian motion, stochastic integral equation.

Mathematics Subject Classification 2000: 60G07, 60G50, 60G51, 30G06, 60H05, 60H20, 60J65.

This book is devoted to new results of investigations of non-archimedean stochastic processes and their transition measures, which is becoming more important nowadays due to the development of non-archimedean mathematical physics, particularly, quantum mechanics, quantum field theory, theory of super-strings and supergravity [4,13,18,40,41,61, 109,110]. On the other hand, quantum mechanics is based on measure theory and probability theory. For comparison references are given below also on works, where real-valued measures on non-archimedean spaces were studied. Stochastic approach in quantum field theory is actively used and investigated especially in recent years [2,41–43].

The terminology a random function $f(t,\omega)$ is used here in more general context, than a stochastic process, because the latter undermines that the variable t is the time parameter.

Any Markov stochastic process with values in the normed space $\mathbf{K}^n$ with $\mathbf{K}$-valued or real-valued transition probabilities can be presented as the limit of Markov processes with finite number of states. Indeed, since any $\mathbf{K}$-valued tight measure on the algebra $Bco(\mathbf{K}^n)$ of clopen (closed and open simultaneously) subsets in $\mathbf{K}^n$ can be presented as the projective limit of $\mathbf{K}$-valued measures on discrete spaces (see also [84,99]). For real-valued measures analogous projective limit decompositions are described by Kolmogoroff's and Prohorov's theorems and due to the total disconnectedness of the normed space $\mathbf{K}^n$. This means that the $\mathbf{K}^n$ valued Markov stochastic process with either $\mathbf{K}$-valued or real-valued transition probabilities can be approximated with some accuracy by Markov processes with finite or countable number of states.

The rational field $\mathbf{Q}$ has the embedding into $\mathbf{K}$, when the field $\mathbf{K}$ is of zero characteristic, $char(\mathbf{K}) = 0$. Therefore, jump or stochastically discontinuous stochastic functions in a topological vector space over $\mathbf{R}$ can be approximated and analyzed with the help of random mappings in non-archimedean spaces.

Random functions with values in non-archimedean spaces and Lie groups modelled on them and their transition probabilities are important for representation theory of groups and stochastic differential equations in them as well.

In this book many recent results in this area are presented. Below representations of stochastic processes with values in finite- and infinite-dimensional vector spaces over infinite fields with non-trivial non-archimedean norms are investigated. Different types of stochastic processes controlled by measures with values in non-archimedean fields of zero characteristic and stochastic integrals are described. Theorems about spectral decompositions of non-archimedean stochastic processes are proved. These studies demonstrate many specific features of the non-archimedean case in comparison with the real or complex classical stochastic analysis. These features arise from many differences between mathematical analysis and functional analysis over $\mathbf{R}$ and $\mathbf{C}$ and that of non-archimedean. General constructions of the book are illustrated in examples.

We remind that locally compact non-archimedean fields have non-archimedean multiplicative norms and their characteristics may be either zero such as for the field of p-adic numbers $\mathbf{Q_p}$ or for its finite algebraic extension, or of a positive characteristic $char(\mathbf{K}) = p > 0$ such as the field $\mathbf{F_p}(\theta)$ of Laurent series over a finite field $\mathbf{F_p}$ with p elements and an indeterminate θ, where $p > 1$ is a prime number, $|\theta| = 1/p$ [111]. Multiplicative norms in such fields $\mathbf{K}$ satisfy stronger inequality, than the triangle inequality, $|x+y| \le \max(|x|,|y|)$ for each $x,y \in \mathbf{K}$. Non-archimedean fields are totally disconnected and balls in them are either non-intersecting or one of them is contained in another.

Very great importance have also branching processes in graphs [1,31,35]. For finite or infinite graphs with finite degrees of vertices there is possible to consider their embeddings into p-adic graphs, which can be embedded into locally compact fields. That is, a consideration of such processes reduces to processes with values in either the field $\mathbf{Q_p}$ of p-adic numbers or $\mathbf{F_p}(\theta)$.

Stochastic differential equations on Banach spaces and manifolds are widely used for solutions of mathematical and physical problems and for construction and investigation of measures on them [16,21,85]. In particular stochastic equations can be used for the constructions of quasi-invariant measures on topological groups. On the other hand, non-archimedean functional analysis fast develops in recent years together with its applications in mathematical physics [3,40,42,99,103,110]. Quasi-invariant transition measures real- or non-archimedean valued induced by random functions on topological groups and their configuration spaces can be used for the investigations of their unitary representations and representations over non-archimedean Banach spaces (see [26,58,60–62] and references therein).

There are different variants for such activity depending on ranges and domains of random functions over different fields. At the same time transition measures may be real, complex or with values in a non-archimedean field.

There is also an interesting interpretation of stochastic processes with values in $\mathbf{Q_p^n}$, for which a time parameter may be either real or p-adic. A random trajectory in $\mathbf{Q_p}^n$ may be continuous relative to the non-archimedean norm in $\mathbf{Q_p}^n$, but its trajectory in $\mathbf{Q}^n$ relative to the usual real metric may be discontinuous. This gives new approach to spasmodic or jump or discontinuous stochastic processes with values in $\mathbf{Q}^n$, when the latter is considered as embedded into $\mathbf{R}^n$. On the other hand, stochastic processes with values in the field $\mathbf{F_p}(\theta)^n$ of the positive characteristic p can naturally take into account cyclic stochastic processes in definite problems.

Stochastic analysis is useful in non-archimedean string dynamics. This is caused at least by three reasons. Integrals and series appearing there which diverge over the real field are converging in the non-archimedean field $\mathbf{Q_p}$. Solutions of string equations have meaning only at distances greater than some threshold value which is compatible with the non-archimedean norm in the p-adic field $\mathbf{Q_p}$. Moreover, the uncertainty Heisenberg principle appears naturally over the p-adic field [12,110].

The approach on manifolds in this book is not restricted by the rigid geometry class [29], since the latter is rather narrow. Below wider classes of functions and manifolds are considered. This is possible with the use of Schikhof's works on classes of functions C^n in the sense of difference quotients, which he had been investigating few years later the published formalism of the rigid geometry.

The first chapter is devoted to random functions and stochastic processes with values in finite-dimensional vector spaces over infinite locally compact fields of zero and positive characteristics with non-trivial non-archimedean norms. Infinitely divisible distributions are investigated. Theorems about their characteristic functionals are proved. Particular cases are demonstrated as applications to non-archimedean analogs of Gaussian and Poisson processes and their generalizations.

The second chapter contains material about random functions and stochastic processes with values in finite- and infinite-dimensional vector spaces over infinite fields $\mathbf{K}$ of zero characteristics with non-trivial non-archimedean norms. For different types of stochastic processes controlled by measures with values in **K** and in complete topological vector spaces over **K** stochastic integrals are investigated. Vector valued measures and integrals in spaces over **K** are studied. Theorems about spectral decompositions of non-archimedean random functions are proved. Applications to totally disconnected topological groups are discussed as well.

Non-Archimedean analogs of Markov quasi-measures and stochastic processes in Banach spaces are described in Chapter 3. They are used for the development of stochastic anti-derivations. The non-archimedean analog of the Itô formula is proved.

Random functions having values in topological vector spaces over non-archimedean fields and with transition measures with values in the real and non-archimedean fields are defined and investigated. For this the non-archimedean analog of the Kolmogorov theorem is proved. The Markov and Poisson processes are studied. For Poisson processes the corresponding Poisson measures are considered and the non-archimedean analog of the Lèvy theorem is proved. Wide classes of stochastic processes are constructed.

Stochastic anti-derivational equations in Banach spaces over local non-archimedean fields are investigated. Theorems about existence and uniqueness of the solutions are proved under definite conditions. In particular Brownian motion processes are considered in relation with the non-archimedean analogs of the Gaussian measures.

Stochastic processes and random functions with ranges in manifolds over non-archimedean fields and with transition measures having values in the field **C** of complex numbers are defined and investigated in Chapter 4. The analogs of Markov, Poisson and Brownian motion processes are studied. For Poisson processes the non-archimedean analog of the Lèvy theorem is proved. Stochastic anti-derivational equations as well as pseudo-differential equations on manifolds are investigated.

In the fifth chapter random functions having ranges in totally disconnected topological groups are described. In particular they are considered for diffeomorphism groups and wrap groups of manifolds modelled on non-archimedean Banach spaces. Theorems about a quasi-invariance and a pseudo-differentiability of transition measures are proved. Transition measures are used for the construction of strongly continuous unitary representations of these groups. In addition stochastic processes in general Banach-Lie groups, wrap monoids, manifolds on Banach spaces over non-archimedean local fields also are investigated. The term a local field means a commutative non-discrete locally compact field. It may be of zero or a positive characteristic.

In Appendix A basic facts about differentiability of functions are written. In Appendix B in more details approximate, global and along curves smoothness of functions $f(x_1,\dots,x_m)$ of variables $x_1,\dots,x_m$ in infinite fields with non trivial non-archimedean norms and relations between them are considered. Theorems about classes of smoothness C^n or C^n_b of functions with continuous or bounded uniformly continuous on bounded domains partial difference quotients up to the order n are investigated. Then classes of smoothness $C^{n,r}$ and $C^{n,r}_b$ and more general in the sense of Lipschitz for partial difference quotients are considered and theorems for them are proved. Moreover, an approximate differentiability of functions relative to measures is defined and investigated. Its relations with lipschitzian property and almost everywhere differentiability are studied. Finally theorems about relations between approximate differentiability by all variables and along curves are proved. This is useful for random functions studies.

In Appendix C Taylor's expansion theorem is recalled. Then Appendices D, E and F contain necessary facts about specific anti-derivation operators, wrap groups and fiber bundles over local fields.

Chapter 1

Infinite Divisible Distributions

1.1. Introduction

We begin with the basic properties of stochastic processes. One of them is infinite divisibility. This notion is so important, because such wide spread processes as Gaussian and Poisson are infinite divisible. Certainly, there exist random vectors having distributions which are not characterized by this property, but the specification of the class of infinitely divisible distributions outlines very important class of them. Their analysis provides also examples of useful stochastic processes.

The role of infinitely divisible distributions is well-known in the theory of stochastic processes over the fields of real and complex numbers [16, 27, 31, 35, 94, 106]. A lot of in this area was done by Khinchin and Levy. The main advantage of their results and works of their followers consists in the fact, that they have taken into account the linear and bilinear functionals on linear spaces. The latter means terms of the first and the second order. Moreover, they have obtained characteristic functionals of infinitely divisible distributions and homogeneous stochastic processes with independent increments.

Later on generalizations on locally compact Abelian groups were also considered [35, 92, 93]. But in the latter cases the obtained results were too general in comparison with those on linear spaces. Indeed, classes of measures corresponding to Gaussian processes on groups appeared to be much wider, than on linear spaces. This means that they do not take into account the field structure, because they operate with the additive group structure only (see [93] and Definition 6.1 and Theorem 6.1 [92]). Moreover, the theory on totally disconnected groups takes into account terms of the first order only and their terms of the second order copied from the real case vanish (see [93] and Section 3 particularly Example 3.4 [92]).

But the terms of the second order are crucial for analysis of the Gaussian and Poisson processes.

Below in this chapter to overcome difficulties met in previous works of others authors later results of fast developing non-archimedean analysis were used [99, 103, 110].

It is worth to mention that limit distributions on non-archimedean local fields (certainly of zero characteristic) were studied in [50, 112]. In these articles results about representations of functionals of infinitely divisible distributions on locally compact Abelian groups from [92,93] were used.

This chapter contains results from the paper [78] as well. It is devoted to infinitely divisible distributions of stochastic processes in vector spaces over locally compact fields **K** with non-trivial non-archimedean norms. Below the approach taking into account the field structure and terms of the first and the second order is presented (see Theorems 5, 7, 8, 10, 12 and 15). This permits to get non-archimedean analogs of the Gaussian and Poisson processes, that is demonstrated below (see Sections 16 and 17).

Earlier in works [8,21]– [24,46,49] stochastic processes on spaces of functions with domains of definition in a non-archimedean linear space and with ranges in the field of real **R** or complex numbers **C** were considered. This means that different variants of non-archimedean stochastic processes are possible depending on a domain of definition, a range of values of functions, values of measures in either the real field or a non-archimedean field [75,84]. Moreover, a time parameter may be real or non-archimedean and so on. That is, such approaches are rather flexible and depending on considered problems different non-archimedean variants arise.

Summarizing we can state that stochastic processes with values in non-archimedean spaces appear while their studies for non-archimedean Banach spaces, totally disconnected topological groups and manifolds [66]– [74].

In this chapter theorems about representations of characteristic functionals of infinitely divisible distributions with values in vector spaces over locally compact infinite fields with zero and positive characteristics with non-trivial non-archimedean norms are formulated and proved. For accomplishing this characteristic functionals are obtained in the specific form. In addition, special non-archimedean classes of mappings are introduced. They are not linear or bilinear, but of the peculiar non-archimedean form. This is caused by the fact that there is not any non-constant linear mapping from the field of real numbers into the field $\mathbf{Q_p}$ or $\mathbf{F_p}(\theta)$ or vice versa. The cases of fields with zero and positive characteristics are considered and not only the terms of the first, but also terms of the second order are taken into account below.

Specific features of the non-archimedean case are elucidated. Therefore, a part of definitions, formulations of theorems and their proofs are changed in comparison with the classical case over the real and complex fields. Some necessary facts from probability theory or non-archimedean analysis are reminded (see, for example, §§2.1–3), that to make reading easier. The main results of the fist chapter are, for example, Theorems 5, 7, 8 and 10.

1.2. Infinitely Divisible Distributions

To avoid misunderstandings we first present our notations and definitions and recall the basic facts.

1. Notations and definitions. Let **K** be a locally compact infinite field with a non-trivial non-archimedean norm, $n \in \mathbf{N}$, let also $\mathbf{Q_p}$ be the field of p-adic numbers, where $1 < p$ is a prime number. Here **K** is either a finite algebraic extension of the field $\mathbf{Q_p}$ or the field $\mathbf{Q_p}$ itself for $char(\mathbf{K}) = 0$, or $\mathbf{K} = \mathbf{F_p}(\theta)$ for $char(\mathbf{K}) = p > 1$, $\mathcal{B}(\mathbf{K^n})$ is the σ-algebra of all Borel subsets in $\mathbf{K^n}$.

We denote by $(\Omega, \mathcal{A}, P)$ a probability space, where Ω is a space of elementary events, $\mathcal{A}$ is a σ-algebra of events in Ω, $P : \mathcal{A} \to [0,1]$ is a probability.

Denote by ξ a random vector (a random variable for $n = 1$) with values in $\mathbf{K^n}$ such that it has the probability distribution $P_\xi(A) = P(\{\omega \in \Omega : \xi(\omega) \in A\})$ for each $A \in \mathcal{B}(\mathbf{K^n})$, where $\xi : \Omega \to \mathbf{K^n}$. Certainly, we suppose that ξ is $(\mathcal{A}, \mathcal{B}(\mathbf{K^n}))$-measurable. This means that $\xi^{-1}(\mathcal{B}(\mathbf{K^n})) \subset \mathcal{A}$.

Random vectors ξ and η with values in $\mathbf{K^n}$ are called independent, if $P(\{\xi \in A, \eta \in B\}) = P(\{\xi \in A\})P(\{\eta \in B\})$ for each $A, B \in \mathcal{B}(\mathbf{K^n})$.

A random vector (a random variable) ξ is called infinitely divisible, if

(1) for each $m \in \mathbf{N}$ there exist random vectors (random variables) $\xi_1, \dots, \xi_m$ such that $\xi = \xi_1 + \dots + \xi_m$ and the probability distributions of $\xi_1, \dots, \xi_m$ are the same.

Let H be a set. A family of subsets $\mathcal{B}(H)$ is called a covering ring if its elements form the covering of H, that is $\bigcup_{B \in \mathcal{B}(H)} = H$, so that the union $A \cup B := \{x : \ x \in A \text{ or } x \in B\} \in \mathcal{B}(H)$, the intersection $A \cap B := \{x : \ x \in A \text{ and } x \in B\} \in \mathcal{B}(H)$ and the difference $A \setminus B := \{x : \ x \in A \text{ and } x \notin B\} = A - B \in \mathcal{B}(H)$ belong to $\mathcal{B}(H)$ for all $A, B \in \mathcal{B}(H)$.

If $H \in \mathcal{B}(H)$, then the covering ring is called an algebra (of subsets in H). The algebra $\mathcal{B}(H)$ is called a σ-algebra if for each sequence $\{A_n \in \mathcal{B}(H) : \ n \in \mathbf{N}\}$ in $\mathcal{B}(H)$ its union $\bigcup_{n=1}^{\infty} A_n \in \mathcal{B}(H)$ belongs to it.

A function $f : T \times \Omega \to H$, where H and T are some sets, $\mathcal{B}(H)$ is a covering ring of subsets in H, is called a random function, if $f(t, \omega)$ is $(\mathcal{A}, \mathcal{B}(H))$-measurable as the function by $\omega \in \Omega$ for each $t \in T$. This means by the definition that $f(t,*)^{-1}(B) \in \mathcal{A}(\Omega)$ for any $B \in \mathcal{B}(H)$, where $f(t,*)^{-1}$ denotes the inverse function by the second argument from H into Ω for a marked $t \in T$.

The terminology a random function $f(t, \omega)$ is used in more general context, than a stochastic process, because the latter undermines the particular case when the variable t is the time parameter. We shall consider as H either a linear space or a manifold over the field $\mathbf{K}$ or a topological group in this and subsequent chapters. For shortening the notation the random function $f(t, \omega)$ is frequently written as $f(t)$.

A stochastic process $\xi = \xi(t) = \xi(t, \omega)$ with the real time, $t \in T$, $T \subset \mathbf{R}$, is called infinitely divisible, if Condition (1) is satisfied for each $t \in T$, where $\xi_j(t)$ are stochastic processes, $j = 1, \dots, m$.

We put $B(X, x, R) := \{y \in X : \rho(x, y) \le R\}$ for the ball in a metric space (X, ρ) with a metric ρ, $0 < R < \infty$.

2. Lemma. *Suppose that ξ and η are two independent random vectors with values in $\mathbf{K^n}$ having probability distributions P_ξ and P_η. Then $\xi + \eta$ has the probability distribution $P_{\xi+\eta}(A) = \int_{\mathbf{K^n}} P_\xi(A - dy) P_\eta(dy)$ for each $A \in \mathcal{B}(\mathbf{K^n})$.*

Proof. Since ξ and η are independent, the probability satisfies the identity $P(\{\omega \in \Omega : \xi(\omega) \in C, \eta(\omega) \in B\}) = P(\{\omega \in \Omega : \xi(\omega) \in C\}) P(\{\omega \in \Omega : \eta(\omega) \in B\})$ for each $C, B \in \mathcal{B}(\mathbf{K^n})$.

Therefore, $P(\{\xi + \eta \in A\}) = P(\{\xi \in A - y, \eta = y, y \in \mathbf{K^n}\})$ for each $A \in \mathcal{B}(\mathbf{K^n})$, consequently, $P_{\xi+\eta}(A) = \int_{\mathbf{K^n}} P_\xi(A - dy) P_\eta(dy)$.

Thus the measure of the sum of these random vectors $P_{\xi+\eta} = P_\xi * P_\eta$ is the convolution of measures P_ξ and P_η.

3. Corollary. *Let ξ be an infinitely divisible random vector. Then $P_\xi = P_{\xi_1}^{*m}$ for each*

$m \in \mathbf{N}$, where P_η^{*m} denotes the m-fold convolution P_η with itself.

Proof. In accordance with Lemma 2 and Definition 1 we have $P_\xi = P_{\xi_1} * P_{\xi_2+\cdots+\xi_m} = \cdots = P_{\xi_1} * P_{\xi_2} * \cdots * P_{\xi_m}$.

On the other hand, the random vectors $\xi_1,\ldots,\xi_m$ have the same probability distributions, hence $P_{\xi_1} * P_{\xi_2} * \cdots * P_{\xi_m} = P_{\xi_1}^{*m}$.

4. Notes and definitions. Certainly, Corollary 3 means, that the equality $P_\xi = P_{\xi_1}^{*m}$ implies the relation: $P_\xi(A) = \int_{\mathbf{K^n}} \cdots \int_{\mathbf{K^n}} P_{\xi_1}(A - dy_2)P_{\xi_2}(dy_2 - dy_3)\cdots P_{\xi_{m-1}}(dy_{m-1} - dy_m)P_{\xi_m}(dy_m)$, where $A \in \mathcal{B}(\mathbf{K^n})$. In the case of $char(\mathbf{K}) = p > 1$ Corollary 3 means, that for $m = kp$, where $k \in \mathbf{N}$, if $P_\xi(\{0\}) = 0$, then $P_{\xi_1}(\{y\}) = 0$ for each singleton $y \in \mathbf{K^n}$. Indeed, the inequalities $P(\xi = 0) \ge P(\xi_1 = \xi_2 = \cdots = \xi_m) \ge P_{\xi_1}(\{y\})^m$ are satisfied. This is the restriction on the atomic property of P_ξ and P_{ξ_1}.

For p-adic numbers x the series decomposition $x = \sum_{k=N}^{\infty} x_k p^k$ exists, where $x_k \in \{0,1,\ldots,p-1\}$, $N \in \mathbf{Z}$, $N = N(x)$, $x_N \neq 0$, $x_j = 0$ for each $j < N$. We put as usually $ord_{\mathbf{Q_p}}(x) = N$ for the order of x so that its norm is $|x|_{\mathbf{Q_p}} = p^{-N}$. We define the function $[x]_{\mathbf{Q_p}} := \sum_{k=N}^{-1} x_k p^k$ for $N < 0$, $[x]_{\mathbf{Q_p}} = 0$ for $N \ge 0$ on $\mathbf{Q_p}$. Therefore, the function $[x]_{\mathbf{Q_p}}$ on $\mathbf{Q_p}$ is considered with values in the segment $[0,1] \subset \mathbf{R}$.

As the norm in the field $\mathbf{F_p}(\theta)$ we put $|x|_{\mathbf{F_p}(\theta)} = p^{-N}$, where $N = ord_{\mathbf{F_p}(\theta)}(x) \in \mathbf{Z}$, $x = \sum_{j=N}^{\infty} x_j \theta^j$, $x_j \in \mathbf{F_p}$ for each j, $x_N \neq 0$, $x_j = 0$ for each $j < N$. Then we define the mapping $[x]_{\mathbf{F_p}(\theta)} = x_{-1}/p$, where we consider elements of the finite field $\mathbf{F_p} = \{0,1,\ldots,p-1\}$ as being embedded into $\mathbf{R}$, hence $[x]_{\mathbf{F_p}(\theta)}$ takes values in $\mathbf{R}$, where $1/p \in \mathbf{R}$, $x_{-1} = 0$ when $N = N(x) \ge 0$.

It is useful to consider a local field of zero characteristic $\mathbf{K}$ as the vector space over the field $\mathbf{Q_p}$, then it is isomorphic with $\mathbf{Q_p^b}$ for some $b \in \mathbf{N}$, since $\mathbf{K}$ is a finite algebraic extension of the field $\mathbf{Q_p}$. For convenience in the case of $\mathbf{K} = \mathbf{F_p}(\theta)$ we take $b = 1$.

We use traditional notations for the probability P and for non-archimedean fields, where also the letter p is used as the lower index, that can be lightly distinguished in different situations. To unify our notation we also put

(i) $\mathbf{F} := \mathbf{Q_p}$ for $char(\mathbf{K}) = 0$ with $\mathbf{K} \supset \mathbf{Q_p}$, while

(ii) $\mathbf{F} := \mathbf{F_p}(\theta)$ for $char(\mathbf{K}) = p > 1$ with $\mathbf{K} = \mathbf{F_p}(\theta)$. Let

$$(x,y) := (x,y)_{\mathbf{F}} := \sum_{j=1}^{b} x_j y_j$$

for $x,y \in \mathbf{F}$, $x = (x_1,\ldots,x_b)$, $x_j \in \mathbf{F}$; while

$$(x,y)_{\mathbf{K}} := \sum_{j=1}^{n} x_j y_j$$

for $x,y \in \mathbf{K^n}$, $x = (x_1,\ldots,x_n)$, $x_j \in \mathbf{K}$.

Then we define the mapping

$$< q >_{\mathbf{F}} := 2\pi[(e,q)]_{\mathbf{F}}$$

for each $q \in \mathbf{K}$, which is considered in (e,q) as the element from $\mathbf{F^b}$, $< q >_{\mathbf{F}} : \mathbf{K} \to \mathbf{R}$, where $e := (1,\ldots,1) \in \mathbf{F^b}$, particularly $e = 1$ for $b = 1$. That is, either in (i) $\mathbf{K} = \mathbf{Q_p}$ or in the case

(ii) for $\mathbf{K} = \mathbf{F_p}(\theta)$. For the additive group $\mathbf{K^n}$ there exists the character

$$\chi_s(z) := \exp(i < (s,z)_{\mathbf{K}} >_{\mathbf{F}})$$

with values in the field of complex numbers $\mathbf{C}$ for each value of the parameter $s \in \mathbf{K^n}$, since

$$s_j(z_j + v_j) = s_j z_j + s_j v_j \quad \text{for each} \quad s_j, z_j, v_j \in \mathbf{K} \quad \text{and}$$

$$(s, z+v)_{\mathbf{K}} = (s,z)_{\mathbf{K}} + (s,v)_{\mathbf{K}}, \quad [x+y]_{\mathbf{F}} - [x]_{\mathbf{F}} - [y]_{\mathbf{F}} \in B(\mathbf{F}, 0, 1)$$

for every $x, y \in \mathbf{F}$, while

$$[x]_{\mathbf{F}} = 0 \quad \text{for each} \quad x \in B(\mathbf{F}, 0, 1),$$

where $i = (-1)^{1/2} \in \mathbf{C}$. In particular, $\chi_0(z) = 1$ for each $z \in \mathbf{K^n}$ for $s = 0$. The character is non-trivial for $s \neq 0$. At the same time

$$\chi_s(z) = \prod_{j=1}^{n} \chi_{s_j}(z_j),$$

where $\chi_{s_j}(z_j)$ are characters of $\mathbf{K}$ as the additive group.

For a σ-additive measure $\mu : \mathcal{B}(\mathbf{K^n}) \to \mathbf{C}$ of a bounded variation the characteristic functional $\hat{\mu}$ is given by the formula:

$$\hat{\mu}(s) := \int_{\mathbf{K^n}} \chi_s(z) \mu(dz),$$

where $s \in \mathbf{K^n}$ is the corresponding continuous $\mathbf{K}$-linear functional on $\mathbf{K^n}$ denoted by the same s.

In general the characteristic functional of the measure μ is defined in the space $C^0(\mathbf{K^n}, \mathbf{K})$ of continuous functions $f : \mathbf{K^n} \to \mathbf{K}$

$$\hat{\mu}(f) := \int_{\mathbf{K^n}} \chi_1(f(z)) \mu(dz), \quad \text{where} \quad 1 \in \mathbf{K}.$$

Let μ be a σ-additive finite non-negative measure on $\mathcal{B}(\mathbf{K^n})$, $\mu(\mathbf{K^n}) < \infty$. We introduce the class $C_1 = C_1(\mathbf{K})$ of continuous functions $A = A_\mu^{\mathbf{K}} : \mathbf{K^n} \to \mathbf{R}$, satisfying Conditions $(F1 - F3)$:

$$(F1) \quad A(y+z) = A(y) + A(z) + 2\pi \int_{\mathbf{K^n}} f_1(y, z; x) \mu(dx)$$

for each $y, z \in \mathbf{K^n}$,

$$(F2) \quad A(\beta y) = [\beta]_{\mathbf{F}} A(y) + 2\pi \int_{\mathbf{K^n}} f_2(\beta, (e, (y,x)_{\mathbf{K}})_{\mathbf{F}}) \mu(dx)$$

for each $y \in \mathbf{K^n}$, $\beta \in \mathbf{F}$, where either

$(F3)$ if $\mathbf{F} = \mathbf{Q_p}$ for $char(\mathbf{K}) = 0$, then $f_1 : (\mathbf{K^n})^3 \to \mathbf{Z}$ and $f_2 : \mathbf{Q_p}^2 \to \mathbf{R}$ are locally constant continuous bounded functions, $f_1(y,z;x) \in \mathbf{Z}$ and $f_2(\alpha, \beta) p^{-N(\alpha,\beta)} \in \mathbf{Z}$ for $N(\alpha, \beta) < 0$ take only integer values, $N(\alpha, \beta) := \min(ord_{\mathbf{Q_p}}(\alpha), ord_{\mathbf{Q_p}}(\beta))$; or

$(F4)$ if $\mathbf{F}=\mathbf{F_p}(\theta)$ for $char(\mathbf{K})=p>0$, then $f_1:(\mathbf{K^n})^3\to\mathbf{R}$ and $f_2:\mathbf{F}^2\to\mathbf{R}$ are locally constant continuous bounded functions, $pf_1(y,z;x)\in\mathbf{Z}$ and $p^2f_2(\alpha,\beta)\in\mathbf{Z}$ for $N(\alpha,\beta)<0$ take only integer values so that

$$N(\alpha,\beta):=\min(ord_{\mathbf{F_p}(\theta)}(\alpha),ord_{\mathbf{F_p}(\theta)}(\beta)).$$

While

$$f_1(y,z;x)=0 \text{ for } \max(|yx|_{\mathbf{K}},|zx|_{\mathbf{K}})\le 1,$$

and

$$f_2(\alpha,\beta)=0 \text{ for } \max(|\alpha|_{\mathbf{F}},|\beta|_{\mathbf{F}})\le 1 \text{ in } (F3,F4).$$

Next we denote by $C_2=C_2(\mathbf{K})$ the class of continuous functions $B=B^{\mathbf{K}}_{\mu}:(\mathbf{K^n})^2\to\mathbf{R}$, satisfying Conditions $(B1-B3)$:

$(B1)$ $B(y,z)=B(z,y)$ for each $y,z\in\mathbf{K^n}$, where $B(y,y)$ is non-negative,

$$(B2)\quad B(q+y,z)=B(q,z)+B(y,z)+2\pi\int_{\mathbf{K^n}}f_1(q,y;x)<(z,x)_{\mathbf{K}}>_{\mathbf{F}}\mu(dx)$$

for each $q,y,z\in\mathbf{K^n}$,

$$(B3)\quad B(\beta y,z)=[\beta]_{\mathbf{F}}B(y,z)+2\pi\int_{\mathbf{K^n}}f_2(\beta,(e,(y,x)_{\mathbf{K}})_{\mathbf{F}})<(z,x)_{\mathbf{K}}>_{\mathbf{F}}\mu(dx),$$

where f_1 and f_2 satisfy Condition either $(F3)$ or $(F4)$ depending on the characteristic $char(\mathbf{K})$.

For $y=z$ we shall also write for short $B(y):=B(y,y)$.

These conditions show that functions of classes $C_1(\mathbf{K})$ and $C_2(\mathbf{K})$ depend on the field $\mathbf{K}$ and its characteristic, but when a field is outlined we shall frequently shorten the notation to C_1 or C_2 correspondingly.

4.1. Lemma. *Let $\chi_s(x):\mathbf{F^n}\to\mathbf{C}$ be a character of the additive group of $\mathbf{F^n}$ as in Section 4, let also $\mu:\mathcal{B}(\mathbf{F^n})\to[0,\infty]$ be the Haar measure such that $\mu(B(\mathbf{F^n},0,1))=1$. Then*

$$\int_{B(\mathbf{F^n},0,p^k)}\chi_s(x)\mu(dx)=J(s,k),$$

where $J(s,k)=p^{kn}$ for $|s|\le p^{-k}$, while $J(s,k)=0$ for $|s|\ge p^{1-k}$.

Proof. The Haar measure μ on the σ-algebra $\mathcal{B}(\mathbf{F^n})$ is the product of the Haar measures μ_1 on the σ-algebras $\mathcal{B}(\mathbf{F})$ of Borel subsets in $\mathbf{F}$. That is $\mu(dx)=\otimes_{j=1}^n\mu_j(dx_j)$, $\mu_j=\mu_1$. Therefore,

$$\int_{B(\mathbf{F^n},0,p^k)}\chi_s(x)\mu(dx)=\prod_{j=1}^n\chi_{s_j}(x_j)\mu_j(dx_j),$$

where $\chi_j=\chi_1$, $\chi_{s_j}(x_j)$ is the character of $\mathbf{F}$.

We consider the case $n=1$. Then

$$K:=\int_{B(\mathbf{F},0,p^k)}\chi_s(x)\mu(dx)=\int_{B(\mathbf{F},y,p^k)}\chi_s(x-y)\mu(dx)$$

for each $y \in B(\mathbf{F},0,p^k)$. Thus

$$K = \chi_s(-y)\int_{B(\mathbf{F},0,p^k)} \chi_s(x)\mu(dx),$$

since the balls $B(\mathbf{F},0,p^k) = B(\mathbf{F},y,p^k)$ coincide for each $y \in B(\mathbf{F},0,p^k)$ due to the ultrametric inequality, while $\mu(A-y) = \mu(A)$ for each $A \in \mathcal{B}(\mathbf{F})$. Then we take $|s|_{\mathbf{F}} \geq p^{-k+1}$ and $|y|_{\mathbf{F}} = p^k$ such that $[sy]_{\mathbf{F}} \neq 0$ is nonzero. Hence $K(1-\chi_s(-y)) = 0$, but $\chi_s(-y) \neq 1$, consequently, $K = 0$.

On the other hand, if $|sx|_{\mathbf{F}} \leq 1$, then $\chi_s(x) = 1$ and inevitably

$$\int_{B(\mathbf{F},0,p^k)} \chi_s(x)\mu(dx) = p^k,$$

when $|s|_{\mathbf{F}} \leq p^{-k}$ (see for comparison the case of the p-adic field $\mathbf{F} = \mathbf{Q_p}$ in Example 6 on page 62 [110]).

5. Theorem. *Suppose that* $\{\psi(v,y) : v \in V\}$ *is a family of characteristic functionals of* σ*-additive non-negative bounded measures on the Borel* σ*-algebra* $\mathcal{B}(\mathbf{K^n})$*, where* V *is a monotonically decreasing sequence of positive numbers converging to zero. Let a limit*

$$g(y) = \lim_{v\downarrow 0}(\psi(v,y)-1)/v$$

exist uniformly on each ball $B(\mathbf{K^n},0,R)$ *for each given* $0 < R < \infty$*. Then a* σ*-additive non-negative bounded measure* ν *exists on* $\{\mathbf{K^n},\mathcal{B}(\mathbf{K^n})\}$*. Moreover, functions* $A(y)$ *and* $B(y)$ *belonging to classes* C_1 *and* C_2 *respectively exist such that*

$$(i)\ \ g(y) = iA(y) - B(y)/2 + \int_{\mathbf{K^n}} (\exp(i<(y,x)_{\mathbf{K}}>_{\mathbf{F}}) - 1 - i<(y,x)_{\mathbf{K}}>_{\mathbf{F}}(1+|x|^2)^{-1}$$
$$+ <(y,x)_{\mathbf{K}}>^2_{\mathbf{F}}(1+|x|^2)^{-1}/2)[(1+|x|^2)/|x|^2]\nu(dx)$$

and $\nu \geq 0$, $\nu(\{0\}) = 0$.

Proof. Let μ_v be a measure corresponding to the characteristic functional $\psi(v,y)$. We define

$$\lambda_v(A) := v^{-1}\int_A |z|^2/[1+|z|^2]\mu_v(dz)$$

for each $A \in \mathcal{B}(\mathbf{K^n})$, where $|z| := \max_{1\leq j\leq n}|z_j|_{\mathbf{K}}$, $z = (z_1,\dots,z_n) \in \mathbf{K^n}$, $z_j \in \mathbf{K}$ for every $j = 1,\dots,n$. Next, we prove that the family of measures $\{\lambda_v : v \in V\}$ is weakly compact. That is, we need to prove that

(i) there exists $L = const > 0$ such that $\sup_{v\in V}\lambda_v(\mathbf{K^n}) \leq L$; also

(ii) $\lim_{R\to\infty}\overline{\lim}_{v\downarrow 0}\lambda_v(\mathbf{K^n}\setminus B(\mathbf{K^n},0,R)) = 0$.

The topologically dual space $\mathbf{K^n}'$ of all continuous $\mathbf{K}$-linear functionals on $\mathbf{K^n}$ is $\mathbf{K}$-linearly and topologically isomorphic with $\mathbf{K^n}$, since $n \in \mathbf{N}$. The field $\mathbf{K}$ is locally compact, consequently, it is spherically complete (see Theorems 3.15, 5.36 and 5.39 [99]). Since $\mathbf{K^n}$ as the linear space over $\mathbf{F}$ is isomorphic with $\mathbf{F^{bn}}$, it is sufficient to verify a weak compactness over the field $\mathbf{F}$, where either $\mathbf{F} = \mathbf{Q_p}$ for $char(\mathbf{K}) = 0$ with $\mathbf{K} \supset \mathbf{Q_p}$ and $b \in \mathbf{N}$, or $\mathbf{F} = \mathbf{F_p}(\theta)$ for $char(\mathbf{K}) = p > 0$ with $\mathbf{K} = \mathbf{F_p}(\theta)$ and $b = 1$. Indeed, apply the non-archimedean variant of the Minlos-Sazonov theorem. Due to this theorem the bijective correspondence between characteristic functionals and measures [66] exists, where characteristic functionals are weakly continuous (see also §IV.1.2 and Theorem IV.2.2 about the Minlos-Sazonov theorem on Hausdorff completely regular (Tychonoff) spaces [108]). Certainly, the characteristic functional is positive definite on $(\mathbf{K^n})'$ or $C^0(\mathbf{K^n},\mathbf{K})$, when μ is non-negative; moreover, $\hat{\mu}(0)=1$ for $\mu(\mathbf{K^n})=1$. In the considered case $\mathbf{K^n}$ is a finite dimensional Banach space over $\mathbf{K}$. Since the multiplication in $\mathbf{K}$ is continuous, this gives the continuous multiplication mapping $f_0:(\mathbf{F^b})^2\to\mathbf{F^b}$ over $\mathbf{F}$. The composition of f_0 with all possible $\mathbf{K}$-linear continuous functionals $s:\mathbf{K^n}\to\mathbf{K}$ separates points in $\mathbf{K^n}$.

Let $|x|\le R_1$, where $0<R_1<\infty$ is an arbitrarily given number. Under the imposed conditions of this theorem for each $\delta>0$ there exists $v_0=v_0(R_1,\delta)>0$ such that for each $\varepsilon>0$ the inequality

$$-Reg(y)+\delta\ge\int_{B(\mathbf{F^{bn}},0,\varepsilon)}[1-\cos<(y,x)_{\mathbf{F}}>_{\mathbf{F}}]|x|^{-2}\lambda_v(dx) \qquad (1)$$

is satisfied for each $0<v\le v_0$, since

$$e^{i\alpha}=\cos(\alpha)+i\sin(\alpha),$$

$$-Re(e^{i\alpha}-1)=1-\cos(\alpha)\quad\text{for each}\quad\alpha\in\mathbf{R},\quad\text{while}$$

$$1+|x|^2\ge1\quad\text{and}\quad[1+|x|^2]|x|^{-2}\ge|x|^{-2}.$$

If $\varepsilon>1$ and $x\in\mathbf{F^{bn}}\setminus B(\mathbf{F^{bn}},0,\varepsilon)$, then from $|x|_{\mathbf{F}}>\varepsilon$ one gets $[1+|x|^2]|x|^{-2}=1+|x|^{-2}\ge1$ and then for each $\delta>0$ there exists $v_0>0$ such that for each $\varepsilon>1$ and each $0<v\le v_0$ the inequality:

$$-Reg(y)+\delta\ge\int_{\mathbf{F^{bn}}\setminus B(\mathbf{F^{bn}},0,\varepsilon)}(1-\cos<(y,x)>_{\mathbf{F}})\lambda_v(dx) \qquad (2)$$

is satisfied.

Then we integrate these inequalities by the variable $y\in B(\mathbf{F^{bn}},0,r)$ and divide on the volume (measure) $\mu(B(\mathbf{F^{bn}},0,r))$, where μ is the nonnegative Haar measure on $\mathbf{F^{bn}}$ such that $\mu(B(\mathbf{F^{bn}},0,1))=1$, $\mu(B(\mathbf{F^{bn}},0,r))=r^{bn}$ for each $r=p^k$ with $k\in\mathbf{Z}$ [10, 111]. Then from (1) we infer:

$$-r^{-bn}\int_{B(\mathbf{F^{bn}},0,r)}Reg(y)\mu(dy)+\delta$$

$$\ge\int_{B(\mathbf{F^{bn}},0,r)}\left(\int_{B(\mathbf{F^{bn}},0,\varepsilon)}|x|_{\mathbf{F}}^{-2}(1-\cos<(y,x)>_{\mathbf{F}})\lambda_v(dx)\right)\mu(dy)r^{-bn}. \qquad (3)$$

From (2) we get:

$$-r^{-bn}\int_{B(\mathbf{F^{bn}},0,r)}Reg(y)\mu(dy)+\delta$$

$$\ge\int_{B(\mathbf{F^{bn}},0,r)}\left(\int_{\mathbf{F^{bn}}\setminus B(\mathbf{F^{bn}},0,\varepsilon)}(1-\cos<(y,x)>_{\mathbf{F}})\lambda_v(dx)\right)\mu(dy)r^{-bn}. \qquad (4)$$

Evidently the equality

$$\cos(<(y,x)>_{\mathbf{F}})=\cos\left(\sum_{j=1}^{bn}<x_jy_j>_{\mathbf{F}}\right)\text{ is satisfied, since}$$

$$(y,x) = \sum_{j=1}^{bn} y_j x_j,$$

also $< a+b >_{\mathbf{F}} = < a >_{\mathbf{F}} + < b >_{\mathbf{F}} + 2w\pi$ for each $a, b \in \mathbf{F}$, where w is an integer number, $w = w(a,b) \in \mathbf{Z}$. Integrals of the characters are known due to Lemma 4.1:

$$\int_{B(\mathbf{F^{bn}},0,p^k)} \chi_s(x)\mu(dx) = \prod_{j=1}^{bn} \int_{B(\mathbf{F},0,p^k)} \chi_{s_j}(x_j)\mu_j(dx_j) = J(s,k),$$

where $J(s,k) = p^{kbn}$ for $|s|_{\mathbf{F}} \leq p^{-k}$, $J(s,k) = 0$ for $|s|_{\mathbf{F}} \geq p^{-k+1}$. Since $(y,x) = (x,y)$ and $\cos(\alpha) = Re(e^{i\alpha})$ for each $\alpha \in \mathbf{R}$, we deduce that

$$\int_{B(\mathbf{F^{bn}},0,p^k)} \cos < (y,x) >_{\mathbf{F}} \mu(dy) = J(x,k),$$

since $J(x,k) \in \mathbf{R}$. Take in $(3,4)$ $r = p^k$, then

$$-p^{-kbn} \int_{B(\mathbf{F^{bn}},0,p^k)} Reg(y)\mu(dy) + \delta$$

$$\geq \left(\int_{B(\mathbf{F^{bn}},0,\varepsilon)} |x|_{\mathbf{F}}^{-2}(1 - J(x,k)p^{-kbn})\lambda_v(dx) \right) \tag{5}$$

$$-p^{-kbn} \int_{B(\mathbf{F^{bn}},0,p^k)} Reg(y)\mu(dy) + \delta$$

$$\geq \int_{\mathbf{F^{bn}} \setminus B(\mathbf{F^{bn}},0,\varepsilon)} (1 - J(x,k)p^{-kbn})\lambda_v(dx)). \tag{6}$$

Since $J(x,k)p^{-kbn} = 1$ for $|x|_{\mathbf{F}} \leq p^{-k}$, while $J(x,k)p^{-kbn} = 0$ for $|x|_{\mathbf{F}} \geq p^{-k+1}$, for $\varepsilon > p^{-k+1}$ with $k \in \mathbf{Z}$, where $p \geq 2$, we get $(1 - J(x,k)p^{-kbn}) = 1$ for $p^{-k+1} \leq |x|_{\mathbf{F}} \leq \varepsilon$, then

$$-p^{-kbn} \int_{B(\mathbf{F^{bn}},0,p^k)} Reg(y)\mu(dy) + \delta \geq \left(\int_{B(\mathbf{F^{bn}},0,\varepsilon) \setminus B(\mathbf{F^{bn}},0,p^{-k})} |x|_{\mathbf{F}}^{-2}\lambda_v(dx) \right)$$

$$\geq \varepsilon^{-2}[\lambda_v(B(\mathbf{F^{bn}},0,\varepsilon)) - \lambda_v(B(\mathbf{F^{bn}},0,p^{-k})],$$

hence

$$[\lambda_v(B(\mathbf{F^{bn}},0,\varepsilon)) - \lambda_v(B(\mathbf{F^{bn}},0,p^{-k})] \leq \varepsilon^2[\delta - p^{-kbn} \int_{B(\mathbf{F^{bn}},0,p^k)} Reg(y)\mu(dy)]. \tag{7}$$

In particular, for $\varepsilon_k = p^{-k+2}$ with $\varepsilon_k \leq \varepsilon$ and $k \to \infty$ Inequality (7) is satisfied. Then the summation of both parts of Inequality (7) by such k gives:

$$\lambda_v(B(\mathbf{F^{bn}},0,\varepsilon)) \leq L_1\delta - \sum_{k=k_0}^{\infty} p^{-kbn-2k+4} \int_{B(\mathbf{F^{bn}},0,p^k)} Reg(y)\mu(dy),$$

where

$$L_1 = p^4 \sum_{k=k_0}^{\infty} p^{-2k} = p^{4-2k_0}/(1-p^{-2}), \tag{8}$$

$k_0 \in \mathbf{Z}$ is fixed. At the same time from (6) it follows that

$$-p^{-kbn}\int_{B(\mathbf{F^{bn}},0,p^k)} Reg(y)\mu(dy) + \delta \geq \lambda_v(\mathbf{F^{bn}} \setminus B(\mathbf{F^{bn}},0,\varepsilon)) \quad (9)$$

for $\varepsilon > p^{-k+1}$. Therefore, due to Inequalities (8,9) there exists $L = const > 0$ such that

$$\lambda_v(\mathbf{F^{bn}}) = \lambda_v(B(\mathbf{F^{bn}},0,\varepsilon)) + \lambda_v(\mathbf{F^{bn}} \setminus B(\mathbf{F^{bn}},0,\varepsilon)) \leq L,$$

for each $v \in (0, v_0]$, where $L = const > 0$.

Due to conditions of this theorem the function $g(y)$ is continuous and $g(0) = 0$, consequently, for each $\delta > 0$ a sufficiently small number $0 < R_1 = p^{k_1} < \infty$ exists such that $R_1^{-bn}|\int_{B(\mathbf{F^{bn}},0,R_1)} Reg(y)\mu(dy)| < \delta$. In view of Inequality (9) for each $\varepsilon > \max(p^{-k_1+1}, 1)$ the inequality

$$\lambda_v(\mathbf{F^{bn}} \setminus B(\mathbf{F^{bn}},0,\varepsilon)) < 2\delta$$

is satisfied for each $v \in (0, v_0]$. Thus, the family of measures $\{\lambda_v : v \in V\}$ is weakly compact.

Choose a sequence $h_n \downarrow 0$ such that λ_{v_n} is weakly convergent to some measure ν on $\mathcal{B}(\mathbf{K^n})$. Due to conditions of this theorem and using the decomposition of exp into the series, we get the following inequality:

$$[\psi(v,y) - 1]/v = \int_{\mathbf{K^n}} (\chi_y(x) - 1)[1 + |x|_{\mathbf{K}}^2]|x|_{\mathbf{K}}^{-2}\lambda_v(dx)$$

$$= iA_v(y) - B_v(y)/2 + \int_{\mathbf{K^n}} f(y,x)\lambda_v(dx),$$

where

$$A_v(y) = \int_{\mathbf{K^n}} < (y,x)_{\mathbf{K}} >_{\mathbf{F}} |x|_{\mathbf{K}}^{-2}\lambda_v(dx),$$
$$B_v(y) = \int_{\mathbf{K^n}} < (y,x)_{\mathbf{K}} >_{\mathbf{F}}^2 |x|_{\mathbf{K}}^{-2}\lambda_v(dx). \quad (10)$$

The function f is given by the expression

$$f(y,x) = (\exp(i < (y,x)_{\mathbf{K}} >_{\mathbf{F}})$$

$$-1 - i < (y,x)_{\mathbf{K}} >_{\mathbf{F}} [1 + |x|_{\mathbf{K}}^2]^{-1} + < (y,x)_{\mathbf{K}} >_{\mathbf{F}}^2 [1 + |x|_{\mathbf{K}}^2]^{-1}/2)[1 + |x|_{\mathbf{K}}^2]|x|_{\mathbf{K}}^{-2}. \quad (10')$$

The multiplier $[1 + |x|_{\mathbf{K}}^2]|x|_{\mathbf{K}}^{-2}$ is continuous and bounded for $|x| \geq R$, where $0 < R < \infty$, $< (y,x)_{\mathbf{K}} >_{\mathbf{F}} = 0$ for $|y|_{\mathbf{K}}|x|_{\mathbf{K}} \leq 1$, hence the function $f(y,x)$ is continuous. The latter function is bounded, when y varies in a bounded subset in $\mathbf{K^n}$, while $x \in \mathbf{K^n}$. Therefore, the limit exists

$$\lim_{k\to\infty} \int_{\mathbf{K^n}} f(y,x)\lambda_{v_k}(dx) = \int_{\mathbf{K^n}} f(y,x)\nu(dx).$$

The functions $< (y,x)_{\mathbf{K}} >_{\mathbf{F}} |x|_{\mathbf{K}}^{-2}$ and $< (y,x)_{\mathbf{K}} >_{\mathbf{F}} < (z,x)_{\mathbf{K}} >_{\mathbf{F}} |x|_{\mathbf{K}}^{-2}$ are locally constant by x for each given value of the parameters y and z. These functions are zero, when $|y|_{\mathbf{K}}|x|_{\mathbf{K}} \leq 1$, that is, they are defined in the continuous manner to be zero at the zero point $x = 0$. There exists the limit in the left hand side of Inequality (10), consequently, the limits

$$\lim_{k\to\infty} A_{v_k}(y) = A(y) = \int_{\mathbf{K^n}} < (y,x)_{\mathbf{K}} >_{\mathbf{F}} |x|_{\mathbf{K}}^{-2}\nu(dx) \quad \text{and}$$

$$\lim_{k\to\infty} B_{\nu_k}(y) = B(y) = \int_{\mathbf{K^n}} < (y,x)_{\mathbf{K}} >_{\mathbf{F}}^2 |x|_{\mathbf{K}}^{-2} \nu(dx)$$

exist. At the same time $B(y) \geq 0$ for each $y \in \mathbf{K^n}$.

For convenience we substitute the measure $\nu(U)$ on $\nu(U \setminus \{0\})$ and denote it by the same symbol, where $U \in \mathcal{B}(\mathbf{K^n})$. Due to the fact that $f(y,0) = 0$, $< (y,0)_{\mathbf{K}} >_{\mathbf{F}} = 0$, one gets that for such substitution of the measure the values of integrals $\int_{\mathbf{K^n}} f(y,x)\nu(dx)$,

$$A(y) = \int_{\mathbf{K^n}} < (y,x)_{\mathbf{K}} >_{\mathbf{F}} |x|_{\mathbf{K}}^{-2} \nu(dx) \quad \text{and}$$

$$B(y,z) = \int_{\mathbf{K^n}} < (y,x)_{\mathbf{K}} >_{\mathbf{F}} < (y,z)_{\mathbf{K}} >_{\mathbf{F}} |x|_{\mathbf{K}}^{-2} \nu(dx)$$

remain unchanged.

It is known that

$$[\alpha+\beta]_{\mathbf{F}} = [\alpha]_{\mathbf{F}} + [\beta]_{\mathbf{F}} + v(\alpha,\beta),$$

where $v(\alpha,\beta) \in \mathbf{Z}$ for $\mathbf{F} = \mathbf{Q_p}$, $pv(\alpha,\beta) \in \mathbf{Z}$ for $\mathbf{F} = \mathbf{F_p}(\theta)$, $0 \leq [\alpha]_{\mathbf{F}} \leq 1$ for each $\alpha, \beta \in \mathbf{F}$. Another useful identity is the following:

$$[\alpha\beta]_{\mathbf{F}} = [\alpha]_{\mathbf{F}}[\beta]_{\mathbf{F}} + u(\alpha,\beta),$$

where $p^{-N(\alpha,\beta)}u(\alpha,\beta) \in \mathbf{Z}$ for $\mathbf{F} = \mathbf{Q_p}$, $p^2u(\alpha,\beta) \in \mathbf{Z}$ for $\mathbf{F} = \mathbf{F_p}(\theta)$, since

$$[\alpha]_{\mathbf{Q_p}}[\beta]_{\mathbf{Q_p}} = \sum_{k=N(\alpha)}^{-1} \sum_{l=N(\beta)}^{-1} \alpha_k \beta_l p^{k+l} \quad \text{and}$$

$$[\alpha\beta]_{\mathbf{Q_p}} = \sum_{N(\alpha)\leq k, N(\beta)\leq l, k+l\leq -1} \alpha_k \beta_l p^{k+l},$$

where

$$\alpha = \sum_{k=N(\alpha)}^{\infty} \alpha_k p^k \in \mathbf{Q_p},$$

$\alpha_k \in \{0,1,\dots,p-1\}$ for each $k \in \mathbf{Z}$, $\alpha_{N(\alpha)} \neq 0$. At the same time we have the identity:

$$[\alpha]_{\mathbf{F_p}(\theta)}[\beta]_{\mathbf{F_p}(\theta)} = \alpha_{-1}\beta_{-1}p^{-2} \quad \text{and}$$

$$[\alpha\beta]_{\mathbf{F_p}(\theta)} = \sum_{N(\alpha)\leq k, N(\beta)\leq l, k+l=-1} \alpha_k \beta_l p^{-1},$$

where

$$\alpha = \sum_{k=N(\alpha)}^{\infty} \alpha_k \theta^k \in \mathbf{F_p}(\theta),$$

$\alpha_k \in \mathbf{F_p}$ for each $k \in \mathbf{Z}$, $\alpha_{N(\alpha)} \neq 0$ [110, 111]. At the same time $[\alpha]_{\mathbf{F}} = 0$, when $|\alpha|_{\mathbf{F}} \leq 1$, hence $v(\alpha,\beta) = 0$ and $u(\alpha,\beta) = 0$ for $\max(|\alpha|_{\mathbf{F}}, |\beta|_{\mathbf{F}}) \leq 1$. Then we infer that

$$< (y+z,x)_{\mathbf{K}} >_{\mathbf{F}} = < (y,x)_{\mathbf{K}} >_{\mathbf{F}} + < (z,x)_{\mathbf{K}} >_{\mathbf{F}} + 2\pi f_1(y,z;x), \tag{11}$$

where $f_1 \in \mathbf{Z}$ for $\mathbf{F} = \mathbf{Q_p}$, $pf_1 \in \mathbf{Z}$ for $\mathbf{F} = \mathbf{F_p}(\theta)$. Since $<(y,x)_{\mathbf{K}}>_{\mathbf{F}}$ is locally constant and $0 \le [\alpha]_{\mathbf{F}} \le 1$ for each $\alpha \in \mathbf{F}$, there is the inequality $-2 \le f_1(y,z;x) \le 1$ for each $x,y,z \in \mathbf{K^n}$ in (11). On the other hand,

$$<(\beta y,x)_{\mathbf{K}}>_{\mathbf{F}} = [\beta]_{\mathbf{F}} <(y,x)_{\mathbf{K}}>_{\mathbf{F}} + 2\pi f_2(\beta,(e,(y,x)_{\mathbf{K}})_{\mathbf{F}}), \qquad (12)$$

where $f_2(\alpha,\beta) = u(\alpha,\beta)$ for each $\alpha,\beta \in \mathbf{F}$, since $\mathbf{F}$ is naturally embedded into $\mathbf{K}$ and $\beta(e,(y,x)_{\mathbf{K}})_{\mathbf{F}} = (e,(\beta y,x)_{\mathbf{K}})_{\mathbf{F}}$. From the inclusion $[\alpha]_{\mathbf{F}} \in [0,1]$ for each $\alpha \in \mathbf{F}$ we deduce that $-1 \le f_2(\alpha,\gamma) \le 1$ for each $\alpha \in \mathbf{F}$ and $\gamma = (e,(y,x)_{\mathbf{K}})_{\mathbf{F}} \in \mathbf{F}$ in (12). In view of the continuity and the locally constant behavior of the mapping $<(y,x)_{\mathbf{K}}>_{\mathbf{F}}$ the continuity and locally constant behaviour of f_1 and f_2 follows. Thus, f_1 and f_2 satisfy Conditions $(F3,F4)$ depending on $char(\mathbf{K})$. Therefore, from $(11,12)$ we get the properties:

$$A(y) = \int_{\mathbf{K^n}} <(y,x)_{\mathbf{K}}>_{\mathbf{F}} |x|_{\mathbf{K}}^{-2}\nu(dx) \quad \text{and} \qquad (13)$$

$$B(y,z) = \int_{\mathbf{K^n}} <(y,x)_{\mathbf{K}}>_{\mathbf{F}}<(z,x)_{\mathbf{K}}>_{\mathbf{F}} |x|_{\mathbf{K}}^{-2}\nu(dx) \qquad (14)$$

with the measure $|x|_{\mathbf{K}}^{-2}\nu(dx)$ here instead of the measure μ in $(F1-F4)$, $(B1-B3)$. In accordance with the construction given above the measures in the definitions of A and B are nonnegative and the functions in integrals are nonnegative, then $A(y)$ and $B(y,z)$ take nonnegative values.

The metric space $\mathbf{K^n}$ is complete separable and hence is the Radon space (see Theorem 1.2 [16]). We remind that this means that the class of compact subsets approximates from below each σ-additive nonnegative finite measure on the Borel σ-algebra $\mathcal{B}(\mathbf{K^n})$. The non-negative measure $|x|_{\mathbf{K}}^{-2}\nu(dx)$ on $\mathbf{K^n} \setminus B(\mathbf{K^n},0,1/|y|_{\mathbf{K}})$ is finite and σ-additive for $|y|_{\mathbf{K}} > 0$, $<(y,x)_{\mathbf{K}}>_{\mathbf{F}} = 0$ for $|(x,y)_{\mathbf{K}}| \le 1$. Therefore, due to the continuity and boundedness of the functions in integrals we have that the mappings $A(y)$ and $B(y,z)$ are continuous.

6. Corollary. *Suppose that the conditions of Theorem* 5 *are satisfied and the integral*

$$J := \int_{\mathbf{K^n}} |x|_{\mathbf{K}}^{-2}\nu(dx) < \infty$$

exists. Then

$$A(y) = -i(\partial\phi(\beta,y)/\partial\beta)|_{\beta=0} \quad \textit{and}$$
$$B(y) = -(\partial^2\phi(\beta,y)/\partial\beta^2)|_{\beta=0},$$

where

$$\phi(\beta,y) = \int_{\mathbf{K^n}} \exp(i<(y,x)>_{\mathbf{F}} \beta)|x|_{\mathbf{K}}^{-2}\nu(dx), \quad -1<\beta<1.$$

Proof. In view of Theorem 5 the functions $A(y)$ and $B(y)$ exist. At the same time the measure ν is nonnegative as the weak limit of a weakly converging sequence of nonnegative measures, consequently, the measure

$$\mu(dx) := |x|_{\mathbf{K}}^{-2}\nu(dx)$$

is nonnegative. In view of the supposition of this Lemma the inequalities $0 \le \mu(\mathbf{K^n}) = J < \infty$ are accomplished. If $J=0$, then $A(y)=0$, $B(y)=0$ and $\phi(\beta,y)=0$, then the statement of this Lemma is evident. Therefore, the case $J > 0$ remains. We consider the random variable $\zeta := < (y,\eta)_{\mathbf{K}} >_{\mathbf{F}}$ with values in $\mathbf{R}$, where η is a random vector in $\mathbf{K^n}$ with the probability distribution $P(dx) := J^{-1}|x|_{\mathbf{K}}^{-2}\nu(dx)$, where $y \in \mathbf{K^n}$ is the given vector.

Then $\phi(\beta,y) = JM\exp(i\beta\zeta)$, where MX denotes the mean value of the random variable X with values in $\mathbf{C}$. This means that

$$M\exp(i\beta\zeta) = \int_{\mathbf{K^n}} \exp(i\beta < (y,x)_{\mathbf{K}} >_{\mathbf{F}})P(dx).$$

For ζ the second moment exists, since the mapping $B(y)$ is defined and has finite values for each $y \in \mathbf{K^n}$. In view of Theorem II.12.1 [106] about relations between moments of the random variable and values of derivatives of their characteristic functions at zero, we get the statement of this Corollary.

7. Theorem. *Suppose that the conditions of Theorem 5 are satisfied and in addition measures $\mu_v(dx)$ posses finite moments of $|x|_{\mathbf{K}}$ of the second order:*

$$\int_{\mathbf{K^n}} |x|^2_{\mathbf{K}}\mu_v(dx) < \infty \quad \forall v \in V,$$

then for $g(y)$ there is the representation:

$$(i) \quad g(y) = i\tilde{A}(y) - \tilde{B}(y)/2 + \int_{B(\mathbf{K^n},0,\varepsilon)} (\exp(i < (y,x)_{\mathbf{K}} >_{\mathbf{F}})$$

$$-1 - i < (y,x)_{\mathbf{K}} >_{\mathbf{F}} + < (y,x)_{\mathbf{K}} >^2_{\mathbf{F}} /2)\eta(dx)$$

$$+ \int_{\mathbf{K^n}\setminus B(\mathbf{K^n},0,\varepsilon)} (\exp(i < (y,x)_{\mathbf{K}} >_{\mathbf{F}}) - 1)\eta(dx),$$

where η is a nonnegative σ-additive measure on $\mathcal{B}(\mathbf{K^n})$, $\eta(\{0\}) = 0$, $\tilde{A}(y) \in C_1$, $\tilde{B}(y,z) \in C_2$.

Proof. We consider the integral

$$\eta_v(A) := v^{-1}\int_A |x|^2_{\mathbf{K}}\mu_v(dx),$$

where $\{\mu_v : v\}$ is the family of measures corresponding to the characteristic functions $\psi(v,y)$. At first we prove the weak compactness of the family of measures $\{\Psi_B(x)\eta_v(dx) : v \in V\}$ for $B = B(\mathbf{K^n},0,R)$, $0 < R < \infty$, where $\Psi_B(x) = 1$ for $x \in B$, $\Psi_B(x) = 0$ for $x \notin B$, $\Psi_B(x)$ is the characteristic function of the set B. Using the non-archimedean analog of the Minlos-Sazonov theorem as in §5 we reduce the proof to the case of measures on $\mathbf{F^{bn}}$. For a marked positive number $0 < R_1 < \infty$ due to the conditions of this theorem for each $\delta > 0$ a positive number $v_0 = v_0(R_1,\delta) > 0$ exists such that for each $\varepsilon > 0$ and each $0 < v \le v_0$ the inequality

$$-Reg(y) + \delta \ge \int_{\mathbf{F^{bn}}} [1 - \cos(< (y,x) >_{\mathbf{F}})]|x|^{-2}\eta_v(dx)$$

is accomplished due to the existence of the limit

$$\lim_{v\downarrow 0}[\psi(v,y) - 1]/v = g(y)$$

uniformly in the ball of the radius $0 < R_1 < \infty$, $\forall y \in \mathbf{F^{bn}} : |y| \le R_1$. After the integration of this inequality by $y \in B(\mathbf{F^{bn}}, 0, r)$ and division on the volume $\mu(B(\mathbf{F^{bn}}, 0, r)) = r^{bn}$ for $r \in \Gamma_{\mathbf{F}} := \{|x| : x \ne 0, x \in \mathbf{F}\} = \{p^k : k \in \mathbf{Z}\}$, where μ is the Haar nonnegative nontrivial measure on $\mathbf{F^{bn}}$ one gets the following inequality:

$$-r^{-bn} \int_{B(\mathbf{F^{bn}},0,r)} Reg(y)\mu(dy) + \delta$$

$$\ge r^{-bn} \int_{B(\mathbf{F^{bn}},0,r)} \left(\int_{\mathbf{F^{bn}}} [1 - \cos < (y,x) >_{\mathbf{F}}] |x|^{-2} \eta_\nu(dx)\mu(dy) \right.$$

$$\ge r^{-bn} \int_{B(\mathbf{F^{bn}},0,r)} \left(\int_{B(\mathbf{F^{bn}},0,\varepsilon)} [1 - \cos < (y,x) >_{\mathbf{F}}] |x|^{-2} \eta_\nu(dx)\mu(dy), \right.$$

since $\eta_\nu \ge 0$ and $\mu \ge 0$ are nonnegative measures. Since $\int_{B(\mathbf{F^{bn}},0,p^k)} \chi_s(x)\mu(dx) = J(s,k)$, where $J(s,k) = p^{kbn}$ for $|s| \le p^{-k}$, $J(s,k) = 0$ for $|s| \ge p^{-k+1}$, we infer that

$$-p^{-bnk} \int_{B(\mathbf{F^{bn}},0,p^k)} Reg(y)\mu(dy) + \delta \ge \int_{B(\mathbf{F^{bn}},0,\varepsilon)} [1 - p^{-bnk} J(x,k)] |x|^{-2} \eta_\nu(dx).$$

For $\varepsilon > p^{-k+1}$ we then get

$$[\eta_\nu(B(\mathbf{F^{bn}},0,\varepsilon) - \eta_\nu(B(\mathbf{F^{bn}},0,p^{-k}))]$$

$$\le \varepsilon^2 [\delta - p^{-bnk} \int_{B(\mathbf{F^{bn}},0,p^k)} Reg(y)\mu(dy)].$$

Then for $\varepsilon = p^{-k_0+2}$ and $\varepsilon_k = p^{-k+2} \le \varepsilon$, $k \to \infty$ the summing of these inequalities leads to:

$$\eta_\nu(B(\mathbf{F^{bn}},0,\varepsilon)) \le L_1\delta - \sum_{k=k_0}^{\infty} p^{-kbn-2k+4} \int_{B(\mathbf{F^{bn}},0,p^k)} Reg(y)\mu(dy),$$

where $L_1 = p^{4-2k_0}/(1-p^{-2})$, $k_0 \in \mathbf{Z}$ is fixed.

From the facts that the function $g(y)$ is continuous and $g(0) = 0$ it follows that for each $\delta > 0$ there exists a positive number $0 < R_1 < \infty$ such that

$$R_1^{-bn} \left| \int_{B(\mathbf{F^{bn}},0,R_1)} Reg(y)\mu(dy) \right| < \delta.$$

Then for $\varepsilon = p^{-k_0+2}$ the inequality: $\eta_\nu(B(\mathbf{F^{bn}},0,\varepsilon)) < 2L_1\delta$ is satisfied for each $\nu \in (0,\nu_0]$. Since

$$\int_{\mathbf{K^n}\setminus B} \Psi_B(x)\eta(dx) = 0,$$

the family of measures $\{\Psi_B\eta_\nu : \nu \in V\}$ is weakly compact for each given $0 < R < \infty$, $B = B(\mathbf{K^n}, 0, R)$.

Take a marked positive number $0 < \varepsilon < \infty$, then

$$\left| \int_{B(\mathbf{K^n},0,\varepsilon)} < (y,x)_{\mathbf{K}} >_{\mathbf{F}} \nu(dx) \right| < \infty \quad \text{and}$$

$$\left| \int_{\mathbf{K^n} \setminus B(\mathbf{K^n},0,\varepsilon)} < (y,x)_{\mathbf{K}} >_{\mathbf{F}} |x|^{-2} \nu(dx) \right| < \infty,$$

hence

$$J_e := \int_{\mathbf{K^n}} (\exp(i < (y,x)_{\mathbf{K}} >_{\mathbf{F}}) - 1 - i < (y,x)_{\mathbf{K}} >_{\mathbf{F}} [1+|x|^2]^{-1}$$

$$+ < (y,x)_{\mathbf{K}} >_{\mathbf{F}}^2 [1+|x|^2]^{-1}/2)[1+|x|^2]|x|^{-2}\nu(dx)$$

$$= \int_{\mathbf{K^n}} (\exp(i < (y,x)_{\mathbf{K}} >_{\mathbf{F}}) - 1 - i < (y,x)_{\mathbf{K}} >_{\mathbf{F}} [1+|x|^2]^{-1}$$

$$+ < (y,x)_{\mathbf{K}} >_{\mathbf{F}}^2 [1+|x|^2]^{-1}/2)\eta(dx),$$

where

$$\eta(A) := \int_A [1+|x|^2]|x|^{-2}\nu(dx)$$

for each $A \in \mathcal{B}(\mathbf{K^n})$. The measure $\eta \geq 0$ is nonnegative, since $\nu \geq 0$ is nonnegative. The equality $\nu(\{0\}) = 0$ implies that $\eta(\{0\}) = 0$. The measure $\eta(A)$ is finite for each $A \in \mathcal{B}(\mathbf{K^n} \setminus B(\mathbf{K^n},0,\varepsilon))$, when $0 < \varepsilon < \infty$, since $\nu(\mathbf{K^n}) < \infty$ and $|x| > \varepsilon$ for $x \in \mathbf{K^n} \setminus B(\mathbf{K^n},0,\varepsilon)$. Therefore,

$$J_e = \left(\int_{B(\mathbf{K^n},0,\varepsilon)} + \int_{\mathbf{K^n} \setminus B(\mathbf{K^n},0,\varepsilon)} \right) (\exp(i < (y,x)_{\mathbf{K}} >_{\mathbf{F}}) - 1)\eta(dx)$$

$$+ \int_{\mathbf{K^n}} (-i < (y,x)_{\mathbf{K}} >_{\mathbf{F}} + < (y,x)_{\mathbf{K}} >_{\mathbf{F}}^2 /2)|x|^{-2}\nu(dx).$$

At the same time

$$\int_{\mathbf{K^n}} (-i < (y,x)_{\mathbf{K}} >_{\mathbf{F}} + < (y,x)_{\mathbf{K}} >_{\mathbf{F}}^2 /2)|x|^{-2}\nu(dx) =$$

$$\int_{B(\mathbf{K^n},0,\varepsilon)} (-i < (y,x)_{\mathbf{K}} >_{\mathbf{F}} + < (y,x)_{\mathbf{K}} >_{\mathbf{F}}^2 /2)\eta(dx)$$

$$- \int_{B(\mathbf{K^n},0,\varepsilon)} (-i < (y,x)_{\mathbf{K}} >_{\mathbf{F}} + < (y,x)_{\mathbf{K}} >_{\mathbf{F}}^2 /2)[(1+|x|^2) - 1]|x|^{-2}\nu(dx)$$

$$+ \int_{\mathbf{K^n} \setminus B(\mathbf{K^n},0,\varepsilon)} (-i < (y,x)_{\mathbf{K}} >_{\mathbf{F}} + < (y,x)_{\mathbf{K}} >_{\mathbf{F}}^2 /2)|x|^{-2}\nu(dx),$$

hence

$$g(y) = i\tilde{A}(y) - \tilde{B}(y)/2$$

$$\int_{B(\mathbf{K^n},0,\varepsilon)} (\exp(i < (y,x)_{\mathbf{K}} >_{\mathbf{F}}) - 1 - i < (y,x)_{\mathbf{K}} >_{\mathbf{F}} + < (y,x)_{\mathbf{K}} >_{\mathbf{F}}^2 /2)\eta(dx)$$

$$+ \int_{\mathbf{K^n} \setminus B(\mathbf{K^n},0,\varepsilon)} (\exp(i < (y,x)_{\mathbf{K}} >_{\mathbf{F}}) - 1)\eta(dx), \tag{1}$$

where

$$\tilde{A}(y) = A(y) + \int_{B(\mathbf{K^n},0,\varepsilon)} < (y,x)_{\mathbf{K}} >_{\mathbf{F}} \nu(dx)$$

$$- \int_{\mathbf{K^n} \setminus B(\mathbf{K^n},0,\varepsilon)} < (y,x)_{\mathbf{K}} >_{\mathbf{F}} |x|^{-2}\nu(dx),$$

$$\tilde{B}(y) = B(y) + \int_{B(\mathbf{K^n},0,\varepsilon)} < (y,x)_{\mathbf{K}} >^2_{\mathbf{F}} \nu(dx) - \int_{\mathbf{K^n}\setminus B(\mathbf{K^n},0,\varepsilon)} < (y,x)_{\mathbf{K}} >^2_{\mathbf{F}} |x|^{-2}\nu(dx).$$

Using the expressions for $A(y)$ and $B(y,z)$ from the proof of Theorem 5, we get

$$\tilde{A}(y) = \int_{B(\mathbf{K^n},0,\varepsilon)} < (y,x)_{\mathbf{K}} >_{\mathbf{F}} \nu(dx)$$

$$+ \int_{B(\mathbf{K^n},0,\varepsilon)} < (y,x)_{\mathbf{K}} >_{\mathbf{F}} |x|^{-2}\nu(dx), \quad (2)$$

$$\tilde{B}(y,z) = \int_{B(\mathbf{K^n},0,\varepsilon)} < (y,x)_{\mathbf{K}} >_{\mathbf{F}} < (z,x)_{\mathbf{K}} >_{\mathbf{F}} \nu(dx)$$

$$+ \int_{B(\mathbf{K^n},0,\varepsilon)} < (y,x)_{\mathbf{K}} >_{\mathbf{F}} < (z,x)_{\mathbf{K}} >_{\mathbf{F}} |x|^{-2}\nu(dx). \quad (3)$$

Due to identities $5(11,12)$ with the measure $[1+|x|^{-2}]\Psi_B\nu(dx)$ here as the measure μ in §4, with $B = B(\mathbf{K^n},0,\varepsilon)$, where $\Psi_B(x)$ is the characteristic function of the set B, $\Psi_B(x) = 1$ for $x \in B$, $\Psi_B(x) = 0$ for $x \in \mathbf{K^n} \setminus B$, we get, that $\tilde{A}$ and $\tilde{B}$ satisfy Conditions $(F1-F4)$ and $(B1-B3)$ respectively. Since the measures in the definition of $\tilde{A}$ and $\tilde{B}$ are nonnegative and the functions in integrals are nonnegative, we get that $\tilde{A}(y)$ and $\tilde{B}(y,z)$ take nonnegative values.

As the metric space $\mathbf{K^n}$ is complete and separable, hence it is the Radon space (see Theorem 1.2 [16]). We remind that this means: the class of compact subsets approximates from below each σ-additive nonnegative finite measure on the Borel σ-algebra $\mathcal{B}(\mathbf{K^n})$. In view of the finiteness and σ-additivity of the nonnegative measure $[1+|x|^{-2}]\Psi_B\nu(dx)$ and the boundedness of the continuous functions in integrals the mappings $\tilde{A}(y)$ and $\tilde{B}(y,z)$ are continuous.

8. Theorem. *For an infinitely divisible distribution in $\mathbf{K^n}$ its characteristic function $\psi(y)$ has the form*

$$\psi(y) = \exp(g(y)),$$

where $g(y)$ is given by Formula $5(i)$. If in addition distributions $\mu_v(dx)$ from Theorem 5 posses finite moments $|x|_{\mathbf{K}}$ of the second order:

$$\int_{\mathbf{K^n}} |x|^2_{\mathbf{K}}\mu_v(dx) < \infty,$$

then $g(y)$ is given by Formula $7(i)$.

Proof. If $h_k = 1/k$, $k \in \mathbf{N}$, we deduce that

$$g(y) = \lim_{k\to\infty}(\psi_k(y)-1)/(1/k) = \lim_{k\to\infty} k(\psi_k(y)-1) = \ln\psi(y),$$

where $\psi_k(y) = \psi(1/k,y)$, $\psi(y) = [\psi_k(y)]^k$. If fix $arg\psi(0) = 0$ and take such a continuous branch of the argument $arg\ \psi(y)$, then $\psi(y) = \exp(g(y))$, where $g(y)$ is given by Theorems 5 or 7.

9. Definitions. Let a random function $\xi(t)$ with values in $\mathbf{K^n}$ be outlined, $t \in T$, where (T,ρ) is a metric space with a metric ρ. Then $\xi(t)$ is called stochastically continuous at a point t_0, if for each $\varepsilon > 0$ there exists

$$\lim_{\rho(t,t_0)\to 0} P(|\xi(t)-\xi(t_0)| > \varepsilon) = 0.$$

If $\xi(t)$ is stochastically continuous at each point of a subset S in T, then it is called stochastically continuous on S. If

$$\lim_{R\to\infty} \sup_{t\in S} P(|\xi(t)| > R) = 0,$$

then a random function $\xi(t)$ is called stochastically bounded on S.

Let $T = [0,a]$ or $T = [0,\infty)$, $a > 0$. A random process $\xi(t)$ with values in $\mathbf{K^n}$ is called a process with independent increments, if $\forall n$, $0 \le t_1 < \dots < t_n$: random vectors $\xi(0)$, $\xi(t_1) - \xi(0), \dots, \xi(t_n) - \xi(t_{n-1})$ are mutually independent. At the same time the vector $\xi(0)$ is called the initial state (value), and its distribution $P(\xi(0) \in B)$, $B \in \mathcal{B}(\mathbf{K^n})$, is called the initial distribution. A process with independent increments is called homogeneous, if the distribution

$$P(t,s,B) := P(\xi(t+s)-\xi(t) \in B), \quad B \in \mathcal{B}(\mathbf{K^n}),$$

of the vector $\xi(t+s)-\xi(t)$ is independent of t, that is, $P(t,s,B) = P(s,B)$ for each $t < t+s \in T$.

10. Theorem. *Let $\psi(t,y)$ be a characteristic function of the vector $\xi(t+s)-\xi(s)$, $t > 0$, $s \ge 0$, where $\xi(t)$ is the stochastically continuous random process with independent increments having values in $\mathbf{K^n}$. Then*

$$\psi(t,y) = \exp(tg(y)),$$

where $g(y)$ is given by Formula 5(i). If in addition $|\xi(t)|_{\mathbf{K}}$ has the second order finite moments, then the function $g(y)$ is written by Formula 7(i).

Proof. We take any homogeneous stochastically continuous process $\xi(t)$ with independent increments having values in $\mathbf{K^n}$, where $t \in T \subset \mathbf{R}$. If $t > s$, then

$$|\psi(t,y)-\psi(s,y)| = |M\exp(i<(y,\xi(t))_{\mathbf{K}}>_{\mathbf{F}}) - M\exp(i<(y,\xi(s))_{\mathbf{K}}>_{\mathbf{F}})|$$

$$= |M(\exp(i<(y,\xi(t)-\xi(s))_{\mathbf{K}}>_{\mathbf{F}})-1)\exp(i<(y,\xi(s))_{\mathbf{K}}>_{\mathbf{F}})|$$

$$\le M|\exp(i<(y,\xi(t)-\xi(s))_{\mathbf{K}}>_{\mathbf{F}})-1|.$$

Therefore, from the stochastic continuity of $\xi(t)$ it follows continuity of $\psi(t,y)$ by t. In view of the stochastic process being homogeneous and independency of its increments the equalities are accomplished:

$$\psi(t_1+t_2,y) = M\exp(i<(y,\xi(t_1+t_2)-\xi(t_1))>_{\mathbf{F}} + i<(y,\xi(t_1)-\xi(0))>_{\mathbf{F}}))$$

$$= M\exp(i<(y,\xi(t_1)-\xi(0))>_{\mathbf{F}}))M\exp(i<(y,\xi(t_2)-\xi(0))>_{\mathbf{F}}) = \psi(t_1,y)\psi(t_2,y)$$

for each $t_1,t_2 \in T$. As it is well-known, the unique continuous solution of the equation $f(v+u) = f(v)f(u)$ for each $v,u \in \mathbf{R}$ has the form $f(v) = \exp(av)$, where $a \in \mathbf{R}$. Thus,

$$\psi(t,y) = \exp(tg(y)),$$

where $g(y) = \lim_{t\downarrow 0}(\psi(t,y)-1)/t$. Applying Theorems 5 and 7, we get the statement of this theorem.

11. Remark. For the illustration of previous constructions we consider an auxiliary random process $\eta := [\xi]_p$ with values in $\mathbf{R^n}$, where $[(q_1,\dots,q_n)]_p := ([q_1]_p, \dots, [q_n]_p)$ for $q=(q_1,\dots,q_n)\in\mathbf{K^n}$. If $\xi(t)$ is a homogeneous process with independent increments, then such is also η. Let $a(t) := M\eta(t)$ is a mean value, while $R(t,s) := M[(\eta(t)-a(t))^*(\eta(s)-a(s))]$ is the correlation matrix, where $\eta=(\eta_1,\dots,\eta_n)$ is the row-vector, A^* denotes the transposed matrix A. For the process with independent increments and finite moments of the second order we have the representation $R(t,s)=B(\min(t,s))$, where the matrix $B(t)$ is symmetric and nonnegative definite. If $\xi(t)$ is the homogeneous process with independent increments, then η has the finite second order moments. As it is known this implies $a(t)=at$, $R(t,s)=B\min(t,s)$, where a is the vector, B is the symmetric nonnegative definite matrix [31].

12. Theorem. *Let P and Q be two nonnegative finite σ-additive measures on the Borel σ-algebra $\mathcal{B}(\mathbf{K^n})$, where $\mathbf{K}$ is a locally compact infinite field with a nontrivial non-archimedean norm, $n\in\mathbf{N}$. Suppose that their characteristic functions are equal $\hat{P}(y)=\hat{Q}(y)$ for each $y\in\mathbf{K^n}$, then $P(A)=Q(A)$ for each $A\in\mathcal{B}(\mathbf{K^n})$.*

Proof. The metric space $\mathbf{K^n}$ is complete and separable, consequently, it is the Radon space. Thus P and Q are the Radon measures (see Theorem 1.2 [16]). Therefore, for each $\delta>0$ the ball $B(\mathbf{K^n},z,R)$ exists, having the finite positive radius $0<R<\infty$, $z\in\mathbf{K^n}$, such that $P(\mathbf{K^n}\setminus B(\mathbf{K^n},z,R))<\delta$ and $Q(\mathbf{K^n}\setminus B(\mathbf{K^n},z,R))<\delta$.

The family of all finite $\mathbf{C}$-linear combinations of characters forms the algebra which is the subalgebra of the algebra of all continuous functions on $B(\mathbf{K^n},z,R_1)$, the complex conjugation preserves this subalgebra. Moreover, this subalgebra contains all complex constants and separates points in $B(\mathbf{K^n},z,R_1)$. In view of the Stone-Weierstrass theorem for each ball $B(\mathbf{K^n},z,R_1)$ having a positive finite radius $0<R_1<\infty$ containing a marked point $z\in\mathbf{K^n}$ and for each $\varepsilon>0$ and each continuous bounded function $f:\mathbf{K^n}\to\mathbf{R}$ there exist $b_1,\dots,b_k\in\mathbf{C}$ and $s_1,\dots,s_k\in\mathbf{K^n}$ such that

$$\sup_{x\in B(\mathbf{K^n},z,R_1)} |b_1\chi_{s_1}(x)+\dots+b_k\chi_{s_k}(x)-f(x)|<\varepsilon,$$

where $\chi_s(x)$ is the character, $k\in\mathbf{N}$ (see Theorem IV.10 [98]).

The characteristic function $\Psi_{B(\mathbf{K^n},z,R)}$ of the set $B(\mathbf{K^n},z,R)$ is continuous on $\mathbf{K^n}$, since $\mathbf{K^n}$ is totally disconnected and the ball $B(\mathbf{K^n},z,R)$ is clopen in $\mathbf{K^n}$ (simultaneously open and closed). We choose a point $z\in\mathbf{K^n}$, positive numbers $0<\delta_k<1/k$ and $0<\varepsilon_k<1/k$, also $R=R(\delta_k)\le R(\delta_{k+1})$ for each k. For an arbitrary vector $z_1\in\mathbf{K^n}$ with $|z-z_1|_{\mathbf{K^n}}<R(\delta_1)$ we take the function of the form $\Psi^\varepsilon(x)=b_1\chi_{s_1}(x)+\dots+b_v\chi_{s_v}(x)$ such that

$$\sup_{x\in B(\mathbf{K^n},z_1,R_1)} |\Psi^{\varepsilon_k}(x)-\Psi_{B(\mathbf{K^n},z_1,R_1)}(x)|<\varepsilon_k.$$

Then we deduce

$$\int_{\mathbf{K^n}}\Psi^{\varepsilon_k}(x)P(dx)=\int_{\mathbf{K^n}}\Psi^{\varepsilon_k}(x)Q(dx) \quad \text{and}$$

$$\int_{\mathbf{K^n}}\Psi_{B(\mathbf{K^n},z_1,R_1)}(x)P(dx)=P(B(\mathbf{K^n},z_1,R_1)),$$

$$\int_{\mathbf{K^n}} \Psi_{B(\mathbf{K^n},z_1,R_1)}(x)Q(dx) = Q(B(\mathbf{K^n},z_1,R_1)).$$

Certainly we have also the following estimates:

$$\left| \int_{\mathbf{K^n}} \Psi_{B(\mathbf{K^n},z_1,R_1)}(x)P(dx) - \int_{\mathbf{K^n}} \Psi_{B(\mathbf{K^n},z_1,R_1)}(x)Q(dx) \right|$$

$$\leq \left| \int_{\mathbf{K^n}} \Psi^{\varepsilon_k}(x)P(dx) - \int_{\mathbf{K^n}} \Psi_{B(\mathbf{K^n},z_1,R_1)}(x)P(dx) \right|$$

$$+ \left| \int_{\mathbf{K^n}} \Psi^{\varepsilon_k}(x)Q(dx) - \int_{\mathbf{K^n}} \Psi_{B(\mathbf{K^n},z_1,R_1)}(x)Q(dx) \right| + \left| \int_{\mathbf{K^n}} \Psi^{\varepsilon_k}(x)P(dx) \right.$$

$$\left. - \int_{\mathbf{K^n}} \Psi^{\varepsilon_k}(x)Q(dx) \right| \leq \varepsilon_k(P(\mathbf{K^n}) + Q(\mathbf{K^n})).$$

The right hand side of the latter inequality tends to zero while $k \to \infty$, consequently,

$$P(B(\mathbf{K^n},z_1,R_1)) = Q(B(\mathbf{K^n},z_1,R_1))$$

for each ball $B(\mathbf{K^n},z_1,R_1)$ in $\mathbf{K^n}$, where $0 < R_1 < \infty$, $z_1 \in \mathbf{K^n}$, since

$$\lim_{k\to\infty} \delta_k = 0$$

and

$$P(\mathbf{K^n} \setminus B(\mathbf{K^n},z,R(\delta_k))) < \delta_k, \;\; Q(\mathbf{K^n} \setminus B(\mathbf{K^n},z,R(\delta_k))) < \delta_k.$$

Since such clopen balls form the base of the topology in $\mathbf{K^n}$, we deduce that $P(A) = Q(A)$ for each $A \in \mathcal{B}(\mathbf{K^n})$.

13. Theorem. *Random vectors $\eta_1,\dots,\eta_k$ in $\mathbf{K^n}$ are independent if and only if*

$$M\exp(i < (y_1,\eta_1)_{\mathbf{K}} + \dots + (y_k,\eta_k)_{\mathbf{K}} >_{\mathbf{F}})$$

$$= M\exp(i < (y_1,\eta_1)_{\mathbf{K}} >_{\mathbf{F}}) \cdots M\exp(i < (y_k,\eta_k)_{\mathbf{K}} >_{\mathbf{F}}) \tag{1}$$

for each $y_1,\dots,y_k \in \mathbf{K^n}$.

Proof. Random vectors $\eta_1,\dots,\eta_k$ are independent, consequently, the random functions $< (y_1,\eta_1)_{\mathbf{K}} >_{\mathbf{F}},\dots,< (y_k,\eta_k)_{\mathbf{K}} >_{\mathbf{F}}$ are independent also. Therefore, there is satisfied the Equality (1), since $\exp(i < (y_1,\eta_1)_{\mathbf{K}} + \dots + (y_k,\eta_k)_{\mathbf{K}} >_{\mathbf{F}}) = \exp(i < (y_1,\eta_1)_{\mathbf{K}} >_{\mathbf{F}}) \cdots \exp(i < (y_k,\eta_k)_{\mathbf{K}} >_{\mathbf{F}})$.

Vice versa let (1) be satisfied. Denote by $P_{\eta_1,\dots,\eta_k}$ the mutual probability distribution of random vectors $\eta_1,\dots,\eta_k$, by P_{η_j} denote the probability distribution of η_j. Then we get

$$\int_{\mathbf{K^n}} \exp(i < (y_1,x_1)_{\mathbf{K}} + \dots + (y_k,x_k)_{\mathbf{K}} >_{\mathbf{F}}) P_{\eta_1,\dots,\eta_k}(dx)$$

$$= M\exp(i < (y_1,\eta_1)_{\mathbf{K}} + \dots + (y_k,\eta_k)_{\mathbf{K}} >_{\mathbf{F}})$$

$$= M\exp(i < (y_1,\eta_1)_{\mathbf{K}} >_{\mathbf{F}}) \dots M\exp(i < (y_k,\eta_k)_{\mathbf{K}} >_{\mathbf{F}})$$

$$= \prod_{j=1}^{k} \int_{\mathbf{K^n}} \exp(i < (y_j, x_j)_{\mathbf{K}} >_{\mathbf{F}}) P_{\eta_j}(dx_j),$$

where $x = (x_1, \dots, x_k), y_1, \dots, y_k, x_1, \dots, x_k \in \mathbf{K^n}$. Therefore, by Theorem 12

$$P_{\eta_1,\dots,\eta_k}(A_1 \times \dots \times A_k) = P_{\eta_1}(A_1) \cdots P_{\eta_k}(A_k)$$

for each $A_1, \dots, A_k \in \mathcal{B}(\mathbf{K^n})$, consequently, $\eta_1, \dots, \eta_k$ are independent.

14. Definitions. If for each continuous bounded function $f : \mathbf{K^n} \to \mathbf{R}$ the limit

$$\lim_{m \to \infty} Mf(\xi_m) = Mf(\xi)$$

exists, then the sequence of random vectors ξ_m in $\mathbf{K^n}$ is called convergent by the distribution to a random vector ξ.

Let a metric space (X, ρ) be given with a metric ρ and a σ-algebra of Borel subsets $\mathcal{B}(X)$.

A sequence $\{P_m : m \in \mathbf{N}\}$ of probability measures P_m is called weakly convergent to a measure P when $m \to \infty$, if for each continuous bounded function $f : X \to R$ the limit

$$\lim_{m \to \infty} \int_X f(x) P_m(dx) = \int_X f(x) P(dx)$$

exists.

The family of probability measures $\mathcal{P} := \{P_\beta : \beta \in \Lambda\}$ on $(X, \mathcal{B}(X))$, where Λ is a set, is called relatively compact, if an arbitrary sequence of measures from $\mathcal{P}$ contains a subsequence weakly converging to some probability measure.

We mention that from the topological point of view the latter corresponds to the sequential relative compactness in the weak topology (see also [19, 87]).

A family of probability measures $\mathcal{P} := \{P_\beta : \beta \in \Lambda\}$ on $(X, \mathcal{B}(X))$ is called dense, if for each $\varepsilon > 0$ there exists a compact subset C in X such that $\sup_{\beta \in \Lambda} P_\beta(E \setminus C) \le \varepsilon$.

15. Theorem. *A random vector ξ in $\mathbf{K^n}$ is a limit by a distribution of sums*

$$\tilde{\xi}_m := \sum_{k=1}^{m} \xi_{m,k}$$

of independent random vectors with the same probability distribution $\xi_{m,k}$, $k = 1, \dots, m$, if and only if ξ is infinitely divisible.

Proof. If ξ is infinitely divisible, then for each $m \ge 1$ there exist independent random vectors with the same distribution $\xi_{m,1}, \dots, \xi_{m,k}$ such that the probability distributions of ξ and of the sum $(\xi_{m,1} + \dots + \xi_{m,k})$ are the same.

Let now $\tilde{\xi}_m$ be a sequence of arbitrary vectors converging by the distribution to ξ when $m \to \infty$. We take $k \ge 1$ and group the summands so that $\tilde{\xi}_{mk}$ takes the form:

$$\tilde{\xi}_{mk} = \zeta_{m,1} + \dots + \zeta_{m,k},$$

where

$$\zeta_{m,1} = \xi_{mk,1} + \dots + \xi_{mk,m}, \dots, \zeta_{mk,k} = \xi_{mk,m(k-1)+1} + \dots + \xi_{mk,mk}.$$

Since the sequence $\tilde{\xi}_{mk}$ converges by the distribution to ξ while $m \to \infty$, the sequence of the probability distributions $P_{\tilde{\xi}_{mk}}$ of random vectors $\tilde{\xi}_{mk}$ is relatively compact. Thus, due to the Prohorov Theorem (see §VI.25 [35] or III.2.1 [106]) the family $\{P_{\tilde{\xi}_{mk}}\}$ is dense.

On the other hand, if $|\tilde{\xi}_{mk}| > R$, then due to non-archimedeanity of the norm in $\mathbf{K^n}$ there exists j such that $|\zeta_{m,j}| > R$. Therefore, we get the estimate

$$P(\zeta_{m,1} \in \mathbf{K^n} \setminus B(\mathbf{K^n}, 0, R)) \leq P(\tilde{\xi}_{mk} \in \mathbf{K^n} \setminus B(\mathbf{K^n}, 0, R)),$$

since $\zeta_{m,j}$ are independent and have the same probability distribution. Thus, $\{P_{\zeta_{m,1}} : m \in \mathbf{N}\}$ is the dense family of probability distributions. We get that there exists the sequence $\{m_j : j \in \mathbf{N}\}$ of natural numbers and random vectors $\eta_1, \dots, \eta_k$ such that $\zeta_{m_j,l}$ converges by the distribution to η_l for each $l = 1, \dots, k$ for $j \to \infty$. In accordance with the definition of convergence by the distribution this means in particular, that for each $b_1, \dots, b_k \in \mathbf{K^n}$ the limit exists

$$\lim_{j \to \infty} M\exp(i < (b_1, \zeta_{m_j,1})_{\mathbf{K}} + \dots + (b_k, \zeta_{m_j,k})_{\mathbf{K}} >_{\mathbf{F}})$$
$$= M\exp(i < (b_1, \eta_1)_{\mathbf{K}} + \dots + (b_k, \eta_k)_{\mathbf{K}} >_{\mathbf{F}}).$$

In view of independency of random vectors $\zeta_{m_j,1}, \dots, \zeta_{m_j,k}$ there is satisfied the equality

$$M\exp(i < (b_1, \zeta_{m_j,1})_{\mathbf{K}} + \dots + (b_k, \zeta_{m_j,k})_{\mathbf{K}} >_{\mathbf{F}})$$
$$= M\exp(i < (b_1, \zeta_{m_j,1})_{\mathbf{K}} >) \cdots M\exp(i(b_k, \zeta_{m_j,k})_{\mathbf{K}} >_{\mathbf{F}}),$$

since $\exp(i < y >_{\mathbf{F}})$ is the character of the additive group of the field $\mathbf{K}$. Therefore,

$$\lim_{j \to \infty} M\exp(i < (b_1, \zeta_{m_j,1})_{\mathbf{K}} + \dots + (b_k, \zeta_{m_j,k})_{\mathbf{K}} >_{\mathbf{F}})$$
$$= M\exp(i < (b_1, \eta_1)_{\mathbf{K}} >_{\mathbf{F}}) \cdots M\exp(i < (b_k, \eta_k)_{\mathbf{K}} >_{\mathbf{F}}).$$

Thus,

$$M\exp(i < (b_1, \eta_1)_{\mathbf{K}} + \dots + (b_k, \eta_k)_{\mathbf{K}} >_{\mathbf{F}})$$
$$= M\exp(i < (b_1, \eta_1)_{\mathbf{K}} >_{\mathbf{F}}) \cdots M\exp(i < (b_k, \eta_k)_{\mathbf{K}} >_{\mathbf{F}})$$

for each $b_1, \dots, b_k \in \mathbf{K^n}$. Then from Theorem 13 it follows, that the random vectors $\eta_1, \dots, \eta_k$ are independent.

We have that the sequence $\tilde{\xi}_{m_jk} = \zeta_{m_j,1} + \dots + \zeta_{m_j,k}$ converges by the distribution to $\eta_1 + \dots + \eta_k$ and $\tilde{\xi}_{m_jk}$ converges by the distribution to ξ. Moreover, the random vector ξ is equal to the sum $\eta_1 + \dots + \eta_k$ by the distribution, since

$$Mf(\xi) = \lim_{j \to \infty} Mf(\tilde{\xi}_{m_jk}) = \lim_{j \to \infty} Mf(\zeta_{m_j,1} + \dots + \zeta_{m_j,k}) = Mf(\eta_1 + \dots + \eta_k)$$

for each continuous bounded function $f : \mathbf{K^n} \to \mathbf{R}$.

16. Particular cases of Theorem 10.

1. At first we consider the case $A(y) = q < (a, y)_{\mathbf{K}} >_{\mathbf{F}}$, $B = 0$, $\nu = 0$, where $a \in \mathbf{K^n}$ is some vector, $q = const > 0$. Then

$$\psi(t, y) = \exp(itq < (a, y)_{\mathbf{K}} >_{\mathbf{F}}).$$

Moreover, the random function $\eta(t) =< (\xi(t),y)_{\mathbf{K}} >_{\mathbf{F}}$ has the form $\eta(t) = \eta(0)+tq$, where ξ is the initial random vector with values in $\mathbf{K^n}$. This means that $\eta(t)$ corresponds to the uniform motion of the point in $\mathbf{R}$ with the velocity q.

In the particular case, when $A(y) = q(v,<y>_{\mathbf{F}})$, $B = 0$, $\nu = 0$, where $v \in \mathbf{R^n}$ is a given vector, $0 \le v_j \le 1$ for each $j = 1,\dots,n$, $v = (v_1,\dots,v_n)$, $q = const > 0$, then

$$\psi(t,y) = \exp(itq(v,<y>_{\mathbf{F}})).$$

Therefore, the random variable $\eta(t) = (<\xi(t)>_{\mathbf{F}},<y>_{\mathbf{F}})_{\mathbf{R}}$ has the form $\eta(t) = \eta(0)+tq$.

2. Now we consider in formulas of functions $A(y)$ and $B(y,z)$ in §§5 and 7 in particular atomic measures, denoting $\tilde{A}$ by A and $\tilde{B}$ by B here for the uniformity. Then we obtain the expressions of the form $\sum_j q_j < (x_j,y)_{\mathbf{K}} >_{\mathbf{F}}$ and $\sum_j q_j < (x_j,y)_{\mathbf{K}} >_{\mathbf{F}} < (x_j,z)_{\mathbf{K}} >_{\mathbf{F}}$, where $q_j = \nu(\{x_j\}) > 0$ or $q_j = \nu(\{x_j\})|x_j|^{-2} > 0$ depending on the considered case, $x_j \ne 0$. In particular, the vectors $x_j = e_j = (0,\dots,0,1,0,\dots) \in \mathbf{K^n}$ with the unity on the j-th place are also acceptable. These expressions may be transformed using conditions $(F1-F4)$ or $(B1-B3)$ (see Formulas $5(i,10,13,14)$ or $7(i,1-3)$). Then there are possible cases

$$A(y) = q < (a,y)_{\mathbf{K}} >_{\mathbf{F}} \quad \text{or} \quad A(y) = (v,<y>_{\mathbf{F}})_{\mathbf{R}},$$

$$B(y,z) = \sum_{j=1}^{n} < s_j y_j z_j >_{\mathbf{F}}, \text{ or } B(y,z) = \sum_{j=1}^{n} q_j < y_j >_{\mathbf{F}} < z_j >_{\mathbf{F}},$$

where $<y>_{\mathbf{F}} = (<y_1>_{\mathbf{F}},\dots,<y_n>_{\mathbf{F}})$, $y = (y_1,\dots,y_n) \in \mathbf{K^n}$, $y_k \in \mathbf{K}$ for each k, $v \in \mathbf{R^n}$, $(*,*)_{\mathbf{R}}$ is the scalar product in $\mathbf{R^n}$, $s_j \in \mathbf{K}$, $a \in \mathbf{K^n}$. The consideration of the transition matrix Y from one basis in $\mathbf{K^n}$ to another or the matrix X of transition from one basis in $\mathbf{R^n}$ into another leads to the more general expressions for $B(y,z)$ such as

$$B(y,z) = (b<y>_{\mathbf{F}},<z>_{\mathbf{F}})_{\mathbf{R}}, \quad B(y,z) =< (hy,z)_{\mathbf{K}} >_{\mathbf{F}},$$

where b is the symmetric nonnegative definite $n \times n$ matrix with elements in the field of real numbers $\mathbf{R}$, h is the symmetric $n \times n$ matrix with elements in the locally compact field $\mathbf{K}$.

3. Let us consider particular expressions of functions:

$$A(y) = q < (a,y)_{\mathbf{K}} >_{\mathbf{F}} \quad \text{and} \quad B(y,z) =< (hy,z)_{\mathbf{K}} >_{\mathbf{F}},$$

where $a \in \mathbf{K^n}$, h is the symmetric $n \times n$ matrix with elements in the field $\mathbf{K}$. We naturally suppose that the correlation term

$$\int_{\mathbf{K^n}} f(y,x)\nu(dx) = 0$$

from §5 or

$$\int_{B(\mathbf{K^n},0,\varepsilon)} (\exp(i < (y,x)_{\mathbf{K}} >_{\mathbf{F}}) - 1 - i < (y,x)_{\mathbf{K}} >_{\mathbf{F}} + < (y,x)_{\mathbf{K}} >_{\mathbf{F}}^2 /2)\eta(dx)$$

$$+ \int_{\mathbf{K^n} \setminus B(\mathbf{K^n},0,\varepsilon)} (\exp(i < (y,x)_{\mathbf{K}} >_{\mathbf{F}}) - 1)\eta(dx) = 0$$

from §7 is zero. Therefore,

$$\psi(t,y) = \exp(itq < (a,y)_{\mathbf{K}} >_{\mathbf{F}} -t < (hy,y)_{\mathbf{K}} >_p /2).$$

Then the random function $\xi(t)$ is one of the non-archimedean variants of the Gaussian process.

4. The next particular case is:

$$A(y) = (v, < y >_{\mathbf{F}})_{\mathbf{R}} \quad \text{and} \quad B(y,z) = (b < y >_{\mathbf{F}}, < z >_{\mathbf{F}})_{\mathbf{R}}$$

(see paragraph 2), while the correlation term is zero. With such data we get:

$$\psi(t,y) = \exp(it(v, < y >_{\mathbf{F}})_{\mathbf{R}} - t(b < y >_{\mathbf{F}}, < y >_{\mathbf{F}})_{\mathbf{R}}/2)$$

and again $\xi(t)$ is one of the analogs of the Gaussian process.

Though Gaussian processes in the non-archimedean case do not exist. That is, we can satisfy a part of properties of the Gaussian type in the non-archimedean case, but not all (see also [75]).

5. When $A = 0$, $B = 0$ (taking into account $(F1 - F4)$ and $(B1 - B3)$; see Formulas $5(i, 10, 13, 14)$ or $7(i, 1-3)$), where ν is the purely atomic measure, concentrated at the point z_0, $\nu(\{z_0\}) = q > 0$, one gets:

$$\psi(t,y) = \exp(qt(\exp(i < (y,z_0)_{\mathbf{K}} >_{\mathbf{F}}) - 1).$$

Therefore, the random function $\xi(t)$ is the non-archimedean analog of the Poisson process.

6. More generally, if

$$\tilde{A}(y) = q < (a,y)_{\mathbf{K}} >_{\mathbf{F}} + \int_{B(\mathbf{K^n},0,\varepsilon)} < (y,x)_{\mathbf{K}} >_{\mathbf{F}} \eta(dx),$$

$$\tilde{B}(y) = -\int_{B(\mathbf{K^n},0,\varepsilon)} < (y,x)_{\mathbf{K}} >_{\mathbf{F}}^2 \eta(dx)/2,$$

$\eta(B(\mathbf{K^n}, 0, \varepsilon)) < \infty$, then the general formula gives:

$$g(y) = i < (a,y)_{\mathbf{K}} >_{\mathbf{F}} + w \int_{\mathbf{K^n}} (\exp(i < (y,x)_{\mathbf{K}} >_{\mathbf{F}}) - 1)\lambda(dx),$$

where λ is the probability measure on $(\mathbf{K^n}, \mathcal{B}(\mathbf{K^n}))$, $0 < w = \eta(\mathbf{K^n}) < \infty$, $\eta(dx) = w\lambda(dx)$ (see Formulas $7(i, 1-3)$ and $(F1 - F4)$, $(B1 - B3)$). Therefore,

$$\psi(t,y) = \exp(itq < (a,y)_{\mathbf{K}} >_{\mathbf{F}}) \sum_{k=0}^{\infty} \exp(-wt)((wt)^k/k!)$$

$$\times \left[\int_{\mathbf{K^n}} \exp(i < (y,x)_{\mathbf{K}} >_{\mathbf{F}})\lambda(dx) \right]^k.$$

This expression of the characteristic function of the random process $\xi(t) = \rho(t) + \xi_1 + \dots + \xi_{\zeta(t)}$, where $\rho(t)$ is the random process in $\mathbf{K^n}$, leads to the characteristic function $\exp(itq < (a,y)_{\mathbf{K}} >_{\mathbf{F}})$. Here $\xi_1, \dots, \xi_k, \dots$ are independent random vectors in $\mathbf{K^n}$ with the same probability distribution $\lambda(dx)$, $\zeta(t)$ is the Poisson process with a parameter w independent of $\rho, \xi_1, \dots, \xi_k, \dots$. Then there arises the non-archimedean analog $\xi(t)$ of the generalized Poisson process.

If put

$$\tilde{A}(y) = (v, <y>_{\mathbf{F}})_{\mathbf{R}} + \int_{B(\mathbf{K^n},0,\varepsilon)} <(y,x)_{\mathbf{K}}>_{\mathbf{F}} \eta(dx),$$

where $\tilde{B}(y)$ is the same as at the beginning of the given paragraph, then

$$\psi(t,y) = \exp(it(v, <y>_{\mathbf{F}})_{\mathbf{R}}) \sum_{k=0}^{\infty} \exp(-wt)((wt)^k/k!)$$

$$\times \left[\int_{\mathbf{K^n}} \exp(i <(y,x)_{\mathbf{K}}>_{\mathbf{F}}) \lambda(dx) \right]^k,$$

where $\rho(t)$ has the characteristic function $\exp(it(v, <y>_{\mathbf{F}})_{\mathbf{R}})$.

17. Remark. Suppose that a branching random process is realized with values in the ring $\mathbf{Z_p}$ of integer p-adic numbers or in the ring $B(\mathbf{F_p}(\theta), 0, 1)$. We denote it by $\mathbf{B}$. In the particular case of the uniform distribution $|x|^{-2}\nu(dx)$ in $\mathbf{B}$ the measure ν is proportional to the Haar measure μ, $|x|^{-2}\nu(dx) = q\mu(dx)$, where $q > 0$, $\mu(\mathbf{B}) = 1$, $\nu(\mathbf{F} \setminus \mathbf{B}) = 0$, and $\mathbf{K} = \mathbf{F} = \mathbf{Q_p}$ or $\mathbf{K} = \mathbf{F} = \mathbf{F_p}(\theta)$ respectively here, $n = 1$. Then it is possible to calculate the functions $A(y)$ and $B(y)$. In view of §5 in this particular case

$$A(y) = q \int_{\mathbf{B}} <yx>_{\mathbf{F}} \mu(dx) \quad \text{and}$$

$$B(y) = q \int_{\mathbf{B}} <yx>_{\mathbf{F}}^2 \mu(dx).$$

If $y = 0$, then $A(0) = 0$ and $B(0) = 0$, therefore, consider the case $y \neq 0$. The function $<yx>_{\mathbf{F}}$ takes the zero value when $|yx|_{\mathbf{F}} \le 1$ and is different from zero when $|x|_{\mathbf{F}} > 1/|y|_{\mathbf{F}}$.

In the considered case the support of the measure ν is contained in $\mathbf{B}$, then $A(y)$ and $B(y)$ are equal to zero when $|y|_{\mathbf{F}} \le 1$. But the Haar measure is invariant relative to shifts $\mu(A+z) = \mu(A)$ for each Borel subset in $\mathbf{F}$ with the finite measure $\mu(A) < \infty$ and each $z \in \mathbf{F}$. Moreover, $\mu(zdx) = |z|_{\mathbf{F}}\mu(dx)$, where $|z|_{\mathbf{F}} = p^{-ord_{\mathbf{F}}(z)}$ (see [111]). Then the functions A and B are given by the formulas:

$$A(y) = q \int_{z \in \mathbf{F}, |y|_{\mathbf{F}} \ge |z|_{\mathbf{F}} > 1} <z>_{\mathbf{F}} \mu(dz)/|y|_{\mathbf{F}} \quad \text{and}$$

$$B(y) = q \int_{z \in \mathbf{F}, |y|_{\mathbf{F}} \ge |z|_{\mathbf{F}} > 1} <z>_{\mathbf{F}}^2 \mu(dz)/|y|_{\mathbf{F}},$$

where $|y|_{\mathbf{F}} > 1$. At the same time

$$z = \sum_{k=N(x)}^{\infty} z_k p^k$$

for $\mathbf{F} = \mathbf{Q_p}$ or

$$z = \sum_{k=N(x)}^{\infty} z_k \theta^k$$

for $\mathbf{F} = \mathbf{F_p}(\theta)$, where $N(z) = ord_p(z)$, $z_k \in \{0, 1, \ldots, p-1\}$ or $z_k \in \mathbf{F_p}$. If $\nu(dx) = q\mu(dx)$, then

$$A(y) = q|y|_{\mathbf{F}} \int_{z \in \mathbf{F}, |y|_{\mathbf{F}} \geq |z|_{\mathbf{F}} > 1} < z >_{\mathbf{F}} |z|_{\mathbf{F}}^{-2} \mu(dz) \quad \text{and}$$

$$B(y) = q|y|_{\mathbf{F}} \int_{z \in \mathbf{F}, |y|_{\mathbf{F}} \geq |z|_{\mathbf{F}} > 1} < z >_{\mathbf{F}}^2 |z|_{\mathbf{F}}^{-2} \mu(dz).$$

These integrals are expressible in the form of finite sums, since $\mu(B(\mathbf{F}, x, p^k)) = p^k$ for each $k \in \mathbf{Z}$ and $z \in \mathbf{F}$, where the functions in the integrals are locally constant.

The measure ν is Borelian, $\nu : \mathcal{B}(\mathbf{K^n}) \to [0, \infty)$, therefore each its atom may be only a singleton. More generally (see Formulas $5(i, 13, 14)$), let us consider the measure μ which has the decomposition

$$\nu = \nu_1 + \nu_2,$$

where ν_2 is the atomic measure, while $\nu_1(dx) = f(x)\mu(dx)$, while $f(x) = g(|x|_{\mathbf{F}}, < x >_{\mathbf{F}})$, $g : \mathbf{R^2} \to [0, \infty)$ is a continuous function. So we deduce from the general formulas that

$$A(y) = \sum_j < yx_j >_{\mathbf{F}} |x_j|^{-2} \nu_2(\{x_j\}) + \int_{\mathbf{F}} < yx >_{\mathbf{F}} f(x)|x|_{\mathbf{F}}^{-2} \mu(dx),$$

$$B(y) = \sum_j < yx_j >_{\mathbf{F}}^2 |x_j|^{-2} \nu_2(\{x_j\}) + \int_{\mathbf{F}} < yx >_{\mathbf{F}}^2 f(x)|x|_{\mathbf{F}}^{-2} \mu(dx),$$

where $\{x_j\}$ are atoms of the measure ν_2, $\nu_2(\{x_j\}) > 0$, each $x_j \neq 0$ is nonzero. At the same time integrals by the Haar measure μ on $\mathbf{F}$ with functions $< yx >_{\mathbf{F}} f(x)|x|_{\mathbf{F}}^{-2}$ and $< yx >_{\mathbf{F}}^2 f(x)|x|_{\mathbf{F}}^{-2}$, where $f(x) = g(|x|_{\mathbf{F}}, < x >_{\mathbf{F}})$, are expressible in the form of series. Indeed, the functions $|x|_{\mathbf{F}}$ and $< x >_{\mathbf{F}}$ are locally constant, consequently, f is locally constant.

Chapter 2

Spectral Expansions of Stochastic Processes

2.1. Introduction

This chapter is devoted to spectral representations of random functions over fields supplied with non-archimedean multiplicative norms. Stochastic integrals and spectral representations of stochastic processes are one of the main instruments in the classical case over the fields of real and complex numbers [16, 27, 31, 35, 106, 108].

Besides locally compact fields in this chapter we consider also non locally compact fields. For example, the algebraic closure of the p-adic field $\mathbf{Q_p}$ can be supplied with the multiplicative non-archimedean norm and its completion relative to this norm gives the field $\mathbf{C_p}$ of complex p-adic numbers, where p is a marked p-adic number. The field $\mathbf{C_p}$ is algebraically closed and complete relative to its norm [48]. Its normalization group $\Gamma_{\mathbf{C_p}} := \{|z| : z \in \mathbf{C_p}, z \neq 0\}$ is isomorphic with the multiplicative group $\{p^x : x \in \mathbf{Q}\}$. Larger fields $\mathbf{U_p}$ are known, which are extensions of $\mathbf{Q_p}$ such that the normalization group $\Gamma_{\mathbf{U_p}} = \{p^x : x \in \mathbf{R}\} = (0, \infty)$ is already the connected multiplicative group. Besides completions relative to multiplicative norms also extensions of fields with the help of the spherical completions are known, when an initial field is not such [17, 20, 99, 103].

In this chapter representations of stochastic processes with values in finite- and infinite-dimensional vector spaces over infinite fields with non-trivial non-archimedean norms are described. This material contains different types of stochastic processes controlled by measures with values in non-archimedean fields of zero characteristic. Theorems about spectral decompositions of non-archimedean stochastic processes are proved below (see, for example, §§20–29, 75–82, Lemmas 27 and 29, Theorems 20, 79, 81). Certainly special features of the non-archimedean stochastic functions are elucidated. These features arise from many differences of the classical over $\mathbf{R}$ and $\mathbf{C}$ analysis and the non-archimedean analysis. General constructions of the paper are illustrated in Examples 9, 31.1, 40, 74, Theorem 41, etc.. Moreover, applications to totally disconnected topological groups are discussed as well.

Some necessary facts from non-archimedean probability theory and non-archimedean analysis are recalled that to make reading easier (see, for example, §§1-6 in section 2). They are written in more details, when it is essential. It is worthwhile to underline that in the second chapter measures and stochastic processes with values not only in non-archimedean fields (see section 2), but also with values in topological linear spaces which may be infinite dimensional over non-archimedean fields are studied (see §§44-74 in section 3).

The random functions with values in $\mathbf{Q_p^n}$ have natural interesting applications, for which a time parameter may be either real or p-adic. A random trajectory in $\mathbf{Q_p^n}$ may be continuous relative to the non-archimedean norm in $\mathbf{Q_p}$, but its trajectory in $\mathbf{Q^n}$ relative to the usual metric induced by the real metric may be discontinuous. This gives new approach to spasmodic or jump or discontinuous stochastic processes with values in $\mathbf{Q^n}$, when the latter is considered as embedded into $\mathbf{R^n}$.

2.2. Scalar Spectral Functions

To avoid misunderstandings we first present our notations and definitions and recall the basic facts.

1. Definitions. Consider a completely regular totally disconnected topological space G and its covering ring $\mathcal{R}$ of subsets in G, that is $\bigcup\{A : A \in \mathcal{R}\} = G$.

We call the ring separating, if for each two distinct points $x, y \in G$ there exists $A \in \mathcal{R}$ such that $x \in A$, $y \notin A$.

A subfamily $\mathcal{S} \subset \mathcal{R}$ is called shrinking, if an intersection of each two elements from $\mathcal{A}$ contains an element from $\mathcal{A}$.

Suppose now that $\mathcal{A}$ is a shrinking family, a mapping $f : \mathcal{R} \to \mathbf{K}$ is given, where $\mathbf{K} = \mathbf{R}$ or $\mathbf{K}$ is the field with the non-archimedean multiplicative norm. If for each $\varepsilon > 0$ there exists $A_0 \in \mathcal{A}$ such that $|f(A)| < \varepsilon$ for each $A \in \mathcal{A}$ with $A \subset A_0$, then by the definition the limit $\lim_{A \in \mathcal{A}} f(A) = 0$ is zero.

A measure $\mu : \mathcal{R} \to \mathbf{K}$ is a mapping with values in the field $\mathbf{K}$ of zero characteristic with the non-archimedean norm satisfying the following properties:

(i) μ is additive;

(ii) for each $A \in \mathcal{R}$ the set $\{\mu(B) : B \in \mathcal{R}, A \subset B\}$ is bounded;

(iii) if $\mathcal{A}$ is the shrinking family in $\mathcal{R}$ and $\bigcap_{A \in \mathcal{A}} A = \emptyset$, then $\lim_{A \in \mathcal{A}} \mu(A) = 0$.

Measures on $\mathsf{Bco}(G)$ are called tight measure, where $\mathsf{Bco}(G)$ is the ring of clopen (simultaneously open and closed) subsets in G.

For each $A \in \mathcal{R}$ there is defined the norm:

$$\|A\|_\mu := \sup\{|\mu(B)| : B \subset A, B \in \mathcal{R}\}$$

For functions $f : G \to X$, where X is a Banach space over $\mathbf{K}$ and $\phi : G \to [0, +\infty)$ the norm

$$\|f\|_\phi := \sup\{|f(x)|\phi(x) : x \in G\}$$

is defined.

More generally for a complete locally $\mathbf{K}$-convex space X with a family of non-archimedean continuous semi-norms $\mathcal{S} = \{u\}$ [87] characterizing its topology, we define the family of the semi-norms

$$\|f\|_{\phi,u} := \sup\{u(f(x))\phi(x) : x \in G\}.$$

Henceforward, we utilize such family $\mathcal{S}$ of continuous semi-norms defining topology of X and frequently for short not repeating that semi-norms u are continuous.

We recall that a subset V in X is called absolutely **K**-convex or a **K**-disc, if $VB+VB \subseteq V$, where $B := \{x \in \mathbf{K} : |x| \leq 1\}$.

Translates $x+V$ of absolutely **K**-convex sets are called **K**-convex, where $x \in X$.

A topological vector space over **K** is called **K**-convex, if it has a base of **K**-convex neighborhoods of zero (see 5.202 and 5.203 [87]). A semi-norm u in X satisfying the inequality

$$u(x+y) \leq \max[u(x), u(y)]$$

for each $x, y \in X$ is called non-archimedean.

A topological vector space X over **K** is locally **K**-convex if and only if its topology is generated by a family of non-archimedean semi-norms.

It is worth to mention that a complete **K**-convex space is the projective limit of Banach spaces over **K** (see 6.204, 6.205 and 12.202 [87]).

We introduce the non-negative mapping:

$$N_\mu(x) := \inf\{\|U\|_\mu : x \in U \in \mathcal{R}\}$$

for each $x \in G$.

If a function f is a finite linear combination over the field **K** of characteristic functions Ch_A of subsets $A \subset G$ from $\mathcal{R}$, then it is called simple. We use this notation Ch_A that to distinguish characteristic functions of subsets from characters of groups or characteristic functionals of measures.

A function $f : G \to X$ is called μ-integrable, if there exists a sequence $f_1, f_2, \dots$ of simple functions for which the limit

$$\lim_{n\to\infty} \|f - f_n\|_{N_\mu, u} = 0$$

exists for each $u \in \mathcal{S}$.

The space $L(\mu, X) = L(G, \mathcal{R}, \mu, X)$ of all μ-integrable functions with values in X is **K**-linear. At the same time the integral is defined by the formula:

$$\int_G \sum_{j=1}^n a_j Ch_{A_j}(x)\mu(dx) := \sum_{j=1}^n a_j \mu(A_j)$$

for any simple function. This integral extends onto $L(\mu, X)$ by the continuity, where $a_j \in X$, $A_j \in \mathcal{R}$ for each j.

Then we define the separating covering ring: $\mathcal{R}_\mu := \{A : A \subset G, Ch_A \in L(\mu, \mathbf{K})\}$ of the topological totally disconnected space G.

For all $A \in \mathcal{R}_\mu$ the extension of the measure is defined with the help of the integral:

$$\bar{\mu}(A) := \int_G \chi_A(x)\mu(dx).$$

For $1 \leq q < \infty$ one usually denotes the non-archimedean norm by

$$\|f\|_q := [\sup_{x \in G} |f(x)|^q N_\mu(x)]^{1/q}$$

for a simple function $f : G \to X$, when X is the Banach space, or the non-archimedean semi-norm by

$$\|f\|_{q,u} := [\sup_{x \in G} u(f(x))^q N_\mu(x)]^{1/q}$$

for each non-archimedean semi-norm $u \in \mathcal{S}$, when X is the complete **K**-convex space. The completion of the space of all simple functions by such norm $\| * \|_q$ or the family of semi-norms $\{\| * \|_{q,u} : u \in \mathcal{S}\}$ is denoted by $L^q(\mu, X)$, where $L(\mu, X) = L^1(\mu, X)$.

Let G be a totally disconnected completely regular space, let also $\mathsf{B_c}(G)$ be its separating covering ring of clopen compact subsets in G, suppose that $\mu : \mathsf{B_c}(G) \to \mathbf{K}$ is a finitely-additive function such that its restriction $\mu|_A$ for each $A \in \mathsf{B_c}(G)$ is a measure on a separating covering ring $\mathcal{R}(G)|_A$, where $\mathsf{B_c}(G)|_A = \mathcal{R}|_A$, $\mathcal{R}|_A := \{E \in \mathcal{R} : E \subseteq A\}$.

A measure $\eta : \mathcal{R} \to \mathbf{K}$ is called absolutely continuous relative to a measure $\mu : \mathcal{R} \to \mathbf{K}$, if there exists a function $f \in L(\mu, \mathbf{K})$ such that

$$\eta(A) = \int_G Ch_A(x) f(x) \mu(dx)$$

for each $A \in \mathcal{R}$. We denote it by $\eta \preceq \mu$.

If simultaneously $\eta \preceq \mu$ and $\mu \preceq \eta$, then we say that η and μ are equivalent $\eta \sim \mu$.

A **K**-valued measure P on $\mathcal{R}(X)$ we call a probability measure if $\|X\|_P =: \|P\| = 1$ and $P(X) = 1$.

The following statements from the non-archimedean functional analysis proved in [99] are useful.

2. Lemma. *Suppose that μ is a measure on a separating covering ring $\mathcal{R}$. Then a unique function $N_\mu : G \to [0, \infty)$ exists satisfying two conditions:*

(1) $\|Ch_A\|_{N_\mu} = \|A\|_\mu$;

(2) if $\phi : G \to [0, \infty)$ and $\|Ch_A\|_\phi \le \|A\|_\mu$ for each $A \in \mathcal{R}$, then $\phi \le N_\mu$; moreover,

$$N_\mu(x) = \inf_{x \in A, A \in \mathcal{R}} \|A\|_\mu$$

for each $x \in X$.

3. Theorem. *Let μ be a measure on $\mathcal{R}$. Then $\mathcal{R}_\mu$ is a separating covering ring of G and $\bar{\mu}$ is a measure on $\mathcal{R}_\mu$ that extends μ.*

4. Lemma. *If μ is a measure on a separating covering ring $\mathcal{R}$, then $N_\mu = N_{\bar{\mu}}$ and $\mathcal{R}_\mu = \mathcal{R}_{\bar{\mu}}$.*

5. Theorem. *Let μ be a measure on a separating covering ring $\mathcal{R}$. Then N_μ is upper semi-continuous and for every $A \in \mathcal{R}_\mu$ and $\varepsilon > 0$ the set $\{x \in A : N_\mu(x) \ge \varepsilon\}$ is $\mathcal{R}_\mu$-compact.*

6. Theorem. *Let μ be a measure on $\mathcal{R}$, let also $\mathcal{S}$ be a separating covering ring of G which is a sub-ring of $\mathcal{R}_\mu$ and let ν be a restriction of μ onto $\mathcal{S}$. Then $\mathcal{S}_\nu = \mathcal{R}_\mu$ and $\bar{\nu} = \bar{\mu}$.*

7. Notations and definitions. We consider a probability space $(\Omega, \mathcal{A}, P)$, where Ω is a space of elementary events, $\mathcal{A}$ is a separating covering ring of events in Ω, $\mathcal{R}(\Omega) \subseteq \mathcal{A} \subseteq \mathcal{R}_P(\Omega)$, $P : \mathcal{A} \to \mathbf{K}$ is a probability. Here **K** denotes a non-archimedean field of zero characteristic, $char(\mathbf{K}) = 0$, complete relative to its multiplicative norm, $\mathbf{K} \supset \mathbf{Q_p}$, $1 < p$ is a prime number, $\mathbf{Q_p}$ is the field of p-adic numbers.

We denote by ξ a random vector (a random variable for $n = 1$) with values in $\mathbf{K^n}$ or in a linear topological space X over $\mathbf{K}$ such that it has the probability distribution

$$P_\xi(A) = P(\{\omega \in \Omega : \xi(\omega) \in A\})$$

for each $A \in \mathcal{R}(X)$, where $\xi : \Omega \to \mathbf{X}$, ξ is $(\mathcal{A}, \mathcal{R}(X))$-measurable, where $\mathcal{R}(X)$ is a separating covering ring of X such that $\mathcal{R}(X) \subset \mathsf{Bco}(X)$, $\mathsf{Bco}(X)$ denotes the separating covering ring of all clopen (simultaneously closed and open) subsets in X. That is, $\xi^{-1}(\mathcal{R}(X)) \subset \mathcal{A}$.

When T is a set and $\xi(t)$ is a random vector for each $t \in T$, this $\xi(t)$ is called a random function (or stochastic function). Particularly, if T is a subset in a field, then $\xi(t)$ is called a stochastic process, while $t \in T$ is interpreted as the time parameter.

As usually one puts

$$M(\xi^k) := \int_\Omega \xi^k(\omega) P(d\omega)$$

for a random variable ξ and $k \in \mathbf{N}$ whenever it exists.

Two random vectors ξ and η with values in X are called independent, if the identity is satisfied:

$$P(\{\xi \in A, \eta \in B\}) = P(\{\xi \in A\}) P(\{\eta \in B\})$$

for each $A, B \in \mathcal{R}(X)$.

8. Definition. Suppose that $\{\Omega, \mathcal{R}, P\}$ is a probability space with a probability measure with values in a non-archimedean field $\mathbf{K}$ complete relative to its multiplicative norm, $\mathbf{K} \supset \mathbf{Q_p}$. We consider a set G and a separating covering ring J of its subsets. We define a $\mathbf{K}$-valued random variable $\xi(A) = \xi(\omega, A)$ for each $A \in J$ and $\omega \in \Omega$ satisfying the following four conditions:

$(M1)$ $\xi(A) \in Y$, $\xi(\emptyset) = 0$, where $Y = L^2(\Omega, \mathcal{R}, P, \mathbf{K})$;

$(M2)$ $\xi(A_1 \cup A_2) = \xi(A_1) + \xi(A_2) \; mod(P)$ for each $A_1, A_2 \in J$ with $A_1 \cap A_2 = \emptyset$;

$(M3)$ $M(\xi(A_1)\xi(A_2)) = \mu(A_1 \cap A_2)$;

$(M4)$ $M(\xi(A_1)\xi(A_2)) = 0$ for each $A_1 \cap A_2 = \emptyset$, $A_1, A_2 \in J$. That is $\xi(A_1)$ and $\xi(A_2)$ are orthogonal random variables, where $\mu(A) \in \mathbf{K}$ for each $A, A_1, A_2 \in J$.

The family of random variables $\{\xi(A) : A \in J\}$ satisfying Conditions $(M1 - M4)$ is called the elementary orthogonal $\mathbf{K}$-valued stochastic measure.

9. Example. If $\xi(A)$ has a zero mean value $M\xi(A) = 0$ for each $A \in J$, while $\xi(A_1)$ and $\xi(A_2)$ are independent random variables for $A_1, A_2 \in J$ with $A_1 \cap A_2 = \emptyset$, then they are orthogonal, since $M(\xi(A_1)\xi(A_2)) = (M\xi(A_1))(M\xi(A_2))$.

10. Lemma. *The function μ from Definition* 8 *is additive.*

Proof. Using the inclusion $\xi(A) \in Y = L^2(\Omega, \mathcal{R}, P, \mathbf{K})$ for each $A \in J$ we get that there exists the integral

$$M\xi(A) = \int_\Omega \xi(\omega, A) P(d\omega),$$

since $\sup_{\omega\in\Omega} |\xi(\omega, A)|^2 N_P^2(\omega) \le \sup_{\omega\in\Omega} |\xi(\omega, A)|^2 N_P(\omega)$ for the probability measure P having the bounded function $N_P(\omega) \le 1$ for each $\omega \in \Omega$. This means that $L^1(\Omega, \mathcal{R}, P, \mathbf{K}) \subset L^2(\Omega, \mathcal{R}, P, \mathbf{K})$.

Therefore, from Conditions $(M2, M4)$ for each $A_1, A_2 \in J$ with the void intersection $A_1 \cap A_2 = \emptyset$ the equalities follow:

$$M(\xi^2(A_1 \cup A_2)) = M[(\xi(A_1) + \xi(A_2))^2]$$

$$= M[\xi^2(A_1) + 2\xi(A_1)\xi(A_2) + \xi^2(A_2)] = M\xi^2(A_1) + M\xi^2(A_2).$$

In view of $(M3)$ one infers the additivity

$$\mu(A_1 \cup A_2) = \mu(A_1) + \mu(A_2).$$

11. Note. Let the measure μ have an extension to a measure on the separating covering ring $\mathcal{R}(G)$, let also G be a totally disconnected completely regular space, where $J \subset \mathcal{R}_\mu(G)$.

12. Definitions. Let a random function $\xi(t)$ be with values in a complete linear locally $\mathbf{K}$-convex space X over $\mathbf{K}$, $t \in T$, where (T, ρ) is a metric space with a metric ρ. The random function $\xi(t)$ is called stochastically continuous at a point t_0, if for each $\varepsilon > 0$ the limit

$$\lim_{\rho(t,t_0) \to 0} P(\{u(\xi(t) - \xi(t_0)) > \varepsilon\}) = 0$$

is zero for each $u \in \mathcal{S}$. If $\xi(t)$ is stochastically continuous at each point of a subset E in T, then it is called stochastically continuous on E.

If $\lim_{R \to \infty} \sup_{t \in E} P(\{u(\xi(t)) > R\}) = 0$ for each $u \in \mathcal{S}$, then a random function $\xi(t)$ is called stochastically bounded on E.

Let $L^0(\mathcal{R}(G), X)$ denotes the class of all step (simple) functions

$$f(x) = \sum_{k=1}^{m} c_k Ch_{A_k}(x),$$

where $c_k \in X$, $A_k \in \mathcal{R}(G)$ for each $k = 1, \dots, m \in \mathbf{N}$, $A_k \cap A_j = \emptyset$ for each $k \neq j$. Then the non-archimedean stochastic integral by the elementary orthogonal stochastic measure $\xi(A)$ of $f \in L^0(\mathcal{R}(G), X)$ is defined by the formula:

$$(SI) \quad \eta(\omega) := \int_G f(x)\xi(\omega, dx) := \sum_{k=1}^{m} c_k \xi(\omega, A_k).$$

13. Lemma. *Suppose that two functions f and g belong to the space $L^0(\mathcal{R}(G), \mathbf{K})$, where*

$$f(x) = \sum_{k=1}^{m} c_k Ch_{A_k}(x) \text{ and } g(x) = \sum_{k=1}^{m} d_k Ch_{A_k}(x).$$

Then

$$M\left(\int_G f(x)\xi(dx) \int_G g(y)\xi(dy)\right) = \sum_{k=1}^{m} c_k d_k \mu(A_k)$$

and a $\mathbf{K}$-linear embedding of $L^0(\mathcal{R}(G), \mathbf{K})$ into $L^2(\mu, \mathbf{K})$ exists.

Proof. In view of Conditions $(M1, M2)$ the integral

$$\int_G f(x)\xi(\omega, dx) \in Y = L(P)$$

exists. Since

$$\int_G f(x)\xi(dx) \int_G g(y)\xi(dy) = \sum_{k,j=1}^{m} c_k d_j \xi(A_k)\xi(A_j),$$

its mean value is

$$M\Big(\int_G f(x)\xi(dx) \int_G g(y)\xi(dy)\Big) = \sum_{k,j=1}^{n} c_k d_j M(\xi(A_k)\xi(A_j)) = \sum_{k=1}^{m} c_k d_k \mu(A_k)$$

due to Conditions $(M3, M4)$, since $A_k \cap A_j = \emptyset$ for each $j \neq k$. This gives the $\mathbf{K}$-linear embedding θ of $L^0(\mathcal{R}(G), \mathbf{K})$ into $L^2(\mu, \mathbf{K})$ such that

$$\theta(f) = \sum_{k=1}^{n} c_k Ch_{A_k}(x) \quad \text{and}$$

$$(i) \quad \|\theta(f)\|_2 = [\max_{k=1}^{m} |c_k|^2 \sup_{x \in A_k} N_\mu(x)]^{1/2} = [\max_{k=1}^{m} |c_k|^2 \|A_k\|_\mu]^{1/2} < \infty$$

due to Lemma 2.

14. Note. We denote by $L^2(\mathcal{R}(G), \mathbf{K})$ the completion of $L^0(\mathcal{R}(G), \mathbf{K})$ by the norm $\| * \|_2$ induced from $L^2(\mu, \mathbf{K})$.

15. Definition. For subsequent needs we extend the non-archimedean stochastic integral by elementary orthogonal stochastic measure $\xi(A)$ from the space of functions $f \in L^0(\xi, X)$ (see §12) on those functions for which the integral exists as the limit of a net in $L^2(\mathcal{R}(G), \mathbf{K})$.

This limit transition is described in details below.

16. Lemma. *Suppose that $f, g \in L^0(\xi, \mathbf{K})$, where*

$$f(x) = \sum_{k=1}^{m} c_k \xi(A_k) \text{ and } g(x) = \sum_{k=1}^{m} d_k \xi(A_k).$$

Then the mean value is:

$$M\Big(\int_G f(x)\xi(dx) \int_G g(y)\xi(dy)\Big) = \sum_{k=1}^{m} c_k d_k \mu(A_k)$$

and a $\mathbf{K}$-linear embedding of $L^0(\mathcal{R}(G), \mathbf{K})$ into $L^2(P, \mathbf{K})$ exists.

Proof. In view of Conditions $(M1, M2)$ the integral $\int_G f(x)\xi(\omega, dx) \in Y = L(P, \mathbf{K})$ exists. But f is the step function, hence

$$(i) \quad \|f\|_{L^2(P,\mathbf{K})} = [\max_{k=1}^{m} |c_k|^2 \sup_{\omega \in \Omega} |\xi^2(\omega, A_k)| N_P(\omega)]^{1/2}$$

$$= [\max_{k=1}^{m} |c_k|^2 \|\xi(*, A_k)\|^2_{L^2(P)}]^{1/2} < \infty$$

and inevitably $f \in L^2(P)$. Thus the mapping

$$\psi(f) := \sum_{k=1}^{m} c_k Ch_{A_k}(x)$$

gives the **K**-linear embedding of $L^0(\xi, \mathbf{K})$ into $L^2(P, \mathbf{K})$. The second statement one can immediately verify as in Lemma 13 due to Formula 12(SI).

17. Note. Henceforth, $L^2(\xi, \mathbf{K})$ denotes the completion of $L^0(\xi, \mathbf{K})$ by the norm $\| * \|_2$ induced from $L^2(P, \mathbf{K})$.

18. Corollary. *The mapping* (SI) *in Definition 12 and Conditions* $(M1 - M4)$ *induce the* **K**-*linear isometry between two spaces* $L^2(\mathcal{R}(G), \mathbf{K})$ *and* $L^2(\xi)$.

Proof. The normalization group $\Gamma_{\mathbf{K}} := \{|z| : z \in \mathbf{K}, z \neq 0\}$ is contained in $(0, \infty)$. In view of Theorem 5, Lemma 10 and Note 11 without loss of generality for a step function f it is possible to take a representation with subsets $A_k \in \mathcal{R}(G)$ such that $\|A_k\|_\mu = |\mu(A_k)|$ for each $k = 1, \dots, m$. The family of all such step functions is everywhere dense in $L^2(\mathcal{R}(G), \mathbf{K})$.

Since $M(\xi^2(A)) = \mu(A)$ for each $A \in \mathcal{R}(G)$, the function

$$N_\mu(x) = \inf_{A \in \mathcal{R}(G), x \in A} \|A\|_\mu$$

is given by such integral, where

$$\|A\|_\mu = \sup\{|\mu(B)| : B \in \mathcal{R}(G), B \subset A\} = \sup\{|M(\xi^2(B))| : B \in \mathcal{R}(G), B \subset A\}.$$

On the other hand, the second moment is

$$M(\xi^2(B)) = \int_\Omega \xi^2(\omega, B) P(d\omega), \text{ consequently,}$$

$$|M(\xi^2(B))| \leq \sup_{\omega \in \Omega} |\xi^2(\omega, B)| N_P(\omega).$$

By our supposition μ is the measure, hence taking a shrinking family $\mathcal{S}$ in $\mathcal{R}(G)$ such that $\bigcap_{A \in \mathcal{S}} A = \{x\}$ we get

$$N_\mu(x) = \inf_{A \in \mathcal{R}(G), x \in A} \left[\sup_{B \in \mathcal{R}(G), B \subset A} \sup_{\omega \in \Omega} |\xi(\omega, B)|^2 N_P(\omega) \right].$$

Thus the equality

$$N_\mu(x) = \inf_{A \in \mathcal{R}(G), x \in A} \left[\sup_{B \in \mathcal{R}(G), B \subset A} \|\xi^2(*, B)\|_{L^2(P)} \right]$$

is satisfied and

$$\|A_k\|_\mu = \|\xi(*, A_k)\|^2_{L^2(P)}$$

for each $k = 1, \dots, m$ due to Lemma 2 and due to the choice $\|A_k\|_\mu = |\mu(A_k)|$ above.

The mapping ψ from §16 also is **K**-linear from $L^0(\xi)$ onto $L^0(\mathcal{R}(G), \mathbf{K})$ such that ψ is the isometry relative to $\| * \|_{L^2(P)}$ and $\| * \|_{L^2(\mu)}$ due to Formulas 13(i) and 16(i) and

Lemma 2, since $\mathcal{R}(G)$ is the separating covering ring on G. Two spaces $L^2(P)$ and $L^2(\mu)$ are complete by their definitions, consequently, ψ has the **K**-linear extension from $L^2(\mathcal{R}(G),\mathbf{K})$ onto $L^2(\xi)$ which is the isometry between $L^2(\mathcal{R}(G),\mathbf{K})$ and $L^2(\xi)$.

19. Definition. For any function $f \in L^2(\mathcal{R}(G),\mathbf{K})$, we put by the definition:

$$\eta = \psi(f) = \int_G f(x)\xi(dx).$$

Such random variable η we call the non-archimedean stochastic integral of the function f by measure ξ.

The operation of taking the limit in the space $L^2(P,X)$ is denoted by $l.i.m.$.

20. Theorems. 1. *For any step function*

$$f(x) = \sum_{k=1}^{n} a_k Ch_{A_k}(x),$$

where $a_k \in \mathbf{K}$, $A_k \in \mathcal{R}(G)$, $n = n(f) \in \mathbf{N}$, *the stochastic integral is given by the formula:*

$$\eta = \int f(x)\xi(dx) = \sum_{k=1}^{n} a_k \xi(A_k).$$

2. *For each* $f,g \in L^2(\mathcal{R}(G),\mathbf{K})$ *the identity*

$$M\Big(\int_G f(x)\xi(dx)\int_G g(y)\xi(dy)\Big) = \int_G f(x)g(x)\mu(dx)$$

is accomplished.

3. *For each* $f,g \in L^2(\mathcal{R}(G),\mathbf{K})$ *and* $\alpha,\beta \in \mathbf{K}$ *the stochastic integral is* **K**-*linear:*

$$\int_G [\alpha f(x) + \beta g(x)]\xi(dx) = \alpha\int_G f(x)\xi(dx) + \beta\int_G g(x)\xi(dx).$$

4. *For each sequence of functions* $f_n \in L^2(G,\mathcal{R}(G),\mu,\mathbf{K})$ *such that*

$$\lim_{n\to\infty} \|f - f_n\|_{L^2(\mu,\mathbf{K})} = 0$$

the limit

$$\int_G f(x)\xi(dx) = l.i.m._{n\to\infty}\int_G f_n(x)\xi(dx)$$

exists.

5. *An extension of the stochastic measure* ξ *from the separating covering ring* $\mathcal{R}$ *onto its completion* $\mathcal{R}_\mu(G)$ *exists.*

Proof. Statements of (1) and (3) follow from the consideration above. To finish the proof of (2) it is sufficient to show that $fg \in L^1(\mu,\mathbf{K})$, if f and $g \in L^2(\mu,\mathbf{K})$, where μ is the measure on G. Since $2|f(x)g(x)| \le |f(x)|^2 + |g(x)|^2$ for each $x \in G$, we infer that $2\sup_{x\in G}|f(x)g(x)|N_\mu(x) \le \sup_{x\in G}(|f(x)|^2 + |g(x)|^2)N_\mu(x) \le \|f\|^2_{L^2(\mu)} + \|g\|^2_{L^2(\mu)}$, consequently, $f(x)g(x)$ is μ-integrable.

4. From

$$\lim_{n\to\infty}[\sup_{x\in G}|f(x)-f_n(x)|^2N_\mu(x)]=0$$

and the formula for the mean value

$$M\left[\int_G(f-f_n)(x)\xi(dx)\int_G(f-f_n)(y)\xi(dy)\right]=\int_G(f-f_n)^2(x)\mu(dx)$$

it follows, that

$$\lim_{n\to\infty}M\left[\left(\int_G(f-f_n)(x)\xi(dx)\right)^2\right]=0.$$

This implies that

$$l.i.m._{n\to\infty}\int_G f_n(x)\xi(dx)=\int_G f(x)\xi(dx)$$

due to Corollary 18.

5. We extend now the stochastic measure ξ from the separating covering ring $\mathcal{R}(G)$ to the stochastic measure $\tilde{\xi}$ on the completed separating covering ring $\mathcal{R}_\mu(G)$. If $A\in\mathcal{R}_\mu(G)$, then $Ch_A\in L(G,\mathcal{R}(G),\mu,\mathbf{K})$. Since $Ch_A\in L(G,\mathcal{R}(G),\mu,\mathbf{K})$, the supremum of the function $N_\mu(x)$ is finite $\sup_{x\in A}N_\mu(x)<\infty$. Putting

$$\tilde{\xi}(A):=\int_G Ch_A(x)\xi(dx)=\int_A\xi(dx)$$

for each $A\in\mathcal{R}_\mu(G)$ implies that

$$\tilde{\xi}\ \text{ is defined on }\ \mathcal{R}_\mu(G). \tag{1}$$

Therefore, $\tilde{\xi}(A)=\xi(A)$ for each $A\in\mathcal{R}(G)$. For each $A,B\in\mathcal{R}_\mu(G)$ there exist sequences of simple functions

$$f_n=\sum_k a_{k,n}Ch_{A_{k,n}}\text{ and }g_m=\sum_l b_{l,m}Ch_{B_{l,m}}$$

with $a_{k,n},b_{l,m}\in\mathbf{K}$, $A_{k,n},B_{l,m}\in\mathcal{R}(G)$ such that

$$\lim_{n\to\infty}\|Ch_A-f_n\|_{L(\mu)}=0\text{ and }\lim_{m\to\infty}\|Ch_B-g_m\|_{L(\mu)}=0.$$

Since $M(a_{k,n}\tilde{\xi}(A_{k,n})b_{l,m}\tilde{\xi}(B_{l,m}))=a_{k,n}b_{l,m}\mu(A_{k,n}\cap B_{l,m})$ for each k,n,l,m, we deduce that

$$M(\tilde{\xi}(A)\tilde{\xi}(B))=\bar{\mu}(A\cap B) \tag{2}$$

for each $A,B\in\mathcal{R}_\mu(G)$, where $\bar{\mu}$ is the extension of the measure μ from the separating covering ring $\mathcal{R}(G)$ onto the completed separating covering ring $\mathcal{R}_\mu(G)$. If $\mathcal{S}\subset\mathcal{R}_\mu(G)$ is a shrinking family such that $\bigcap_{A\in\mathcal{S}}A=\emptyset$, then

$$l.i.m._{A\in\mathcal{S}}\tilde{\xi}(A)=0 \tag{3}$$

in accordance with Corollary 18, since $M[\tilde{\xi}(A)]^2=\bar{\mu}(A)$ and

$$\lim_{A\in\mathcal{S}}\bar{\mu}(A)=0$$

due to Theorem 3.

21. Definition. A random function of sets satisfying conditions 20.5(1 – 3) is called the orthogonal stochastic measure.

22. Corollary. *Let* ξ *and* $\tilde{\xi}$ *be as in Theorem* 20.5, *then two spaces coincide* $L^2(\xi, \mathbf{K}) = L^2(\tilde{\xi}, \mathbf{K})$.

23. Note. If ξ is an orthogonal stochastic measure with a structure measure μ on $\mathcal{R}_\mu(G)$ and $g \in L^2(\mu, \mathbf{K})$, then we define two set functions

$$\rho(A) := \int_G Ch_A(x) g(x) \xi(dx)$$

and

$$\nu(A) := \int_A g^2(x) \mu(dx)$$

for each $A \in \mathcal{R}_\mu(G)$.

24. Lemma. *If* $f \in L^2(\nu, \mathbf{K})$, *then* $f(x)g(x) \in L^2(\mu, \mathbf{K})$ *and*

$$\int_G f(x) \rho(dx) = \int_G f(x) g(x) \xi(dx).$$

Proof. In view of Theorems 20 for each $A, B \in \mathcal{R}_\mu(G)$ there is the equality

$$M[\rho(A)\rho(B)] = M \int_G Ch_A(x) g(x) \xi(dx) \int_G Ch_B(y) g(y) \xi(dy)$$

$$= \int_{A \cap B} g^2(x) \mu(dx) = \nu(A \cap B).$$

Since $g \in L^2(\mu, \mathbf{K})$, the function ν is the measure absolutely continuous relative to μ on $\mathcal{R}_\mu(G)$. If

$$f(x) = \sum_k a_k Ch_{A_k}(x)$$

is a simple function with expansion coefficients $a_k \in \mathbf{K}$ and subsets $A_k \in \mathcal{R}_\mu(G)$, then its integral is:

$$\int_G f(x) \rho(dx) = \sum_k a_k \int_G Ch_{A_k}(x) g(x) \xi(dx) = \sum_k a_k \rho(A_k) = \int_G f(x) g(x) \xi(dx),$$

since

$$\sup_{x \in G} |f(x)g(x)|^2 N_\mu(x) \le [\max_k |a|_k^2] \sup_{x \in G} |g(x)|^2 N_\mu(x) < \infty.$$

If f_n is a fundamental sequence of simple functions in $L^2(\nu, \mathbf{K})$, then the mean value is

$$M\left[\left(\int_G (f_n - f_m)(x) \rho(dx)\right)^2\right] = \int_G [(f_n - f_m)(x)]^2 g^2(x) \mu(dx),$$

hence $f_n g$ is the fundamental sequence in $L^2(\mu, \mathbf{K})$. Therefore, the limit exists

$$\lim_{n\to\infty}\int_G f_n(x)\rho(dx) = \lim_{n\to\infty}\int_G f_n(x)g(x)\xi(dx),$$

consequently,

$$\int_G f(x)\rho(dx) = \int_G f(x)g(x)\xi(dx).$$

25. Lemma. *If $A \in \mathcal{R}_\mu(G)$, then*

$$\xi(A) = \int_G [Ch_A(x)/g(x)]\rho(dx).$$

Proof. Since $\nu(\{x : g(x) = 0\}) = 0$, the reciprocal function $1/g(x)$ is defined ν-almost everywhere on G, consequently,

$$\int_G [Ch_A(x)/g^2(x)]\nu(dx) = \int_G [g^2(x)/g^2(x)]\mu(dx) = \mu(A).$$

In view of Lemma 24

$$\int_G [Ch_A(x)/g(x)]\rho(dx) = \int_G [Ch_A(x)g(x)/g(x)]\xi(dx) = \xi(A).$$

26. Notation and Remark. Let T be a totally disconnected Hausdorff topological space with a separating covering ring $\mathcal{R}(T)$ and with a non-trivial measure $h : \mathcal{R}(T) \to \mathbf{K}$. We denote by $B(X,x,R) := \{y \in X : \rho(x,y) \leq R\}$ the ball in a metric space (X,ρ) supplied with a metric ρ, of radius $0 < R < \infty$, containing a point x. Particularly, the set T may be either the clopen subset in $\mathbf{K_r}$ or the segment in $\mathbf{R}$. The measure h may be the non-trivial $\mathbf{K}$-valued measure on the separating covering ring $\mathcal{R}(B(T,t_0,R))$ for each $t_0 \in T$ and every $0 < R < \infty$, where $\mathcal{R}(B(T,t_0,R_1)) \subset \mathcal{R}(B(T,t_0,R_2))$ for each $0 < R_1 < R_2 < \infty$ and each $t_0 \in T$. We consider the fields $\mathbf{K}$ and $\mathbf{K_r}$ complete relative to their multiplicative non-archimedean norms so that $\mathbf{K} \supset \mathbf{Q_p}$ and $\mathbf{K_r} \supset \mathbf{Q_{p'}}$, where $r = p'$, r and p are primes.

As it is well-known, a continuous mapping θ from a clopen subset G in $\mathbf{Q'_p}$ onto the segment $[a,b]$ in $\mathbf{R}$ exists, where $-\infty < a < b < \infty$ (see [19]). This mapping θ induces the separating covering ring $\mathcal{R}([a,b]) := \{A \subset [a,b] : \theta^{-1}(A) \in \mathcal{R}(G)\}$ of the segment $[a,b]$ and the measure $h(A) = \mu(\theta^{-1}(A))$ on $[a,b]$, where $\mathcal{R}(G)$ is any separating covering ring of G and $\mu : \mathcal{R}(G) \to \mathbf{K}$ is any measure, $h = \mu^\theta$.

We recall that a measure $h : \mathsf{B_c}(\mathbf{K_r}) \to \mathbf{K}$ is called the Haar measure, if

$$h(t+B) = h(B)$$

for each clopen compact subset B in $\mathbf{K_r}$ and each $t \in \mathbf{K_r}$. The Haar measure exists due to the Monna-Springer Theorem 8.4 [99], when $r \neq p$ are mutually prime, $(r,p) = 1$, since the additive group $B(\mathbf{K_r},0,R)$ is p-free. For example, one can take a non-trivial $\mathbf{K}$-valued Haar measure h on the algebra $\mathsf{B_c}(\mathbf{K_r})$ of clopen compact subsets in $\mathbf{K_r}$ such that $h(B(\mathbf{K_r},0,1)) = 1$. But generally we do not demand, that either h is the Haar measure or $r \neq p$.

Let a function $g(t,x)$ on $T \times G$ be $\mathcal{R}_h \times \mathcal{R}_\mu$-measurable and let this function g belong to the space $L^2(T \times G, \mathcal{R}_h \times \mathcal{R}_\mu, h \times \mu, \mathbf{K})$, where $\mathcal{R}_h := \mathsf{B}_{ch}(\mathbf{K_r})$.

27. Lemma. *The stochastic integral*

$$\rho(t) = \int_G g(t,x)\xi(dx) \tag{1}$$

is defined for each $t \in T$ for P-almost all $\omega \in \Omega$ and it can be defined such that the stochastic function $\rho(t)$ would be measurable.

Proof. If

$$g(t,x) = \sum_k a_k Ch_{B_k}(t) Ch_{A_k}(x)$$

is a simple function with subsets $A_k \in \mathcal{R}_\mu$ and $B_k \in \mathcal{R}_h$ and coefficients $a_k \in \mathbf{K}$ for each $k = 1,\dots,m$, $m \in \mathbf{N}$, then the function

$$\rho(t) = \sum_k a_k Ch_{B_k}(t)\xi(A_k)$$

is $\mathcal{R}_h \times \mathcal{A}$-measurable by variables $(t,\omega) \in T \times \Omega$ (see also Definitions 7). For each $g \in L^2(h \times \mu, \mathbf{K})$ there exists a sequence of simple functions $g_n(t,x)$ such that

$$\lim_{n\to\infty} \sup_{t\in T, x\in G} |g(t,x) - g_n(t,x)|^2 N_h(t) N_\mu(x) = 0.$$

Let the sequence of stochastic functions be given by

$$\rho_n(t) := \int_G g_n(t,x)\xi(dx),$$

then there exists the stochastic function $\tilde{\rho}(t)$ so that

$$\lim_{n\to\infty} \sup_{t\in T} |M[(\tilde{\rho}(t) - \rho_n(t))^2]| N_h(t) = 0.$$

Then the equality

$$\int_T M[(\tilde{\rho}(t) - \rho_n(t))^2] h(dt) = \int_T \int_G [g(t,x) - g_n(t,x)]^2 \mu(dx) h(dt)$$

is satisfied, consequently,

$$\sup_{t\in T} |M[(\tilde{\rho}(t) - \rho_n(t))^2]| N_h(t) \le \sup_{t\in T, x\in G} |g(t,x) - g_n(t,x)|^2 N_h(t) N_\mu(x) < \infty$$

and inevitably the stochastic function $\tilde{\rho}(t)$ is $\mathcal{R}_h \times \mathcal{A}$-measurable by $(t,\omega) \in T \times \Omega$. Moreover, this function exists with the unit probability. Thus

$$h(\{t \in A : M[(\rho(t) - \tilde{\rho}(t))^2] = 0\}) = h(A)$$

for each $A \in \mathcal{R}(T)$.

Finally we put

$$\eta(t) = \tilde{\rho}(t) \quad \text{if} \quad P(\{\rho(t) \neq \tilde{\rho}(t)\}) = 0, \quad \text{while}$$

$$\eta(t) = \rho(t) \quad \text{if} \quad P(\{\rho(t) \neq \tilde{\rho}(t)\}) \neq 0.$$

Therefore, the stochastic function η is $\mathcal{R}_h \times \mathcal{A}$-measurable, since η differs from $\mathcal{R}_h \times \mathcal{A}$-measurable function $\tilde{\rho}(t)$ on a set of zero $h \times P$-measure and η is stochastically equivalent with ρ.

28. Remark. Henceforth, due to Lemma 27 we shall suppose that the stochastic integrals 27(1) are $\mathcal{R}_h \times \mathcal{A}$-measurable.

29. Lemma. *Suppose that $g(t,y)$ and $z(t)$ are $\mathcal{R}_h \times \mathcal{R}_\mu$ and $\mathcal{R}_h$-measurable functions, $g \in L^2(T \times \mathbf{K}, \mathcal{R}_h \times \mathcal{R}_\mu, h \times \mu, \mathbf{K})$ and $z \in L^2(T, \mathcal{R}_h, h, \mathbf{K})$, ξ is an orthogonal stochastic measure on $(\mathbf{K}, \mathcal{R}_\mu)$. Then*

$$\int_T z(t) \int_{\mathbf{K}} g(t,y)\xi(dy)h(dt) = \int_{\mathbf{K}} q(y)\xi(dy),$$

where

$$q(y) = \int_T z(t)g(t,y)h(dt). \tag{1}$$

Proof. Since $z \in L^2(h)$ and $g \in L^2(h \times \mu)$, the estimates are satisfied:

$$\sup_{t\in T, y\in \mathbf{K}} |z(t)g(t,y)|^2 N_h^2(t)N_\mu(y)$$

$$\leq [\sup_{t\in T} |z(t)|^2 N_h(t)] \sup_{t\in T, y\in \mathbf{K}} |g(t,y)|^2 N_h(t)N_\mu(y) < \infty. \tag{2}$$

We consider $g \in L^2(h \times \mu)$ and a sequence

$$g_n(t,y) = \sum_k a_{k,n} Ch_{B_{k,n}}(t) Ch_{A_{k,n}}(y)$$

of step functions converging to g in $L^2(h \times \mu, \mathbf{K})$ with expansion coefficients $a_{k,n} \in \mathbf{K}$ and subsets $A_{k,n} \in \mathcal{R}_\mu$, $B_{k,n} \in \mathcal{R}_h$ for each k,n. The mean value of the square of the left side of Equation (1) is:

$$M\left(\left[\int_T z(t) \int_{\mathbf{K}} g(t,y)\xi(dy)h(dt)\right]^2\right)$$

$$= \left(\int_T \int_T z(t_1)z(t_2) \int_{\mathbf{K}} g(t_1,y)g(t_2,y)\mu(dy)h(dt_1)h(dt_2)\right)$$

$$= \int_{\mathbf{K}} \left[\int_T z(t)g(t,y)h(dt)\right]^2 \mu(dy). \tag{3}$$

Equation (1) is satisfied for step functions. In view of (3) the left an the right sides of (1) are continuous relative to taking a limit by $g_n(t,y)$ in $L^2(h \times \mu, \mathbf{K})$ relative to the probability P as well as in the space $L^2(P,\mathbf{K})$. Finally, the statement of this Lemma follows from the fact that the family of step functions is dense in $L^2(h\times\mu,\mathbf{K})$.

30. Remark and Notation. If conditions of Lemma 29 are satisfied for each $T = B(\mathbf{K_r},0,R)$, or $T=[-R,R]$ respectively, $0<R<\infty$ and if the integral

$$\int_{\mathbf{K_r}} z(t)g(t,y)h(dt) = \lim_{R\to\infty}\int_{B(\mathbf{K_r},0,R)} z(t)g(t,y)h(dt)$$

exists in $L^2(\mu)$, then

$$\int_{\mathbf{K_r}} z(t)\int_{\mathbf{K}} g(t,y)\xi(dy)h(dt) = \int_{\mathbf{K}} s(y)\xi(dy), \tag{1}$$

where

$$s(y) := \int_{\mathbf{K_r}} z(t)g(t,y)h(dt).$$

This conclusion follows from Lemma 29, since the left side of 30(1) is the limit of the left side of 29(1), when R tends to the infinity. In the right side of 29(1) it is possible to take the limit under the sign of the stochastic integral in the mean square sense relative the function N_P for the probability P as well as in the space $L^2(P)$.

Next we describe the generalization of the above construction onto $\mathbf{K}$-linear spaces X, which may be infinite-dimensional over $\mathbf{K}$.

2.3. Vector Spectral Functions

1. Remark. We consider a complete locally $\mathbf{K}$-convex space X over an infinite field $\mathbf{K}$ of zero characteristic, $char(\mathbf{K})=0$, with a non-archimedean multiplicative norm relative to which $\mathbf{K}$ is complete. Then the space $Lin(X,X)=Lin(X)$ of all $\mathbf{K}$-linear continuous operators $F: X\to X$ is locally $\mathbf{K}$ convex and complete.

A $\mathbf{K}$-linear continuous operator F is called compact, if for each $\varepsilon>0$ there exists a finite-dimensional over $\mathbf{K}$ vector subspace X_ε such that it has a complement $Z_\varepsilon := X\ominus X_\varepsilon$ in X and $u(Fx)\le \varepsilon u(x)$ for each $x\in Z_\varepsilon$ and each semi-norm u in X, where Z_ε is the $\mathbf{K}$-vector subspace in X such that $Z_\varepsilon\cap X_\varepsilon=\{0\}$, $Z_\varepsilon\oplus X_\varepsilon = X$.

Henceforth, also the subspace $Lc(X)$ in $Lin(X)$ of all compact operators on X is considered.

Let $W_1 : X\to X^T$ and $W_2 : Lin(X)\to Lin(X)$ be linear isomorphisms of transposition denoted simply by W such that a restriction of W on each finite-dimensional subspace $\mathbf{K^n}$ in X or $Mat_n(\mathbf{K})$ in $Lin(\mathbf{K})$ gives $W(F)$ a transposed vector or matrix, where $Mat_n(\mathbf{K})$ denotes the $\mathbf{K}$-linear space of all $n\times n$ matrices with entries in $\mathbf{K}$. Let two $\mathbf{K}$-bilinear continuous multiplications be given:

$(T1)$ $X\times X\ni \{a,b^T\}\mapsto (a,b)\in\mathbf{K}$ and

$(T2)$ $X\times X\ni\{a^T,b\}\mapsto [a,b]\in Lc(X)$,

so that they are jointly continuous by two variables, where $b^T := W(b)$.

1.1. Examples. As usually $X=c_0(\alpha,\mathbf{K})$ denotes the Banach space consisting of all vectors $x=(x_j : j\in\alpha, x_j\in\mathbf{K})$ such that for each $\varepsilon>0$ the set $\{j : |x_j|>\varepsilon\}$ is finite with the norm $\|x\|_{c_0} := \sup_{j\in\alpha}|x_j|$, where α is a set. Due to the Zermelo Theorem (see [19]) as α one can take an ordinal. This Banach space $c_0(\alpha,\mathbf{K})$ has the standard basis $\{e_j : j\in\alpha\}$, where $e_j = (0,\dots,0,1,0,\dots)$ with 1 in the j-th place and others entries zero. Such basis is orthonormal in the non-archimedean sense [99].

Therefore, for each $F\in Lin(X)$ linear continuous operator there are numbers $F_{i,j}\in\mathbf{K}$ such that

$$Fe_i = \sum_{j\in\alpha} F_{i,j}e_j \tag{1}$$

for each $i\in\alpha$.

If $F\in Lc(X)$ is any compact operator, then for each $\varepsilon>0$ the set
$\beta(F) := \{(i,j) : |F_{i,j}|>\varepsilon, i,j\in\alpha\}$ is finite, where

$$X_\varepsilon = span_{\mathbf{K}}\{e_j : \exists (i,j)\vee(j,i)\in\beta(F)\},$$

$$span_{\mathbf{K}}\{y_j : j\in\beta\} := \{z = a_1y_{j_1}+\cdot+a_ky_{j_k} : a_1,\dots,a_k\in\mathbf{K}, k\in\mathbf{N}, j_1,\dots,j_k\in\beta\}$$

denotes the $\mathbf{K}$-linear span of vectors.

If x is a row-vector, then $W(x)$ is a column-vector. If $F\in Lin(c_0(\alpha,\mathbf{K}))$, then $[W(F)]_{i,j} = F_{j,i}$ for all $i,j\in\alpha$. Taking $\varepsilon_n = p^{-n}$, $n\in\mathbf{N}$, gives that x has non-zero entries only in a countable subset $\beta(x)\subset\alpha$ and the limit

$$\lim_{j\in\alpha} x_j = 0$$

exists. If $a\in X$ and $b\in X$, then the series

$$(a,b) = \sum_{j\in\alpha} a_jb_j$$

converges due to the non-archimedean inequality for the norm and the limit

$$\lim_{j\in\alpha} a_jb_j = 0$$

is zero. If $a,b\in X$, then $[a,b] = F$ with $F_{l,j} = a_lb_j$ for all $l,j\in\alpha$, consequently, $F\in Lc(c_0(\alpha,\mathbf{K}))$.

If the field $\mathbf{K}$ is spherically complete, then a Banach space over $\mathbf{K}$ is isomorphic with $c_0(\alpha,\mathbf{K})$ for some set α and each closed $\mathbf{K}$-linear subspace Z in X is complemented (see Theorems 5.13 and 5.16 [99]). Further, certain closed $\mathbf{K}$-linear subspaces of products of Banach spaces $c_0(\alpha,\mathbf{K})$ can serve as further examples.

2. Note. Henceforth, we shall suppose that

(D) a complete $\mathbf{K}$-convex space X (see §1) has an everywhere dense linear subspace X_0 isomorphic with $c_0(\alpha,\mathbf{K})$ such that a topology τ_0 in X_0 inherited from the topology τ in X is weaker or equal to that of the norm topology τ_c in $c_0(\alpha,\mathbf{K})$.

In the particular case of $\tau_0 = \tau_c$ we can take $X = c_0(\alpha,\mathbf{K})$.

3. Lemma. *Suppose that X is a complete locally $\mathbf{K}$-convex space satisfying Condition 32(D). Then there exists a continuous linear mapping $Tr : Lc(X)\mapsto\mathbf{K}$.*

Proof. Consider an arbitrary compact operator $F \in Lc(X)$ and a semi-norm u in X. If a finite-dimensional over $\mathbf{K}$ subspace X_ε is complemented in X, then the $\mathbf{K}$-linear space $X_{0,\varepsilon} := X_0 \cap X_\varepsilon$ exists so that $X_0 \ominus X_\varepsilon = X_0 \cap Z_\varepsilon =: Z_{0,\varepsilon}$, where $Z_\varepsilon = X \ominus X_\varepsilon$ and $X_{0,\varepsilon} \cap Z_{0,\varepsilon} = \{0\}$, since $X_\varepsilon \cap Z_\varepsilon = \{0\}$. Thus the decomposition into the direct sum $X_0 = X_{0,\varepsilon} \oplus Z_{0,\varepsilon}$ is valid.

Hence there exists the continuous compact restriction of F on X_0. Then for each $\varepsilon > 0$ a finite-dimensional over $\mathbf{K}$ subspace $X_{0,\varepsilon}$ in X_0 exists such that $u(Fx) \le \varepsilon u(x)$ for each vector $x \in Z_{0,\varepsilon}$ and each semi-norm u in X_0. Therefore,

$$\lim_{i\in\alpha} \sup_{j\in\alpha} |F_{i,j}| = 0, \tag{1}$$

since the family of semi-norms $\{u\}$ separates points in X, where $\{e_j : j \in \alpha\}$ is the basis in X_0 inherited from $c_0(\alpha, \mathbf{K})$ and by the Zermelo theorem we take as α an ordinal (see Example 1.1). Thus the series $\sum_{j\in\alpha} F_{j,j}$ converges in $\mathbf{K}$, since $\mathbf{K}$ is complete relative to its non-archimedean norm. Next we put

$$TrF|_{X_0} := \sum_{j\in\alpha} F_{j,j}. \tag{2}$$

The space $X_{0,\varepsilon}$ is isomorphic with $\mathbf{K}^{\mathbf{m}}$ for some $m \in \mathbf{N}$, where the norm in $\mathbf{K}^{\mathbf{m}}$ is equivalent to that of inherited from $c_0(\alpha, \mathbf{K})$. Then each basic vector v_k in $X_{0,\varepsilon}$ has an expansion over $\mathbf{K}$ by the basis $\{e_j : j \in \alpha\}$, consequently, for each $\delta > 0$ a finite subset β in α exists such that $\|v_k - y_k\|_{c_0} < \delta$ for each k, where a vector y_k belongs to the $\mathbf{K}$-linear span $span_{\mathbf{K}}\{e_j : j \in \beta\}$.

Thus for a suitable finite subset β in α for each $x \in X_{0,\varepsilon}$ a vector $y \in span_{\mathbf{K}}\{e_j : j \in \beta\}$ in the $\mathbf{K}$-linear span exists such that $u(x-y) \le \varepsilon u(x)$, since

$$0 \le u(ax+by) \le \max(|a|u(x), |b|u(y))$$

for all vectors $x, y \in X$ and numbers $a, b \in \mathbf{K}$. Therefore, we get the continuous $\mathbf{K}$-linear mapping $Tr : Lc(X_0) \to \mathbf{K}$ relative to the topology τ_0 in X_0 provided by the family of semi-norms $\{u\}$ and inevitably Tr has the continuous $\mathbf{K}$-linear extension on the completion X of X_0 relative to the locally $\mathbf{K}$-convex topology τ in X.

4. Corollary. *Let the conditions of Lemma* 3 *be satisfied and let* $F \in Lc(X)$ *be a compact operator. Then* $\|F|_{X_0}\|_{c_0(\alpha,\mathbf{K})} < \infty$.

Proof. In view of 3(1) the supremum $\sup_{i,j\in\alpha} |F_{i,j}| < \infty$ is finite. Applying the equality

$$\sup_{i,j\in\alpha} |F_{i,j}| = \|F|_{X_0}\|_{c_0(\alpha,\mathbf{K})}$$

finishes the proof.

5. Definition. Suppose that for each subset $A \in \mathcal{R}(G)$ in the separating covering ring a random vector $\xi(A) \in X$ is given. Let the conditions be satisfied:

$$(M1) \quad \xi(A) \in Y,$$

$\xi(\emptyset) = 0$, where $Y = L^2(\Omega, \mathcal{R}, P, X)$;

$$(M2) \quad \xi(A_1 \cup A_2) = \xi(A_1) + \xi(A_2) \quad mod(P)$$

for each $A_1, A_2 \in \mathcal{R}(G)$ with $A_1 \cap A_2 = \emptyset$;

$$(M3) \quad M[\xi(A_1), \xi(A_2)] = \mu(A_1 \cap A_2);$$

$$(M4) \quad M[\xi(A_1), \xi(A_2)] = 0$$

for each $A_1 \cap A_2 = \emptyset$, $A_1, A_2 \in \mathcal{R}(G)$. This means that $\xi(A_1)$ and $\xi(A_2)$ are orthogonal random variables, where $\mu(A) \in Lc(X)$ for each $A, A_1, A_2 \in \mathcal{R}(G)$.

The family of random vectors $\{\xi(A) : A \in \mathcal{R}(G)\}$ satisfying Conditions $(M1-M4)$ is called the (elementary) orthogonal X-valued stochastic measure. The compact operator $\mu(A)$ is called the structural operator.

6. Lemma. *If subsets $A_1, A_2 \in \mathcal{R}(G)$ belong to the separating covering ring, $A_1 \cap A_2 = \emptyset$, then*

$$\mu(A_1 \cup A_2) = \mu(A_1) + \mu(A_2).$$

Proof. Generalizing the proof of Lemma 10 we get the statement of this lemma, since the product in X with values in $Lc(X)$ is continuous and $Lc(X)$ is the locally **K**-convex space having also the structure of the algebra over **K**. At the same time $L^1(P,X) \subset L^2(P,X)$. The equalities

$$\mu(A_1 \cup A_2) = M[\xi(A_1 \cup A_2), \xi(A_1 \cup A_2)] = M[\xi(A_1) + \xi(A_2), \xi(A_1) + \xi(A_2)]$$

$$= M[\xi(A_1), \xi(A_1)] + M[\xi(A_2), \xi(A_2)] = \mu(A_1) + \mu(A_2)$$

are satisfied, since

$$M[\xi(A_1), \xi(A_2)] = 0 \text{ and } M[\xi(A_2), \xi(A_1)] = 0 \quad \text{for } A_1 \cap A_2 = \emptyset.$$

7. Note. Definitions 2.1 can be generalized in the following manner. Let Z be a locally **K**-convex space supplied with a family of semi-norms $\mathcal{S}(Z)$ defining its topology. If $\mathcal{A}$ is a shrinking family, $f : \mathcal{R} \to Z$, then the limit

$$\lim_{A \in \mathcal{A}} f(A) = 0$$

is zero, if for each $\varepsilon > 0$ and each $u \in \mathcal{S}(Z)$ there exists a subset $A_0 \in \mathcal{A}$ such that $u(f(A)) < \varepsilon$ for each $A \in \mathcal{A}$ with $A \subset A_0$.

A measure $\mu : \mathcal{R} \to Z$ is a mapping with values in Z satisfying the following properties:
(i) μ is additive;
(ii) for each $A \in \mathcal{R}$ the set $\{\mu(B) : B \in \mathcal{R}, A \subset B\}$ is bounded;
(iii) if $\mathcal{A}$ is the shrinking family in $\mathcal{R}$ and $\bigcap_{A \in \mathcal{A}} A = \emptyset$, then the limit

$$\lim_{A \in \mathcal{A}} \mu(A) = 0$$

is zero. Henceforth, we suppose that μ has an extension to a $Lc(X)$-valued measure on the separating covering ring $\mathcal{R}(G) = \mathcal{R}$.

8. Lemma. *If X is a complete locally* **K**-*convex space satisfying Condition $2(D)$ and a measure μ is as in §7, then there exists the trace $Tr\mu(A)$ of μ for each $A \in \mathcal{R}(G)$. Moreover, $Tr\mu$ is the* **K**-*valued measure.*

Proof. In view of Lemma 3 there exists the continuous mapping $Tr: Lc(X) \to \mathbf{K}$, hence $Tr\mu(A) \in \mathbf{K}$ is a finite number for each $A \in \mathcal{R}(G)$, since $\mu(A)$ is the compact operator. Then the semi-norm

$$\|\mu(A)\|_u := \sup_{u(x)\neq 0, x\in X,} u(\mu(A)x)/u(x) \leq \sup_{i,j\in\alpha} |[\mu(A)]_{i,j}| = \|[\mu(A)]|_{X_0}\|_{c_0} < \infty$$

is finite due to Corollary 34. Therefore,

$$|Tr\mu(A)| \leq \|[\mu(A)]|_{X_0}\|_{c_0} \qquad (1)$$

for every $A \in \mathcal{R}(G)$. Thus if $\mathcal{A}$ is the shrinking family in the separating covering ring $\mathcal{R}$ and $\bigcap_{A\in\mathcal{A}} A = \emptyset$, then the limit $\lim_{A\in\mathcal{A}} \mu(A) = 0$ is zero, consequently, $\lim_{A\in\mathcal{A}} \|\mu(A)\|_u = 0$ for each semi-norm u in X and inevitably $\lim_{A\in\mathcal{A}} Tr\mu(A) = 0$ due to Inequality (1) and $\mu(A) \in Lc(X)$ and $3(1)$.

9. Definition. Let g be a complete locally **K**-convex algebra with a unit 1 and X be a complete locally **K**-convex space satisfying Condition $2(D)$ and let simultaneously X be a unital left g-module, where g also satisfies Conditions $1(T1, T2)$. This means, that there exists a mapping $g \times X \to X$ satisfying the following conditions $(1-7)$:

(1) $b(x_1 + x_2) = bx_1 + bx_2$,

(2) $(b_1 + b_2)x = b_1x + b_2x$,

(3) $b_1(b_2x) = (b_1b_2)x$,

(4) $1x = x$,

(5) there exists a family of consistent semi-norms $\mathcal{S} = \{u\}$ in g and X defining their Hausdorff topologies such that $u(bx) \leq u(b)u(x)$, $u(ab) \leq u(a)u(b)$,

(6) $(ax, by) = (b^T ax, y) = (x, a^T by) \in \mathbf{K}$ and

(7) $[ax, by] = [a,b][x,y] \in Lc(X)$ such that $Lc(X)$ is the left $Lc(g)$-module for each $a, b, b_1, b_2 \in g$ and every $x, y, x_1, x_2 \in X$.

For each simple function $f \in L^0(\mathcal{R}, g)$ we define the stochastic integral:

$$(SI) \quad \eta = \int_G f(x)\xi(dx) := \sum_k a_k \xi(A_k),$$

where $f(x) = \sum_k a_k Ch_{A_k}(x)$, $a_k \in g$.

Henceforward, $L^0(\xi, g) = L^0(\xi)$ denotes the family of all random vectors η of the form (SI).

10. Examples. At first we consider either $g = \mathbf{K}$ or a subalgebra g in $Lin(X)$, where X is a complete locally **K**-convex space. Each semi-norm v in X induces the consistent semi-norm

$$\|F\|_v := \sup_{x\in X, v(x)\neq 0} v(Fx)/v(x)$$

for each $F \in L(X)$. For simplicity of the notation we can denote $\|F\|_v$ also by $v(F)$ and these semi-norms in X and in $Lin(X)$ are consistent, since $v(Fx) \leq \|F\|_v v(x)$, where we distinguish $v(F)$ and $v(Fx)$. Evidently, each multiple bI of the unit operator I also belongs to $Lin(X)$ for $b \in \mathbf{K}$.

We take now a group H with a **K**-valued measure ν on the separating covering ring $\mathcal{R}(H)$ or particularly ν on $\mathsf{B_c}(H)$. Here a topological totally disconnected group H may be such that $\sup_{x\in H} N_\nu(x) = 1$. Let $L^q_b(H,\mathsf{B_c}(H),\nu,\mathsf{b})$ be a completion of the family of all step functions $f: H \to \mathsf{b}$ with supports in $A \in \mathsf{B_c}(H)$ on which $\nu|_A$ is the measure relative to the family of all non-archimedean semi-norms

$$\|f\|_{q,b,v} = [\sup_{x\in H, y\in H} v[f(y^{-1}x)]^q N_\nu(x)]^{1/q} < \infty,$$

where $1 \le q < \infty$, b is a complete locally convex algebra over $\mathbf{K}$ with a family of semi-norms $\{v\}$ in it and b satisfies Conditions $1(T1,T2)$. Certainly $v(xy) \le v(x)v(y)$ for each $x,y \in \mathsf{b}$ and every semi-norm v. In particular, this space is defined for the measure ν on $\mathcal{R}(H)$.

If $f_1, f_2 \in L^1_b(H,\nu,\mathsf{b})$, the convolutions are defined by the formulas:

$$conv\{f_1,f_2\} := \{f_1 * f_2\}(x) := \int_H f_1(y^{-1}x)f_2(y)\nu(dy), \tag{1}$$

$$conv[f_1,f_2] := [f_1 * f_2](x) := \int_H [f_1(y^{-1}x), f_2(y)]\nu(dy), \tag{2}$$

$$conv(f_1,f_2) := (f_1 * f_2)(x) := \int_H (f_1(y^{-1}x), f_2(y))\nu(dy). \tag{3}$$

They are defined for simple functions. If they exist, then

$$\sup_{x\in H, z\in H} v[\{f_1 * f_2\}(z^{-1}x)]N_\nu(x)$$

$$\le \sup_{x\in H, y\in H, z\in H} v[f_1(y^{-1}z^{-1}x)]v[f_2(y)]N_\nu(x)N_\nu(y)$$

$$\le \|f_1\|_{1,b,v}\|f_2\|_{1,b,v} < \infty, \tag{4}$$

$$\sup_{x\in H, z\in H} v[[f_1 * f_2](z^{-1}x)]N_\nu(x)$$

$$\le \sup_{x\in H, y\in H, z\in H} C_v v[f_1(y^{-1}z^{-1}x)]v[f_2(y)]N_\nu(x)N_\nu(y)$$

$$\le C_v\|f_1\|_{1,b,v}\|f_2\|_{1,b,v} < \infty, \tag{5}$$

$$\sup_{x\in H, z\in H} |(f_1 * f_2)(z^{-1}x)|N_\nu(x)$$

$$\le \sup_{x\in H, y\in H, z\in H} J_v v[f_1(y^{-1}z^{-1}x)]v[f_2(y)]N_\nu(x)N_\nu(y)$$

$$\le J_v\|f_1\|_{1,b,v}\|f_2\|_{1,b,v} < \infty, \tag{6}$$

since the mappings $1(T1,T2)$ are continuous, $y^{-1}z^{-1} = (zy)^{-1}$, where the semi-norm in $Lin(\mathsf{b})$ induced by the semi-norm v in b is also denoted by v,

$$J_v = \sup_{a,b,v(a)>0,v(b)>0} |(a,b)|/[v(a)v(b)] \quad \text{and}$$

$$C_v = \sup_{a,b,v(a)>0,v(b)>0} v([a,b])/[v(a)v(b)]$$

are finite constants equal to semi-norms of the jointly continuous mappings 1 $(T1, T2)$ correspondingly relative to the semi-norm v on g.

This implies that the convolutions have the continuous extensions on the space $L^1_b(H, \nu, \mathsf{b}) =: X$ such that

$$conv\{f_1, f_2\} \in L^1_b(H, \nu, \mathsf{b}), \quad conv[f_1, f_2] \in L^1_b(H, \nu, Lc(\mathsf{b})),$$

$$conv(f_1, f_2)k \in L^1_b(H, \nu, \mathbf{K}).$$

The space X is **K**-linear and complete and it is the algebra with the multiplication being the convolution $conv\{f_1, f_2\}$. If the constant function 1 is not in this space, then it may be adjoined to get a complete locally **K**-convex algebra with a unit $1(x) = 1$ for each $x \in H$. Such function permits to encompass the case of the algebra b with the unit element. This constant function generally need not be a unit relative to the multiplication in this algebra X of functions. Certainly, to such algebra X the unit element relative to the multiplication may be attached.

This is the group algebra X of H over b. Particularly, we can take $\mathsf{b} = Mat_m(\mathbf{K})$ also or more general algebras as above. Then the transposition in b induces it in the space $X = L^1_b(H, \nu, \mathsf{b})$ also so that

$$(f_1, f_2) := \int_H (f_1(x), f_2(x))\nu(dx) \in \mathbf{K} \quad \text{and}$$

$$[f_1, f_2] := [f_1 * f_2] \in Lin(X)$$

can be considered as the unique continuous linear operator F on X such that $F \in Lin(X)$,

$$Ff(x) =< [f_1 * f_2] * f > (x), \tag{7}$$

where

$$< g * f > (x) := \int_H g(y^{-1}x) f(y)\nu(dy) \tag{8}$$

for each $g \in L^1_b(H, \mathcal{R}, \nu, Lin(\mathsf{b}))$ and each $f \in X$ and every $x \in H$. If b satisfies Condition $9(1-7)$, then X also satisfies them. Particularly, if b is the Banach algebra, then X also is the Banach algebra.

11. Theorem. *Let a mapping F be given by the formula*

$$Ff(x) =< [f_1 * f_2] * f > (x),$$

where $f, f_1, f_2 \in L^1_b(H, \nu, \mathsf{b}) = X$ and H is the topological group, $\mathcal{R}(H) \subset \mathsf{Bco}(H)$ as in Example 10, $\sup_{x \in H} N_\nu(x) < \infty$ is finite. Then F is the compact operator $F \in Lc(X)$ and the mapping $X^2 \ni \{f_1, f_2\} \mapsto < [f_1 * f_2] * > \in Lc(X)$ is continuous.

Proof. As it was demonstrated in Example 10, the convolution $conv[f_1, f_2]$ is the continuous mapping from X^2 into $L^1_b(H, \nu, Lc(\mathsf{b}))$, where ν is the **K**-valued measure. The space $L^1_b(H, \nu, Lc(\mathsf{b}))$ is the completion of the family of all step functions

$$g(x) = \sum_k Ch_{A_k}(x) a_k$$

relative to the family of semi-norms

$$\|g\|_{1,b,v} = [\sup_{x\in H, y\in H} v[g(y^{-1}x)]N_\nu(x)] < \infty,$$

where $\|Y\|_{v,\mathsf{b}} := \sup_{v(t)\neq 0, t\in \mathsf{b}} v(Yt)/v(t)$ is the semi-norm in $Lin(\mathsf{b})$ denoted also by $v(Y)$, with subsets $A_k \in \mathcal{R}(H)$ and expansion coefficients $a_k \in Lc(\mathsf{b})$. Therefore, it is sufficient to demonstrate that $F \in Lc(X_s)$ and the mapping $X_s^2 \ni \{f_1, f_2\} \mapsto [f_1 * f_2] \in Lc(X_s)$ is continuous, where $X_s := L^1_b(H, \nu, \mathbf{K})$.

Theorem 7.12 [99] states, that $f \in L(H, \mathcal{R}, \nu, \mathbf{K})$ if and only if it has two properties:

(i) f is $\mathcal{R}_\nu$-continuous,

(ii) for every $\varepsilon > 0$ the set $\{x : |f(x)|N_\nu(x) \geq \varepsilon\}$ is $\mathcal{R}_\nu$-compact, consequently, contained in $\{x : N_\nu(x) \geq \delta\}$ for some $\delta > 0$.

For vector valued functions its generalization is given in Theorem 16 below, which is proved independently of §§10, 11. Thus if $f_1, f_2 \in X_s$, then $f_1 * f_2$ is $\mathcal{R}_\nu$-continuous and for each $\varepsilon > 0$ and every semi-norm v in b a positive number exists $\delta > 0$ so that the inclusion

$$\{x : v(f_1 * f_2(x))N_{\nu,v}(x) \geq \varepsilon\} \subset \{x : N_{\nu,v}(x) \geq \delta\}$$

is accomplished, where the sets $\{x : v(f_1 * f_2(x))N_{\nu,v}(x) \geq \varepsilon\}$ and $\{x : N_{\nu,v}(x) \geq \delta\}$ are $\mathcal{R}_\nu$-compact, consequently, they are $\mathcal{R}$-compact sets.

If $f \in L(H, \mathcal{R}, \nu, \mathsf{b})$, then for each $\varepsilon > 0$ and each continuous semi-norm v in b and every $x \in H$ there exists an open symmetric neighborhood U_x of the unit element e in the topological group H such that $v(f(y^{-1}x) - f(x)) < \varepsilon$ for each $y \in U_x^3$. From the covering $\{xU_x : x \in H, N_{\nu,v}(x) \geq \delta\}$ of $\{x : N_{\nu,v}(x) \geq \delta\}$ one can a finite covering $\{x_j U_{x_j} : j = 1, \dots, q\}$ extract and take $U = \bigcap_{j=1}^q U_{x_j}$, since $\mathcal{R} \subset \mathsf{Bco}(H)$. Then the set U is open symmetric $U^{-1} = U$ and the unit element $e \in U$ belongs to U.

If $y \in U$ and $N_{\nu,v}(x) \geq \delta$, then there exists a natural number j such that $x \in x_j U_{x_j}$, consequently,

$$v(f(y^{-1}x) - f(x)) \leq \max(v(f(y^{-1}x) - f(y^{-1}x_j)),$$
$$v(f(y^{-1}x_j) - f(x_j)), v(f(x_j) - f(x))) < \varepsilon,$$

since $(y^{-1}x)(y^{-1}x_j)^{-1} \in U^3 \subset U^3_{x_j}$. Hence

$$v(f(y^{-1}z) - f(z)) \leq \max(v(f(y^{-1}z) - f(y^{-1}t)), v(f(y^{-1}t) - f(t))) < \varepsilon$$

for each $z \in [U\{x \in H : N_{\nu,v}(x) \geq \delta\}]$, where $t \in \{x \in H : N_{\nu,v}(x) \geq \delta\}$ is such that $zt^{-1} \in U$, since $(y^{-1}z)(y^{-1}t)^{-1} \in U^3$. At the same time

$$\{x \in H : N_{\nu,v}(x) \geq \delta\} \subset [U\{x \in H : N_{\nu,v}(x) \geq \delta\}]$$

and $v(f(z))N_{\nu,v}(z) < \varepsilon$ for each $z \in H \setminus \{x \in H : N_{\nu,v}(x) \geq \delta\}$. If $z \in H \setminus [U\{x \in H : N_{\nu,v}(x) \geq \delta\}]$, then $Uz \cap \{x \in H : N_{\nu,v}(x) \geq \delta\} = \emptyset$. Thus for each function $f \in L(H, \mathcal{R}, \nu, \mathsf{b})$ and every positive number $\varepsilon > 0$ and each semi-norm v in b an open symmetric neighborhood U of the unit element e in H exists such that $v(f(y^{-1}x) - f(x)) < \varepsilon$ for each $x \in H$ and $y \in U$.

Let us take the functions $f, f_1, f_2 \in X_s$. Then for each $\varepsilon > 0$ an open symmetric neighborhood U of the unit element e in H exists such that the estimate (*iii*) $\nu(<[f_1 * f_2] * f > (y^{-1}x) - <[f_1 * f_2] * f > (x))N_{\nu,v}(x) < \varepsilon \sup_{x\in H} \nu(f(x))$ $N_{\nu,v}(x)$ is fulfilled for each $x \in H$ and every $y \in U$. Then we take the sets

$$\{x : v([f_1 * f_2](x))N_{\nu,v}(x) \geq \varepsilon\} \subset \{x : N_{\nu,v}(x) \geq \delta\} =: A$$

for $[f_1 * f_2]$ as above. It is useful to make the decomposition: $F = F_1 + F_2$, where $F_1 f :=< [f_1 * f_2] * (Ch_A f) >$ and $F_2 f :=< [f_1 * f_2] * ((1 - Ch_A) f >$. Hence $\|F_2\|_v \leq \varepsilon$, since $\sup_{x\in H} N_\nu(x) < \infty$ is finite.

In view of Inequality (*iii*) for each $\varepsilon > 0$ a finite covering family of subsets $A_1, \dots, A_m \in \mathcal{R}$ in A exists, $\bigcup_{j=1}^m A_j = A$, such that for each $f \in X_s$ a simple function

$$g(x) = \sum_{j=1}^{m} b_k Ch_{A_k}(x)$$

exists for which

$$\|F_1(f - g)\|_v < \varepsilon.$$

But the family of such simple functions g is finite-dimensional over **K**, consequently, F is the compact operator.

In view of Formulas 10(2,5) the mapping $X^2 \ni \{f_1, f_2\} \mapsto < [f_1 * f_2] * > \in Lc(X)$ is continuous.

12. Lemma. *Suppose that for a natural number k the moment exists $M\xi^k$, where ξ is a random vector with values in a locally* **K***-convex algebra* g. *Then for each $1 \leq l < k$ the moment of the l-th order $M\xi^l$ exists.*

Proof. If ξ is a random vector with values in g, then the random vector ξ by the definition is $(\mathcal{A}, \mathcal{R}(\mathrm{g}))$-measurable and for each semi-norm u in g the supremum

$$\sup_{\omega\in\Omega} [u(\xi)^k N_P(\omega)]^{1/k} = \|\xi\|_{k,u} < \infty$$

exists, where $\mathcal{R}(\mathrm{g})$ is a separating covering ring of g such that $\mathcal{R}(\mathrm{g}) \subset \mathsf{Bco}(\mathrm{g})$. Therefore,

$$\sup_{\omega\in\Omega} [u(\xi)^l N_P(\omega)]^{1/l} \leq \sup_{\omega\in\Omega} [u(\xi)^k N_P(\omega)]^{1/k}$$

for each $1 \leq l < k$, since $N_P(\omega) \leq 1$ for each $\omega \in \Omega$.

13. Remark. By $L^2(\xi, \mathrm{g})$ the completion of $L^0(\xi, \mathrm{g})$ is denoted. The completion is done by the family of continuous semi-norms

$$\|f\|_{2,P,u} := [\sup_{x\in G, \omega\in\Omega} u^2(f(x)\xi(\omega, x))N_P(\omega)]^{1/2} \tag{1}$$

induced from $L^2(\Omega, \mathcal{A}, P, X)$, where $L^0(\xi, \mathrm{g})$ is the space of all step functions of the form $\sum_{k=1}^l a_k \xi(A_k)$ with expansion coefficients $a_k \in \mathrm{g}$ and subsets $A_k \in \mathcal{R}(G)$, $A_j \cap A_k = \emptyset$ for each $k \neq j$, $l \in \mathbf{N}$.

14. Definition. Let Z be a locally **K**-convex space and let $\mu : \mathcal{R} \to Z$ be a measure, where $\mathcal{R} = \mathcal{R}(G)$ is a separating covering ring of a set G. For each semi-norm u in Z and any $A \in \mathcal{R}(G)$ we define the semi-norm

$$\|A\|_{\mu,u} := \sup\{u(\mu(B)) : B \in \mathcal{R}, B \subset A\}. \tag{1}$$

15. Lemma. *If $\mu : \mathcal{R} \to Z$ is a measure, then for each semi-norm u in Z there exists a unique function $N_{\mu,u} : G \to [0,\infty)$ such that*

$$\|Ch_A\|_{N_{\mu,u}} = \|A\|_{\mu,u} \tag{1}$$

for each $A \in \mathcal{R}$ (see Definition 44);

$$\text{if } \phi : G \to [0,\infty) \text{ and } \|Ch_A\|_\phi \leq \|A\|_{\mu,u} \text{ for all } A \in \mathcal{R}, \text{ then } \phi \leq N_{\mu,u}; \tag{2}$$

moreover, this function can be evaluated as

$$N_{\mu,u}(x) = \inf_{A : x \in A \in \mathcal{R}} \|A\|_{\mu,u} \quad \text{for each } x \in G. \tag{3}$$

Proof. If the function $N_{\mu,u}(x)$ is defined by the formula (3), then the second statement is evident, since $\|Ch_A\|_\phi \leq \|A\|_{\mu,u}$ for all $A \in \mathcal{R}$. We take any $A \in \mathcal{R}$ and consider the family

$$\mathcal{E} := \{B \in \mathcal{R} : B \subset A, \|A \setminus B\|_{\mu,u} \leq \|Ch_A\|_{N_{\mu,u}} + \varepsilon\}.$$

On the other hand, from Formula 14(1) it follows that

$$\|A_1 \cup A_2\|_{\mu,u} \leq \max(\|A_1\|_{\mu,u}, \|A_2\|_{\mu,u}\}$$

for each $A_1, A_2 \in \mathcal{R}$, since μ is additive and for each $B \subset A_1 \cup A_2$ we have $B = B_1 \cup B_2$, while

$$u(\mu(A_1 \cup A_2)) \leq \max\{u(\mu(A_1 \setminus A_2)), u(\mu(A_2 \setminus A_1)), u(\mu(A_1 \cap A_2))\},$$

where $B_1 := A_1 \cap B$ and $B_2 := A_2 \cap B$. Therefore, $\mathcal{E}$ is the shrinking family.

Then for each $x \in A$ there exists $B \in \mathcal{R}$ such that $x \in B$ and

$$\|B\|_{\mu,u} \leq N_{\mu,u}(x) + \varepsilon \leq \|Ch_B\|_{N_{\mu,u}} + \varepsilon,$$

consequently, $A \setminus B \in \mathcal{E}$. Since $\bigcap_{A \in \mathcal{E}} A = \emptyset$, a subset $B \in \mathcal{E}$ exists so that

$$\|B\|_{\mu,u} \leq \varepsilon$$

and inevitably

$$\|A\|_{\mu,u} \leq \max\{\|B\|_{\mu,u}, \|A \setminus B\|_{\mu,u} + \varepsilon\}.$$

16. Definitions. Suppose that μ is an X-valued measure as in §§17, 19. For a function $f : G \to \mathrm{g}$ we consider the family of semi-norms

$$\|f\|_{q,\mu,u} := [\sup_{x \in G} u^q(f) N_{\mu,u}(x)]^{1/q}$$

whenever it exists, where $1 \leq q < \infty$ and u is a semi-norm in X.

The space $L^q(G, \mathcal{R}, \mu, X; \mathrm{g})$ is defined as the completion of the family of all step (simple) functions f relative to the family of semi-norms $\{\| * \|_{q,\mu,u} : u \in \mathcal{S}\}$.

If $f \in L^1(G, \mathcal{R}, \mu, X; \mathrm{g})$, then it is called μ-integrable.

In the particular case when the algebra coincides with the field $\mathrm{g} = \mathbf{K}$ we can omit it from the notation. When G, $\mathcal{R}$, μ, X and g are specified it also can be written shortly $L^q(\mu)$. For $q = 1$ this upper left index can be omitted writing $L(\mu)$.

A function f is called μ-integrable, if $f \in L(\mu)$. The completed separating covering ring is defined by the formula:

$$\mathcal{R}_\mu := \{A \subset G : Ch_A \in L(\mu)\}$$

for the X-valued measure μ. The measure μ is extended with the help of the integral

$$\bar{\mu}(A) := \int_G Ch_A(x)\mu(dx),$$

since $1 \in \mathrm{g}$.

17. Lemma. *If the measure μ and the separating covering ring $\mathcal{R}_\mu$ are as in §16, then $A \in \mathcal{R}_\mu$ if and only if for each $\varepsilon > 0$ and each semi-norm u in X a subset $B \in \mathcal{R}$ exists such that $N_{\mu,u}(x) \le \varepsilon$ for every $x \in A \triangle B$.*

Proof. If for each $\varepsilon > 0$ and a semi-norm u in X a subset $B \in \mathcal{R}$ exists so that

$$\|Ch_A - Ch_B\|_{N_{\mu,u}} < \varepsilon,$$

then taking a sequence $\varepsilon_n = p^{-n}$ we get, that the characteristic function Ch_A belongs to the space $L(\mu)$, hence $A \in \mathcal{R}_\mu$.

On the other hand, if $A \in \mathcal{R}_\mu$, then for each $1 > \varepsilon > 0$ there exists a simple function $f \in L(\mu)$ such that $\|Ch_A - f\|_{N_{\mu,u}} < \varepsilon$. We take the subset $B := \{x : u(f(x) - 1) < 1\}$ belonging to the separating covering ring $\mathcal{R}$, $B \in \mathcal{R}$. Then

$$u(f(x) - Ch_B(x)) \le \min[u(f(x)), u(f(x) - 1)] \le u(f(x) - Ch_A(x))$$

for each $x \in G$ and inevitably

$$\|Ch_A - Ch_B\|_{N_{\mu,u}} \le \max[\|f - Ch_A\|_{N_{\mu,u}}, \|f - Ch_B\|_{N_{\mu,u}}]$$

$$= \|f - Ch_A\|_{N_{\mu,u}} \le \varepsilon.$$

18. Lemma. *If a measure $\mu : \mathcal{R} \to X$ satisfies Conditions $7(i, ii)$, then Property $7(iii)$ is equivalent to:*

(iii') if $\mathcal{A} \subset \mathcal{R}$ is a shrinking family and $\bigcap_{A \in \mathcal{A}} A = \emptyset$, then

$$\lim_{A \in \mathcal{A}} \|A\|_{\mu,u} = 0$$

for each semi-norm $u \in \mathcal{S}$ in X.

Proof. Since $\|A\|_{\mu,u} \ge u(\mu(A))$ for each subset A in the separating covering ring $\mathcal{R}$, Condition (iii') implies Property $7(iii)$.

Prove now the converse statement supposing that μ satisfies Conditions $7(i - iii)$. Suppose that $\mathcal{A}$ is a shrinking family with the void intersection $\bigcap_{A \in \mathcal{A}} A = \emptyset$. For each $\varepsilon > 0$ and every $u \in \mathcal{S}$ there exist an element $E \in \mathcal{A}$ such that $u(\mu(A)) < \varepsilon$ for each $A \in \mathcal{A}$ such that $A \subset E$. Next we take an arbitrary continuous semi-norm $u \in \mathcal{S}$ in X. For every $A \in \mathcal{A}$ choose $V_A \in \mathcal{R}$ such that $V_A \subset A$ and $u(\mu(V_A)) > \min(\varepsilon, \|A\|_{\mu,u}/2)$. If $A \in \mathcal{A}$, then the family

$$\mathcal{C} := \{V_A \cap B : B \in \mathcal{A}, B \subset A\}$$

is shrinking and $\bigcap_{U \in \mathcal{C}} U = \emptyset$, hence for each $A \in \mathcal{A}$ there exists $W_A \in \mathcal{A}$ such that $W_A \subset A$ and $u(\mu(V_A \cap W_A)) < \varepsilon$.

We consider the family of subsets

$$\mathcal{V} := \{V_A \cup W_A : A \in \mathcal{A}, A \subset E\}.$$

If $A, B \in \mathcal{A}$, then there exists $C \in \mathcal{A}$ with $C \subset W_A \cap W_B$. Then $V_C \cup W_C \subset C \subset W_A \subset V_A \cup W_A$, also $V_C \cup W_C \subset V_B \cup W_B$, consequently, the family $\mathcal{V}$ is shrinking. Moreover, its intersection is void $\bigcap_{C \in \mathcal{V}} C = \emptyset$. Thus a subset $A \in \mathcal{A}$ exists with $A \subset E$ and $u(\mu(V_A \cup W_A)) < \varepsilon$, also $u(\mu(V_A \cup W_A)) < \varepsilon$ and $u(\mu(V_A \cap W_A)) < \varepsilon$ due to the definition of W_A, as well as $u(\mu(W_A)) < \varepsilon$, since $W_A \in \mathcal{A}$ and $W_A \subset A \subset E$. Hence the estimate

$$u(\mu(V_A)) = u(\mu(V_A \cup W_A) + \mu(V_A \cap W_A) - \mu(W_A)) < \varepsilon$$

is satisfied. Therefore,

$$\|A\|_{\mu,u} \leq 2\varepsilon,$$

since $u(\mu(V_A)) > \min(\varepsilon, \|A\|_{\mu,u}/2)$.

19. Theorem. *Suppose that μ is an X-valued measure on a separating covering ring $\mathcal{R}$. Then $\mathcal{R}_\mu$ is a separating covering ring of G and $\bar{\mu}$ is an X-valued measure extending μ.*

Proof. In view of Lemma 17 $\mathcal{R}_\mu$ is the separating covering ring of the set G and the mapping $\bar{\mu} : \mathcal{R}_\mu \to X$ is additive. If $A \in \mathcal{R}_\mu$, then for each $B \subset A$ such that $B \in \mathcal{R}$ and each semi-norm u in X the inequalities are satisfied:

$$u(\bar{\mu}(B)) \leq \|Ch_B\|_{N_{\mu,u}} \leq \|Ch_A\|_{N_{\mu,u}} < \infty,$$

hence $\bar{\mu}$ has property $17(ii)$, where Ch_A denotes as usually the characteristic function of the set A.

Consider now any shrinking family $\mathcal{A} \subset \mathcal{R}_\mu$ having empty intersection. For $\varepsilon > 0$ and a semi-norm u in X we take the family of subsets

$$\mathcal{E} := \{B \in \mathcal{R} : \exists A \in \mathcal{A} \text{ such that } A \cap G_{\varepsilon,u} = B \cap G_{\varepsilon,u}\},$$

where $G_{\varepsilon,u} := \{x \in G : N_{\mu,u}(x) \geq \varepsilon\}$. This implies that the family $\mathcal{E}$ is shrinking. If $x \notin G_{\varepsilon,u}$, then there exists $V \in \mathcal{R}$ such that $x \in V$ and $\varepsilon > \|V\|_{\mu,u}$, hence $B \setminus V \in \mathcal{E}$ for each $B \in \mathcal{E}$, consequently, $V \cap (\bigcap_{B \in \mathcal{E}} B) = \emptyset$ and inevitably $\bigcap_{B \in \mathcal{E}} B \subset G_{\varepsilon,u}$. Then by the construction of $\mathcal{E}$ we get:

$$\bigcap_{B \in \mathcal{E}} B = \bigcap_{B \in \mathcal{E}} B \cap G_{\varepsilon,u} = \bigcap_{A \in \mathcal{A}} A \cap G_{\varepsilon,u} = \emptyset.$$

In view of Lemma 18 there exists s subset $B \in \mathcal{E}$ such that $\|B\|_{\mu,u} < \varepsilon$, hence $B \cap G_{\varepsilon,u} = \emptyset$. Then there exists $A \in \mathcal{A}$ such that

$$A \cap G_{\varepsilon,u} = B \cap G_{\varepsilon,u} = \emptyset,$$

consequently, $\|A\|_{\mu,u} < \varepsilon$. Again in accordance with Lemma 18 the limit

$$\lim_{A\in\mathcal{A}} \bar{\mu}(A) = 0$$

is zero. Thus $\bar{\mu}$ is the X-valued measure.

20. Lemma. *If μ is an X-valued measure on $\mathcal{R}$, then $N_{\mu,u} = N_{\bar{\mu},u}$ for each semi-norm u in X. Moreover,*

$$\| * \|_{N_{\mu,u}} = \| * \|_{N_{\bar{\mu},u}}, \quad L(\bar{\mu}) = L(\mu),$$

$$\int_G f d\bar{\mu} = \int_G f d\mu, \quad \mathcal{R}_{\bar{\mu}} = \mathcal{R}_{\mu}.$$

Proof. We take an arbitrary continuous semi-norm u in X and a point $x \in G$ and a number $b > N_{\mu,u}(x)$. Then there exists a subset $A \in \mathcal{R} \subset \mathcal{R}_{\mu}$ such that $x \in A$ and $\|A\|_{N_{\bar{\mu},u}} \leq b$. Then for every $B \in \mathcal{R}$ such that $B \subset A$ the inequalities

$$u(\bar{\mu}(B)) \leq \|B\|_{N_{\bar{\mu},u}} \leq \|A\|_{N_{\bar{\mu},u}} \leq b$$

are satisfied , hence

$$\|A\|_{N_{\bar{\mu},u}} \leq b, \tag{1}$$

consequently, the inequality

$$N_{\bar{\mu},u}(x) \geq \inf\{\|A\|_{\bar{\mu},u} : A \in \mathcal{R}_{\mu}, x \in A\}$$

is fulfilled.

We take now a positive number $0 < d < N_{\mu,u}(x)$ and a subset $A \in \mathcal{R}_{\mu}$ with a point $x \in A$ in it. In view of Lemma 17 a subset $B \in \mathcal{R}$ exists so that $N_{\bar{\mu},u}(y) \leq d$ for each $y \in A \triangle B$. Therefore,

$$\|B\|_{N_{\bar{\mu},u}} \geq N_{\bar{\mu},u}(x) > d,$$

so $u(\mu(E)) > d$ for some $E \in \mathcal{R}$ such that $E \subset B$. Then we infer the inequalities

$$u(\mu(E) - \bar{\mu}(E \cap A)) = u(\bar{\mu}(E \setminus A)$$

$$\leq \|E \setminus A\|_{N_{\bar{\mu},u}} \leq \|E \setminus B\|_{N_{\bar{\mu},u}} \leq d < u(\mu(E)).$$

Thus $u(\bar{\mu}(E \cap A)) = u(\mu(E))$, consequently, the semi-norm of the set A is subordinated to the inequalities:

$$\|A\|_{\bar{\mu},u} \geq u(\bar{\mu}(E \cap A)) = u(\mu(E)) > d \tag*{(2).}$$

Finally, from these two Inequalities $(1,2)$ the equality $N_{\bar{\mu},u}(x) = N_{\mu,u}(x)$ for each $x \in G$ follows.

21. Theorem. (1). *If μ is a measure on a separating covering ring $\mathcal{R}$, then the function $N_{\mu,u}$ is upper semi-continuous for each semi-norm $u \in \mathcal{S}$ in X. Moreover, it is $\mathcal{R}_{\mu}$-upper semi-continuous and for every $A \in \mathcal{R}_{\mu}$ and $\varepsilon > 0$ the set $\{x \in A : N_{\mu,u}(x) \geq \varepsilon\}$ is $\mathcal{R}_{\mu}$-compact, hence it is $\mathcal{R}$-compact.*

(2). *Conversely, let* $\mu : \mathcal{R} \to X$ *satisfies* $7(i)$ *and let for every* $u \in \mathcal{S}$ *there exist an* $\mathcal{R}$*-upper-semi-continuous function* $\phi_u : G \to [0,\infty)$ *such that*

$$u(\mu(A)) \leq \sup_{x\in A} \phi_u(x)$$

for each $A \in \mathcal{R}$ *and let the set* $\{x \in A : \phi_u(x) \geq \varepsilon\}$ *be* $\mathcal{R}$*-compact for each* $\varepsilon > 0$. *Then* μ *is an* X*-valued measure and*

$$N_{\mu,u}(x) \leq \phi_u(x)$$

for each $x \in G$ *and each* $u \in \mathcal{S}$.

Proof. (1). Let us consider the set

$$G_{\varepsilon,u} := \{x \in G : N_{\mu,u}(x) \geq \varepsilon\},$$

where $\varepsilon > 0$. Then for each $x \in G \setminus G_{\varepsilon,u}$ there exists $A \in \mathcal{R}$ such that $x \in A$ and $\|A\|_{\mu,u} < \varepsilon$, hence $A \subset G \setminus G_{\varepsilon,u}$ and inevitably $G_{\varepsilon,u}$ is $\mathcal{R}$-closed and $N_{\mu,u}$ is $\mathcal{R}$-upper semi-continuous. Then we take an arbitrary subset $A \in \mathcal{R}_\mu$ and a covering $\mathcal{V}$ of $A \cap G_{\varepsilon,u}$ by elements of $\mathcal{R}_\mu$. Then the sets $A \setminus (V_1 \cup \cdots \cup V_n \cup V)$, where $n \in \mathbf{N}$, $V_1, \dots, V_n \in \mathcal{V}$, $V \in \mathcal{R}_\mu$ and $V \subset G \setminus G_{\varepsilon,u}$, form a shrinking subfamily $\mathcal{A}$ in the completed separating covering ring $\mathcal{R}_\mu$ with the empty intersection $\bigcap_{E\in\mathcal{A}} E = \emptyset$. In accordance with Property $18(iii')$ of an X-valued measure there exist $V_1, \dots, V_n \in \mathcal{V}$ and $V \subset G \setminus G_{\varepsilon,u}$ such that the semi-norm fulfills the inequality

$$\|A \setminus (V_1 \cup \cdots \cup V_n \cup V)\|_{\bar{\mu},u} < \varepsilon,$$

hence $A \setminus (V_1 \cup \cdots \cup V_n \cup V) \subset G \setminus G_{\varepsilon,u}$. Since $V \subset G \setminus G_{\varepsilon,u}$, the formula is valid $A \cap G_{\varepsilon,u} \subset V_1 \cup \cdots \cup V_n$. Thus the topological space $A \cap G_{\varepsilon,u}$ is $\mathcal{R}_\mu$-compact.

(2). Each function of the form ϕ_u is $\mathcal{R}$-upper semi-continuous and ϕ_u is bounded from above on each $\mathcal{R}$-compact set, also $\|A\|_{\mu,u} \leq \|Ch_A\|_{\phi_u}$ for each $A \in \mathcal{R}$. Thus for each $u \in \mathcal{S}$ and every subset $A \in \mathcal{R}$ the semi-norm $\|A\|_{\mu,u} < \infty$ is finite, consequently, $7(ii)$ is satisfied.

We consider a positive number $\varepsilon > 0$ and a shrinking subfamily $\mathcal{A}$ in $\mathcal{R}$ such that $\bigcap_{E\in\mathcal{A}} = \emptyset$. Then the sets $\{x \in A : \phi_u(x) \geq \varepsilon\}$ form a family $\mathcal{E}$ of $\mathcal{R}$-compact sets closed under finite intersections and $\bigcap_{B\in\mathcal{E}} B = \emptyset$. Therefore, a subset $E \in \mathcal{A}$ exists so that

$$\{x \in E : \phi_u(x) \geq \varepsilon\} = \emptyset.$$

Hence the inclusion $E \subset \{x \in G : \phi_u(x) < \varepsilon\}$ is satisfied and inevitably the semi-norm $\|A\|_{\mu,u} < \varepsilon$ is finite.

22. Corollary. *If* μ *is an* X*-valued measure on a separating covering ring* $\mathcal{R}$ *then for every* $\varepsilon > 0$ *and each continuous semi-norm* $u \in \mathcal{S}$ *the set* $\{x \in G : N_{\mu,u}(x) \geq \varepsilon\}$ *is* $\mathcal{R}$*-locally compact.*

23. Theorem. *Let* μ *be a measure on a separating covering ring* $\mathcal{R}$ *and let also* $\mathcal{U}$ *be a separating covering ring of a set* G *being a sub-ring of* $\mathcal{R}_\mu$ *and let* ν *be a restriction of the measure* μ *on this sub-ring* $\mathcal{U}$. *Then* $\mathcal{U}_\nu = \mathcal{R}_\mu$ *and* $\bar{\nu} = \bar{\mu}$.

Proof. Let $u \in \mathcal{S}$ be a continuous semi-norm in X. At first we prove that the inequality $N_{\mu,u}(x) \geq N_{\nu,u}(x)$ is fulfilled for each point $x \in G$. Suppose the contrary, that there exists

$y \in G$ such that the opposite inequality $N_{\mu,u}(y) < N_{\nu,u}(y)$ is satified. This implies that a subset $V \in \mathcal{R}$ exists so that $\|V\|_{\mu,u} < N_{\nu,u}(y)$.

In accordance with Lemma 20 $\|V\|_{\mu,u} = \|V\|_{\bar{\mu},u}$. Then for every $x \in G \setminus V$ there exists $B \in \mathcal{U}$ such that $y \in B$ and $x \notin B$, since $\mathcal{U}$ is the separating covering ring. Therefore, the family $\{B \setminus V : B \in \mathcal{U}, y \in B\}$ is a shrinking sub-collection of $\mathcal{R}_\mu$ whose intersection is empty. In view of $18(iii')$ a subset $B \in \mathcal{U}$ exists so that $y \in B$ and

$$\|B \setminus V\|_{\bar{\mu},u} < N_{\nu,u}(y),$$

since $\|V\|_{\bar{\mu},u} < N_{\mu,u}(y)$. But the measure ν is the restriction of the measure $\bar{\mu}$ on the sub-ring $\mathcal{U}$ and $B \in \mathcal{U}$. This implies the inequality

$$\|B\|_{\nu,u} \leq \|B\|_{\bar{\mu},u} < N_{\nu,u}(y)$$

giving the contradiction, since $y \in B$.

We demonstrate now the inclusion $\mathcal{R}_\mu \subset \mathcal{U}_\nu$. Let $A \in \mathcal{R}_\mu$ and $\varepsilon > 0$. In view of Lemma 7 it is sufficient for each $u \in \mathcal{S}$ to construct $B \in \mathcal{U}$ such that $N_{\nu,u}(x) < \varepsilon$ for every $x \in A \triangle B$. To prove this we take the subset

$$W := \{x \in A : N_{\mu,u}(x) \geq \varepsilon\}.$$

In view of Theorem 51 the set W is $\mathcal{R}_\mu$-compact, consequently, it is $\mathcal{U}$-compact as well. One can mention that for each $x \in G \setminus W$ a subset $B \in \mathcal{U}$ exists such that $W \subset B$ with $x \notin B$. Then the shrinking sub-family $\{B \setminus A : B \in \mathcal{U}, B \supset W\}$ of the separating covering ring $\mathcal{R}_\mu$ has the empty intersection and there exists $B \in \mathcal{U}$ such that $B \supset W$ for which $\|B \setminus A\|_{\bar{\mu},u} < \varepsilon$. Thus

$$N_{\bar{\mu},u} = N_{\mu,u} < \varepsilon$$

on the set $B \setminus A$. On the other hand, $N_{\mu,u}(x) < \varepsilon$ on the set $A \setminus W$ hence on $A \triangle B$ as well.

Next we show that μ can be obtained as the restriction of the measure $\bar{\nu}$. For A, B, ε, u as in the preceding paragraph we have

$$u(\bar{\nu}(A) - \bar{\mu}(A)) = u(\bar{\nu}(A) - \nu(B) + \bar{\mu}(B) - \bar{\nu}(A))$$

$$= u(\bar{\nu}(A \setminus (A \cap B)) - \bar{\nu}(B \setminus (A \cap B)) + \bar{\mu}((B \setminus (A \cap B)) - \bar{\mu}(B \setminus (A \cap B)))$$

$$\leq \max(\|A \triangle B\|_{\bar{\nu},u}, \|A \triangle B\|_{\bar{\mu},u}) = \|A \triangle B\|_{\bar{\mu},u} = \sup_{x \in A \triangle B} N_{\mu,u}(x).$$

Therefore, these two measures coincide $\bar{\nu} = \bar{\mu}$ on the separating covering ring $\mathcal{R}_\mu$.

This implies the inclusion $\mathcal{R} \subset \mathcal{U}_\nu$ and the measure μ is the restriction of the measure $\bar{\nu}$ on the separating covering ring $\mathcal{R}$. Symmetrically interchanging two measures μ and ν in the proof above one also obtains $\mathcal{U}_\nu = \mathcal{R}_\mu$ and $\bar{\mu} = \bar{\nu}$.

24. Lemma. *Suppose that μ is an X-valued measure on a separating covering ring $\mathcal{R}$. For a continuous semi-norm u in X and for a positive number $\varepsilon > 0$ we put $G_{\varepsilon,u} := \{x \in G : N_{\mu,u}(x) \geq \varepsilon\}$. Then the restriction of the $\mathcal{R}$ and $\mathcal{R}_\mu$-topologies to $G_{\varepsilon,u}$ coincide. A function $f : G \to \mathbf{g}$ is $\mathcal{R}_\mu$-continuous if and only if for each $u \in \mathcal{S}$ and $\varepsilon > 0$ the restriction $f|_{G_{\varepsilon,u}}$ is $\mathcal{R}$-continuous.*

Proof. In view of Lemma 17 the $\mathcal{R}$-topology and $\mathcal{R}_\mu$-topology induce the same topology on $G_{\varepsilon,u}$. Therefore, if $f: G \to \mathsf{g}$ is $\mathcal{R}_\mu$-continuous, then it is $\mathcal{R}$-continuous on $G_{\varepsilon,u}$.

Suppose that $f: G \to \mathsf{g}$ has $\mathcal{R}$-continuous restrictions $f|_{G_{\varepsilon,u}}$ for each $u \in \mathcal{S}$ and every $\varepsilon > 0$. Take any clopen subset V in g. If $A \in \mathcal{R}_\mu$, then the intersection $A \cap G_{\varepsilon,u}$ is $\mathcal{R}$-compact due to Theorem 21. Moreover, the set $f^{-1}(V) \cap A \cap G_{\varepsilon,u}$ is $\mathcal{R}$-clopen as a subset of $G_{\varepsilon,u}$. For each $x \in f^{-1}(V) \cap A \cap G_{\varepsilon,u}$ we take a subset $U_x \in \mathcal{R}$ with $x \in U_x$ so that $U_x \cap G_{\varepsilon,u} \subset f^{-1}(V) \cap A \cap G_{\varepsilon,u}$. From this covering of the compact set $G_{\varepsilon,u}$ we choose a finite sub-covering such that $f^{-1}(V) \cap A \cap G_{\varepsilon,u} \subset U$, where $U := \bigcup_{j=1}^{k} U_{x_j}$, consequently, $U \in \mathcal{R}$. Therefore, the equality $f^{-1}(V) \cap A \cap G_{\varepsilon,u} = U \cap G_{\varepsilon,u}$ is satisfied. In view of Lemma 17 the set $f^{-1}(V) \cap A$ belongs to the completed separating covering ring $\mathcal{R}_\mu$ for each $A \in \mathcal{R}$. Thus $f^{-1}(V)$ is $\mathcal{R}_\mu$-clopen and inevitably f is $\mathcal{R}_\mu$-continuous.

25. Corollary. *If a function $f: G \to \mathsf{g}$ is $\mathcal{R}_\mu$-continuous on each $\mathcal{R}_\mu$-compact set, then f is $\mathcal{R}_\mu$-continuous on G. If $u \in \mathcal{S}$ is a continuous semi-norm and $E \subset G$ is $\mathcal{R}_\mu$-compact, then the set $H := \{x \in E : N_{\mu,u}(x) = 0\}$ is finite and there exists $\delta > 0$ such that $N_{\mu,u} > \delta$ on the set $E \setminus H$.*

Proof. In accordance with Lemma 24 each $\mathcal{R}$-compact set $G_{\varepsilon,u}$ is $\mathcal{R}_\mu$-compact and f is continuous on every $\mathcal{R}$-compact subset of $G_{\varepsilon,u}$. On the other hand, $G_{\varepsilon,u}$ is $\mathcal{R}$-locally compact due to Corollary 22. Therefore, f is $\mathcal{R}$-continuous on $G_{\varepsilon,u}$ and $\mathcal{R}_\mu$-continuous on G.

We have that each subset A of the set $\{x \in G : N_{\mu,u}(x) = 0\}$ is $\mathcal{R}_\mu$-clopen, since the characteristic function Ch_A belongs to the space $L(\mu)$. Take an $\mathcal{R}_\mu$-compact subset E in G. Hence H is finite. Let a number $\pi \in \mathbf{K}$ be with the norm $0 < |\pi| < 1$. If the infimum

$$\inf\{N_{\mu,u}(x) : x \in E \setminus H\} = 0$$

is zero, then there exists a sequence $\{x_k \in E : k \in \mathbf{N}\}$ in E such that $N_{\mu,u}(x_k) < |\pi|^k$ and $N_{\mu,u}(x_k) < N_{\mu,u}(x_{k-1})$ for each k. We choose $A_k \in \mathcal{R}$ such that $x_k \in A_k$ and $N_{\mu,u}(x) < |\pi|^k$ for each $x \in A_k$ and $A_k \cap A_l = \emptyset$ for each $k \neq l$. Without loss of generality we can consider the family of non-archimedean semi-norms $\mathcal{S}$ in X such that if $u, q \in \mathcal{S}$, then the functional

$$x \mapsto t(x) := \max(u(x), q(x))$$

belongs to the family $\mathcal{S}$, where $x \in X$. This is natural, because $t(x)$ is the semi-norm in X (see also [87]). Therefore,

$$\|A\|_{\mu,t} = \max(\|A\|_{\mu,u}, \|A\|_{\mu,q})$$

for each $A \in \mathcal{R}$, as well as

$$N_{\mu,t}(x) = \max(N_{\mu,u}(x), N_{\mu,q}(x))$$

for each $x \in G$. Hence $G_{\varepsilon,q} \cap G_{\varepsilon,u} \supset G_{\varepsilon,t}$ for each $\varepsilon > 0$ and $t = \max(u,q)$, $u, q, t \in \mathcal{S}$. If a function $f|_U$ is continuous on a set U and W is a subset of U, $W \subset U$, then evidently $f|_W$ is continuous.

In view of Lemma 24 and Theorem 21 the function

$$g(x) := \sum_k \pi^{-k} Ch_{A_k \cap \{x \in G: N_{\mu,u}(x) > 0\}}(x) v_k$$

is $\mathcal{R}_\mu$-continuous, where $v_k \in X$, $u(v_k) = 1$ for each k, since the restriction of g on each $G_{\gamma,q}$ is continuous for each $q \in \mathcal{S}$ and $\gamma > 0$. But g appears to be not bounded on the $\mathcal{R}_\mu$-compact set E. This gives the contradiction, consequently,
$\inf\{N_{\mu,u}(x) : x \in E \setminus H\} > 0$.

26. Theorem. *Let μ be an X-valued measure on a separating covering ring $\mathcal{R}$. A function $f : G \to \mathrm{g}$ is μ-integrable if and only if it satisfies two conditions (1) and (2):*

(1) f is $\mathcal{R}_\mu$-continuous;

(2) for each $u \in \mathcal{S}$ and $\varepsilon > 0$ the set $\{x : x \in G, u(f(x))N_{\mu,u}(x) \geq \varepsilon\}$ is $\mathcal{R}_\mu$-compact, consequently, contained in some $\{x : N_{\mu,u}(x) \geq \delta\}$ with $\delta > 0$.

Proof. If u is a continuous semi-norm in X and $\varepsilon > 0$, then the set

$$G_{\varepsilon,u} := \{x \in G : N_{\mu,u}(x) \geq \varepsilon\}$$

is $\mathcal{R}$-compact by Theorem 21. Without loss of generality we consider complete X and g. For each function f belonging to the space $L(G,\mathcal{R},\mu,X;\mathrm{g})$ and each continuous semi-norm u in X there exists a sequence of simple functions $\{f_k : k \in \mathbf{N}\}$ such that the limit

$$\lim_{k\to\infty} \|f - f_k\|_{\mu,u} = 0$$

is zero. Then each mapping f_k is $\mathcal{R}$-continuous and the sequence $\{f_k : k\}$ converges uniformly on $G_{\varepsilon,u}$ to f. Hence f is $\mathcal{R}$-continuous on $G_{\varepsilon,u}$. In view of Corollary 25 the function f is $\mathcal{R}_\mu$-continuous.

We take a step function g such that $\|f - g\|_{\mu,u} < \varepsilon$, consequently, the set $\{x : u(f(x))N_{\mu,u}(x) \geq \varepsilon\} = \{x : u(g(x))N_{\mu,u}(x) \geq \varepsilon\}$ is compact by Theorem 21. Thus from $f \in L(G,\mathcal{R},\mu,X;\mathrm{g})$ Properties $(1,2)$ follow.

Let now Properties $(1,2)$ be satisfied for a mapping $f : G \to X$. For $\delta > 0$ and a continuous semi-norm u in X we take a $\mathcal{R}_\mu$-step function g such that $\|f - g\|_{\mu,u} < \delta$. It is useful to consider the set

$$V := \{x \in G : u(f(x))N_{\mu,u}(x) \geq \delta\}.$$

The function $N_{\mu,u}$ is $\mathcal{R}_\mu$-upper semi-continuous by Theorem 21 and

$$\sup_{x\in C} N_{\mu,u}(x) =: w < \infty.$$

Since the set V is compact, there exists a finite clopen covering $B_1,\dots,B_k$ of V such that $u(f(x) - f(y))w < \delta$ for each $x,y \in B_j \cap V$ with the same j, $j = 1,\dots,k$. Let us choose now points $b_j \in B_j$, consequently, the set

$$\{x \in G : u(f(x) - f(b_j))N_{\mu,u}(x) < \delta\}$$

is $\mathcal{R}_{\mu,u}$-open and contains B_j. Therefore, there exist disjoint sets $W_1,\dots,W_k$ such that

$$W_j \subset \{x \in G : u(f(x) - f(b_j))N_{\mu,u}(x) < \delta\}$$

and $B_j := W_j \cap V$, since G is $\mathcal{R}$ totally disconnected with the clopen base of its topology. Let $q(x)$ be a step function such that

$$g(x) := \sum_{j=1}^{k} f(b_j)Ch_{W_j}(x).$$

For $x \in W_j$ we have the estimate

$$u(f(x)-g(x))N_{\mu,u}(x) = u(f(x)-f(b_j))N_{\mu,u}(x) < \delta.$$

At the same time for $x \notin \bigcup_{j=1}^k W_j$ we have the inequality

$$u(f(x)-g(x))N_{\mu,u}(x) = u(f))N_{\mu,u}(x) < \delta,$$

consequently, $\|f-g\|_{\mu,u} \le \delta$.

27. Corollary. *Let μ be a X-valued measure on a separating covering ring $\mathcal{R}$, $g \in L(G,\mathcal{R},\mu,X;\mathrm{g})$, let also a function $f: G \to \mathrm{g}$ be $\mathcal{R}_\mu$-continuous and $u(f(x)) \le u(g(x))$ for each semi-norm u on an algebra* g *and for every $x \in G$, then $f \in L(G,\mathcal{R},\mu,X;\mathrm{g})$.*

28. Corollary. *The space $L(G,\mathcal{R},\mu,X;\mathrm{g})$ is complete and locally* **K***-convex. If X and* g *are normed spaces, then $L(G,\mathcal{R},\mu,X;\mathrm{g})$ is the Banach space.*

Proof. By the construction above the space $L(G,\mathcal{R},\mu,X;\mathrm{g})$ is the completion of the space of step functions relative to the family of semi-norms $\|*\|_{\mu,u}$, where $u \in \mathcal{S}$ is a continuous semi-norm in X, g. Therefore, the space $L(G,\mathcal{R},\mu,X;\mathrm{g})$ is isomorphic with $L(G,\mathcal{R},\mu,\tilde{X};\tilde{\mathrm{g}})$ and complete, where $\tilde{X}$ denotes the completion of X and $\tilde{\mathrm{g}}$ is the completion of g as the **K**-convex spaces. Particularly, when X and g are normed spaces, then $\tilde{X}$ and $\tilde{\mathrm{g}}$ and $L(G,\mathcal{R},\mu,\tilde{X};\tilde{\mathrm{g}})$ are the Banach spaces.

29. Definition. If X is a normed space over an infinite field **K** with a multiplicative non-archimedean norm and $x_1,x_2,\dots$ is a sequence in X such that the inequality

$$\|a_1x_1+\dots+a_nx_n\| \ge t\max\{\|a_jx_j\| : j=1,\dots,n\}$$

is fulfilled for each numbers $a_1,\dots,a_n \in \mathbf{K}$ and a natural number $n \in \mathbf{N}$ not exceeding the length of the sequence, where $0 < t \le 1$ is a marked number, then the vectors of the family $\{x_1,x_2,\dots\}$ is called t-orthogonal. If $t=1$, then the vectors of the family $\{x_1,x_2,\dots\}$ are called orthogonal.

Naturally in the case $t=1$ the inequality, $\ge$, reduces to the equality, $=$, due to the non-archimedean property of the norm.

30. Theorem. *Let μ be a X-valued measure on a measure space $(G,\mathcal{R})$ and let a Banach algebra* g *has a t_0-orthogonal basis, where $0 < t_0 \le 1$. Then the space $L(G,\mathcal{R},\mu,X;\mathrm{g})$ has a t-orthogonal basis with $0 < t < t_0$. If the normalization group of the field* **K** *is discrete in $(0,\infty)$, then the space $L(G,\mathcal{R},\mu,X;\mathrm{g})$ has an orthogonal basis.*

Proof. If the normalization group of the field **K** is discrete, then the Banach space over **K** has an orthogonal basis by Theorem 5.16 [99], particularly, for the **K**-linear space $L(G,\mathcal{R},\mu,X;\mathrm{g})$ due to Corollary 28.

In general, suppose that the normalization group of the field **K** is dense in $(0,\infty)$ and $N_\mu(x) > 0$ for each $x \in G$. We choose a marked number t so that $0 < t < t_0$, where the algebra g has the t_0-orthogonal basis as the **K**-linear normed space. Then we choose a number $\pi \in \mathbf{K}$ so that $t_1 < |\pi| < 1$, where $t_1 = t/t_0$, and define the function $h: G \to \mathbf{K}$ such that $h(x) = \pi^n$, when $|\pi|^{n+1} < N_\mu(x) \le |\pi|^n$ and $n \in \mathbf{Z}$. Therefore, $|\pi||h(x)| < N_\mu(x) \le |h(x)|$ for each $x \in G$. Denote by G_d the set G in the discrete topology and define the mapping

$$q: L(G,\mathcal{R},\mu,X;\mathrm{g}) \ni f \mapsto hf \in BC(G_d,\mathrm{g}),$$

where $BC(Y,W)$ denotes the space of all bounded continuous mappings from a topological space Y into a **K**-linear normed space W. The space $BC(G_d,\mathrm{g})$ is supplied with the norm

$$\|v\|_\infty := \sup_{x\in G_d} \|v(x)\|,$$

where $\|*\|$ is the norm in g. Therefore, the inequality

$$|\pi|\|qf\|_\infty < \|f\|_\mu \leq \|qf\|_\infty \tag{1}$$

is satisfied for each $f\in L(G,\mathcal{R},\mu,X;\mathrm{g})$. If $A\in\mathcal{R}$ and $b\in \mathrm{g}$, then the function $hbCh_A$ belongs to the space $BC(G_d,\mathrm{g})$, since

$$\|h(x)bCh_A(x): x\in G\}\subset\{\pi^n : n\in\mathbf{Z}; |\pi|^{n+1} < \|b\||Ch_A(x)|\}.$$

By the **K** linearity and Property (1) of the mapping q the range $q(L(G,\mathcal{R},\mu,X;\mathrm{g}))$ is the closed **K**-linear subspace in the normed space $BC(G_d,\mathrm{g})$. In the space $BC(G_d,\mathrm{g})$ the product $BC(G_d,\mathbf{K})\times \mathrm{g}$ is everywhere dense, while the space $L(G,\mathcal{R},\mu,X;\mathbf{K})\times\mathrm{g}$ is everywhere dense in the space $L(G,\mathcal{R},\mu,X;\mathrm{g})$. In view of Corollaries 5.23 and 5.25 [99] the space $BC(G_d,\mathbf{K})$ has an orthogonal basis, hence $BC(G_d,\mathrm{g})$ has a t_0-orthogonal basis.

In accordance with the Gruson Theorem 5.9 [99] if E is a Banach space with an orthogonal basis, then each its closed **K**-linear subspace has an orthogonal basis. The space $q(L(G,\mathcal{R},\mu,X;\mathbf{K}))$ is closed in $BC(G_d,\mathbf{K})$, consequently, it has an orthogonal basis $\{e_j : j\}$. Thus the **K**-linear normed space $L(G,\mathcal{R},\mu,X;\mathrm{g})$ has the t-orthogonal basis $q^{-1}(e_j\times s_k)$, where $\{s_k : k\}$ is any t_0-orthogonal basis in g.

31. Theorems. *Let X and* g *be as in §9 and in addition let X be an algebra over the non-archimedean field* **K** *with a family of multiplicative semi-norms in X. Suppose that μ and ν are X-valued measures on separating covering rings $\mathcal{R}$ of a set G and $\mathcal{T}$ of a set H. Then*

(1) *the finite unions of the sets $A\times B$, $A\in\mathcal{R}$, $B\in\mathcal{T}$ form the separating covering ring $\mathcal{R}\otimes\mathcal{T}$ of the product set $G\times H$;*

(2) *there exists a unique measure $\mu\times\nu$ on $\mathcal{R}\times\mathcal{T}$ such that $\mu\times\nu(A\times B)=\mu(A)\nu(B)$ for each $A\in\mathcal{R}$ and $B\in\mathcal{T}$, $N_{\mu\times\nu,u}(x,y)=N_{\mu,u}(x)N_{\nu,u}(y)$ for each $x\in G$, $y\in H$ and every semi-norm $u\in\mathcal{S}$ in X;*

(3) *if $f\in L(G\times H,\mathcal{R}\times\mathcal{T},\mu\times\nu,X;\mathrm{g})$, then $H\ni y\mapsto \int_G f(x,y)\mu(dx)$ is a ν-almost everywhere defined ν-integrable function on H and the mapping $G\ni x\mapsto\int_H f(x,y)\nu(dy)$ is μ-almost everywhere defined μ-integrable on G and*

$$\int_{G\times H} f(x,y)\mu\times\nu(dx,dy) = \int_H\left(\int_G f(x,y)\mu(dx)\right)\nu(dy).$$

Moreover, if X is commutative, then

$$\int_H\left(\int_G f(x,y)\mu(dx)\right)\nu(dy) = \int_G\left(\int_H f(x,y)\nu(dx)\right)\mu(dy);$$

(4) *in particular, if* g *and* X *are commutative,* $f \in L(G,\mathcal{R},\mu,X;\mathrm{g})$ *and* $h \in L(H,\mathcal{T},\nu,X;\mathrm{g})$, *then* $f(x)h(y) \in L(G\times H,\mathcal{R}\times\mathcal{T},\mu\times\nu,X;\mathrm{g})$ *and*

$$\int_{G\times H} f(x)g(y)\mu(dx)\nu(dy) = \left(\int_G f(x)\mu(dx)\right)\left(\int_H g(y)\nu(dy)\right);$$

(5) *the algebra* $L(G\times H,\mathcal{R}\times\mathcal{T},\mu\times\nu,X;\mathrm{g})$ *is* **K**-*linearly topologically isomorphic with the tensor product* $L(G,\mathcal{R},\mu,X;\mathrm{g})\hat{\otimes}L(H,\mathcal{T},\nu,X;\mathrm{g})$.

Proof. (1). If $\mathcal{U}$ and $\mathcal{V}$ are coverings the sets of G and H by elements from the separating covering rings $\mathcal{R}$ and $\mathcal{T}$ respectively, then

$$\bigcup_{A\in\mathcal{U},B\in\mathcal{V}} A\times B = G\times H.$$

If $(x_1,y_1)\neq(x_2,y_2)\in G\times H$, then either $x_1\neq x_2$ or $y_1\neq y_2$. In the first case take $A\in\mathcal{R}$ such that $x_1\in A$ and $x_2\in G\setminus A$ and $x_1\in B$ with $B\in\mathcal{T}$, then $A\times B$ separates them: $(x_1,y_1)\in A\times B$ and $(x_2,y_2)\notin A\times B$.

(2). We define the product measure by the formula:

$$\mu\times\nu(E) := \int_{G\times H} Ch_E(x,y)\mu(dx)\nu(dy)$$

for each $E\in\mathcal{R}\times\mathcal{T}$, consequently, $\mu\times\nu$ is additive. For $N_u(x,y) := N_{\mu,u}(x)N_{\nu,u}(y)$ and each $A\in\mathcal{R}$, $B\in\mathcal{T}$ we get

$$u((\mu\times\nu))(A\times B)) \leq \|Ch_{A\times B}\|_{N_u},$$

consequently, $\|C\|_{\mu\times\nu,u}\leq\|Ch_C\|_{N_u}$.

Naturally G and H and $G\times H$ are supplied with the $\mathcal{R}$ and $\mathcal{T}$ and $\mathcal{R}\times\mathcal{T}$ topologies. Then the inclusion

$$\{(x,y)\in A\times B : N_u(x,y)\geq\varepsilon\}\subset\{x\in A : N_{\mu,u}(x)\geq\varepsilon\}\times\{y\in B : N_{\nu,u}(y)\geq\varepsilon\}$$

is satisfied for each $A\in\mathcal{R}$ and $B\in\mathcal{T}$ and $\varepsilon>0$. The function N_u is upper semi-continuous and the subset $\{x\in A : N_{\mu,u}(x)\geq\varepsilon\}\times\{y\in B :\ N_{\nu,u}(y)\geq\varepsilon\}$ is compact, consequently, the subset $\{(x,y)\in A\times B :\ N_u(x,y)\geq\varepsilon\}$ is compact and inevitably the subset $\{(x,y)\in C :\ N_u(x,y)\geq\varepsilon\}$ is also compact for each $C\in\mathcal{R}\times\mathcal{T}$ and every $\varepsilon>0$. Therefore, by Theorem 21 $\mu\times\nu$ is the measure and $N_{\mu\times\nu,u}(x,y)\leq N_u(x,y)$ for each $u\in\mathcal{S}$ and each $x\in G$ and every $y\in H$. Each u is the multiplicative semi-norm in X and the formula

$$u((\mu\times\nu)(A\times B)) = u(\mu(A))u(\nu(B))$$

is fulfilled and hence

$$\sup\{u((\mu\times\nu)(C)) :\ C\in\mathcal{R}\times\mathcal{T}, C\subset A\times B\} = \|A\times B\|_{\mu\times\nu,u}\geq\|A\|_{\mu,u}\|B\|_{\nu,u}$$

for each $A\in\mathcal{R}$ and every $B\in\mathcal{T}$. Therefore, the inequality

$$N_{\mu\times\nu,u}(x,y)\geq N_u(x,y)$$

is accomplished for each u, x and y.

(3). If f is a step function, then Property (3) is satisfied due to (2). If a function f belongs to the space $L(G,\mathcal{R},\mu,X;\mathsf{g})$, then for each continuous semi-norm u there exists a sequence of step functions $f_1, f_2, \dots$ converging to f such that $\|f - f_n\|_{N_u} \le 1/n$ for each $n \in \mathbf{N}$. Hence we see that the estimate

$$u(f(x,y) - f_n(x,y))N_{\mu,u}(x)N_{\nu,u}(y) \le 1/n$$

is satisfied for all $x, y \in G$. Thus $f(*,y) \in L(G,\mathcal{R},\mu,X;\mathsf{g})$ for ν-almost every $y \in H$, consequently,

$$u\left(\int_G f(x,y)\mu(dx) - \int_G f_n(x,y)\mu(dx)\right)N_{\nu,u}(y) \le 1/n.$$

Therefore, the function $H \ni y \mapsto \int_G f(x,y)\mu(dx)$ is defined for ν-almost all $y \in H$ and is ν-integrable. Since

$$\int_H \left(\int_G f_n(x,y)\mu(dx)\right)\nu(dy) = \int_{G\times H} f_n(x,y)\mu\times\nu(dx,dy)$$

for each n, we infer the inequality

$$u(\int_H \left(\int_G f_n(x,y)\mu(dx)\right)\nu(dy) - \int_{G\times H} f_n(x,y)\mu\times\nu(dx,dy)) \le 1/n$$

for each $n \in \mathbf{N}$ and each $u \in \mathcal{S}$, consequently, the integral equality

$$\int_H \left(\int_G f(x,y)\mu(dx)\right)\nu(dy) = \int_{G\times H} f(x,y)\mu\times\nu(dx,dy)$$

is fulfilled.

Part (4) follows from Properties (2,3) as the particular case.

(5). The bilinear continuous operator

$$\mathsf{m} : L(G,\mathcal{R},\mu,X;\mathsf{g}) \times L(H,\mathcal{T},\nu,X;\mathsf{g}) \ni (f,h) \mapsto fh \in L(G\times H,\mathcal{R}\times\mathcal{T},\mu\times\nu,X;\mathsf{g})$$

exists. If X is a Banach space, then the norm of the operator m is $\|\mathsf{m}\| = 1$.

Let Y be a complete locally $\mathbf{K}$-convex space over the non-archimedean field $\mathbf{K}$ and let

$$F : L(G,\mathcal{R},\mu,X;\mathsf{g}) \times L(H,\mathcal{T},\nu,X;\mathsf{g}) \to Y$$

be a continuous $\mathbf{K}$-bilinear mapping. Then the mapping

$$F_{\mathsf{m}}(Ch_{A\times B}) = F_{\mathsf{m}}(Ch_A \otimes Ch_B) = F(Ch_A, Ch_B) \in Y$$

exists for each $A \in \mathcal{R}$ and $B \in \mathcal{T}$. As usually a step function $f : G\times H \to \mathsf{g}$ is written in the form:

$$f(x,y) = \sum_j b_j Ch_{A_j\times B_j}(x,y),$$

where expansion coefficients b_j belong to the algebra g and subsets A_j are from the separating covering ring $\mathcal{R}$, $B_j \in \mathcal{T}$ for each j, $(A_j\times B_j)\cap(A_i\times B_i) = \emptyset$ for each $i \ne j$.

For each semi-norm v in Y and each continuous semi-norm u in X there exists k such that

$$\sup_{x\in G,y\in H} v(F_{\mathrm{m}}(b_kCh_{A_k\times B_k})(x,y))N_{\mu\times\nu,u}(x,y)$$

$$=\max_j v(F_{\mathrm{m}}(b_jCh_{A_j\times B_j})(x,y))N_{\mu\times\nu,u}(x,y),$$

consequently, the inequalities

$$v(F_{\mathrm{m}}f(x,y))N_{\mu\times\nu,u}(x,y)\leq \sup_{x\in G,y\in H} v(F_{\mathrm{m}}(b_kCh_{A_k\times B_k})(x,y))N_{\mu\times\nu,u}(x,y)$$

$$\leq u(b_k)\|F\|_{u,v}\|Ch_{A_k}\|_{\mu,u}\|Ch_{B_k}\|_{\nu,u}\leq \|F\|_{u,v}\|f\|_{\mu\times\nu,u},$$

are satisfied, where the semi-norms are defined by the equation $\|F\|_{u,v}:=\sup_{w\in X,u(w)>0} v(Fw)/u(w)$.

Therefore, the mapping F_{m} has the continuous extension F_{m} onto the space $L(G\times H,\mathcal{R}\times\mathcal{T},\mu\times\nu,X;\mathrm{g})$ with values in Y. Thus $F_{\mathrm{m}}(f\otimes h)=F(f,h)$ for each functions $f\in L(G,\mathcal{R},\mu,X;\mathrm{g})$ and $h\in L(H,\mathcal{T},\nu,X;\mathrm{g})$ and inevitably the algebra $L(G\times H,\mathcal{R}\times\mathcal{T},\mu\times\nu,X;\mathrm{g})$ is **K**-linearly topologically isomorphic with the tensor product $L(G,\mathcal{R},\mu,X;\mathrm{g})\hat{\otimes}L(H,\mathcal{T},\nu,X;\mathrm{g})$.

32. Theorem. *Let g_j and X_j be a family of algebras and locally convex spaces over an infinite field* **K** *with a multiplicative non-trivial non-archimedean norm satisfying Conditions of §9, $j\in\beta$, where β is a set. Suppose that for each j there is a measure $\mu_j:\mathcal{R}\to X_j$, $X=\bigotimes_{j\in\beta}X_j$ and $\mathrm{g}=\bigotimes_{j\in\beta}\mathrm{g}_j$ are supplied with the product topologies. Then the product measure*

$$\mu=\bigotimes_{j\in\beta}\mu_j$$

on the separating covering ring $\mathcal{R}$ exists with values in X so that the space $L(G,\mathcal{R},\mu,X;\mathrm{g})$ is the completion of the direct sum $\bigoplus_{j\in\beta}L(G,\mathcal{R},\mu_j,X_j;\mathrm{g}_j)$.

Proof. Since each X_j and every g_j are complete, the space X and the algebra g are complete (see [19,87]). Naturally X is the locally **K**-convex space and g is the algebra such that $x+y=(x_j+y_j:j)$ and $ab=(a_jb_j:j)$, where $x,y\in X$, $a,b\in\mathrm{g}$, $x=(x_j:j)$, $a=(a_j:j)$, $ax=(a_jx_j:j)$ (see Theorem 5.6.1 [87]). Thus X is the unital left g-module. For each $j\in\beta$ there the projectors $\pi_j(x)=x_j$ and $\pi_j(a)=a_j$ are defined on X and g. Therefore, topologies of the space X and the algebra g are characterized by the families of semi-norms u so that

$$u(x)=\max\{u_j(x_j):j\in\alpha\}\quad\text{and}$$

$$u(b)=\max\{u_j(b_j):j\in\alpha\},$$

where α is a finite subset in β, $u_j\in\mathcal{S}_j$, where $\mathcal{S}_j$ is a consistent family of semi-norms u_j in X_j and g_j denoted by the same symbol for shortening the notation.

If $A,B\in\mathcal{R}$ and $A\cap B=\emptyset$, then the measure *mu* satisfies the equalities:

$$\mu(A\cup B)=(\mu_j(A\cup B):j)=(\mu_j(A)+\mu_j(B):j)$$

$$=(\mu_j(A):j)+(\mu_j(B):j)=\mu(A)+\mu(B),$$

consequently, μ is additive. If u is any continuous semi-norm in X and $A \in \mathcal{R}$, then for each $C \subset A, C \in \mathcal{R}$ the inequality

$$u(\mu(C)) = \max\{u_j(\mu_j(C)) : j \in \alpha\} < \infty$$

is fulfilled, since α is a finite set and each μ_j is bounded. If $\mathcal{A}$ is a shrinking family in $\mathcal{R}$ and $\bigcap_{A \in \mathcal{A}} A = \emptyset$, then the limit

$$\lim_{A \in \mathcal{A}} u_j(\mu_j(A)) = 0$$

is zero for each j and each $u_j \in \mathcal{S}_j$, consequently,

$$\lim_{A \in \mathcal{A}} u(\mu(A)) = 0$$

for each semi-norm $u \in \mathcal{S}$ in X. Thus μ is the measure on $\mathcal{R}$ with values in X.

If $f \in L(G, \mathcal{R}, \mu, X; \mathsf{g})$, then for each $u \in \mathcal{S}$ there exists a sequence $\{f_n : n \in \mathbf{N}\}$ of simple functions such that the upper estimate

$$\|f - f_n\|_{\mu,u} \leq 1/n$$

is fulfilled. We have that the direct sum $\bigoplus_j \mathsf{g}_j$ is everywhere dense in the topological algebra g, where elements of the direct sum as usually are $b = (b_j : j \in \beta, b_j \in \mathsf{g}_j\}$ such that the set $\{j : b_j \neq 0\}$ is finite (see Example 5.10.6 in [87]). Thus each simple function can be chosen taking values in the direct sum $\bigoplus_{j \in \beta} \mathsf{g}_j$. Therefore, the space $L(G, \mathcal{R}, \mu, X; \mathsf{g})$ is the completion of the direct sum $\bigoplus_{j \in \beta} L(G, \mathcal{R}, \mu_j, X_j; \mathsf{g}_j)$.

Certainly if β is finite, then the direct sum and the direct product coincide.

33. Corollary. *If suppositions of Theorem 32 are satisfied and $f \in L(G, \mathcal{R}, \mu, X; \mathsf{g})$, then*

$$\int_G f(x)\mu(dx) = \left(\int_G f_j(x)\mu_j(dx_j) : j \in \beta \right),$$

where a function f_j belongs to the space $L(G, \mathcal{R}, \mu, X_j; \mathsf{g}_j)$ for each j.

Proof. The formula of this Corollary is satisfied for each f in the direct sum $\bigoplus_{j \in \beta} L(G, \mathcal{R}, \mu_j, X_j; \mathsf{g}_j)$, but the latter space is everywhere dense in $L(G, \mathcal{R}, \mu, X; \mathsf{g})$ in accordance with Theorem 31. The mapping $\int_G : L(G, \mathcal{R}, \mu, X; \mathsf{g}) \to X$ is continuous: $u(\int_G f(x)\mu(dx)) \leq \|f\|_{\mu,u}$ due to Lemma 15, consequently, the statement of this Corollary follows by the continuity.

34. Theorem. *Let g, X, G, $\mathcal{R}$, μ be as in §9 and let F be a continuous homomorphism of a left unital g-module X into a uniformly complete left unital h-module Y. Then F induces a continuous homomorphism $\hat{F} : L(G, \mathcal{R}, \mu, X; \mathsf{g}) \to L(G, \mathcal{R}, \nu, Y; \mathsf{h})$ such that*

$$F\left(\int_G f(x)\mu(dx) \right) = \int_G \hat{F}(f)\nu(dx)$$

for each $f \in L(G, \mathcal{R}, \mu, X; \mathsf{g})$, where $\nu = F(\mu)$ is an Y-valued measure. If $F(\mathsf{g}) = \mathsf{h}$, then $\hat{F}$ is epimorphic.

Proof. If A and B are two elements of the separating covering ring $\mathcal{R}$ with $A \cap B = \emptyset$, then we have the additivity

$$\nu(A \cup B) := F(\mu(A \cup B)) = F(\mu(A)) + F(\mu(B)) = \nu(A) + \nu(B) \in Y.$$

If $A \in \mathcal{R}$ and u is a continuous semi-norm in X, then the supremum

$$\sup_{C \subset A, C \in \mathcal{R}} u(\mu(C)) < \infty$$

is finite, consequently, for each continuous semi-norm v in Y the supremum

$$\sup_{C \subset A, C \in \mathcal{R}} v(\nu(C)) < \infty$$

is also finite, since F is continuous. If $\mathcal{A}$ is a shrinking family in the separating covering ring $\mathcal{R}$ with the void intersection $\bigcap_{A \in \mathcal{A}} A = \emptyset$, then the limit $\lim_{\mathcal{A}} \mu(A) = 0$ is zero, consequently,

$$0 = \lim_{A \in \mathcal{A}} F(\mu(A)) = \lim_{A \in \mathcal{A}} \nu(A),$$

since F is continuous. Thus ν is the Y-valued measure.

If v is a continuous semi-norm in Y, then $v_F(q) := v(F(q))$ is the continuous semi-norm in X, where $q \in X$, consequently, $v_F(\mu(A)) = v(\nu(A))$ for each $A \in \mathcal{R}$ and inevitably $N_{\mu, v_F}(x) = N_{\nu, v}(x)$ for each $x \in G$.

If $f : G \to \mathrm{g}$ is a step function having the form

$$f(x) = \sum_j a_j Ch_{A_j}(x),$$

where each coefficient a_j belong to the algebra g, every subset A_j is from the separating covering ring $\mathcal{R}$, then

$$\hat{F}(f) = F(f) = \sum_j F(a_j) Ch_{A_j}(x),$$

since $F(a_1 b_1 + a_2 b_2) = F(a_1)F(b_1) + F(a_2)F(b_2)$ for each $a_j, b_j \in \mathrm{g}$, also $0, 1 \in \mathrm{g}$, $F(0) = 0$, $F(1) = 1 \in \mathsf{h}$. Moreover, $\|F(f)\|_{\nu,v} = \|f\|_{\mu,v_F}$ for each step function f and each continuous semi-norm v in Y, consequently, the mapping $\hat{F}$ is continuous and $\mathbf{K}$-linear and has the continuous extension $\hat{F} : L(G, \mathcal{R}, \mu, X; \mathrm{g}) \to L(G, \mathcal{R}, \nu, Y; \mathsf{h})$.

For each $s \in L(G, \mathcal{R}, \nu, Y; \mathsf{h})$ and each continuous semi-norm v in Y we take a sequence s_n of simple functions $s_n : G \to \mathsf{h}$ converging to s such that $\|s - s_n\|_{\nu,v} \le 1/n$ for each n. If $F(\mathrm{g}) = \mathsf{h}$, then for each

$$s_n = \sum_j b_{j,n} Ch_{A_j}$$

a simple function

$$f_n = \sum_j a_{j,n} Ch_{A_j}$$

exists such that $f_n : G \to \mathrm{g}$ and $\hat{F}(f_n) = s_n$, $a_{j,n} \in \mathrm{g}$ for each j, n, where $b_{j,n} \in \mathsf{h}$, $A_j \in \mathcal{R}$. But $\{f_n : n\}$ is the fundamental sequence relative to each semi-norm $\| * \|_{\mu, v_f}$, consequently, there exists $f \in L(G, \mathcal{R}, \mu, X; \mathrm{g})$ such that $\hat{F}(f) = s$.

By the conditions of this theorem $F(a_1w_1+a_2w_2)=F(a_1)F(w_1)+F(a_2)F(w_2)$ for each $a_1,a_2\in \mathsf{g}$, $w_1,w_2\in X$, where $F(a_j)\in \mathsf{h}$ and $F(w_j)\in Y$. Therefore, for each step function $f(x)=\sum_j a_jCh_{A_j}(x)$ we have the equalities

$$F\Big(\int_G f(x)\mu(dx)\Big)=F\Big(\sum_j a_j\mu(A_j)\Big)=\sum_j F(a_j)\nu(A_j)=\int_G \hat{F}(f)(x)\nu(dx)$$

$$\text{and}\quad v_F\Big(\int_G f(x)\mu(dx)\Big)=v\Big(\int_G \hat{F}(f)(x)\nu(dx)\Big)$$

for each continuous semi-norm v in Y, consequently, by the continuity we get the equality

$$F\Big(\int_G f(x)\mu(dx)\Big)=\int_G \hat{F}(f)(x)\nu(dx)$$

for each function f in the space $L(G,\mathcal{R},\mu,X;\mathsf{g})$.

35. Definition. Let $\mathcal{R}=\mathsf{Bco}(G)$ be the ring of all clopen subsets of a zero-dimensional Hausdorff space. A measure $\mu:\mathsf{Bco}(G)\to X$ is called a tight measure, where X is as in §9. The family $M=M(G,X)$ of all such tight measures form the **K**-linear space supplied with the family of semi-norms

$$\|\mu\|_u=\sup_{A\in \mathsf{Bco}(G)} u(\mu(A))=\|G\|_{\mu,u}=\sup_{x\in G}N_{\mu,u}(x),$$

where $u\in\mathcal{S}$ is a continuous semi-norm in X. In particular, if X is the normed space, then $M(G,X)$ is the normed space.

The closure of the set $\{x\in G:\exists u\in\mathcal{S},N_{\mu,u}(x)>0\}$ is traditionally called the support of the measure μ.

36. Theorem. *If $\mathcal{R}$ is a covering ring of G being the base of the zero-dimensional Hausdorff topology in G and if μ is a X-valued measure on $\mathcal{R}$, then the formula*

$$(f\mu)(A):=\int_G Ch_A(x)f(x)\mu(dx)$$

provides the tight measure for each $f\in L(G,\mathcal{R},\mu,X;\mathbf{K})$ and the mapping $\psi_\mu:=\psi:L(G,\mathcal{R},\mu,X;\mathbf{K})\ni f\mapsto (f\mu)$ is the **K***-linear topological embedding of the space $L(G,\mathcal{R},\mu,X;\mathbf{K})$ into $M(G,X)$.*

Proof. For each continuous semi-norm $u\in\mathcal{S}$ the set $\{x\in G:N_{\mu,u}(x)>0\}$ is σ-compact, that is, this set is the countable union of compact subsets by Theorem 21. Therefore, if A is a clopen subset in G, then $Ch_A\in L(G,\mathcal{R},\mu,X;\mathbf{K})$, since for each continuous semi-norm $u\in\mathcal{S}$ and $\varepsilon>0$ a sequence $f_n\in L(G,\mathcal{R},\mu,X;\mathbf{K})$ exists satisfying the restriction $\|Ch_A-f_n\|_{\mu,u}\le 1/n$ and with supports $supp(f_n)$ contained in the set $\{x\in G:N_{\mu,u}(x)\ge 1/n\}$. Thus $f\mu$ is defined on the separating covering sub-ring $\mathcal{R}_{f\mu}\supset \mathsf{Bco}(G)$, consequently, $f\mu\in M(G,X)$. Evidently the mapping ψ is linear $\psi(af+bg)=a\psi(f)+b\psi(g)=af\mu+bg\mu$. On the other hand, $N_{f\mu,u}(x)\le\|f\|_{\mu,u}N_{\mu,u}(x)$, consequently, ψ is continuous. In view of Theorem 26 $u(f(x))N_{\mu,u}(x)=N_{f\mu,u}(x)$ for each continuous semi-norm $u\in\mathcal{S}$ and a point $x\in G$, hence ψ is the topological embedding. If X is a Banach space, then ψ is the isometric embedding.

37. Let Y^* denote the topological dual space of all continuous **K**-linear functionals on a **K**-linear space Y. Traditionally $Mat_n(\mathbf{K})$ denotes the algebra of all square $n \times n$ matrices, $n \in \mathbf{N}$, while $BC(G,Y)$ denotes the space of all continuous bounded functions from G into Y.

Theorem. *If G is a zero-dimensional Hausdorff space and $\mu \in M(G, Mat_n(\mathbf{K}))$, then $BC(G, Mat_n(\mathbf{K})) \subset L(G, \mathcal{R}, \mu, Mat_n(\mathbf{K}); Mat_n(\mathbf{K}))$ and the mapping*

$$\lambda_\mu(f) := \int_G f(x)\mu(dx)$$

provides the **K***-linear isometric embedding*

$$\lambda : M(G, Mat_n(\mathbf{K})) \hookrightarrow BC(G, Mat_n(\mathbf{K}))^*.$$

If G is compact, then λ is the isomorphism.

Proof. In view of Theorem 26 we get the inclusion $BC(G, Mat_n(\mathbf{K})) \subset L(G, \mathcal{R}, \mu, Mat_n(\mathbf{K}); Mat_n(\mathbf{K}))$. Moreover, the mapping λ is defined and it is **K**-linear.

Since the algebra $Mat_n(\mathbf{K})$ is finite dimensional over the field **K**, we certainly get that its topological dual space is isomorphic with $Mat_n(\mathbf{K})$. It has the natural norm topology

$$\|b\| := \max_{1 \le i,j \le n} |b_{i,j}|.$$

If $\mu \in M(G, Mat_n(\mathbf{K}))$, then

$$\|\mu\| = \sup_{A \in \mathsf{Bco}(G)} \max_{i,j} |\mu_{i,j}(A)| = \sup_{A \in \mathsf{Bco}(G)} |\lambda_\mu(Ch_A)| \le \|\lambda_\mu\|,$$

consequently, λ is the isometric embedding.

If $q \in BC(G, Mat_n(\mathbf{K}))^*$, then $\mu(A)$ having matrix elements $q(E_{i,j}Ch_A) =: \mu^q_{i,j}(A)$ is the additive function on the separating covering ring $\mathsf{Bco}(G)$ with values in $Mat_n(\mathbf{K})$, where $E_{i,j}$ is the $n \times n$ matrix with 1 at the (i,j)-th place and zeros at others places. For each $A \in \mathsf{Bco}(G)$ and $b \in Mat_n(\mathbf{K})$ we have $bCh_A \in BC(G, Mat_n(\mathbf{K}))$. Therefore, μ^q is defined on $\mathsf{Bco}(G)$.

If $\mathcal{A} \subset \mathsf{Bco}(G)$ is a shrinking family and G is compact, then from $\bigcap_{A \in \mathcal{A}} A = \emptyset$ it follows, that $\emptyset \in \mathcal{A}$, since each $A \in \mathcal{A}$ is closed in G. Since q is continuous, the mapping μ^q is the measure for the compact topological space G. In this particular case $BC(G, Mat_n(\mathbf{K}))$ is isomorphic with the space $C(G, Mat_n(\mathbf{K}))$ of all continuous functions from G into $Mat_n(\mathbf{K})$.

38. Theorem. *A function $f : G \to \mathsf{g}$ is μ-integrable for each tight measure $\mu \in M(G,X)$ if and only if f is bounded and for each compact subset V in G the restriction $f|_V$ of f to V is continuous.*

Proof. Suppose that f is μ-integrable for each tight measure $\mu \in M(G,X)$. Take a semi-norm $u \in \mathcal{S}$ in X and a number $\pi \in \mathbf{K}$ such that $0 < |\pi| < 1$. If the case when the function f is not bounded, a sequence $b_j \in \mathbf{G}$ exists so that $b_i \ne b_j$ for each $i \ne j$ and

$$\lim_{n \to \infty} u(\pi^n f(b_n)) = \infty.$$

We define the measure by the formula:

$$\mu := \sum_n \pi^n x_n \delta_{b_n},$$

where $\delta_b(A) := 1$ if $b \in A$, $\delta_b(A) = 0$ if $b \notin A$, $x_n \in X$, $u(x_n) = 1$. Therefore, μ is the tight measure $\mu \in M(G,X)$ and $N_{\mu,u}(b_n) = |\pi|^n$ for every $n \in \mathbf{N}$, consequently, $\|f\|_{\mu,u} = \infty$ and $f \notin L(G, \mathsf{Bco}(G), \mu, X; \mathrm{g})$. In view of Theorem 7.9 [99] if V is a compact subset in G, then there exists a measure $\lambda : \mathsf{Bco}(G) \to \mathbf{K}$ such that $N_\lambda(x) = 1$ for each $x \in V$ and $N_\lambda(x) = 0$ for each $x \in G \setminus V$. One can take a point $y \in X$ with $u(y) = 1$, then $\mu = y\lambda$ is the X-valued measure and $N_{\mu,u} = Ch_V$. Therefore, due to Lemma 24 and Theorem 26 the restriction $f|_V$ of the function f is continuous.

Suppose now that a function f is bounded and its restriction to each compact subset of G is continuous. Taking any tight measure $\mu \in M(G,X)$ we get due to Corollary 25 that the mapping f is $\mathsf{Bco}(G)_\mu$-continuous. If $z \in \mathrm{g}$, then $q_z \in L(G, \mathsf{Bco}(G), \mu, X; \mathrm{g})$, where $q_z(x) = z$ for every $x \in G$. We choose an element $z \in \mathrm{g}$ such that the inequality $u(f(x)) \le u(z)$ is fulfilled for each $x \in G$, consequently, in view of Corollary 27 the function f belongs to the space $L(G, \mathsf{Bco}(G), \mu, X; \mathrm{g})$.

39. Definition. Let G be a zero-dimensional Hausdorff topological space. If for each subset $U \subset G$ the property is satisfied: it is clopen in G if and only if $U \cap V$ is clopen in V for each compact subset V in G, then G is called the k_0-space.

40. Corollary. *If G is a k_0-space, then*

$$BC(G, \mathrm{g}) = \bigcap_{\mu \in M(G,X)} L(G, \mathsf{Bco}(G), \mu, X; \mathrm{g}).$$

Proof. We recall that a topological space is called a k-space if it is Hausdorff and the image of some locally compact Hausdorff space under the quotient mapping. In view of Theorem 3.3.21 [19] a mapping $f : G \to Y$ from a k-space G into a topological space Y is continuous if and only if for each compact subset V in G the restriction $f|_V$ of the function f to V is continuous. Therefore, due to Theorem 37 we get the statement of this corollary.

41. Definition. A functional $J \in BC(G, \mathrm{g})^*$ is said to have a compact support, if there exists a compact subset V in G such that $J(f) = 0$ for each $f \in BC(G, \mathrm{g})$ with $f(x) = 0$ for every $x \in V$.

A hood on G is a mapping $h : G \to [0, \infty)$ such that the set $\{x \in G : h(x) \ge \varepsilon\}$ is compact for each $\varepsilon > 0$.

A subset $W \subset BC(G, \mathrm{g})$ is called strictly open if for each $f \in W$ and a continuous semi-norm $u \in \mathcal{S}$ in g there exists a hood h such that the inclusion

$$W \supset \{g \in BC(G, \mathrm{g}) : \sup_{x \in G} u(f(x) - g(x))h(x) \le 1\}$$

is fulfilled.

Strictly open subsets in $BC(G, \mathrm{g})$ form a topology τ_{str} in $BC(G, \mathrm{g})$ called the strict topology. A functional J continuous on $(BC(G, \mathrm{g}), \tau_{str})$ is called strictly continuous.

42. Theorem. *The following conditions on a functional $J \in BC(G, Mat_n(\mathbf{K}))^*$ are equivalent:*

(1) *a tight measure* $\mu \in M(G, Mat_n(\mathbf{K}))$ *exists such that* $J = \lambda_\mu$ *(see Theorem* 37*);*

(2) *for each* $\varepsilon > 0$ *a compact subset* V *in* G *exists so that the inequality*

$$|J(f)| \leq \max\{\|J\| \sup_{x\in V} \|f(x)\|, \varepsilon\|f\|)$$

is fulfilled for every continuous bounded function $f \in BC(G, Mat_n(\mathbf{K}))$*;*

(3) *the functional* J *is the limit of elements in* $BC(G, Mat_n(\mathbf{K}))^*$ *having compact supports;*

(4) *the functional* J *is strictly continuous.*

Proof. If W_1 and W_2 are strictly open subsets in $BC(G, \mathrm{g})$, $f \in W = W_1 \cap W_2$, $u \in S$, then hoods h_j exist such that the inclusion

$$W_j \supset \{g \in BC(G, \mathrm{g}) : \sup_{x\in G} u(f(x) - g(x))h_j(x) \leq 1\}$$

is fulfilled for $j = 1, 2$. Therefore,

$$W \supset \{g \in BC(G, \mathrm{g}) : u(f(x) - g(x))h(x) \leq 1\},$$

where the mapping $h(x) = \max(h_1(x), h_2(x))$ for each x is the hood such that the set $\{x \in G : h(x) \geq \varepsilon\} = \{x \in G : h_1(x) \geq \varepsilon\} \cup \{x \in G : h_2(x) \geq \varepsilon\}$ is compact for each $\varepsilon > 0$ as the union of two compact sets. Thus strictly open subsets form the base of a topology.

The $\mathbf{K}$-algebra $Mat_n(\mathbf{K})$ is normed. Since $Mat_n(\mathbf{K})$ is the finite dimensional space over the field $\mathbf{K}$, its topologically dual space is isomorphic with the algebra $Mat_n(\mathbf{K})$. For a compact subset W in G let R_W denote the restriction mapping $R_W : BC(G, Mat_n(\mathbf{K})) \to C(W, Mat_n(\mathbf{K}))$. In view of Theorem 5.24 [99] there exists a $\mathbf{K}$-linear isometric embedding

$$T_W : C(W, Mat_n(\mathbf{K})) \hookrightarrow PC(G, Mat_n(\mathbf{K}))$$

so that $R_W \circ T_W = I$, since $n \in \mathbf{N}$, where $PC(G,X)$ denotes the closed $\mathbf{K}$-linear hull in $BC(G,X)$ of the subset $\{Ch_A : A \in \mathsf{Bco}(G), A \text{ is compact}\}$ of characteristic functions.

If the functional is $J = \lambda_\mu$ for some tight measure $\mu \in M(G, Mat_n(\mathbf{K}))$, then N_μ is the hood and $|J(f)| \leq \|f\|_{N_\mu}$ for every $f \in BC(G, Mat_n(\mathbf{K}))$, consequently, J is strictly continuous, that is, $(1) \Rightarrow (4)$.

If J is strictly continuous take $\pi \in \mathbf{K}$ with $0 < |\pi| < 1$. There exists a hood h for which the inclusion

$$\{f : \|f\|_h < 1\} \subset \{f : |J(f)| \leq |\pi|\}$$

is satisfied. Therefore, $|J(f)| \leq \|f\|_h$ for each f. For $\varepsilon > 0$ we take the subset $W := \{x \in G : h(x) \geq \varepsilon\}$. If $f \in BC(G, Mat_n(\mathbf{K}))$, then we choose the function $g := T_W R_W f$, hence $J(f) = J(f - g) + J(g)$ and the inequalities

$$|J(f-g)| \leq \sup_{x\in G} \|f(x) - g(x)\| h(x) \leq \sup_{x\in G\setminus W} \|f(x) - g(x)\|\varepsilon \leq \|f\|\varepsilon$$

are accomplished as well as the following

$$|J(g)| \leq \|J\|\|g\| \leq \|J\| \sup_{x\in W} \|f(x)\|,$$

consequently, we have demonstrated the implication (4) $\Rightarrow$ (3).

Suppose that (2) is satisfied, then the functional J_W has the compact support. Therefore, we get the estimate

$$|J(f) - J_W(f)| = |J(f - T_W R_W f)| \leq \varepsilon \|f - T_W R_W f\| < \varepsilon \|f\|,$$

consequently, $\|J - J_W\| \leq \varepsilon$. Thus we have proved the implication (2) $\Rightarrow$ (3) also.

Let Condition (3) be satisfied. We have that the space $M(G, Mat_n(\mathbf{K}))$ is complete and the mapping λ is isometric, consequently, the range of λ is closed in the topologically dual space $BC(G, Mat_n(\mathbf{K}))^*$. Therefore, without loss of generality we can consider a functional J with the compact support. Suppose that $W \subset G$ and $J(f)$ for f vanishing identically on W. We define the mapping by the formula

$$\mu^J_{i,j}(A) := J(E_{i,j}\ Ch_A)$$

for each $A \in \mathsf{Bco}(G)$ and all $i, j = 1, \dots, n$, consequently, this mapping $\mu : \mathsf{Bco}(G) \to Mat_n(\mathbf{K})$ is additive. Here as above $E_{i,j}$ is the matrix with the element 1 at the (i, j)-th place and zeros at others places, $\mu(A) = \mu^J(A)$ is the matrix with matrix elements $\mu^J_{i,j}(A)$. Then $N_\mu(x) = 0$ for every $x \in G \setminus W$ and N_μ is bounded on G, consequently, μ is the measure on $\mathsf{Bco}(G)$.

The normed space $C(W, Mat_n(\mathbf{K}))$ has the orthonormal base consisting of functions $E_{i,j}Ch_A$, where $i, j = 1, \dots, n$ and $A \in \mathsf{Bco}(G)$. Suppose that a continuous bounded function $f \in BC(G, Mat_n(\mathbf{K}))$ is given, then sequences $A_k \in \mathsf{Bco}(G)$ and $b_k \in Mat_n(\mathbf{K})$ exist so that $b_k = a_k E_{i(k),j(k)}$ and $\lim_{k\to\infty} a_k = 0$ and the series $f = \sum_k b_k Ch_{A_k}$ converges uniformly on W. Therefore, the functional has the decomposition:

$$J(f) = J\Big(\sum_k b_k Ch_{A_k}\Big) = \sum_k a_k \mu_{i(k),j(k)}(A_k) = \int_G \sum_k a_k Ch_{A_k} \mu_{i(k),j(k)}(dx)$$

$$= \int_G \sum_k b_k \mu(A_k) = \int_G f(x)\mu(dx),$$

since $\mu(A) = \sum_{i,j=1}^n E_{i,j} J(E_{i,j}\ Ch_A)$. This finishes the proof of the implication (3) $\Rightarrow$ (1).

43. Corollary. *Let G and H be zero-dimensional Hausdorff spaces, let also X be a complete topological algebra over* $\mathbf{K}$. *If $\mu \in M(G,X)$ and $\nu \in M(H,X)$ are tight measures, then $\mu \times \nu$ is a tight measure on the product topological space $G \times H$, $\mu \times \nu \in M(G \times H, X)$.*

Proof. This follows from Theorem 32(2).

44. Example. In this section convolutions of tight measures are considered. Suppose that G is a zero-dimensional Hausdorff topological semi-group and X is a topological algebra over an infinite field $\mathbf{K}$ with a multiplicative non-trivial non-archimedean norm. For $\mu, \nu \in M(G,X)$ and $f \in BC(G, \mathsf{g})$ we define the functional

$$Jf := \int_G f(xy)(\mu(dx) \times \nu(dy)).$$

If $u \in \mathcal{S}$ is a consistent continuous semi-norm in X and g, then the inequality

$$u(Jf) \leq \sup_{x,y \in G} u(f(xy)) N_{\mu,u}(x) N_{\nu,u}(y)$$

is accomplished. For each $\varepsilon > 0$ the sets $G_{\mu,u,\varepsilon} := \{x \in G : N_{\mu,u}(x) \geq \varepsilon\}$ and $G_{\nu,u,\varepsilon} := \{y \in G : N_{\nu,u}(y) \geq \varepsilon\}$ are compact, consequently, their product $G_{\mu,u,\varepsilon}G_{\nu,u,\varepsilon}$ is compact. Moreover, the inequality

$$u(Jf) \leq \max\{\sup\{u(f(z)) : z \in G_{\mu,u,\varepsilon}G_{\nu,u,\varepsilon}\}\|\mu\|_u\|\nu\|_u; \|f\|_u\|\mu\|_u\varepsilon; \|f\|_u\|\nu\|_u\varepsilon\}$$

is fulfilled. Therefore, J is induced by a tight measure denoted by $\mu * \nu$ so that it satisfies the condition:

$$\int_G f(x)[\mu * \nu](dx) = \int_G f(xy)(\mu(dx) \times \nu(dy))$$

for each continuous bounded function $f \in BC(G,X)$. In particular, for $f = Ch_A$ with $A \in \mathsf{Bco}(G)$ we get

$$(i) \quad [\mu * \nu](A) = (\mu \times \nu)(\{(x,y) \in G \times G, xy \in A\}).$$

The tight measure $\mu * \nu$ is called the convolution product of tight measures μ and ν. Evidently, the convolution has properties:

$$(a\mu + b\zeta) * \nu = (a\mu * \nu) + (b\zeta * \nu) \quad \text{and}$$

$$\mu * (a\nu + b\zeta) = (a\mu * \nu) + (b\mu * \zeta)$$

for each $a, b \in \mathbf{K}$ and $\mu, \nu, \zeta \in M(G,X)$, since $\mathbf{K}$ is the commutative field. Hence $M(G,X)$ is the algebra with the addition $(\mu + \nu)(A) = \mu(A) + \nu(A)$ for each $A \in \mathsf{Bco}(G)$ and the multiplication given by the convolution product of measures.

From (i) it follows, that

$$N_{\mu*\nu,u}(z) = \sup_{x,y \in G, xy = z} N_{\mu,u}(x) N_{\nu,u}(y) < \infty,$$

consequently, $M(G,X)$ is the topological algebra relative to the family of semi-norms

$$\|\mu\|_u := \sup_{x \in G} N_{\mu,u}(x),$$

$u \in \mathcal{S}$, such that $\|\mu * \nu\|_u \leq \|\mu\|_u \|\nu\|_u$ for each $\mu, \nu \in M(G,X)$.

45. Lemma. *The mapping 9(SI) and Conditions 5(M1 – M4) induce an isometry between two spaces* $L^2(\mathcal{R}(G), \mathrm{g})$ *and* $L^2(\xi, \mathrm{g})$.

Proof. At first we demonstrate, that there exists a linear isometric mapping of the space $L^0(\mathcal{R}, \mathrm{g})$ onto the space $L^0(\xi, \mathrm{g})$. Let $f(x) = \sum_k a_k Ch_{A_k}(x)$ and $g(x) = \sum_l b_l Ch_{A_l}(x)$ be two simple functions in $L^0(\mathcal{R}, \mathrm{g})$, where $a_k, b_l \in \mathrm{g}$ are expansion coefficients (see Definition 9). Then due to Conditions $5(M1 - M4)$ and $9(1 - 7)$ the equalities:

$$M\Big[\Big(\int_G [f(x)\xi(dx)), \Big(\int_G g(x)\xi(dx)\Big)\Big] = \sum_k [a_k, b_k]\mu(A_k)$$

$$= \int_G [f(x), g(x)]\mu(dx) \tag{1}$$

are satisfied, since $\mathrm{g}^2 \ni \{a,b\} \mapsto [a,b] \in Lc(\mathrm{g})$ is the continuous mapping, while $\mu(A) \in Lc(X)$ for each $A \in \mathcal{R}$.

In view of Lemma 8 the **K**-valued measure $Tr\mu$ exists. Therefore, the mean value is:

$$M\Big(\Big(\int_G f(x)\xi(dx)\Big),\Big(\int_G g(x)\xi(dx)\Big)\Big) = \sum_k Tr(b_k^T a_k\mu)(A_k)$$

$$= \int_G Tr(g^T(x)f(x)\mu)(dx), \quad (2)$$

since $g^2 \ni \{a,b\} \mapsto (a,b) \in \mathbf{K}$ is the continuous mapping from g^2 into the field **K**, where $(Tr\mu)(A) := Tr\mu(A)$ for each $A \in \mathcal{R}(G)$.

If $F_1 \in Lin(X)$ and $F_2 \in Lc(X)$, then $F_1F_2 \in Lc(X)$. To each element $b \in g$ the **K**-linear continuous operator $X \ni x \mapsto bx \in X$ corresponds. Naturally $Lin(X)$ and $Lc(X)$ are the left g-modules, since $(bF) \in Lin(X)$ for each $F \in Lin(X)$ and $(bF) \in Lc(X)$ for every $F \in Lc(X)$ and each $b \in g$, where $(bF)(x) := b(Fx)$ for all $x \in X$. Since $\mu(A) \in Lc(X)$ and $g^T(x)f(x) \in g$, we deduce that $g^T(x)f(x)\mu(A) \in Lc(X)$ for each $x \in G$ and $A \in \mathcal{R}(G)$ and all $f,g \in L^0(\mathcal{R},g)$.

We supply the space $L^0(\mathcal{R},g)$ with the semi-norm

$$\|f\|_{2,\mu} = [\sup_{x\in G} N_{Trf^Tf\mu}(x)] \quad (3)$$

and in $L^0(\xi,g)$ put

$$\|\eta_f\|_{2,N_P} := [\sup_{x\in G} |(f(x)\xi(x), f(x)\xi(x))|N_P(x)]^{1/2} \quad (4)$$

for each $\eta_f = \eta = \int_G f(x)\xi(dx)$. The semi-norm given by Formula (4) is continuous relative to the family of semi-norms 13(1).

The semi-norm defined by Formula (3) is continuous relative to the family of semi-norms:

$$\|f\|_{\mu,u} := \sup_{x\in G} N_{f^Tf\mu,u}(x). \quad (5)$$

Since $M((a\xi(A), b\xi(A))) = Tr(b^Ta\mu(A))$ for each subset A in the separating covering ring $\mathcal{R}(G)$, the equality

$$N_{Trf^Tf\mu}(x) = \inf_{A\in\mathcal{R}(G), x\in A} \|A\|_{Trf^Tf\mu}$$

is satisfied, where $(f^Tf\mu)(dx) = f^T(x)f(x)\mu(dx)$. At the same time we have

$$M(a\xi(B), b\xi(B)) = \int_\Omega (a\xi(\omega,B), b\xi(\omega,B))P(d\omega)$$

and

$$|M(a\xi(B), b\xi(B))| \le \sup_{\omega\in\Omega} |(a\xi(\omega,B), b\xi(\omega,B))|N_P(\omega)$$

for each $a,b \in g$.

In view of Lemma 8 $Trg^Tf\mu$ is the measure for each functions $f,g \in L^0(\mathcal{R},g)$, consequently, taking a shrinking family S in the separating covering ring $\mathcal{R}(G)$ such that $\bigcap_{A\in S} A = \{x\}$ gives the equality

$$N_{Trg^Tf\mu}(x) = \inf_{A\in\mathcal{R}(G), x\in A} [\sup_{B\in\mathcal{R}(G), B\subset A_k, k} \sup_{\omega\in\Omega, k} |(a_k\xi(\omega,B), b_k\xi(\omega,B))|N_P(\omega)].$$

Form this we deduce that

$$N_{Trg^T f\mu}(x) = \inf_{A\in\mathcal{R}(G), x\in A}[\sup_{B\in\mathcal{R}(G), B\subset A_k, k} \|(a_k\xi(*,B), b_k\xi(*,B))\|_{L^2(P,\mathbf{K})}] \quad \text{and}$$

$$\|A_k\|_{Trg^T f\mu} = \|(a_k\xi(*,A_k), b_k\xi(*,A_k))\|_{L^2(P)}]$$

for each $k = 1,\dots,m$ due to Lemma 2 and due to the choice of subsets A_k satisfying the equality

$$\|A_k\|_{Trb_k^T a_k\mu} = |Trb_k^T a_k\mu(A_k)|$$

which is sufficient for our consideration.

On the other hand, $M[a\xi(A), b\xi(A)] = b^T a\mu(A) \in Lc(X)$ for each $A \in \mathcal{R}(G)$. Then the estimate

$$N_{b^T a\mu,u}(x) = \inf_{A\in\mathcal{R}(G), x\in A}[\sup_{B\in\mathcal{R}(G), B\subset A} u(b^T a\mu(B))]$$

$$\le u(b^T a)N_{\mu,u}(x) < \infty$$

follows, where the function $N_{\mu,u}(x)$ is defined by the equality

$$N_{\mu,u}(x) := \inf_{A\in\mathcal{R}(G), x\in A}[\sup_{B\in\mathcal{R}(G), B\subset A} u(\mu(B)).$$

Therefore, the follwing semi-norms are equal:

$$\|f\|_{\mu,u} = \|f\|_{2,P,u}$$

for each step function $f \in L^0(\mathcal{R}, \mathrm{g})$ and each consistent semi-norm u in g, X and $Lin(X)$.

The mapping ψ from $9(SI)$ also is **K**-linear from the space $L^0(\xi, \mathrm{g})$ into $L^0(\mathcal{R}(G), \mathrm{g})$ such that ψ is the isometry relative to the consistent semi-norms $13(1)$ and $45(5)$ due to Formula (1) and Lemmas 2, 15 and Theorem 26.

Two spaces $L^2(P, \mathrm{g})$ and $L^2(\mu, \mathrm{g})$ are complete by their definitions, consequently, ψ has the **K**-linear extension from the space $L^2(\mathcal{R}(G), \mathrm{g})$ onto the space $L^2(\xi, \mathrm{g})$ which is the isometry between these two spaces $L^2(\mathcal{R}(G), \mathrm{g})$ and $L^2(\xi, \mathrm{g})$.

46. Definition. For each function $f \in L^2(\mathcal{R}(G), \mathrm{g})$ the random vector η is defined by the equality:

$$\eta = \psi(f) = \int_G f(x)\xi(dx).$$

The random vector η is called the non-archimedean stochastic integral of the function f by an orthogonal stochastic measure ξ.

47. Remark. It is possible to consider random vectors of the form:

$$\eta(t) = \int_G g(t,x)\xi(dx),$$

where ξ is an orthogonal stochastic measure on a measurable space $(G, \mathcal{R}(G))$ with values in X and a structural measure μ with values in the space of **K**-linear compact operators $Lc(X)$ as above, $t \in T$, $g(t,x) \in L^2(G, \mathcal{R}(G), \mu, \mathrm{g})$ as the function by $x \in G$ for each $t \in T$, where T is a set.

The covariance operator of a random vector η is

$$B(t_1,t_2) = M\{\eta^T(t_1),\eta(t_2)\} = \int_G \{g^T(t_1,x),g(t_2,x)\}\mu(dx), \quad (1)$$

moreover,

$$M\{\eta(t_1),\eta^T(t_2)\} = \int_G \{g(t_1,x),g^T(t_2,x)\}Tr\mu(dx) \quad (2)$$

in the notation of §§1, 9, where X is the left g-module, while $Lc(X)$ is supplied with the natural structure of the left $Lc(\mathrm{g})$-module, $\{a^T,b\} \in Lin(\mathrm{g})$, $\{a,b^T\} \in \mathrm{g}$ for each $a,b \in \mathrm{g}$. We denote by $L^2\{g\}$ the closure in the space $L^2(G,\mathcal{R}(G),\mu,\mathrm{g})$ of the **K**-linear span of the family of functions $\{g(t,x) : t \in T\}$. Therefore, $L^2\{g\}$ is the **K**-linear closed subspace in $L^2(G,\mathcal{R}(G),\mu,\mathrm{g})$. If $L^2\{g\} = L^2(G,\mathcal{R}(G),\mu,\mathrm{g})$, then the system of functions $\{g(t,x) : t \in T\}$ is called complete in $L^2(G,\mathcal{R}(G),\mu,\mathrm{g})$.

Let $\{\eta(t) : t \in T\}$ be an X-valued random vector. By $L^0\{\eta\}$ the family of all random vectors of the form:

$$\zeta = \sum_{k=1}^{l} a_k\eta(t_k)$$

is denoted, where $l \in \mathbf{N}$, $t_k \in T$, $a_k \in \mathrm{g}$. Then $L^2\{\eta\}$ denotes the closure of $L^0\{\eta\}$ in the space $L^2(\Omega,\mathcal{R},P,X)$.

A family of random vectors $\{\zeta_\beta : \zeta_\beta \in L^2(\Omega,\mathcal{R},P,X); \beta \in \Lambda\}$ is called subordinated to the random X-valued function $\{\eta(t) : t \in T\}$, if $\zeta_\beta \in L^2\{\eta\}$ for each $\beta \in \Lambda$.

48. Lemma. *If X is a left* g*-module, then $Lin(X)$ is a left $Lin(\mathrm{g})$-module, $Lc(X)$ is a left $Lin(\mathrm{g})$-module as well as left $Lc(\mathrm{g})$-module.*

Proof. Let x be a vector in X, let also y be an element of the algebra g, let also A and B be linear operators $A \in Lin(\mathrm{g})$ and $B \in Lin(X)$, then $Bx \in X$, $y(Bx) \in X$ and $Ay \in \mathrm{g}$, consequently, $AyBx \in X$, since the left module is associative. Particularly, for $1 \in \mathrm{g}$ it gives the inclusion $AB \in Lin(X)$. Therefore, the multiplication $Lin(\mathrm{g}) \times Lin(X) \ni (A,B) \mapsto AB \in Lin(X)$ exists. If $B \in Lc(X)$, then $AyB \in Lc(X)$, since $vB \in Lc(X)$ for each $v \in \mathrm{g}$. In particular, it is valid for $y = 1$, consequently, $Lc(X)$ is the left $Lin(\mathrm{g})$-module and inevitably left $Lc(\mathrm{g})$-module, since $Lc(\mathrm{g}) \subset Lin(\mathrm{g})$.

49. Theorem. *Let a covariance operator $B(t_1,t_2)$ of a random X-valued function $\{\eta(t) : t \in T\}$ admits Representation* 46(1)*, where X and* g *are as in* §§2.30 *and* 3.9*, μ is a $Lc(X)$-valued measure on a measure space $(G,\mathcal{R}(G))$, $\mu^T = \mu$, $g(t,x) \in L^2(G,\mathcal{R}(G),\mu,Lc(X);\mathrm{g})$ for each $t \in T$ and the family $\{g(t,x) : t \in T;\}$ is complete in $L^2(G,\mathcal{R}(G),\mu,Lc(X);\mathrm{g})$. Then the random function $\eta(t)$ can be presented in the form:*

$$\eta(t) = \int_G g(t,x)\xi(dx) \quad (1)$$

with probability 1 *for each $t \in T$, where ξ is a stochastic orthogonal X-valued measure subordinated to the random function $\eta(t)$ and with a structure function μ.*

Proof. Consider functions of the form:

$$f(x) = \sum_{k=1}^{l} b_k g(t_k,x), \quad (2)$$

where $t_k \in T$, $b_k \in \mathrm{g}$, $l \in \mathbf{N}$.

We recall that a family Ψ of vectors in a topological vector space H over a field $\mathbf{K}$ is called complete if its linear span $span_{\mathbf{K}}\Psi$, consisting of all finite linear combinations of vectors from Ψ over $\mathbf{K}$, is everywhere dense in the space H. We put

$$(3) \quad \psi(f) = \zeta = \sum_{k=1}^{l} b_k \eta(t_k).$$

As above $L^0\{g\}$ denotes the family of all vectors of Form (2). In $L^0\{g\}$ the $\mathbf{K}$-bi-linear functional defined by the equation:

$$(f_1, f_2) := \int_G \{f_1(x), f_2^T(x)\} Tr\mu(dx) \tag{4}$$

exists. In view of Lemma 45 the mapping $\zeta = \psi(f)$ is the $\mathbf{K}$-linear topological isomorphism of the space $L^0\{g\}$ onto $L^0\{\eta\}$. When particularly X and g are normed, then the mapping ψ is the isometry. Thus ψ has the continuous extension up to the $\mathbf{K}$-linear topological isomorphism of the space $L^2\{g\}$ onto the space $L^2\{\eta\}$.

If a subset A belongs to the separating covering ring $\mathcal{R}(G)$, then $Ch_A \in L^2(G, \mathcal{R}(G), \mu, \mathrm{g})$, since $1 \in \mathrm{g}$. But the equality $L^2\{g\} = L^2(G, \mathcal{R}(G), \mu, \mathrm{g})$ is valid due to completeness of the family $\{g(t,x) : t \in T\}$. Therefore, $Ch_A \in L^2(G, \mathcal{R}(G), \mu, \mathrm{g})$. Put $\xi(A) := \psi(Ch_A)$, then $\xi(A)$ is the orthogonal stochastic measure with the structure function μ due to Lemma 38, since

$$M\{\xi^T(A), \xi(B)\} = \int_G \{Ch_A^T(x), Ch_B(x)\} \mu(dx) = \mu(A \cap B) \tag{5}$$

for each $A, B \in \mathcal{R}(G)$.

Let now the function $g(t)$ has the form

$$\gamma(t) := \int_G g(t,x) \xi(dx).$$

Since

$$M\{\eta^T(t), \xi(A)\} = \int_G \{g^T(t,x), Ch_A(x)\} \mu(dx) \quad \text{and}$$

$$M\{\xi^T(A), \eta(t)\} = \int_G \{Ch_A^T(x), g(t,x)\} \mu(dx)$$

and ψ is the $\mathbf{K}$-linear topological isomorphism, we infer the formulas

$$M\{\eta^T(t), \gamma(t)\} = M\{\gamma^T(t), \eta(t)\} = \int_G \{g^T(t,x), g(t,x)\} \mu(dx).$$

Therefore, the equalities

$$M\{Ch_A(\eta(t) - \gamma(t))^T, Ch_A(\eta(t) - \gamma(t))\} = M\{Ch_A \eta^T(t), Ch_A \eta(t)\}$$

$$- M\{Ch_A \eta^T, Ch_A \gamma(t)\} - M\{Ch_A \gamma^T(t), Ch_A \eta(t)\} + M\{Ch_A \gamma^T(t), Ch_A \gamma(t)\} = 0$$

are satisfied for each subset A in the separating covering ring $\mathcal{R}(G)$, consequently, Formula 49(1) is accomplished with probability 1 for each $t \in T$.

50. Definition. Let $\eta(t)$ be a g-valued stochastic process or stochastic function such that for each natural number $n \in \mathbf{N}$ and each values of the time variable $t_1, \dots, t_n \in \mathbf{T}$ with $t, t+t_1, \dots, t+t_n \in \mathbf{T}$ the mutual distribution of $\eta(t+t_1), \dots, \eta(t+t_n)$ is independent of t, where $\mathbf{T}$ is an additive semigroup. Then $\eta(t)$ is called the stationary stochastic function, where $P : \mathcal{R} \to \mathbf{K}$ is a probability measure.

Suppose that $\mathbf{T}$ is a uniform space. A g-valued stochastic function $\eta(t) \in L^b(\Omega, \mathcal{R}, P, \mathbf{K}; g)$, $t \in \mathbf{T}$, $1 \le b < \infty$, is called mean-b-continuous at $t_0 \in \mathbf{T}$, if the limit

$$\lim_{t \to t_0} \eta(t) = \eta(t_0)$$

exists in the sense of convergence in the space $L^b(\Omega, \mathcal{R}, P, \mathbf{K}; g)$, when t tends to t_0 in $\mathbf{T}$. In particular, for $b = 2$ it is mean-square continuity and convergence respectively. If $\eta(t)$ is mean-b-continuous at each point of $\mathbf{T}$, then $\eta(t)$ is called mean-b-continuous on $\mathbf{T}$.

51. Suppose that an algebra g over the complex non-archimedean field $\mathbf{C_p}$ has a uniformity τ relative to which it is complete. Let g have a $\mathbf{Q_p}$-linear embedding into the Banach space $c_0(\gamma, \mathbf{Q_p})$ for some set γ such that the norm uniformity n_u in g inherited from the Banach space $c_0(\gamma, \mathbf{Q_p})$ with the standard norm $\| * \|$ is not stronger, than τ, that is $n_u \subset \tau$. Let also the algebra g be everywhere dense in $(c_0(\gamma, \mathbf{Q_p}), \| * \|)$.

Theorem. *Let $\eta(t)$ be a stationary mean-square-continuous stochastic process with values in the algebra* g, $t \in \mathbf{T} = \mathbf{C_r}$, *where r and p are mutually prime numbers,* $M\eta(t) = 0$, *then there exists an orthogonal* g*-valued stochastic measure* $\xi(A)$ *on* $\mathsf{Bco}(\mathbf{C_r})$ *subordinated to* $\eta(t)$ *such that*

$$\eta(t) = \int_{\mathbf{C_r}} g(t,x)\xi(dx), \tag{1}$$

where $g(t,x)$ *is a* $\mathbf{C_p}$*-valued character from the additive group* $(\mathbf{C_r}, +)$ *into the multiplicative group* $(\mathbf{C_p}, \times)$. *Between the spaces* $L^2\{\eta\}$ *and* $L^2\{\mu\}$ *a* $\mathbf{K}$*-linear topological isomorphism* ψ *exists such that*

$$\psi(\eta(t)) = g(t,*) \text{ and } \psi(\xi(A)) = Ch_A \text{ if} \tag{2}$$

$$\zeta_j = \psi(f_j). \tag{3}$$

Moreover,

$$\zeta_j = \int_{\mathbf{C_r}} f_j(x)\xi(dx) \quad \text{and}$$

$$M\{\zeta_1^T, \zeta_2\} = \int_{\mathbf{C_r}} \{f_1(x)^T, f_2(x)\}\mu(dx).$$

52. Definition. Formula 51(1) is called the spectral decomposition of the stationary stochastic process $t \mapsto \eta(t)$, $t \in \mathbf{T}$. A measure $\xi(A)$ is called a stochastic spectral measure of the stationary stochastic process $\eta(t)$.

Proof of Theorem 51. Since $\eta(t)$ is a stationary stochastic process, the mean value $Mf(\eta(t+t_1), \dots, \eta(t+t_n))$ is independent of t for each continuous function $f : g^n \to \mathbf{K}$, where $n \in \mathbf{N}$. By the condition of this theorem $\eta(t) \in L^2(\Omega, \mathcal{R}, P, \mathbf{C_p}; \mathrm{g})$, consequently, the mean value $M\eta(t) = m$ exists and

$$M\{[\xi(t) - m]^T, [\xi(q) - m]\} = B(t-q) \in Lc(\mathrm{g}) \quad (4)$$

for each t and q in the complex non-archimedean field $\mathbf{C_r}$, where $m = 0$. Evidently $B^T(t - q) = B(q-t)$ for each $t, q \in \mathbf{C_r}$.

Consider now $\mathbf{C_p}$-valued characters of $(\mathbf{C_r}, +)$ as the additive group, where $r = p'$, $p \neq p'$ are prime numbers. For p-adic numbers the decomposition

$$x = \sum_{k=N}^{\infty} x_k p^k$$

is valid, where $x \in \mathbf{Q_p}$, $x_k \in \{0, 1, \dots, p-1\}$, $N \in \mathbf{Z}$, $N = N(x)$, $x_N \neq 0$, $x_j = 0$ for each $j < N$. The order the number x is defined as $ord_p(x) = N$, consequently, its norm is $|x|_{\mathbf{Q_p}} = p^{-N}$.

The function

$$[x]_{\mathbf{Q_p}} := \sum_{k=N}^{-1} x_k p^k$$

for $N < 0$, $[x]_{\mathbf{Q_p}} = 0$ for $N \geq 0$ on $\mathbf{Q_p}$ is used below. Therefore, the function $[x]_{\mathbf{Q_p}}$ on the field $\mathbf{Q_p}$ is considered with values in the segment $[0,1] \subset \mathbf{R}$.

Consider the non-archimedean complex field $\mathbf{C_r}$ as the vector space over the field $\mathbf{Q_r}$. It can be supplied with the multiplicative non-archimedean norm $|*|_{\mathbf{C_r}} = |*|$ in $\mathbf{C_r}$, which gives the uniformity in it. The equivalent uniformity can be given by a norm $|*|_r$ so that $|x|_r \in \{r^l : l \in \mathbf{Z}\} \cup \{0\}$ for each $x \in \mathbf{C_r}$. It is the following:

$$|x|_r := \min\{r^l : |x|_{\mathbf{C_r}} \leq r^l, l \in \mathbf{Z}\}$$

if $x \neq 0$, while $|0|_r = 0$, consequently,

$$(i) \quad |x|_r / r \leq |x|_{\mathbf{C_r}} \leq |x|_r$$

for each $x \in \mathbf{C_r}$ and inevitably $\mathbf{C_r}$ is the topological vector space over $\mathbf{Q_r}$ relative to $|x|_r$. Since the field $\mathbf{C_r}$ is the extension of $\mathbf{Q_r}$, the restriction of this norm $|*|_r$ on $\mathbf{Q_r}$ is the r-adic norm. On the entire $\mathbf{C_r}$ this norm $|*|_r$ in general need not be multiplicative.

We verify, that it is indeed the non-archimedean norm. At first it is evident, that $|x|_r \geq 0$ for each $x \in \mathbf{C_r}$, while $|x|_r = 0$ if and only if $x = 0$ due to Formula (i). If $x, y \in \mathbf{C_r}$, $x \neq 0$, $y \neq 0$, then $|x| = r^a$, $|y| = r^b$, $|x+y| = r^c$ with $a, b, c \in \mathbf{R}$, $c \leq \max(a,b)$, where $r \geq 2$ is the prime number. Therefore, $|x|_r = r^A$, $|y|_r = r^B$, $|x+y|_r = r^C$, where $a \leq A$, $b \leq B$, $c \leq C$, $A, B, C \in \mathbf{Z}$ are the least integers satisfying these inequalities. Therefore, $C \leq max(A,B)$, consequently, $|x+y|_r \leq \max(|x|_r, |y|_r)$ for each $x, y \in \mathbf{C_r}$.

In view of Theorems 5.13 and 5.16 [99] the $\mathbf{Q_r}$-linear space $(\mathbf{C_r}, |*|_r)$ is isomorphic with the Banach space $c_0(\alpha, \mathbf{Q_r})$, where α is a set. This set is convenient to consider as an ordinal due to Zermelo theorem [19].

Let

$$(x,y) := (x,y)_{\mathbf{Q_r}} := \sum_{j \in \alpha} x_j y_j$$

for all $x, y \in \mathbf{C_r}$, $x = (x_j : j \in \alpha, x_j \in \mathbf{Q_r})$. This series (x, y) converges in $\mathbf{Q_r}$, since for each $\varepsilon > 0$ the set $\{j : |x_j|_r \geq \varepsilon\}$ is finite.

If X is a complete locally $\mathbf{C_r}$-convex space, then it is the projective limit of Banach spaces $V_u := X/Y_u$ over $\mathbf{C_r}$, where $Y_u := \{x \in X : u(x) = 0\}$, while u is a continuous semi-norm in X, $u \in \mathcal{S}$ [87]. Each V_u can be supplied with the structure of a Banach space over the field $\mathbf{Q_r}$. Therefore, X can be supplied with the structure X_r of the complete locally $\mathbf{Q_r}$-convex space with the coressponding topology denoted here by τ_r.

We consider the case of such space X, when X_r has an embedding into $c_0(\beta, \mathbf{Q_r})$ for some $\beta \geq \alpha$. Let naturally the norm topology n_r of $|*|_r$ in X_r inherited from $c_0(\beta, \mathbf{Q_r})$ be such that $\tau_r \supset n_r$. Then each $\mathbf{Q_r}$-linear continuous functional on $(c_0(\beta, \mathbf{C_r}), |*|_r)$ is also continuous on (X_r, τ_r).

Now it is possible to define a character with values in $\mathbf{C_p}$ for $(X, +)$ as the additive group, $r \neq p$. For this it is sufficient to put

$$\chi_{r,p;s}(x) = \varepsilon^{[(s,z)_{\mathbf{Q_r}}]_{\mathbf{Q_r}}/z},$$

where $\varepsilon = 1^z$ is a root of unity in $\mathbf{C_p}$,

$$z = r^{ord_r[(s,z)_{\mathbf{Q_r}}]_{\mathbf{Q_r}}},$$

$s, z \in X_r$ or we can consider s, z as elements in X as well (see above).

For a tight measure $\mu : \mathcal{R}(X) \to Lc(\mathrm{g})$ or $\mu : \mathcal{R}(X) \to \mathbf{C_p}$ the characteristic functional $\hat{\mu}$ is given by the formula:

$$\hat{\mu}(s) := \int_X \chi_{r,p;s}(z)\mu(dz),$$

where $s \in X_r$, X is a non-archimedean space over $\mathbf{C_r}$.

In general the characteristic functional of the measure $\mu : \mathcal{R}(G) \to Lc(\mathrm{g})$ or $\mu : \mathcal{R}(G) \to \mathbf{C_p}$ is defined in the space $C^0(G, \mathbf{C_r})$ of continuous functions $f : G \to \mathbf{C_r}$

$$\hat{\mu}(f) := \int_G \chi_{r,p;1}(f(z))\mu(dz),$$

where $1 \in \mathbf{C_r}$, G is the totally disconnected topological Hausdorff space with the separating covering ring $\mathcal{R}(G)$.

In view of Theorems 2.21 and 2.30 [76] and Theorem 24 and (4) above a $Lc(\mathrm{g})$-valued measure μ on $\mathsf{Bco}(\mathbf{C_r})$ exists so that

$$B(t) = \int_{\mathbf{C_r}} \chi_{r,p;1}(ty)\mu(dy).$$

Functions $g(t,y) := \chi_{r,p;1}(ty)$ are continuous and uniformly bounded. Since $|g(t,y)|_{C_p}$
$= 1$ for each $t, y \in \mathbf{C_r}$, the inclusion is satisfied $g(t,y) \in L^2\{\mu\}$.

In view of the Kaplansky Theorem A.4 [103] and Theorem 26 above the family of functions $\{g(t,y) : t, y \in \mathbf{C_r}\}$ is complete in the space $L^2(\mathrm{g}, \mathsf{Bco}(\mathrm{g}), \mu, Lc(\mathrm{g}); \mathbf{C_p})$. Thus statements $51(1-3)$ follow from Theorem 49.

Chapter 3

Random Functions and Stochastic Differential Equations in Banach Spaces

3.1. Introduction

This chapter is devoted to stochastic analysis in Banach spaces. Stochastic processes in finite- and infinite-dimensional Banach spaces over a field **K** are used in many problems and applications. This also is the base for subsequent studies of random functions in locally **K**-convex, ultra-metric and ultra-uniform spaces. Stochastic differential equations, Lévy and diffusion processes with ranges in Banach spaces and manifolds are widely used for solutions of mathematical, physical, economic, etc. problems and for construction and investigation of transition probabilities on them.

In Section 2 some necessary facts from the measure theory are recalled. Random functions with Markov and Poisson properties are considered in Sections 3 and 4. The existence of random functions is proved below. Poisson and Markov processes play very important role in stochastic analysis.

In the classical stochastic analysis indefinite integrals are widely used. But in the non-archimedean field **K**, for example, of p-adic numbers $\mathbf{Q_p}$, any linear order structure compatible with the addition or multiplication in **K** does not exist apart from the real field **R**.

In this and subsequent chapters spaces of functions with values in Banach spaces over non-archimedean fields are considered, in particular, with values in the field $\mathbf{Q_p}$ of p-adic numbers. That is stochastic processes are considered on spaces of functions with values in the non-archimedean field such that a parameter analogous to the time is either real, p-adic or more generally can take values in any group. Certainly this encompasses cases of the time parameter with values in the adele ring $\mathcal{A} := \mathbf{R} \times \prod_{p \in \mathbf{P}} \mathbf{Q_p}$ for **Q** and the idele I, which is the multiplicative subgroup in $\mathcal{A}$, where **P** denotes the family of all prime numbers (see also [111]). Certainly having random functions or transition measures with values in $\mathbf{Q_p}$ or **R** one can construct their products with values in the adele ring $\mathcal{A}$ or in the idele group I. This may have advantages due to the availability of different multiplicative norms on multipliers in the adele ring.

Specific anti-derivation operators generalizing Schikhof anti-derivation operators on spaces of functions C^n are described in Section 5. Their continuity and differentiability properties are given. Also operators analogous to nuclear operators are studied. The non-archimedean stochastic integral is defined in Section 6. Its continuity as the operator on the corresponding spaces of functions is proved. Functions of stochastic processes are considered as well and analogs of the Itô formula are proved. Spaces of analytic functions lead to simpler expressions of the Itô formula analog. But the space of analytic functions is very narrow and though it is helpful in non-archimedean mathematical physics it is insufficient for solutions of all mathematical and physical problems. For example, in many cases of topological groups for non-archimedean manifolds spaces of analytic functions are insufficient. On the other hand, for spaces C^n rather simple formulas are found. Particularly, processes like Brownian motion controlled by transition measures with values in the real and non-archimedean fields are presented in Sections 6.1 and 6.2.

There are not differentiable functions from the field $\mathbf{Q_p}$ into $\mathbf{R}$ or in another non-archimedean field $\mathbf{Q}_{p'}$ with $p \neq p'$ besides locally constant. Therefore, for such mappings instead of differentiability their pseudo-differentiability is considered.

Then in Section 7 of this chapter stochastic anti-derivational equations are considered. In the non-archimedean case anti-derivational equations are used instead of stochastic integrals or differential equations in the classical case.

For this suitable analogs of Gaussian transition measures are considered. They are used for the definition of the standard Brownian motion. Integration by parts formula for the non-archimedean stochastic processes is studied. Some particular cases of the general Itô formula are discussed here more concretely. Analogs of theorems about existence and uniqueness of solutions of stochastic anti-derivational equations are proved. Generating operators of solutions of stochastic equations are investigated.

In this chapter notations and definitions of two previous chapters are used.

3.2. p-Adic Probability Measures

At first we remind some necessary facts from the measure theory, which are written in details in the book [81].

We consider a non-void topological space X. A topological space is called zero-dimensional if it has a base of its topology consisting of clopen subsets. A topological space X is called a T_0-space if for each two distinct points x and y in X there exists an open subset U in X such that either $x \in U$ and $y \in X \setminus U$ or $y \in U$ and $x \in X \setminus U$.

A covering ring $\mathcal{R}$ of a space X defines on it a base of zero-dimensional topology $\tau_{\mathcal{R}}$ such that each element of $\mathcal{R}$ is considered as a clopen subset in X. If $\pi : X \to Y$ is a mapping such that $\pi^{-1}(\mathcal{R}_Y) \subset \mathcal{R}_X$, then a measure μ on $(X, \mathcal{R}_X)$ induces a measure $\nu := \pi(\mu)$ on $(Y, \mathcal{R}_Y)$ by the formula

$$\nu(A) = \mu(\pi^{-1}(A)) \quad \text{for each } A \in \mathcal{R}_Y.$$

1. Proposition. *Let $(X, \mathcal{R}, \mu)$ be a measure space. Then there exists a quotient mapping $\pi : X \to Y$ on a Hausdorff zero-dimensional space $(Y, \tau_{\mathcal{G}})$ and $\pi(\mu) := \nu$ is a measure on Y such that $\mathcal{G} = \pi(\mathcal{R})$, where $(Y, \mathcal{G}, \nu)$ is a measure space.*

2. Note. In view of Proposition 1 we consider henceforth Hausdorff zero-dimensional measurable $(X, \mathcal{R})$ spaces if another is not specified.

We consider now a family of probability measure spaces $\{(X_j, \mathcal{R}_j, \mu_j) : j \in \Lambda\}$, where Λ is a set. Suppose that each covering ring $\mathcal{R}_j$ is complete relative to a measure μ_j. This means that $\mathcal{R}_j = \mathcal{R}_{\mu_j}$, where $\mathcal{R}_{\mu_j}$ denotes a completion of $\mathcal{R}_j$ relative to the measure μ_j. Let

$$X := \prod_{j\in\Lambda} X_j$$

be the product of topological spaces supplied with the product (Tychonoff) topology τ_X, where each X_j is considered in its $\tau_{\mathcal{R}_j}$-topology. There is the natural continuous projection $\pi_j : X \to X_j$ for each $j \in \Lambda$. Let $\mathcal{R}$ be the ring of the form $\bigcup_{j_1,\dots,j_n\in\Lambda, n\in\mathbf{N}} \bigcap_{l=1}^{n} \pi_{j_l}^{-1}(\mathcal{R}_{j_l})$.

3. Definition. A triple $(X, \mathcal{R}, \mu)$ is called a cylindrical distribution if it satisfies the following condition:

$$\mu|_{\bigcap_{l=1}^{n} \pi_{j_l}^{-1}(\mathcal{R}_{j_l})} = \prod_{l=1}^{n} \tilde{\mu}_{j_l}$$

for each $j_1, \dots, j_n \in \Lambda$ and $n \in \mathbf{N}$, where $\tilde{\mu}_j(\pi_j^{-1}(A)) := \mu_j(A)$ for each $A \in \mathcal{R}_j$; $\tilde{\mu}_j$ is the measure on $(X, \pi_j^{-1}(\mathcal{R}_j))$.

4. Theorem. *A cylindrical distribution μ on $(X, \mathcal{R})$ has an extension up to a probability measure μ on $(X, \mathcal{R}_\mu)$, where μ and X are the same as in §3.*

5. Note. Theorem 4 has an evident generalization for bounded measures μ_j if two products $\prod_{j\in\Lambda_0} \mu_j(X_j) \in \mathbf{K}$ and $\prod_{j\in\Lambda} \|X_j\|_{\mu_j} < \infty$ converge, where $\Lambda_0 := \{j : j \in \Lambda, \mu_j(X_j) \neq 0\}$, when $\Lambda \setminus \Lambda_0$ is finite. Since μ is defined on the separating covering ring $\mathcal{R}$ and bounded on it, the measure μ has an extension to the bounded measure μ on $\mathcal{R}_\mu$ such that

$$\mu(X) = \prod_{j\in\Lambda} \mu_j(X_j) \quad \text{and}$$

$$\|X\|_\mu = \prod_{j\in\Lambda} \|X_j\|_{\mu_j},$$

where $\mathcal{R}_\mu$ is the completion of $\mathcal{R}$ relative to μ.

6. Note. A set Λ is called directed if there exists a relation $\leq$ on it satisfying the following conditions:

$(D1)$ If $j \leq k$ and $k \leq m$, then $j \leq m$;

$(D2)$ For every $j \in \Lambda$, $j \leq j$;

$(D3)$ For each j and k in Λ there exists $m \in \Lambda$ such that $j \leq m$ and $k \leq m$. A subset Υ of Λ directed by $\leq$ is called cofinal in Λ if for each $j \in \Lambda$ there exists $m \in \Upsilon$ such that $j \leq m$. Suppose that Λ is a directed set and $\{(X_j, \mathcal{R}_j, \mu_j) : j \in \Lambda\}$ is a family of probability measure spaces, where $\mathcal{R}_j$ is the covering ring (not necessarily separating). Supply each X_j with a topology τ_j such that its base is a ring $\mathcal{R}_j$. Let this family be consistent in the following sense:

(1) there exists a mapping $\pi_j^k : X_k \to X_j$ for each $k \geq j$ in Λ such that $(\pi_j^k)^{-1}(\mathcal{R}_j) \subset \mathcal{R}_k$, $\pi_j^j(x) = x$ for each $x \in X_j$ and each $j \in \Lambda$, $\pi_k^m \circ \pi_l^k = \pi_l^m$ for each $m \geq k \geq l$ in Λ;

(2) $\pi_l^k(\mu_k) = (\mu_l)$ for each $k \geq l$ in Λ. Such family of measure spaces is called consistent.

7. Theorem. *Let $\{(X_j,\mathcal{R}_j,\mu_j) : j\in\Lambda\}$ be a consistent family of measure spaces as in §6. Then there exists a probability measure space $(X,\mathcal{R}_\mu,\mu)$ and a mapping $\pi_j : X\to X_j$ for each $j\in\Lambda$ such that $(\pi_j)^{-1}(\mathcal{R}_j)\subset\mathcal{R}$ and $\pi_j(\mu)=\mu_j$ for each $j\in\Lambda$.*

8. Note. Theorem 7 has an evident generalization to the following case:

$$\|X_j\|_{\mu_j}<\infty$$

for each j and two limits

$$\lim_{j\in\Lambda_0}\mu_j(X_j)\in\mathbf{K} \quad\text{and}\quad \lim_{\mathbf{j}\in\blacksquare}\|\mathbf{X_j}\|_{\mu_\mathbf{j}}<\infty$$

exist, where $\Lambda_0:=\{j : j\in\Lambda,\mu_j(X_j)\neq 0\}$ and $\Lambda\setminus\Lambda_0$ is bounded in Λ. We have

$$\|X_j\|_{\mu_j}\leq\|X_k\|_{\mu_k}$$

for each $j\leq k$ in Λ, since $\pi_j^k(\mu_k)=\mu_j$ and $(\pi_j^k)^{-1}(\mathcal{R}_k)\subset\mathcal{R}_j$. Since Λ is directed, the limit

$$\lim_{j\in\Lambda}\|X_j\|_{\mu_j}=\sup_{j\in\Lambda}\|X_j\|_{\mu_j}$$

exists. Since a cylindrical distribution μ is defined on $\mathcal{R}$ and bounded on it, this cylindrical distribution μ has an extension to the bounded measure μ on $\mathcal{R}_\mu$ such that

$$\mu(X)=\lim_{j\in\Lambda}\mu_j(X_j) \quad\text{and}\quad \|X\|_\mu=\lim_{j\in\Lambda}\|X_j\|_{\mu_j},$$

where $\mathcal{R}_\mu$ is the completion of $\mathcal{R}$ relative to μ.

Let now X be a set with a covering ring $\mathcal{R}$ such that $X\in\mathcal{R}$. Let also $\{(X,\mathcal{G}_j,\mu_j) : j\in\Lambda\}$ be a family of measure spaces such that Λ is directed and $\mathcal{G}_j\subset\mathcal{G}_k$ for each $j\leq k\in\Lambda$, $\mathcal{R}=\bigcup_{j\in\Lambda}\mathcal{G}_j$. Suppose that the measure $\mu:\mathcal{R}\to\mathbf{K}$ is such that its restrictions are

$$\mu|_{\mathcal{G}_j}=\mu_j \quad\text{and}\quad \mu_k|_{\mathcal{G}_j}=\mu_j$$

for each $j\leq k$ in Λ. Then the triple $(X,\mathcal{R},\mu)$ is called the cylindrical distribution. For each $A\in\mathcal{R}$ there exists $j\in\Lambda$ such that $A\in\mathcal{G}_j$, consequently,

$$\|A\|_{\mu_j}=\|A\|_{\mu_k} \quad\text{for each}\quad k\geq j \quad\text{in}\quad \Lambda.$$

Thus

$$\|A\|_\mu:=\lim_{k\in\Lambda}\|A\|_{\mu_k}$$

is correctly defined. Let also the cylindrical distribution μ be bounded, that is,

$$\|X\|_\mu<\infty.$$

9. Theorem. *Let $(X,\mathcal{R},\mu)$ be a bounded cylindrical distribution as in §8. Then μ has an extension to a bounded measure μ on a completion $\mathcal{R}_\mu$ of $\mathcal{R}$ relative to μ.*

10. Note. Let $X:=\prod_{t\in T}X_t$ be a product of sets X_t and let a covering ring $\mathcal{R}$ be given on X such that for each $n\in\mathbf{N}$ and pairwise distinct points $t_1,\dots,t_n$ in a set T there exists a

$\mathbf{K}$-valued measure $P_{t_1,\dots,t_n}$ on a covering ring $\mathcal{R}_{t_1,\dots,t_n}$ of the Cartesian product $X_{t_1}\times\cdots\times X_{t_n}$ such that $\pi^{t_1,\dots,t_{n+1}}_{t_1,\dots,t_n}(\mathcal{R}_{t_1,\dots,t_{n+1}})=\mathcal{R}_{t_1,\dots,t_n}$ for each $t_{n+1}\in T$ and

$$P_{t_1,\dots,t_{n+1}}(A_1\times\cdots\times A_n\times X_{n+1})=P_{t_1,\dots,t_n}(A_1\times\cdots\times A_n)$$

$$\text{for each } A_1\times\cdots\times A_n\in\mathcal{R}_{t_1,\dots,t_n},$$

where

$$\pi^{t_1,\dots,t_{n+1}}_{t_1,\dots,t_n}:X_{t_1}\times\cdots\times X_{t_{n+1}}\to X_{t_1}\times\cdots\times X_{t_n}$$

is the natural projection, $A_l\subset X_{t_l}$ for each $l=1,\dots,n$. Suppose that this cylindrical distribution is bounded, that is,

$$\sup_{t_1,\dots,t_n\in T,n\in\mathbf{N}}\|P_{t_1,\dots,t_n}\|<\infty$$

and the limit

$$\lim_{t_1,\dots,t_n\in T_0;n\in\mathbf{N}}P_{t_1,\dots,t_n}(X_{t_1}\times\cdots\times X_{t_n})\in\mathbf{K}$$

exists, where $T_0:=\{t\in T:P_t(X_t)\neq 0\}$, $T\setminus T_0$ is finite.

11. Theorem. *A cylindrical distribution $P_{t_1,\dots,t_n}$ from §10 has an extension to a bounded measure P on a completion $\mathcal{R}_P$ of the product separation covering ring*

$$\mathcal{R}:=\bigcup_{t_1,\dots,t_n\in T,n\in\mathbf{N}}\mathcal{G}_{t_1,\dots,t_n}$$

relative to P, where $\mathcal{G}_{t_1,\dots,t_n}:=(\pi_{t_1,\dots,t_n})^{-1}(\mathcal{R}_{t_1,\dots,t_n})$ and $\pi_{t_1,\dots,t_n}:X\to X_{t_1}\times\cdots\times X_{t_n}$ is the natural projection.

The proofs of these theorems are given in details in [81].

3.3. Markov Distributions for a Non-Archimedean Banach Space

1. Remark. Markov processes are characterized by the absence of a posterity (aftereffect). Therefore, they play very important role in stochastic processes.

Let $H=c_0(\alpha,\mathbf{K})$ be the Banach space. Suppose that a field $\mathbf{K}$ is complete as the ultrametric space. For example, the field $\mathbf{K}$ is such that $\mathbf{Q_p}\subset\mathbf{K}\subset\mathbf{C_p}$ or $\mathbf{F_p}(\theta)\subset\mathbf{K}$, where p is a prime number, $\mathbf{Q_p}$ is the field of p-adic numbers, $\mathbf{C_p}$ is the field of complex p-adic numbers, $\mathbf{F_p}(\theta)$ is the field of formal power series by an indeterminate θ over the finite field $\mathbf{F_p}$ consisting of p elements.

Let also $\mathsf{U^P}$ be a cylindrical ring generated by projections $\pi_F:H\to F$ on finite dimensional over $\mathbf{K}$ subspaces F in H and rings $Bco(F)$ of clopen subsets. This ring $\mathsf{U^P}$ is the base of the weak topology $\tau_{H,w}$ in H. Each vector $x\in H$ is considered as continuous linear functional on H by the formula

$$x(y)=\sum_j x^jy^j$$

for each $y\in H$, so there is the natural embedding $H\hookrightarrow H^*=l^\infty(\alpha,\mathbf{K})$, where $x=\sum_j x^je_j$, $x^j\in\mathbf{K}$,

$$l^\infty(\alpha,\mathbf{K}):=\{x:x=(x_i:i\in\alpha,x_i\in\mathbf{K}),$$

$\sup_{i\in\alpha}|x_i| =: \|x\| < \infty\}$ is the Banach space topologically dual to H. This justifies the following generalization.

2. Notes and definitions. Suppose that ■ is an additive group. For example, ■ is contained in $\mathbf{R}$ or $\mathbf{C}$ or in a non-archimedean field. Let T be a subset in ■ and containing a point t_0 and let $X_t = X$ be a locally $\mathbf{K}$-convex space for each $t \in T$. We put

$$(\tilde{X}_T, \tilde{\mathsf{U}}) := \prod_{t\in T}(X_t, \mathsf{U}_t)$$

for the product of measurable spaces. Here U_t denotes a ring of clopen subsets of X_t, $\tilde{\mathsf{U}}$ is a ring of cylindrical subsets of $\tilde{X}_T$ generated by projections $\tilde{\pi}_q : \tilde{X}_t \to X^q$, where $X^q := \prod_{t\in q} X_t$, $q \subset T$ is a finite subset of T (see §I.1.3 [16]). We take a subfield $\mathbf{K_s}$ of $\mathbf{C_s}$ such that $\mathbf{K_s}$ is complete as an ultra-metric space, where s is a prime number. A function $P(t_1,x_1,t_2,A)$ with values in $\mathbf{K_s}$ for each $t_1 \ne t_2 \in T$, $x_1 \in X_{t_1}$, $A \in \mathsf{U}_{t_2}$ is called a transition measure if it satisfies the following conditions:

(i) the set function $\nu_{x_1,t_1,t_2}(A) := P(t_1,x_1,t_2,A)$ is a measure on $(X_{t_2}, \mathsf{U}_{t_2})$;

(ii) the function $\alpha_{t_1,t_2,A}(x_1) := P(t_1,x_1,t_2,A)$ of the variable x_1 is U_{t_1}-measurable, that is, $\alpha^{-1}_{t_1,t_2,A}(Bco(\mathbf{K_s})) \subset \mathsf{U}_{t_1}$;

$$(iii)\ P(t_1,x_1,t_2,A) = \int_{X_z} P(t_1,x_1,z,dy)P(z,y,t_2,A) \text{ for each } t_1 \ne t_2 \in T,$$

that is, $P(z,y,t_2,A)$ as the function by y is in $L((X_z,\mathsf{U}_z),\nu_{x_1,t_1,z},\mathbf{K_s})$. A transition measure $P(t_1,x_1,t_2,A)$ is called normalized if

$$(iv)\ P(t_1,x_1,t_2,X_{t_2}) = 1 \text{ for each } t_1 \ne t_2 \in T.$$

For each set $q = (t_0,t_1,\dots,t_{n+1})$ of pairwise distinct points in T there is defined a measure in in the product $X^g := \prod_{t\in g} X_t$ of measurable spaces by the formula

$$(v)\ \mu^q_{x_0}(E) = \int_E \prod_{k=1}^{n+1} P(t_{k-1},x_{k-1},t_k,dx_k),\ E \in \mathsf{U}^g := \prod_{t\in g}\mathsf{U}_t,$$

where $g = q\setminus\{t_0\}$, variables $x_1,\dots,x_{n+1}$ are such that $(x_1,\dots,x_{n+1}) \in E$, $x_0 \in X_{t_0}$ is fixed.

Suppose that the transition measure $P(t,x_1,t_2,dx_2)$ is normalized. For the product of two sets $E = E_1 \times X_{t_j} \times E_2$, where $E_1 \in \prod_{i=1}^{j-1}\mathsf{U}_{t_i}$, $E_2 \in \prod_{i=j+1}^{n+1}\mathsf{U}_{t_i}$, the equality

$$(vi)\quad \mu^q_{x_0}(E) = \int_{E_1\times E_2}\left[\prod_{k=1}^{j-1} P(t_{k-1},x_{k-1},t_k,dx_k)\right]$$

$$\times\left[\int_{X_{t_j}} P(t_{j-1},x_{j-1},t_j,dx_j)\prod_{k=j+1}^{n+1} P(t_{k-1},x_{k-1},t_k,dx_k)\right] = \mu^r_{x_0}(E_1\times E_2)$$

is fulfilled, where $r = q\setminus\{t_j\}$. From Equation (vi) it follows, that

$$(vii)\ [\mu^q_{x_0}]^{\pi^q_v} = \mu^v_{x_0}$$

for each $v < q$ (that is, $v \subset q$), where $\pi_v^q : X^g \to X^w$ is the natural projection, $g = q \setminus \{t_0\}$, $w = v \setminus \{t_0\}$. We consider the family Υ_T of all finite subsets q in T such that $t_0 \in q \subset T$, $v \leq q \in \Upsilon_T$, $\pi_q : \tilde{X}_T \to X^g$ is the natural projection, $g = q \setminus \{t_0\}$. Therefore, due to Conditions (iv, v, vii) : $\{\mu_{x0}^q; \pi_v^q; \Upsilon_T\}$ is the consistent family of measures. It induces a cylindrical distribution $\tilde{\mu}_{x_0}$ on the measurable space $(\tilde{X}_T, \tilde{\mho})$ such that

$$\tilde{\mu}_{x0}(\pi_q^{-1}(E)) = \mu_{x0}^q(E)$$

for each $E \in \mho^g$.

The cylindrical distribution given by Equations $(i - v, vii)$ is called the Markov distribution (with time $t \in T$).

3. Proposition. *If a normalized transition measure P satisfies the condition*

$$(i) \quad C := \sup_q \left[\sum_{k=1}^{n} ln(\sup_x \|\nu_{x,t_{k-1},t_k}\|) \right] < \infty,$$

where $q = (t_0, t_1, \ldots, t_n)$ with pairwise distinct points $t_0, \ldots, t_n \in T$ and $n \in \mathbf{N}$, then the Markov cylindrical distribution $\tilde{\mu}_{x_0}$ is bounded and it has an extension to a bounded measure $\tilde{\mu}_{x_0}$ on a completion $\tilde{\mho}_{\tilde{\mu}_{x_0}}$ of $\tilde{\mho}$ relative to $\tilde{\mu}_{x_0}$.

4. Proposition. *If*

$$(ii) \quad C_x := \sup_q \left[\sum_{k=1}^{n} ln\|\nu_{x,t_{k-1},t_k}\| \right] = \infty$$

for each x, where $q = (t_0, t_1, \ldots, t_n)$ with pairwise distinct points $t_0, \ldots, t_n \in T$ and $n \in \mathbf{N}$, then the Markov cylindrical distribution $\tilde{\mu}_{x0}$ has an unbounded variation on each non-void set $E \in \tilde{\mho}$.

Proof. (1). If $E \in \tilde{\mho}$, then $E \in \mho^g$ for some set $q = (t_0, t_1, \ldots, t_n)$ with pairwise distinct points $t_0, \ldots, t_n \in T$ and $n \in \mathbf{N}$ and $g = q \setminus \{t_0\}$, consequently,

$$|\mu_{x0}^q(E)| \leq \prod_{k=1}^{n} \sup_x \|\nu_{x,t_{k-1},t_k}\| \leq exp(C) < \infty,$$

since $t_k \in T$ for each $k = 0, 1, \ldots, n$, hence

$$\sup_{q,E} |\mu_{x0}^q(E)| = \|\tilde{\mu}_{x_0}\| \leq exp(C).$$

In view of Theorem 11 we get an extension of $\tilde{\mu}_{x_0}$ to a bounded measure on $\tilde{\mho}_{\tilde{\mu}_{x0}}$.

(2). For each (t_1, t_2, x) with x in $\pi_{t_0,t_2}(E)$ there exists a set $\delta(t_1, t_2, x) \in \mho_{t_2} \cap \pi_{t_0,t_2}(E)$ such that

$$\|\delta(t_1, t_2, x)\|_{\nu_{x_1,t_1,t_2}} > 1 + \varepsilon(t_1, t_2, x_1, x),$$

where $\varepsilon(t_1, t_2, x_1, x) > 0$. In view of Condition (ii) for each $R > 0$ and x we choose q such that

$$\sum_{k=1}^{n} \varepsilon(t_k, t_{k+1}, x_1, x) > R.$$

For chosen points $u \neq u_1 \in T$ and $x \in \pi_{t_0,u}(E) \subset X_u$ we represent the set $\delta(u,u_1,x)$ as a finite union of disjoint subsets $\gamma_{j_1} \in \mathsf{U}_{u_1}$ such that for each γ_{j_1} and $u_2 \neq u_1$ there is a set $\delta_{j_1} \in \mathsf{U}_{u_2} \cap \pi_{t_0,u_2}(E)$ satisfying the restriction

$$\|\delta_{j_1}\|_{\nu_{x_1,u_1,u_2}} \geq 1+\varepsilon(u_1,u_2,x_1,x)$$

for each $x \in \gamma_{j_1}$. Then by induction

$$\delta_{j_1,\dots,j_n} = \bigcup_{j_{n+1}=1}^{m_{n+1}} \gamma_{j_1,\dots,j_{n+1}}$$

so that for $u_{n+2} \neq u_{n+1} \in T$ there is a set $\delta_{j_1,\dots,j_{n+1}} \in \mathsf{U}_{u_{n+1}} \cap \pi_{t_0,u_{n+1}}(E)$ for which

$$\|\delta_{j_1,\dots,j_{n+1}}\|_{\nu_{x_{n+1},u_{n+1},u_{n+2}}} \geq 1+\varepsilon(u_{n+1},u_{n+2},x_{n+1},x)$$

for each $x \in \gamma_{j_1,\dots,j_{n+1}}$. We define the sets

$$\Gamma^{u,x_0}_{j_1,\dots,j_n} = \{x : x(u)=x_0, x(u_1) \in \gamma_{j_1},\dots,x(u_n) \in \delta_{j_1,\dots,j_n}, x(u_{n+1}) \in \gamma_{j_1,\dots,j_n}\}$$

and $\Gamma^{u,x_0} := (\bigcup_{j_1,\dots,j_n} \Gamma^{u,x_0}{}_{j_1,\dots,j_n})$. The latter set belongs to the separation covering ring $\tilde{\mathsf{U}}$, since $m_1 \in \mathbf{N},\dots,m_n \in \mathbf{N}$. Then we get the inequality

$$\|\Gamma^{u,x_0}\|_{\tilde{\mu}_{x_0}} \geq \sup_{j_1,\dots,j_n} \left\| \left(\prod_{k=1}^{n+1} \nu_{u_{k-1},x_{k-1},u_k}(dx_k)\right)_{\gamma_{j_1}\times\dots\times\gamma_{j_1,\dots,j_n}\times\delta_{j_1,\dots,j_n}} \right\|$$

$$\geq \prod_{k=1}^{n}[1+\varepsilon(u_{k-1},u_k,x_{k-1},x_k)] > R,$$

consequently, $\|E\|_{\tilde{\mu}_{x_0}} = \infty$, since $\|E\|_{\tilde{\mu}_{x_0}} \geq \sup_{\Gamma^{u,x_0}} \|\Gamma^{u,x_0}\|_{\tilde{\mu}_{x_0}}$ and $R>0$ is arbitrary.

5. Note. Let $X_t = X$ for each $t \in T$, $\tilde{X}_{t_0,x_0} := \{x \in \tilde{X}_T : x(t_0) = x_0\}$. We define a projection operator $\bar{\pi}_q : x \mapsto x_q$, where x_q is defined on $q = (t_0,\dots,t_{n+1})$ such that $x_q(t) = x(t)$ for each $t \in q$, that is, $x_q = x|_q$. For every function $F : \tilde{X}_T \to \mathbf{C_s}$ there corresponds a function $(S_qF)(x) := F(x_q) = F_q(y_0,\dots,y_n)$, where $y_j = x(t_j)$, $F_q : X^q \to \mathbf{C_s}$. We put

$$\mathsf{F} := \{F | F : \tilde{X}_T \to \mathbf{C_s}, S_qF \text{ are } \mathsf{U}^q-\text{measurable}\}.$$

If $F \in \mathsf{F}$, $\tau = t_0 \in q$, then the integral

$$(i)\ J_q(F) = \int_{X^q} (S_qF)(x_0,\dots,x_n) \prod_{k=1}^{n+1} P(t_{k-1},x_{k-1},t_k,dx_k)$$

exists.

6. Definition. A function F is called integrable with respect to a Markov cylindrical distribution μ_{x_0} if the limit

$$(ii)\ \lim_q J_q(F) =: J(F)$$

along the generalized net by finite subsets q of T exists. This limit is called a functional integral with respect to the Markov cylindrical distribution:

$$(iii)\ J(F) = \int_{\tilde{X}_{t_0,x_0}} F(x)\mu_{x_0}(dx).$$

7. Remark. Consider a $\mathbf{K_s}$-valued measure $P(t,A)$ on (X,U) for each $t \in T$ such that $A - x \in \mathsf{U}$ for each $A \in \mathsf{U}$ and $x \in X$, where $A \in \mathsf{U}$, X is a locally $\mathbf{K}$-convex space, U is a covering ring of X, where $\mathbf{K_s}$ denotes a finite algebraic extension of the field $\mathbf{Q_s}$. Suppose that P is a spatially homogeneous transition measure (see also Chapter 1 §2.9), that is,

$$(i)\ P(t_1,x_1,t_2,A) = P(t_2 - t_1, A - x_1)$$

for each $A \in \mathsf{U}$, $t_1 \ne t_2 \in T$ and $t_2 - t_1 \in T$ and every $x_1 \in X$, where $P(t,A)$ satisfies the following condition:

$$(ii)\ P(t_1 + t_2, A) = \int_X P(t_1,dy)P(t_2,A-y)$$

for each t_1 and t_2 and $t_1 + t_2 \in T$. Such a transition measure $P(t_1,x_1,t_2,A)$ is called homogeneous. In particular for $T = \mathbf{Z_p}$ we have

$$(iii)\ P(t+1,A) = \int_X P(t,dy)P(1,A-y).$$

If $P(t,A)$ is a continuous function by $t \in T$ for each fixed $A \in \mathsf{U}$, then Equation (iii) defines $P(t,A)$ for each $t \in T$, when $P(1,A)$ is given, since $\mathbf{Z}$ is dense in $\mathbf{Z_p}$. One can mention that §§2.9, 2.11 and 3 provide examples of Markov distributions. Since a non-archimedean field is not ordered, we consider the condition $t_1 \ne t_2$ in $7(i)$ instead of $t_1 < t_2$ used in the classical case. In the case $\Lambda \subset \mathbf{R}$ and in particular $\Lambda = \mathbf{Z}$ it is possible to consider the condition $t_1 < t_2$. Examples of Markov distributions are also Poisson and Gaussian distributions.

8. Notes. Let X be a locally $\mathbf{K}$-convex space and P satisfies Conditions $7(i-iii)$. For x and $z \in \mathbf{Q_p^n}$ we denote by (z,x) the following sum $\sum_{j=1}^n x_j z_j$, where $x = (x_j : j = 1,\dots,n)$, $x_j \in \mathbf{Q_p}$. Each number $y \in \mathbf{Q_p}$ has a decomposition $y = \sum_l a_l p^l$, where $a_l \in (0,1,\dots,p-1)$, $\min(l : a_l \ne 0) =: ord_p(y) > -\infty$ for $y \ne 0$ and $ord(0) := \infty$ [87, 103]. We define the following symbol

$$\{y\}_p := \sum_{l<0} a_l p^l \quad \text{for } |y|_p > 1 \quad \text{and}$$

$$\{y\}_p = 0 \quad \text{for } |y|_p \le 1.$$

It is useful to consider a character of X, $\chi_\gamma : X \to \mathbf{C_s}$ given by the following formula:

$$(i)\ \chi_\gamma(x) = \varepsilon^{z^{-1}\{(e,\gamma(x))\}_p}$$

for each $\{(e,\gamma(x))\}_p \ne 0$, $\chi_\gamma(x) := 1$ for $\{(e,\gamma(x))\}_p = 0$, where $\varepsilon = 1^z$ is a root of unity, $z = p^{ord(\{(e,\gamma(x))\}_p)}$, $\gamma \in X^*$. Traditionally X^* denotes the topologically dual space of continuous $\mathbf{K}$-linear functionals on X. Mention, that the field $\mathbf{K}$ as the $\mathbf{Q_p}$-linear space is n-dimensional, that is, $dim_{\mathbf{Q_p}}\mathbf{K} = n$, $\mathbf{K}$ as the Banach space over $\mathbf{Q_p}$ is isomorphic with $\mathbf{Q_p^n}$, $e = (1,\dots,1) \in \mathbf{Q_p^n}$, where $s \neq p$ are prime numbers. Then

$$(ii)\ \phi(t_1,x_1,t_2,y) := \int_X \chi_y(x) P(t_1,x_1,t_2,dx)$$

is the characteristic functional of the transition measure $P(t_1,x_1,t_2,dx)$ for each $t_1 \neq t_2 \in T$ and each $x_1 \in X$. In [81] were considered conditions, when a measure is completely characterized by ϕ. In the particular case of P satisfying Conditions $7(i,ii)$ with $t_0 = 0$ its characteristic functional satisfies the equalities:

$$(iii)\quad \phi(t_1,x_1,t_2,y) = \psi(t_2 - t_1,y)\chi_y(x_1),$$

where

$$(iv)\quad \psi(t,y) := \int_X \chi_y(x) P(t,dx) \quad \text{and}$$

$$(v)\quad \psi(t_1+t_2,y) = \psi(t_1,y)\psi(t_2,y)$$

for each $t_1 \neq t_2 \in T$ and $t_2 - t_1 \in T$ and $t_1 + t_2 \in T$ respectively and $y \in X^*$, $x_1 \in X$.

9. Remark and notation. We recall the following. A measurable space (Ω,F) with a probability $\mathbf{K_s}$-valued measure λ on a covering ring F of a set Ω is called a probability space and it is denoted by $(\Omega,\mathsf{F},\lambda)$. Points $\omega \in \Omega$ are called elementary events and values $\lambda(S)$ are probabilities of events $S \in \mathsf{F}$. A measurable map $\xi : (\Omega,\mathsf{F}) \to (X,\mathsf{B})$ is called a random variable with values in X, where B is a covering ring such that $\mathsf{B} \subset Bco(X)$, $Bco(X)$ is the ring of all clopen subsets of a locally $\mathbf{K}$-convex space X, $\xi^{-1}(\mathsf{B}) \subset \mathsf{F}$, where $\mathbf{K}$ is a non-archimedean field complete as an ultra-metric space.

The random variable ξ induces a normalized measure $\nu_\xi(A) := \lambda(\xi^{-1}(A))$ in X and a new probability space (X,B,ν_ξ).

Let T be a set with a covering ring $\mathcal{R}$ and a measure $\eta : \mathcal{R} \to \mathbf{K_s}$. Consider the following Banach space $L^q(T,\mathcal{R},\eta,H)$ as the completion of the set of all $\mathcal{R}$-step functions $f : T \to H$ relative to the following norm:

(1) $\|f\|_{\eta,q} := \sup_{t\in T} \|f(t)\|_H N_\eta(t)^{1/q}$ for $1 \le q < \infty$ and

(2) $\|f\|_{\eta,\infty} := \sup_{1\le q<\infty} \|f(t)\|_{\eta,q}$, where H is a Banach space over $\mathbf{K}$. For $0 < q < 1$ this is the metric space with the metric

(3) $\rho_q(f,g) := \sup_{t\in T} \|f(t) - g(t)\|_H N_\eta(t)^{1/q}$.

If H is a complete locally $\mathbf{K}$-convex space, then H is a projective limit of Banach spaces $H = \lim\{H_\alpha, \pi^\alpha_\beta, \Upsilon\}$, where Υ is a directed set, $\pi^\alpha_\beta : H_\alpha \to H_\beta$ is a $\mathbf{K}$-linear continuous mapping for each $\alpha \ge \beta$, $\pi_\alpha : H \to H_\alpha$ is a $\mathbf{K}$-linear continuous mapping such that $\pi^\alpha_\beta \circ \pi_\alpha = \pi_\beta$ for each $\alpha \ge \beta$ (see §6.205 [87]). Each norm p_α on H_α induces a prednorm $\tilde{p}_\alpha$ on H. If $f : T \to H$, then $\pi_\alpha \circ f =: f_\alpha : T \to H_\alpha$. In this case $L^q(T,\mathcal{R},\eta,H)$ is defined as a completion of a family of all step functions $f : T \to H$ relative to the family of prednorms

(1′) $\|f\|_{\eta,q,\alpha} := \sup_{t\in T} \tilde{p}_\alpha(f(t)) N_\eta(t)^{1/q}$,

$\alpha \in \Upsilon$, for $1 \le q < \infty$ and

(2′) $\|f\|_{\eta,\infty,\alpha} := \sup_{1\le q<\infty} \|f(t)\|_{\eta,q,\alpha}$, $\alpha \in \Upsilon$, or pseudo-metrics

(3′) $\rho_{q,\alpha}(f,g) := \sup_{t\in T} \tilde{p}_\alpha(f(t) - g(t)) N_\eta(t)^{1/q}$,

$\alpha \in \Upsilon$, for $0 < q < 1$. Therefore, $L^q(T,\mathcal{R},\eta,H)$ is isomorphic with the projective limit

$\lim\{L^q(T,\mathcal{R},\eta,H_\alpha),\pi^\alpha_\beta,\Upsilon\}$. For $q=1$ we write simply $L(T,\mathcal{R},\eta,H)$ and $\|f\|_\eta$. This definition is correct, since $\lim_{q\to\infty} a^{1/q}=1$ for each $\infty > a > 0$.

For example, T may be a subset of the real field $\mathbf{R}$. Let $\mathbf{R_d}$ be the field $\mathbf{R}$ supplied with the discrete topology. Since the cardinality $card(\mathbf{R}) = \mathsf{c} = 2^{\aleph_0}$, there are bijective mappings of $\mathbf{R}$ on $Y_1 := \{0,\dots,b\}^{\mathbf{N}}$ and also on $Y_2 := \mathbf{N}^{\mathbf{N}}$, where b is a positive integer number. We supply the sets $\{0,\dots,b\}$ and $\mathbf{N}$ with the discrete topologies and Y_1 and Y_2 with the product topologies. Then zero-dimensional spaces Y_1 and Y_2 supply $\mathbf{R}$ with covering separating rings $\mathcal{R}_1$ and $\mathcal{R}_2$ contained in $Bco(Y_1)$ and $Bco(Y_2)$ respectively. Certainly such construction is not related with the standard (Euclidean) metric in $\mathbf{R}$. Therefore, for the space $L^q(T,\mathcal{R},\eta,H)$ we can consider $t\in T$ as the real time parameter. If $T\subset \mathbf{F}$ with a non-archimedean field $\mathbf{F}$, then we can consider the non-archimedean time parameter.

If T is a zero-dimensional T_1-space, then denote by $C^0_b(T,H)$ the Banach space of all continuous bounded functions $f: T\to H$ supplied with the norm:

(4) $\|f\|_{C^0} := \sup_{t\in T}\|f(t)\|_H < \infty$.

If T is compact, then $C^0_b(T,H)$ is isomorphic with the space $C^0(T,H)$ of all continuous functions $f: T\to H$.

For a set T and a complete locally $\mathbf{K}$-convex space H over $\mathbf{K}$ consider the product $\mathbf{K}$-convex space $H^T := \prod_{t\in T} H_t$ in the product topology, where $H_t := H$ for each $t\in T$.

Then we take on the space either $X := X(T,H) = L^q(T,\mathcal{R},\eta,H)$ or $X := X(T,H) = C^0_b(T,H)$ or on $X = X(T,H) = H^T$ a separating covering ring B such that $\mathsf{B}\subset Bco(X)$. Consider a random variable $\xi:\omega\mapsto \xi(t,\omega)$ with values in (X,B), where $t\in T$.

Events $S_1,\dots,S_n$ are called independent in total if $P(\prod_{k=1}^n S_k) = \prod_{k=1}^n P(S_k)$. Sub-rings $\mathsf{F}_k\subset\mathsf{F}$ are said to be independent if all collections of events $S_k\in\mathsf{F}_k$ are independent in total, where $k=1,\dots,n$, $n\in\mathbf{N}$. To each collection of random variables ξ_γ on (Ω,F) with $\gamma\in\Upsilon$ is related the minimal ring $\mathsf{F}_\Upsilon\subset\mathsf{F}$ with respect to which all ξ_γ are measurable, where Υ is a set. Collections $\{\xi_\gamma:\gamma\in\Upsilon_j\}$ are called independent if such are F_{Υ_j}, where $\Upsilon_j\subset\Upsilon$ for each $j=1,\dots,n$, $n\in\mathbf{N}$.

Consider a set T such that $card(T) > n$. For $X = C^0_b(T,H)$ or $X = H^T$ define $X(T,H;(t_1,\dots,t_n);(z_1,\dots,z_n))$ as a closed sub-manifold in X of all $f: T\to H$, $f\in X$ such that $f(t_1)=z_1,\dots,f(t_n)=z_n$, where $t_1,\dots,t_n$ are pairwise distinct points in T and $z_1,\dots,z_n$ are points in H. For $X = L^q(T,\mathcal{R},\eta,H)$ and pairwise distinct points $t_1,\dots,t_n$ in T with $N_\eta(t_1)>0,\dots,N_\eta(t_n)>0$ define $X(T,H;(t_1,\dots,t_n);(z_1,\dots,z_n))$ as a closed sub-manifold which is the completion relative to the norm $\|f\|_{\eta,q}$ of a family of $\mathcal{R}$-step functions $f: T\to H$ such that $f(t_1)=z_1,\dots,f(t_n)=z_n$. In these cases $X(T,H;(t_1,\dots,t_n);(0,\dots,0))$ is the proper $\mathbf{K}$-linear subspace of $X(T,H)$ such that $X(T,H)$ is isomorphic with $X(T,H;(t_1,\dots,t_n);(0,\dots,0))\oplus H^n$, since if $f\in X$, then $f(t)-f(t_1) =: g(t)\in X(T,H;t_1;0)$ (in the third case we use that $T\in\mathcal{R}$ and hence there exists the embedding $H\hookrightarrow X$). For $n=1$ and $t_0\in T$ and $z_1=0$ we denote $X_0 := X_0(T,H) := X(T,H;t_0;0)$.

10. Definition. We define a (non-archimedean) stochastic process $w(t,\omega)$ with values in the space H as a random variable such that:

(i) the differences $w(t_4,\omega)-w(t_3,\omega)$ and $w(t_2,\omega)-w(t_1,\omega)$ are independent for each chosen (t_1,t_2) and (t_3,t_4) with $t_1\ne t_2$, $t_3\ne t_4$, such that neither t_1 nor t_2 is in the two-element set $\{t_3,t_4\}$, where $\omega\in\Omega$;

(ii) the random variable $\omega(t,\omega)-\omega(u,\omega)$ has a distribution $\mu^{F_{t,u}}$, where μ is a probability $\mathbf{K_s}$-valued measure on $(X(T,H),\mathsf{B})$, $\mu^g(A) := \mu(g^{-1}(A))$ for $g : X \to H$ such that $g^{-1}(\mathcal{R}_H) \subset \mathsf{B}$ and each $A \in \mathcal{R}_H$. By $F_{t,u}$ a continuous linear operator $F_{t,u} : X \to H$ is denoted which is given by the formula

$$F_{t,u}(w) := w(t,\omega) - w(u,\omega)$$

for each $w \in L^q(\Omega,\mathsf{F},\lambda;X)$, where $1 \le q \le \infty$, $\mathcal{R}_H$ is a separating covering ring of H such that $F_{t,u}^{-1}(\mathcal{R}_H) \subset \mathsf{B}$ for each $t \ne u$ in T;

(iii) we also put $w(0,\omega) = 0$, that is, we consider a $\mathbf{K}$-linear subspace $L^q(\Omega,\mathsf{F},\lambda;X_0)$ of $L^q(\Omega,\mathsf{F},\lambda;X)$, where $\Omega \ne \emptyset$, X_0 is the closed subspace of X as in §9.

The stochastic process $w(t,\omega)$ satisfying conditions $(i-iii)$ possesses the Markovian property with the transition measure $P(u,x,t,A) = \mu^{F_{t,u}}(A-x)$.

The realization of stochastic processes is described in the following theorem.

11. Theorem. *Let either $X = C^0_b(T,H)$ or $X = H^T$ or $X = L^g(T,\mathcal{R},\eta,H)$ with $1 \le g \le \infty$ be the same spaces as in §9, where the normalization group $\Gamma_{\mathbf{K}}$ is discrete in $(0,\infty)$. Then there exists a family Ψ of pairwise inequivalent (non-archimedean) stochastic processes on X of a cardinality $card(\Psi) \ge card(T)card(H)$ or $card(\Psi) \ge card(\mathcal{R})card(H)$ respectively.*

Proof. Each complete locally $\mathbf{K}$-convex space H is the projective limit of Banach spaces H_α. Therefore, due to §10 it is sufficient to consider the case of the Banach space H. Since H is over the field $\mathbf{K}$ with the normalization group $\Gamma_{\mathbf{K}}$ discrete in $(0,\infty)$, the space H is isomorphic with the Banach space $c_0(\alpha,\mathbf{K})$ (see Theorems 5.13 and 5.16 [99]), where α is an ordinal.

Let $\mathcal{R}_{\mathbf{K}}$ be a covering separating ring of $\mathbf{K}$ much that elements of $\mathcal{R}_{\mathbf{K}}$ are clopen subsets in $\mathbf{K}$. Then a lot of different probability measures m on $(\mathbf{K},\mathcal{R}_{\mathbf{K}})$ with values in $\mathbf{K_s}$ exist. For example, such measure can be realized as an atomic measure with atoms a_j such that for each $U \in \mathcal{R}_{\mathbf{K}}$ either $a_j \subset U$ or $a_j \subset \mathbf{K}\setminus U$. Particularly, atoms may be singletons. Let the family Υ of a_j be countable and

$$\lim_j m(a_j) = 0,$$

when Υ is infinite. Then

$$m(S) := \sum_{a_j \subset S} a_j$$

for each $S \in \mathcal{R}_{\mathbf{K}}$ and

$$\|m\| = \sup_j |m(a_j)|.$$

If $\mathbf{K}$ is infinite and contains a locally compact infinite subfield $\mathbf{F}$ with a nontrivial normalization, then $\mathbf{K}$ can be considered as a locally $\mathbf{F}$-convex space. As the locally $\mathbf{F}$-convex space $\mathbf{K}$ in its weak topology is isomorphic with $\mathbf{F}^\gamma$, since $\Gamma_{\mathbf{F}}$ is discrete in $(0,\infty)$ and there is the non-archimedean variant of the Hahn-Banach theorem, where γ is a set (see [87,99]). Having a measure on $\mathbf{F}$ we can construct a probability measure on $\mathbf{K}$ due to Theorem 2.7 and Note 2.8 and Remarks 2.5.

Therefore, we consider the particular case of the locally compact field $\mathbf{K}$. If $\mathbf{K}$ is infinite, then either $\mathbf{K} \supset \mathbf{Q_p}$ or $\mathbf{K} = \mathbf{F_p}(\theta)$ with the corresponding prime number p, since $\mathbf{K}$ is with the nontrivial normalization [111]. If $\mathbf{K}$ is finite, then $\mathbf{K} = \mathbf{F_p}$. Let s be a prime number such that $s \neq p$, then $\mathbf{K}$ is s-free as the additive topological group (see the Monna-Springer theorem in §8.4 [99]). Therefore, there exists the $\mathbf{K_s}$-valued Haar measure w on $\mathbf{K}$. This means the bounded measure on each clopen compact subset of $\mathbf{K}$ with

$$w(B(\mathbf{K},0,1)) = 1 \text{ and } w(y+A) = w(A)$$

for each $A \in Bco(\mathbf{K})$ and each $y \in \mathbf{K}$, where $B(Y,y,r) := \{z : z \in Y, d(y,z) \leq r\}$ denotes the ball in an ultra-metric space Y with an ultra-metric d and a point $y \in Y$.

We have the following isomorphisms: $L^g(T,\mathcal{R},\eta,H) = L^g(T,\mathcal{R},\eta,\mathbf{K}) \otimes H$ and $C^0_b(T,H) = C^0_b(T,\mathbf{K}) \otimes H$, moreover, $L^g(T,\mathcal{R},\eta,\mathbf{K})$ is isomorphic with $c_0(\beta_L,\mathbf{K})$ and $C^0_b(T,\mathbf{K})$ is isomorphic with $c_0(\beta_C,\mathbf{K})$, where β_L and β_C are ordinals, since $\Gamma_\mathbf{K}$ is discrete (see Chapter 5 [99]).

We consider also the space $Y_1 := \mathbf{K}^T$ and $Y_1 \otimes H$ means the completion of the product of these spaces. As above we consider the product space H^T supplied with the Tychonoff product topology of copies of the space H. For the Banach space H infinite dimensional over the locally compact field $\mathbf{K}$ the space H^T is $\mathbf{K}$-linearly topologically isomorphic with $Y_1 \otimes H$. If H is finite dimensional over $\mathbf{K}$, then H^T is isomorphic with Y_1, since T is infinite, $card(T) \geq \aleph_0$. For the Banach space $c_0(\alpha,\mathbf{K})$ the Hahn-Banach theorem is satisfied, since the normalization group of $\mathbf{K}$ is discrete in $(0,\infty)$ (see §8.203 in [87]). Therefore, the rings of clopen subsets in $(c_0(\alpha,\mathbf{K}),\tau_w)$ and in Y_1 supply these spaces with the covering separating rings, where τ_w denotes the weak topology in the space $c_0(\alpha,\mathbf{K})$. On the other hand, evidently the field $\mathbf{K}$ has $\mathbf{K}$-linear embeddings into the space H over $\mathbf{K}$.

If μ_1 and μ_2 are $\mathbf{K_s}$-valued measures on Banach spaces Z_1 and Z_2 with covering rings $\mathcal{R}_1$ and $\mathcal{R}_2$ respectively, then $\mu_1 \otimes \mu_2$ is the $\mathbf{K_s}$-valued measure on $(Z_1 \otimes Z_2, \mathcal{R}_1 \times \mathcal{R}_2)$.

In the Banach space $c_0(\beta,\mathbf{K})$ the canonical base $(e_j : j \in \beta)$ is useful, where $e_j := (0,\dots,0,1,0,\dots)$ with 1 on the j-th place. With this standard base are associated projections

$$\pi_{j_1,\dots,j_n}(x) := \sum_{l=1}^{n} x^{j_l} e_{j_l}$$

for each $j_1,\dots,j_n \in \beta$ and each $n \in \mathbf{N}$ and for each vector $x \in c_0(\beta,\mathbf{K})$ with coordinates $x^j \in \mathbf{K}$ in the standard base. Consider a covering separating ring $\mathcal{R}$ of $c_0(\beta,\mathbf{K})$ such that

$$\mathcal{R} := \bigcup_{j_1,\dots,j_n \in \beta; n \in \mathbf{N}} (\pi_{j_1,\dots,j_n})^{-1}(Bco(span_\mathbf{K}(e_{j_1},\dots,e_{j_n}))),$$

where $span_\mathbf{K}(z_l : l \in \gamma) := \{x : x \in c_0(\beta,\mathbf{K}); x = \sum_{j \in \zeta} a^j z_j; a^j \in \mathbf{K}; card(\zeta) < \aleph_0, \zeta \subset \gamma\}$ denotes the $\mathbf{K}$-linear span for each $\gamma \subset \beta$. On the completion $\mathcal{R}_\mu$ there exists a probability $\mathbf{K_s}$-valued measure μ generated by a bounded cylindrical distribution as in §2.7, §2.9 or §2.11. For example, each

$$\mu_j(dx) := f_j(x)w(dx)$$

is a measure on $\mathbf{K}$, where $f_j \in L(\mathbf{K},\mathcal{R}(\mathbf{K}),w,\mathbf{K_s})$, w is either the Haar measure or any other probability measure on $\mathbf{K}$, $\mu_j = \pi_j(\mu)$ for each $j \in \beta$.

We take in the Banach space $c_0(\beta,\mathbf{K}))$ the standard basis $\{e_j : j\}$ and projection operators

$$\pi_{j_1,\dots,j_n} : c_0(\beta,\mathbf{K}) \to span_\mathbf{K}\{e_{j_1},\dots,e_{j_n}\}$$

so that $\pi_{j_1,\dots,j_n}(x) = (x_{j_1},\dots,x_{j_n})$, where $x \in c_0(\beta,\mathbf{K})$, $x = x_1e_1 + \dots + x_ne_n + \cdots$, $x_j \in \mathbf{K}$ for each j. In particular, for $card(\beta) \le \aleph_0$ and a locally compact non-archimedean infinite field $\mathbf{K}$ with nontrivial normalization there exists μ such that $\mathcal{R}_\mu \supset Bco(c_0(\beta,\mathbf{K}))$. For this consider on the Banach space $c_0 := c_0(\omega_0,\mathbf{K})$ a linear operator $J \in L_0(c_0)$, where $L_0(H)$ denotes the Banach space of compact $\mathbf{K}$-linear operators on the Banach space H, such that $Je_i = v_ie_i$ with $v_i \ne 0$ for each i and a measure

$$\nu(dx) := f(x)w(dx),$$

where $f : \mathbf{K} \to B(\mathbf{K},0,r)$ with $r \ge 1$ is a function belonging to the space $L(\mathbf{K},\mathcal{R}_w, w,\mathbf{K_s})$ such that

$$\lim_{|x|\to\infty} f(x) = 0 \quad \text{and} \quad \nu(\mathbf{K}) = 1,$$

$$\|S\|_\nu > 0 \quad \text{for each clopen subset } S \text{ in } \mathbf{K},$$

for example, when $f(x) \ne 0$ w-almost everywhere. In particular we can choose ν with $\|\nu\| = 1$. The sequence $\{v_j : j\}$ satisfies the condition

$$\lim_{j\to\infty} v_j = 0,$$

since the operator J is compact. In view of Lemma 2.3 and Theorem 2.30 from Chapter 2 of the book [81] and Theorem 2.7 and 2.11 above there exists the product measure

$$(i) \quad \mu(dx) := \prod_{i=1}^{\infty} \nu_i(dx^i)$$

on the ring $Bco(c_0)$ of clopen subsets of c_0, where

$$(ii) \quad \nu_i(dx^i) := f(x^i/v_i)\nu(dx^i/v_i).$$

Formula (i) means that each measure $\pi_{j_1,\dots,j_n}(\mu) = \nu_{j_1} \times \cdots \times \nu_{j_n}$ is the probability measure on the covering separating ring

$$Bco(span_{\mathbf{K}}\{e_{j_1},\dots,e_{j_n}\}).$$

Moreover, the family of cylindrical measures $\mu_{j_1,\dots,j_n}$ on cylindrical covering rings $\pi^{-1}_{j_1,\dots,j_n}[Bco(span_{\mathbf{K}}\{e_{j_1},\dots,e_{j_n}\})]$ is consistent, where

$$\mu_{j_1,\dots,j_n}(\pi^{-1}_{j_1,\dots,j_n}(A_{j_1} \times \cdots \times A_{j_n})) = \nu_{j_1}(A_{j_1}) \times \cdots \times \nu_{j_n}(A_{j_n})$$

for each $A_j \in Bco(\mathbf{K})$.

Consider, for example, the particular case of $X = C^0(T,H)$ with compact T. If $t_0 \in T$ is an isolated point, then $C^0(T,H) = C^0(T \setminus \{t_0\},H) \oplus H$, so we consider the case of T dense in itself. Let Z be a compact subset without isolated points in a locally compact field $\mathbf{K}$. Then the Banach space $C^0(Z,\mathbf{K})$ has the Amice polynomial orthonormal base $Q_m(x)$, where $x \in Z$, $m \in \mathbf{N_o} := \{0,1,2,\dots\}$ [3]. Each $f \in C^0$ has a decomposition into the converging series

$$f(x) = \sum_m a_m(f)Q_m(x)$$

such that

$$\lim_{m\to\infty} a_m = 0$$

with coefficients $a_m \in \mathbf{K}$. These decompositions establish the isometric isomorphism

$$\theta : C^0(T,\mathbf{K}) \to c_0(\omega_0,\mathbf{K})$$

such that

$$\|f\|_{C^0} = \max_m |a_m(f)| = \|\theta(f)\|_{c_0}.$$

If u_i are zeros of basic polynomials Q_m as in [3], then

$$Q_m(u_i) = 0 \quad \text{for each} \quad m > i.$$

The set $\{u_i : i\}$ is dense in T.

The locally $\mathbf{K}$-convex space $X = X(T,H)$ is isomorphic with the tensor product $X(T,\mathbf{K}) \otimes H$ (see §4.R [99] and [87]). If $J_i \in L_0(Y_i)$ is a compact $\mathbf{K}$-linear operator on the Banach space Y_i nondegenerate for each $i = 1,2$, that is, $ker(J_i) = \{0\}$, then their direct product $J := J_1 \otimes J_2 \in L_0(Y_1 \otimes Y_2)$ is nondegenerate (see also Theorem 4.33 [99]). If the spaces $X(T,\mathbf{K})$ and H are of separable type over the locally compact infinite field $\mathbf{K}$ with the nontrivial normalization, then we can construct a measure μ on X such that $\mathcal{R}_\mu \supset Bco(X)$. The case H^T is considered analogously. Put $Y_1 := X(T,\mathbf{K})$ and $Y_2 := H$ and $J := J_1 \otimes J_2 \in L_0(Y_1 \otimes Y_2)$, where $J_1 e_m := \alpha_m e_m$ such that $\alpha_m \neq 0$ for each m and $\lim_i \alpha_i = 0$. Take J_2 also nondegenerate. Then J induces a product measure μ on $X(T,H)$ such that $\mu = \mu_1 \otimes \mu_2$, where μ_i are measures on Y_i induced by J_i due to Formulas (i,ii). Analogously considering the following subspace $X_0(T,H)$ and operators $J := J_1 \otimes J_2 \in L_0(X_0(T,\mathbf{K}) \otimes H)$ we get the measures μ on it also, where $t_0 \in T$ is a marked point.

We recall, that a topological space is called Lindelöf, if if it is regular and from each its open covering one can choose a countable sub-covering. On the other hand, the space $X(T,H)$ is Lindelöf (see §3.8 [19]), consequently, each subset U open in $X(T,H)$ is a countable union of clopen subsets. Hence the characteristic function Ch_U of U belongs to $L(X,\mathcal{R},\mu,\mathbf{K_s})$, since

$$\|\mu\| = \sup_x N_\mu(x) < \infty,$$

consequently, $\mathcal{R}_\mu \ni U$. In general the condition $\mathcal{R}_\mu \supset Bco(X_0)$ is not imposed, so $\mathcal{R}_\mu$ may be any covering separating ring of X_0.

For each finite number of pairwise distinct points $(t_0,t_1,\dots,t_n)$ in T and points $(0,z_1,\dots,z_n)$ in H there exists a closed subset

$$X(T,H;(t_0,t_1,\dots,t_n);(0,z_1,\dots,z_n)) \quad \text{in} \quad X(T,H)$$

such that

$$X(T,H;(t_0,t_1,\dots,t_n);(0,z_1,\dots,z_n))$$
$$= (0,z_1,\dots,z_n) + X(T,H;(t_0,t_1,\dots,t_n);(0,\dots,0)),$$

where $X(T,H;(t_0,t_1,\dots,t_n);(0,\dots,0))$ is the $\mathbf{K}$-linear subspace in $X(T,H)$. Therefore,

(iii) rings $F^{-1}_{t_2,t_1}(\mathcal{R}(H))$ and $F^{-1}_{t_4,t_3}(\mathcal{R}(H))$ are independent sub-rings in the ring $\mathcal{R}(X(T,H))$, when (t_1,t_2) and (t_3,t_4) satisfy Condition $10(i)$, where the covering rings $\mathcal{R}(H)$ and $\mathcal{R}(X)$ and the measures μ, μ_1 and μ_2 are as above. We put

$$P(t_1,x_1,t_2,A) := \mu(\{f : f \in X(T,H;(t_0,t_1);(0,x_1)), \quad f(t_2) \in A\})$$

for each $t_1 \neq t_2 \in T$, $x_1 \in H$ and $A \in \mathcal{R}(H)$. In view of (iii) we get, that P satisfies Conditions $2(i-iv)$. By the above construction (and Proposition 3 also) the Markov cylindrical distribution $\tilde{\mu}_{x_0}$ induced by μ is bounded, since μ is bounded, where $x_0 = 0$ for $X_0(T,H)$. Let Υ be the set of all elementary events

$$\omega := \{f : f \in X(T,H;(t_0,t_1,\dots,t_n);(0,x_1,\dots,x_n))\},$$

where Λ_ω is a finite subset of $\mathbf{N}$, $x_i \in H$, $(t_i : i \in \Lambda_\omega)$ is a subset of $T \setminus \{t_0\}$ of pairwise distinct points. There exists the ring $\tilde{\mathsf{U}}$ of cylindrical subsets of $X_0(T,H)$ induced by projections $\pi_s : X_0(T,H) \to H^s$, where

$$H^s := \prod_{t \in s} H_t,$$

$s = (t_1,\dots,t_n)$ are finite subsets of T, $H_t = H$ for each $t \in T$. This induces the covering ring $\mathcal{R}(\Upsilon)$ of Υ, where $(\Upsilon,\mathcal{R}(\Upsilon),\nu)$ is the image of $(X_0(T,H),\tilde{\mathsf{U}},\tilde{\mu}_{x_0})$ due to Proposition 2.1. In view of Theorem 2.11 and §2.5 $\tilde{\mu}_{x_0}$ on $(X_0(T,H),\tilde{\mathsf{U}})$ induces the probability measure ν on $(\Upsilon,\mathcal{R}_\nu(\Upsilon))$.

Thus for each probability space $(\Omega,\mathsf{F},\lambda)$ and a measurable mapping $\pi : \Omega \to \Upsilon$, that is, $\pi^{-1}(\mathcal{R}(\Upsilon)) \subset \mathsf{F}$, such that $\pi(\lambda) = \nu$ we get the space $L^q(\Omega,\mathsf{F},\lambda,X_0)$ and the realization of the stochastic process $\xi(t,\omega)$. In particular, we can take $\Omega = \Upsilon$ and $\pi = id$.

In the case $X(T,H) = H^T$ (apart from $C^0(T,H)$ and $L^q(T,\mathcal{R},\eta,H)$) it is sufficient to take any bounded linear operator J_1 on $Y_1 = \mathbf{K}^T$, that is, $J_1 \in L(Y_1)$, that brings the difference, when $card(T) \geq \aleph_0$.

Therefore, using cylindrical distributions we get examples of such measures μ for which stochastic processes exist. Hence to each such measure on $X_0(T,H)$ there corresponds the stochastic process.

Evidently, on the field $\mathbf{K}$ there exists a family $\Psi_{\mathbf{K}}$ of inequivalent $\mathbf{K_s}$-valued measures of the cardinality $card(\Psi_{\mathbf{K}}) \geq card(\mathbf{K})$, since the subfamily of atomic measures satisfies this inequality. In the particular case of $\mathbf{C_p} \supset \mathbf{F} \supset \mathbf{Q_p}$ or $\mathbf{F} = \mathbf{F_p}(\theta)$ we use the Haar measure w also for which $card(L^q(\mathbf{F},w,Bco(\mathbf{F}),\mathbf{K_s})) = \mathsf{c} := card(\mathbf{R})$ for each $1 \leq q \leq \infty$. In view of the non-archimedean variant of the Kakutani theorem (see 3.5 in Chapter 2 of the book [81]) we get the inequalities for $card(\Psi)$, since $card(H) = card(\beta_H)card(\mathbf{K})$. In particular, for $T = B(\mathbf{K},0,r)$ with $r > 0$ and a locally compact field $\mathbf{K}$ either $\mathbf{K} \supset \mathbf{Q_p}$ or $\mathbf{K} = \mathbf{F_p}(\theta)$ considering all operators $J := J_1 \otimes J_2 \in L_0(Y_1 \otimes Y_2)$ and the corresponding measures as above we get $\mathsf{c}^{\aleph_0} = \mathsf{c}$ inequivalent measures for each chosen f.

12. Note. Evidently, this theorem is also true for $C^0(T,H)$, that follows from the proof.

If take ν with $supp(\nu) = B(\mathbf{K},0,1)$, then repeating the proof it is possible to construct μ with $supp(\mu) \subset B(C^0(T,\mathbf{K}),0,1) \times B(H,0,1)$. Certainly such measure μ can not be quasi-invariant relative to shifts from a dense $\mathbf{K}$-linear subspace in $C^0(T,H)$, but (starting from the Haar measure w on $\mathbf{F}$) μ can be constructed quasi-invariant relative to a dense additive subgroup G' of $B(C^0(T,\mathbf{K}),0,1) \times B(H,0,1)$, moreover, there exists μ for which G' is also $B(\mathbf{K},0,1)$-absolutely convex.

Properties $9(i,ii)$ serve for a definition of a specific class of stochastic processes, which are analogous to the classical Brownian motion processes.

3.4. Poisson Processes

1. Definition. Let T be an additive group such that $T \subset B(\mathbf{K_s},0,r)$ and $0 \neq \rho \in \mathbf{K_s}$ with $|\rho| r < s^{1/(1-s)}$, where $\mathbf{K_s}$ is a field such that $\mathbf{Q_s} \subset \mathbf{K_s} \subset \mathbf{C_s}$, $\mathbf{K_s}$ is complete as an ultra-metric space. We consider a stochastic process $\xi \in L^q(\Omega, \mathsf{F}, \lambda, X_0(T,H))$ such that a transition measure has the form

$$P(t_1,x,t_2,A) := P(t_2-t_1,x,A) := Exp(-\rho(t_2-t_1))P(A-x)$$

(see also §3.2 and §3.9) for each $x \in H$ and $A \in \mathcal{R}_H$ and t_1 and t_2 in T, where

$$Exp(x) := \sum_{n=0}^{\infty} x^n/n!.$$

Then such process is called the Poisson process.

We remind that the series defining the exponential function over the field $\mathbf{K}$ has a positive finite radius of convergence. In accordance with Theorem 25.6 [103] this series converges on the set $E = \{x \in \mathbf{K} : \ |x| < p^{1/(1-p)}\}$ if $char(k) = p$; while $E = \{x \in \mathbf{K} : \ |x| < 1\}$ if $char(k) = 0$, where k denotes the residue class field of $\mathbf{K}$. Proposition 45.6 [103] gives the locally analytic extension EXP of the exponential function Exp from E onto $\mathbf{C}_p$ with values in $\mathbf{C}_p^+ := \{x \in \mathbf{C}_p : \ |x-1|_p < 1\}$, when $\mathbf{K} = \mathbf{C}_p$; $EXP : \mathbf{C}_p \to \mathbf{C}_p^+$. For such extension the property of $\mathbf{C}_p^+$ being commutative divisible group is used.

2. Proposition. *Let the identity*

$$P(A-x) = \int_H P(-x+dy)P(A-y)$$

be satisfied for each $x \in H$ and $A \in \mathcal{R}_H$ and let also properties

$$P(H) = 1 \quad and \quad \|P\| = 1$$

be fulfilled, then there exists a measure μ on $X_0(T,H)$ for which the Poisson process exists.

Proof. The exponential function converges if $|x| < s^{1/(1-s)}$, since $|n!|_s^{-1} \leq s^{(n-1)/(s-1)}$ for each $0 < n \in \mathbf{Z}$. Hence $|Exp(-\rho(t_2-t_1)) - 1| < 1$ for each t_1 and t_2 in T, consequently, $|Exp(-\rho(t_2-t_1))| = 1$. We take the measure

$$\mu_{t_1,\dots,t_n} := P(t_2-t_1,0,*) \cdots P(t_n-t_{n-1},0,*) \quad \text{on } \mathcal{R}_{t_1,\dots,t_n} := \mathcal{R}_{t_1} \times \cdots \times \mathcal{R}_{t_n}$$

for each pairwise distinct points $t_1,\dots,t_n \in T$, where

$$\mu_{t_1,\dots,t_n} = \pi_{t_1,\dots,t_n}(\mu) \quad \text{and}$$

$$\pi_{t_1,\dots,t_n} : X_0(T,H) \to H_{t_1} \times \cdots \times H_{t_n}$$

is the natural projection, $H_t = H$ for each $t \in T$. Here $\mathcal{R}_t = \mathcal{R}_H$ is the separating covering ring on H for each $t \in T$. There is a family Λ of all finite subsets of T directed by inclusion. In view of Theorem 2.9 the cylindrical distribution μ generated by the family $P(t_2 - t_1, x, A)$ has an extension to a measure on $X_0(T,H)$. All others conditions are satisfied in accordance with §3.10 and §1. Finally, we construct a space of elementary events Ω and a random function analogously to §3.11.

3. Note. Let K be a complete ultra-metric space with an ultra-metric d, that is,

$$d(x,y) \le \max(d(x,z), d(y,z)) \quad \text{for each} \quad x,y,z \in X.$$

Let

$$d(x,y) := \max_{1 \le i \le n} d(x_i, y_i)$$

be the ultra-metric in K^n, where $x = (x_i : i = 1, \dots, n) \in K^n$, $x_i \in K$. We put

$$\tilde{K}^n := (x \in K^n : x_i \ne x_j \quad \text{for each} \quad i \ne j).$$

Then we supply $\tilde{K}^n$ with an ultra-metric

$$\delta^n_K(x,y) := d^n_K(x,y)/[\max(d^n_K(x,y), d^n_K(x, (\tilde{K}^n)^c), d(y, (\tilde{K}^n)^c)]$$

where $A^c := K^n \setminus A$ for a subset $A \subset K^n$. Therefore, $(\tilde{K}^n, \delta^n_K)$ is the complete ultra-metric space. Let also B^n_K denotes the collection of all n-point subsets of K. Then the ultra-metric δ^n_K is equivalent with the following ultra-metric

$$d^{(n)}_K(\gamma, \gamma') := \inf_{\sigma \in \Sigma_n} d^n_K((x_1, \dots, x_n), (x'_{\sigma(1)}, \dots, x'_{\sigma(n)})).$$

Here as usually Σ_n denotes the symmetric group of $(1, \dots, n)$, $\sigma \in \Sigma_n$, $\sigma : (1, \dots, n) \to (1, \dots, n)$; $\gamma, \gamma' \in B^n_K$. For each subset $A \subset K$ a number mapping $N_A : B^n_K \to \mathbf{N_o}$ is defined by the following formula: $N_A(\gamma) := card(\gamma \cap A)$, where $\mathbf{N} := \{1,2,3,\dots\}$, $\mathbf{N_o} := \{0,1,2,3,\dots\}$. It remains to show, that δ^n_K is the ultra-metric for the ultra-metric space (K,d). For this we mention, that

$$(i) \quad \delta^n_K(x,y) > 0,$$

when $x \ne y$, and $\delta^n_K(x,x) = 0$.

$$(ii) \quad \delta^n_K(x,y) = \delta^n_K(y,x),$$

since this symmetry is true for d^n_K and for $[*]$ in the denominator in the formula defining δ^n_K. To prove

$$(iii) \quad \delta^n_K(x,y) \le \max(\delta^n_K(x,z), \delta^n_K(z,y))$$

we consider the case $\delta^n_K(x,z) \ge \delta^n_K(y,z)$, hence it is sufficient to show, that $\delta^n_K(x,y) \le \delta^n_K(x,z)$. If the inequality

$$(a) \quad d^n_K(x,z) \ge \max(d^n_K(z, (\tilde{K}^n)^c), d^n_K(x, (\tilde{K}^n)^c)),$$

is satisfied, then we get $\delta^n_K(x,z) = 1$, hence $\delta^n_K(x,y) \le \delta^n_K(x,z)$, since $\delta^n_K(x,y) \le 1$ for each $x,y \in \tilde{K}^n$. In the case

$$(b) \quad d^n_K(x,(\tilde{K}^n)^c) > \max(d^n_K(x,z), d^n_K(z,(\tilde{K}^n)^c)),$$

we infer that

$$\delta^n_K(x,z) = d^n_K(x,z)/d^n_K(x,(\tilde{K}^n)^c) \le 1.$$

Since $d^n_K(z,A) := \inf_{a\in A} d^n_K(z,a)$, the inequality

$$d^n_K(z,(\tilde{K}^n)^c) \le \max(d^n_K(y,(\tilde{K}^n)^c), d^n_K(y,z))$$

follows for each y.

If the case when inequalities $d^n_K(x,z) < d^n_K(z,(\tilde{K}^n)^c)$ and $d^n_K(x,y) \le d^n_K(x,z)$ are satisfied, then we deduce that

$$d^n_K(z,(\tilde{K}^n)^c) \le d^n_K(x,(\tilde{K}^n)^c).$$

Hence the estimate

$$d^n_K(x,y)\max(d^n_K(x,z), d^n_K(x,(\tilde{K}^n)^c), d^n_K(z,(\tilde{K}^n)^c))$$
$$\le d^n_K(x,z)\max(d^n_K(x,y), d^n_K(x,(\tilde{K}^n)^c), d^n_K(y,(\tilde{K}^n)^c))$$

is fulfilled. With the help of (ii) the remaining cases can be written using the transposition $(x,y) \mapsto (y,x)$.

4. Notes and definitions. As usually the direct sum of topological spaces is considered

$$B_K := \bigoplus_{n=0}^{\infty} B^n_K, \tag{1}$$

where $B^0_K := \{0\}$ is a singleton, $B_K \ni x = (x_n : x_n \in B^n_K, n = 0,1,2,\dots)$. If a complete ultra-metric space X is not compact, then there exists an increasing sequence of subsets $K_n \subset X$ such that

$$X := \bigcup_n K_n \tag{2}$$

and K_n are complete spaces in the uniformity induced from X. Moreover, K_n can be chosen clopen in X. Then the following space

$$\Gamma_X := \{\gamma : \gamma \subset X \text{ and } card(\gamma \cap K_n) < \infty \text{ for each } n\} \tag{3}$$

is called the configuration space and it is isomorphic with the projective limit $pr-\lim\{B_{K_n}, \pi^n_m, \mathbf{N}\}$, where $\pi^n_m(\gamma_m) = \gamma_n$ for each $m > n$ and $\gamma_n \in B_{K_n}$. If d_n denotes the ultra-metric in B_{K_n}, then their sequence is consistent $d_{n+1}|_{B_{K_n}} = d_n$, since $K_n \subset K_{n+1}$. Then $\prod_{n=1}^{\infty} B_{K_n} =: Y$ in the Tychonoff product topology is ultra-metrizable. This induces the ultra-metric in Γ_X, for example, by the formula:

$$\rho(x,y) := d_n(x_n,y_n)p^{-n} \text{ is the ultra-metric in } \Gamma_X, \tag{4}$$

where $n = n(x,y) := \min_{(x_j \ne y_j)} j$, $x = (x_j : j \in \mathbf{N}, x_j \in B_{K_j})$, $1 < p \in \mathbf{N}$.

Let $K \in \{K_n : n \in \mathbf{N}\}$, then m_K denotes the restriction $m|_K$, where $m : \mathcal{R} \to \mathbf{K_s}$ is a measure on a covering separating ring $\mathcal{R}_m$ of X, $K_n \in \mathcal{R}_m$ for each $n \in \mathbf{N}$. Suppose that $K_l^n \in \mathcal{R}_{m^n}$ for each n and l in $\mathbf{N}$, where $\mathcal{R}_m$ is the completion of the covering ring $\mathcal{R}^n$ of X^n relative to the product measure $m^n = \bigotimes_{j=1}^n m_j$, $m_j = m$ for each j. Then $m_K^n := \bigotimes_{j=1}^n (m_K)_j$ is a measure on K^n and hence on $\tilde{K}^n$, when m is such that $\|m|_{(K^n \setminus \tilde{K}^n)}\| = 0$. For example, we can take a non-atomic m, where $(m_K)_j = m_K$ for each j. Let $m(K_l) \neq 0$ for each $l \in \mathbf{N}$, $m(X) \neq 0$ and $\|m\| < s^{1/(1-s)}$. Therefore, the formula

$$(i) \quad P_{K,m} := Exp(-m(K)) \sum_{n=0}^{\infty} m_{K,n}/n!$$

provides the measure on the covering separating ring $\mathcal{R}(B_K)$, where

$$\mathcal{R}(B_K) = B_K \cap \left(\bigoplus_{n=0}^{\infty} \mathcal{R}_{m^n} \right),$$

$m_{K,0}$ is a probability measure on the singleton B_K^0, and $m_{K,n}$ are images of m_K^n under the following mappings:

$$p_K^n : (x_1, \ldots, x_n) \in \tilde{K}^n \to \{x_1, \ldots, x_n\} \in B_K^n.$$

Such system of measures $P_{K,n}$ is consistent, that is,

$$\pi_l^n(P_{K_l,m}) = P_{K_n,m} \quad \text{for each } n \leq l.$$

This defines the unique measure P_m on $\mathcal{R}(\Gamma_X)$, which is called the Poisson measure, where

$$\pi_n : Y \to B_{K_n}$$

is the natural projection for each $n \in \mathbf{N}$. For each natural numbers $n_1, .., n_l \in \mathbf{N_o}$ and disjoint subsets $B_1, \ldots, B_l$ in the topological space X given by Equality 4(2) belonging to the covering separating ring $\mathcal{R}_m$ the following equality:

$$(ii) \quad P_m \left(\bigcap_{j=1}^{l} \{\gamma : card(\gamma \cap B_i) = n_i\} \right) = \prod_{i=1}^{l} m(B_i)^{n_i} Exp(-m(B_i))/n_i!$$

is valid.

There exists the following embedding

$$\Gamma_X \hookrightarrow S_X,$$

where

$$S_X := \lim \{E_{K_n}, \pi_m^n, \mathbf{N}\}$$

is the limit of an inverse mapping sequence,

$$E_K := \bigoplus_{l=0}^{\infty} K^l$$

for each $K \in \{K_n : n = 0, 1, 2, \dots\}$. The Poisson measure P_m on the covering separating ring $\mathcal{R}(\Gamma_X)$ considered above has an extension on the covering separating ring $\mathcal{R}(S_X)$ such that

$$\|P_m|_{S_X \setminus \Gamma_X}\| = 0.$$

The latter equality mean that the measure P_m is zero on the excess space $S_X \setminus \Gamma_X$. If each K_n is a complete $\mathbf{K}$-linear space, then E_K and S_X are complete $\mathbf{K}$-linear spaces, since

$$S_X \subset \left(\prod_{n=1}^{\infty} E_{K_n} \right).$$

Then on the covering separating ring $\mathcal{R}(S_X)$ the Poisson measure P_m exists, but without the restriction $\|m_K^n|_{K^n \setminus \tilde{K}^n}\| = 0$, where

$$(iii) \quad P_{K,m} := Exp(-m(K)) \sum_{n=0}^{\infty} m_K^n / n!,$$

$$\pi_l^n(P_{K_l,m}) = P_{K_n,m} \quad \text{for each} \quad n \le l.$$

5. Corollary. *Let suppositions of Proposition 2 be satisfied with $H = S_X$ for a complete $\mathbf{K}$-linear space X and $P(A) = P_m(A)$ for each $A \in \mathcal{R}(S_X)$, then there exists a measure μ on the space $X_0(T,H)$ for which the Poisson process exists.*

6. Definition. The stochastic process of Corollary 5 is called the Poisson process with values in X.

7. Note. If $\xi \in L^q(\Omega, \mathsf{F}, \lambda; X_0(T,H))$ is a stochastic process, then its mean value at the moment $t \in T$ is defined by the following formula:

$$(i) \quad M_t(\xi) := \int_{\Omega} \xi(t,\omega)\lambda(d\omega).$$

Let $H = \mathbf{K}$ be a field, where $\mathbf{Q_p} \subset \mathbf{K} \subset \mathbf{C_p}$, let also a measure λ be with values in $\mathbf{K}$. Suppose that

$$\|\{\omega : |\xi(t,\omega)| > R\}\|_{\lambda} = 0 \quad \text{for each} \quad t \in T \quad \text{for some} \quad R > 0.$$

Let $\rho \in B(\mathbf{K}, 0, c)$ and let T be a subgroup of $B(\mathbf{K}, 0, r)$, where $R\max(c,r) < p^{1/(1-p)}$, $\mathbf{K}$ is the locally compact field. Henceforth, the anti-derivation operator

$$\tilde{P}^1 : C^0(B(\mathbf{K},0,c),\mathbf{K}) \to C^1(B(\mathbf{K},0,c),\mathbf{K})$$

is used. This operator is given in accordance with §§54, 80 [103].

We recall the following. For any natural number $n \in \mathbf{N}$ and a subset X dense in itself contained in the field $\mathbf{K}$ one puts

$$\nabla^n X := \{x = (x_1, \dots, x_n) \in X^n : \ x_i \neq x_j \ \forall i \neq j\}.$$

The n-th order difference quotient $\Phi^n f$ of a function $f : X \to \mathbf{K}$ is defined by induction:

$$\Phi^0 f = f, \quad \Phi^1 f(x_1, x_2) = [f(x_1) - f(x_2)]/(x_1 - x_2),$$

$$\Phi^{n+1}f(x_1,\dots,x_{n+1}) = [\Phi^n f(x_1,x_3,\dots,x_{n+1}) - \Phi^n(x_2,x_3,\dots,x_{n+1})]/(x_1-x_2).$$

The space $C^n(X,\mathbf{K})$ is defined as consisting of all continuous functions f on X for which the n-th order difference quotient $\Phi^n f$ has a continuous extension from $\nabla^n X$ onto X^n. Then the space $C^\infty(X,\mathbf{K}) := \bigcap_{n=1}^\infty C^n(X,\mathbf{K})$ of infinite differentiable functions is defined. The subspace $BC^n(X,\mathbf{K})$ of $C^n(X,\mathbf{K})$ is defined as:

$$BC^n(X,\mathbf{K}) := \{f \in C^n(X,\mathbf{K}) : \|f\|_n := \max_{j=0,\dots,n} \|\Phi^j f\| < \infty\},$$

where $\|\Phi^j f\| := \sup_{x\in\nabla^n X} |\Phi^j f(x)|$.

An approximation of the identity on X is a sequence $\sigma_0,\dots,\sigma_m,\dots$ of mappings $\sigma_m : X \to X$ satisfying the following four conditions:

$(A1)$ σ_0 is constant,

$(A2)$ $\sigma_m \circ \sigma_n = \sigma_n \circ \sigma_m = \sigma_n$ for all $m \ge n \ge 0$,

$(A3)$ $\sigma_n(x) = \sigma_n(y)$ for each $|x-y| < \rho^n$,

$(A4)$ $|\sigma_n(x) - x| < \rho^n$ for each n,

where $0<\rho<1$ is some marked positive number less than one. For a natural number $n \in \mathbf{N}$ for $char(k) = 0$ or so that $n < char(k)$ for $char(k) > 0$, where k is the residue class field of the field $\mathbf{K}$, the anti-derivation operator $\tilde{P}^n : C^{n-1}(X,\mathbf{K}) \to C^n(X,\mathbf{K})$ is defined by the formula:

$$(AD)\quad \tilde{P}^n f(x) := \sum_{m=0}^{\infty}\sum_{j=0}^{n-1} \bar{\Phi}^j f(x_m; x_{m+1}-x_m,\dots,x_{m+1}$$

$$-x_m;0,\dots,0)(x_{m+1}-x_m)/(j+1),$$

where $f^{(j)}(x).(v_1,\dots,v_j) = j!\bar{\Phi}^j f(x;v_1,\dots,v_j;0,\dots,0)$ denotes the j-th derivative of f at a point x along $(v_1,\dots,v_j)$, $(v,\dots,v) =: v^{\otimes j}$. In the scalar case $f^{(j)}(x).(v_1,\dots,v_j) = f^{(j)}(x)v_1\dots v_j$ and $f^{(j)}(x) \in \mathbf{K}$, when $X \subset \mathbf{K}$ and $v_1,\dots,v_j \in \mathbf{K}$. The anti-derivation operator is characterized by the equality

$$(D)\quad d\tilde{P}^n f(x)/dx = f(x)$$

and each function has the Taylor expansion

$$(T)\quad \tilde{P}^n f(x) - \tilde{P}^n f(y) = \sum_{j=1}^{n} (x-y)^j f^{(j-1)}(y)/j! + (x-y)^n R_n(x,y)$$

for all $x,y \in X$, where $R_n(x,y)$ is a continuous function vanishing on the diagonal $\{(x,x) : x \in X\} \subset X^2$.

8. Theorem. *Let ψ be a continuously differentiable function, from T into $\mathbf{K}$ belonging to the space $\tilde{P}^1(C^0(B(\mathbf{K},0,c),\mathbf{K}))$ and $\psi(0)=0$. Then a stochastic process $\xi(t,\omega)$ exists such that*

$$M_t(Exp(-\rho\xi(t,\omega))) = Exp(-t\psi(\rho))$$

for each t in T and each $\rho \in B(\mathbf{K},0,c)$.

Proof. For the construction of the stochastic process ξ we consider solution of the following equation

$$M_t[Exp(-\rho\xi(t,\omega))] = Exp(-t\psi(\rho)).$$

Then $e(t) = e(t-s)e(s)$ for each t and s in T and each $\rho \in B(\mathbf{K}, 0, c)$, where

$$e_\rho(t) := e(t) := M_t(Exp(-\rho\xi(t,\omega))).$$

Therefore, we deduce the differential equation

$$\partial e_\rho(t)/\partial\rho = -t\psi'(\rho)Exp(-t\psi(\rho)),$$

consequently,

$$\psi'(\rho) = t^{-1}\int_{\mathbf{K}} l \quad EXP(-\rho l)P(\{\omega : \xi(t,\omega) \in dl\})$$

for each $t \neq 0$, where EXP is the locally analytic extension of Exp on $\mathbf{K}$ (with values in $\{x : x \in \mathbf{C_p}, \quad |x-1| < 1\}$, see §1 above). In particular, we get the equality

$$\psi'(\rho) = \lim_{t\to 0, t\neq 0} t^{-1}\int_{\mathbf{K}} l \quad EXP(-\rho l)P(\{\omega : \xi(t,\omega) \in dl\}).$$

By the conditions of this theorem we have

$$\psi(\rho) = \tilde{P}^1_\beta \psi'(\beta)|_0^\rho.$$

Consider a measure m on a separating covering ring $\mathcal{R}(\mathbf{K})$ such that $\mathcal{R}(\mathbf{K}) \supset Bco(\mathbf{K}) \cup \{0\}$ with values in $\mathbf{K}$ for which

$$m(dl) := \lim_{t\to 0, t\neq 0} lP(\{\omega : \xi(t,\omega) \in dl\})/t.$$

Therefore, we infer the equality

$$\psi(\rho) = \tilde{P}^1_\beta \left(\int_{\mathbf{K}} EXP(-\beta l)m(dl) \right)|_0^\rho.$$

From $\psi(0) = 0$ we have $e_\rho(1) = 1$ for each ρ, consequently, the formula

$$\psi(\rho) = \rho m_0 + \int_{\mathbf{K}} [1 - EXP(-\rho l)]l^{-1}m(dl)$$

is valid, where $m_0 := m(\{0\})$, since the limits

$$\lim_{l\to 0, l\neq 0} [1 - EXP(-\rho l)]/l = \rho \quad \text{and}$$

$$\lim_{\rho\to 0, \rho\neq 0} \int_{B(\mathbf{K},0,k)} [1 - EXP(-\rho l)]l^{-1}m(dl) = 0$$

exist for each $k > 0$. Next we define a measure $n(dl)$ such that $n(\{0\}) = 0$ and $n(dl) = l^{-1}m(dl)$ on $\mathbf{K} \setminus \{0\}$, consequently,

$$\psi(\rho) = \rho m_0 + \int_{\mathbf{K}} [1 - EXP(-\rho l)]n(dl).$$

We search a solution of the problem in the form

$$\xi(t,\omega) = tm_0 + \int_{\mathbf{K}} l\eta(t, dl, \omega),$$

where $\eta(t,dl,\omega)$ is the measure on $\mathcal{R}(\mathbf{K})$ for each fixed $t \in T$ and $\omega \in \Omega$ such that its moments satisfy the Poisson distribution with the Poisson measure P_{tn}. The latter means that

$$M_t[\eta^k(t,dl,\omega)] = \sum_{s\le k} a_{s,k}(tn)^s(dl)/s!$$

for each $t \in T$, where $a_{0,j} = 0$, $a_{1,j} = 1$ and recurrently we deduce

$$a_{k,j} = k^j - \sum_{s=1}^{k} \binom{k}{s} a_{k-s,j}$$

for each $k \le j$, in particular, $a_{j,j} = j!$, that is,

$$a_{k,j} = \sum_{s_1+\dots+s_k=j, s_1\ge 1,\dots,s_k\ge 1} [j!/(s_1!\cdots s_k!)].$$

Using the fact that the set of step functions is dense in the space $L(\mathbf{K},\mathcal{R}(\mathbf{K}),n,\mathbf{C_p})$ we get the following equalities:

$$M_t\left[EXP(-\rho\int_{\mathbf{K}} l\eta(t,dl,\omega))\right] = \lim_{\mathcal{Z}} M_t\left[\prod_j EXP(-\rho l_j\eta(t,\delta_j,\omega))\right]$$

$$= \lim_{\mathcal{Z}}\prod_j M_t[EXP(-\rho l_j\eta(t,\delta_j,\omega))] = \lim_{\mathcal{Z}} EXP\Big(-\rho t\sum_j (1-EXP(-\rho l_j))n(\delta_j)\Big)$$

$$= EXP\left[-\rho t\int_{\mathbf{K}}(1-EXP(-\rho l)n(dl))\right],$$

where $\mathcal{Z}$ is an ordered family of partitions $\mathcal{U}$ of $\mathbf{K}$ into disjoint union of elements of $\mathcal{R}(\mathbf{K})$, $\mathcal{U} \le \mathcal{V}$ in $\mathcal{Z}$ if and only if each element of the disjoint covering $\mathcal{U}$ is a union of elements of $\mathcal{V}$, $l_j \in \delta_j \in \mathcal{U} \in \mathcal{Z}$. The limit

$$\lim_{\mathcal{U}\in\mathcal{Z}} f(\mathcal{U}) =: \lim_{\mathcal{Z}} f = a$$

means that for each $\varepsilon > 0$ there exists $\mathcal{U}$ such that for each $\mathcal{V}$ with $\mathcal{U} \le \mathcal{V}$ we have

$$|a - f(\mathcal{V})| < \varepsilon,$$

where $f(\mathcal{U})$ is one of the functions defined as above with $l_j \in \delta_j \in \mathcal{U}$, that is,

$$f(\mathcal{U}) = M_t[g\circ h(\eta)],$$

where $g\circ h(\eta)$ is the composition of the continuous function g and of the function

$$h(\eta) = \int_{\mathbf{K}} \zeta(y)\eta(t,dy,\omega)$$

with the step function ζ. Then we get the equation

$$M_t\left[EXP(-\rho\xi(t,\omega))\right] = EXP(-\rho t m_0)M_t\left[EXP(-\rho\int_{\mathbf{K}} l\eta(t,dl,\omega))\right].$$

In view of Corollary 5 the latter equation defines the stochastic process with the probability space $(\Omega,\mathsf{F},\lambda)$, the existence of which follows from §3.11.

3.5. Specific Anti-Derivations of Operators

1. Let $X := c_0(\alpha, \mathbf{K_p})$ be a Banach space over a local field $\mathbf{K_p}$ of zero characteristic or of the positive characteristic. This means that $\mathbf{K_p}$ is locally compact and either a finite algebraic extension of the p-adic field such that $\mathbf{K_p} \supset \mathbf{Q_p}$ for $char(\mathbf{K_p}) = 0$ or $\mathbf{K_p} \supset \mathbf{F}_p(\theta)$ for $char(\mathbf{K_p}) = p > 0$. Take the ball $B_r := B(\mathbf{K_p}, t_0, r)$ in the field $\mathbf{K_p}$ containing t_0 and of radius r. Let F be a continuous function on $B_r \times C^0(B_r, X)^{\otimes k}$ with values in $C^0(B_r, X)$:

$$F \in C^0(B_r \times C^0(B_r, X)^{\otimes k}, C^0(B_r, X)), \tag{1}$$

where $Z^{\otimes k} = Z \otimes \cdots \otimes Z$ is the product of k copies of a normed space Z and $Z^{\otimes k}$ is supplied with the box (i.e. maximum norm) topology ([19, 87]):

$$\|z\| = \sup_{j \in k} \|z_j\|, \quad z = (z_j \in Z : j \in k).$$

Classes of differentiable functions are described in Appendix A. The space $C^t(M, X)$ consists of all mappings $f : M \to X$ from a C^∞-manifold M with clopen charts modelled on a Banach space Y over $\mathbf{K_p}$ into X of class of smoothness C^t with $0 \le t < \infty$ (see also [58, 61, 62]). When M is closed and bounded in the corresponding Banach space, this space $C^t(M, X)$ is Banach and it can be supplied with the supremum-norm of $\bar{\Phi}^v f$, $0 \le v \le t$. Each continuous function F prescribed by (1) above can be written in the following form:

$$F(v, \xi_1, \dots \xi_k) = \sum_{j \in \alpha} F^j(v, \xi_1, \dots, \xi_k) e_j, \tag{2}$$

where $F^j \in C^0(B_r \times C^0(B_r, X)^{\otimes k}, \mathbf{K_p})$ for each $j \in \alpha$; $\xi_1, \dots, \xi_k \in C^0(B_r, X)$ are continuous functions. In particular let

$$F(v; \xi_1, \dots, \xi_k) = G(v; \xi_1, \dots, \xi_l).(A_{l+1}(v)\xi_{l+1}, \dots, A_k(v)\xi_k). \tag{3}$$

As usually $L(X, Y)$ denotes the Banach space of all continuous linear operators $A : X \to Y$ supplied with the operator norm

$$\|A\| := \sup_{0 \ne x \in X} \|Ax\|_Y / \|x\|_X$$

and for brevity $L(X) := L(X, X)$, $A_i(v)$ are continuous linear operators for each $v \in B_r$ such that $A_i \in C^0(B_r, L(X))$,

$$G(v, \xi_1, \dots, \xi_l) \in L_{k-l}(X^{\otimes(k-l)}; X)$$

for each fixed $v \in B_r$ and $\xi_1, \dots \xi_l \in C^0(B_r, X)$. This means that F is the $(k-l)$-linear operator by $\xi_{l+1}, \dots, \xi_k$, where $G = G(v, \xi_1, \dots, \xi_l)$ is the short notation of $G(v, \xi_1(v), \dots, \xi_l(v))$. Then $L_k(X_1, \dots, X_k; Y)$ denotes the Banach (normed) space of k-linear continuous operators from $X_1 \otimes \cdots \otimes X_k$ into Y for Banach (normed) spaces $X_1, \dots, X_k, Y$ over $\mathbf{K}$ and $L_k(X^{\otimes k}; Y) := L_k(X_1, \dots, X_k; Y)$ for the particular case $X_1 = \cdots = X_k = X$. When $l = 0$ put $G = G(v)$. There exists the following anti-derivation of operators given by Equation (3):

$$\hat{P}_{(\xi_{l+1}, \dots, \xi_k)}[G(s; \xi_1, \dots, \xi_l) \circ (A_{l+1} \otimes \cdots \otimes A_k)](v):$$

$$= \sum_{n=0}^{\infty} G(v_n;\xi_1,\dots,\xi_l).(A_{l+1}(v_n)[\xi_{l+1}(v_{n+1})$$

$$-\xi_{l+1}(v_n)],\dots,A_k(v_n)[\xi_k(v_{n+1})-\xi_k(v_n)]), \quad (4)$$

where $v_n=\sigma_n(t)$, $\{\sigma_n : n=0,1,2,..\}$ is an approximation of the identity in B_r, satisfying conditions $4.7(A1-A4)$ given above.

2. Lemma. (1). *If $G\in C^0(B_r\times X^{\otimes l},L_{k-l}(X^{\otimes(k-l)};X))$ is a continuous operator-valued mapping, $\xi_i\in C^0(B_r,X)$ is a continuous function for each $i=1,\dots,k$, while $A_{l+i}\in C^0(B_r,L(X))$ is a continuous operator-valued mapping for each $i=1,\dots,k-l$, then the anti-derivation mapping $\hat{P}_{(\xi_{l+1},\dots,\xi_k)}[G(s;\xi_1,\dots,\xi_l)\circ(A_{l+1}\otimes\cdots\otimes A_k)](v)$ is of class $C^0(B_r\times C^0(B_r,X)^{\otimes l},C^0(B_r,X))$ as the function by variables $v,\xi_1,\dots,\xi_l$ for each fixed $\xi_{l+1},\dots,\xi_k$ and $\hat{P}$ is of smoothness class C^∞ by variables $\xi_{l+1},\dots,\xi_k$.*

(2). *Moreover, if the mapping G is of smoothness class C^m by the arguments $\xi_1,\dots,\xi_l$, then $\hat{P}_{(\xi_{l+1},\dots,\xi_k)}G$ is also of smoothness class C^m by the arguments $\xi_1,\dots,\xi_l$.*

Proof. Since the ball B_r is compact, the functions ξ_i are uniformly continuous together with the mapping $A_{l+i}(v)[\xi_{l+i}(v)]$. In addition the anti-derivation operator $\hat{P}$ is linear by the variables $\xi_{l+1},\dots,\xi_k$.

Each **K** linear span of a finite family of vectors $v_1,\dots,v_n$ is a finite-dimensional linear space over the field **K**. The composition $f\circ g$ of k times continuously differentiable functions (i.e. of class C^k) $f:U\to Z$ and $g:V\to Y$ is of class C^k on V, when U and V are open subsets in Banach spaces X and Y correspondingly over the same field **K**, Z is also a Banach space over **K** and $g(V)\subset U$. This can be demonstrated by induction using formulas for the partial difference quotients of the composite mapping $f\circ g$. Using embeddings of finite dimensional spaces $\mathbf{K}^n$ and $\mathbf{K}^m$ into X and Y this can also be demonstrated with the help of the formulas from Lemma B.9 and Corollary B.10.

From this, Formula 1(4) and Conditions $4.7(A1-A4)$ the first statement follows. The last statement follows from the linearity of $\hat{P}$ by G, Formulas $4.7(D,T)$ and applying the operator of difference quotients $\bar{\Phi}^m$ by the variables $\xi_1,\dots,\xi_l$ (see Appendix A or [58,62]).

3. Lemma. *If $\xi_i\in C^1(B_r,X)$ is a continuous function for each $i=1,\dots,k$ and Conditions (1) of Lemma 2 are satisfied, then the anti-derivation mapping is of smoothness class*

$$\hat{P}_{(\xi_{l+1},\dots,\xi_k)}[G(s;\xi_1,\dots,\xi_l)\circ(A_{l+1}\otimes\cdots\otimes A_k)](x)\in C^1(B_r,X)$$

as the function by the argument $x\in B_r$ and its partial derivative by x is:

$$\partial(\hat{P}_{(\xi_{l+1},\dots,\xi_k)}[G(s;\xi_1,\dots,\xi_l)\circ(A_{l+1}\otimes\cdots\otimes A_k)](x))/\partial x$$

$$= \sum_{q=l+1}^{k} \hat{P}_{(\xi_{l+1},\dots,\xi_{q-1},\xi_{q+1},\dots,\xi_k)} G(x;\xi_1,\dots,\xi_l).(A_{l+1}(x)\xi_{l+1}(x),\dots,A_{q-1}(x)\xi_{q-1}(x),$$

$$A_q(x)\xi'_q(x),A_{q+1}(x)\xi_{q+1}(x),\dots,A_k(x)\xi_k(x))$$

such that the estimate for norms

$$\|\hat{P}_{(\xi_{l+1},\dots,\xi_k)}[G(s;\xi_1,\dots,\xi_l)\circ(A_{l+1}\otimes\cdots\otimes A_k)](x)\|_{C^1(B_r,X)}$$

$$\leq \|G\|_{C^0(B_r\times X^{\otimes l},L_{k-l}(X^{\otimes(k-l)};X))}\prod_{i=l+1}^{k}[\|A_i\|_{C^0(B_r,L(X))}\|\xi_i\|_{C^1(B_r,X)}]$$

is satisfied.

Proof. Let us consider the difference

$$\gamma := \gamma(x,y) := \hat{P}_{(\xi_{l+1},\dots,\xi_k)}[G(z;\xi_1,\dots,\xi_l)\circ(A_{l+1}\otimes\cdots\otimes A_k)](x) - \hat{P}_{(\xi_{l+1},\dots,\xi_k)}$$

$$[G(z;\xi_1,\dots,\xi_l)\circ(A_{l+1}\otimes\cdots\otimes A_k)](y) - (x-y)\sum_{q=l+1}^{k}\hat{P}_{(\xi_{l+1},\dots,\xi_{q-1},\xi_{q+1},\dots,\xi_k)}$$

$$[G(y;\xi_1,\dots,\xi_l).(A_{l+1}(y)\xi_{l+1}(y),\dots,A_{q-1}(y)\xi_{q-1}(y),A_q(y){\xi'}_q(y),$$
$$A_{q+1}(y)\xi_{q+1}(y),\dots,A_k(y)\xi_k(y))]$$

and $\rho^{s+1}\leq|x-y|<\rho^s$, where $s\in\mathbf{N}$ is a natural number. Therefore, $x_0=y_0,\dots,x_s=y_s$, $x_{s+1}\neq y_{s+1}$ and we infer the expression:

$$\gamma=\left[\sum_{q=l+1}^{k}E(x_s)(v_{l+1},\dots,v_{q-1},h_q,z_{q+1},\dots,z_k)\right]+E(x_s)(h_{l+1},h_{l+2},z_{l+3},\dots,z_k)$$

$$+E(x_s)(h_{l+1},v_{l+2},h_{l+3},z_{l+4},\dots,z_k)+\cdots+E(x_s)(v_{l+1},\dots,v_{k-2},h_{k-1},h_k)+\cdots$$

$$+E(x_s)(h_{l+1},\dots,h_k)+\sum_{j=s+1}^{\infty}\{E(x_j)(\xi_{l+1}(x_{j+1})-\xi_{l+1}(x_j)),\dots,(\xi_k(x_{j+1})-\xi_k(x_j)))$$

$$-E(y_j)(\xi_{l+1}(y_{j+1})-\xi_{l+1}(y_j)),\dots,(\xi_k(y_{j+1})-\xi_k(y_j)))$$

$$-(x-y)\sum_{q=l+1}^{k}\hat{P}_{(\xi_{l+1},\dots,\xi_{q-1},\xi_{q+1},\dots,\xi_k)}$$

$$\times E(y)(\xi_{l+1}(y),\dots,\xi_{q-1}(y),{\xi'}_q(y),\xi_{q+1}(y),\dots,\xi_k(y)),$$

where $v_j=\xi_j(x_{s+1})-\xi_j(x_s)$ is the increment of the function ξ_j, $h_j=\xi_j(x_{s+1})-\xi_j(y_{s+1})$, $z_j=\xi_j(y_{s+1})-\xi_j(y_s)$ for $j=l+1,\dots,k$ and

$$(i)\ E:=E(x):=G(x;\xi_1,\dots,\xi_l).(A_{l+1}(x)\otimes\cdots\otimes A_k(x))\quad\text{and}$$

$$(ii)\ E(x)(\xi_{l+1},\dots,\xi_k):=G(x;\xi_1,\dots,\xi_l).(A_{l+1}(x)\xi_{l+1},\dots,A_k(x)\xi_k)$$

in accordance with Formula 1(3). On the other hand, the inequality

$$\|\xi_i(y_{j+1})-\xi_i(y_j)-(y_{j+1}-y_j)\xi_i(y)\|=\|(y_{j+1}-y_j)[(\bar{\Phi}^1\xi_i)(y_j;1;y_{j+1}-y_j)-\xi_i(y)]\|$$
$$\leq|y_{j+1}-y_j|\|\xi_i\|_{C^1(B_r,X)}$$

is fulfilled and

$$E(x).(a_{l+1}+b_{l+1},\dots,a_k+b_k)-E(y).(a_{l+1},\dots,a_k)$$

$$=E(x).(a_{l+1}+b_{l+1},\dots,a_k+b_k)-E(x)(a_{l+1},\dots,a_k)$$
$$+[E(x)-E(y)].(a_{l+1},\dots,a_k)=E(x).(b_{l+1},a_{l+2},\dots,a_k)$$

$$+\cdots+E(x).(a_{l+1},\ldots,a_{k-1},b_k)+E(x).(b_{l+1},b_{l+2},a_{l+3},\ldots,a_k)+\cdots$$
$$+E(x).(a_{l+1},\ldots,a_{k-2},b_{k-1},b_k)+\cdots+E(x).(b_{l+1},\ldots,b_k)$$
$$+[E(x)-E(y)].(a_{l+1},\ldots,a_k)$$

for each functions $a_{l+1},\ldots,a_k,b_{l+1},\ldots,b_k\in C^0(B_r,X)$. Thus the estimate follows

$$\left\|\left[\sum_{q=l+1}^{k}E(x_s)(v_{l+1},\ldots,v_{q-1},h_q,z_{q+1},\ldots,z_k)\right]\right.$$
$$-(x-y)\sum_{q=l+1}^{k}\hat{P}_{(\xi_{l+1},\ldots,\xi_{q-1},\xi_{q+1},\ldots,\xi_k)}$$
$$\left.\times E(y)(\xi_{l+1}(y),\ldots,\xi_{q-1}(y),\xi'_q(y),\xi_{q+1}(y),\ldots,\xi_k(y))\right\|$$
$$\leq\|E\|_{C^0}\rho^s\prod_{q=l+1}^{k}\|\xi_q\|_{C^1}\alpha(s)$$

and

$$\|E(x_j)(\xi_{l+1}(x_{j+1})-\xi_{l+1}(x_j)),\ldots,(\xi_k(x_{j+1})-\xi_k(x_j)))$$
$$-E(y_j)(\xi_{l+1}(y_{j+1})-\xi_{l+1}(y_j)),\ldots,(\xi_k(y_{j+1})-\xi_k(y_j)))\|$$
$$\leq\|E\|_{C^0}\rho^s\prod_{q=l+1}^{k}\|\xi_q\|_{C^1}\alpha(s)$$

for each $j\geq s+1$, where the limit

$$\lim_{s\to\infty}\alpha(s)=0$$

is zero, consequently, $\lim_{|x-y|\to 0}\gamma(x,y)=0$. Therefore, the partial difference quotient $\bar{\Phi}^1(\hat{P}_{(\xi_{l+1},\ldots,\xi_k)}E)(x)$ belongs to the space $C^0(B_r,X)$, where

$$\Phi^1\eta(x;h;\zeta)=\{\eta(x+\zeta h)-\eta(x)\}/\zeta$$

for $0\neq\zeta\in\mathbf{K}$, $h\in H$, $\eta\in C^1(U,Y)$, U is an open subset in X, X and Y are Banach spaces over the field $\mathbf{K}$, $\bar{\Phi}^1\eta$ denotes the continuous extension of $\Phi^1\eta$ on $U\times V\times B(\mathbf{K},0,1)$ for a neighborhood V of 0 in X (see also Appendix A or §2.3 [58] or §I.2.3 [62]). Then the anti-derivation operator takes the form:

$$(iii)\ (\hat{P}_{(\xi_{l+1},\ldots,\xi_k)}E)(x)=\sum_{n=0}^{\infty}(x_{n+1}-x_n)^{k-l}G(x_n;\xi_1,\ldots,\xi_l)$$
$$.(A_{l+1}(x_n)(\bar{\Phi}^1\xi_{l+1})(x_n;1;x_{n+1}-x_n),\ldots,(A_k(x_n)(\bar{\Phi}^1\xi_k)(x_n;1;x_{n+1}-x_n)).$$

Let us consider the increment $\eta:=\eta(x,y):=(\hat{P}_wE)(x)-(\hat{P}_wE)(y)$, then it can be written in the form:

$$\eta=E(x_s)(w(x_{s+1})-w(y_{s+1}))$$

$$+\sum_{n=s+1}^{\infty}\{E(x_n)(w(x_{n+1})-w(x_n))-E(y_n)(w(y_{n+1})-w(y_n))\},$$

consequently, the estimate for the norm

$$\|\eta\|\leq\|E\|_{C^0(B_r\times X^{\otimes l},L_{k-l}(X^{\otimes(k-l)};X))}\left(\prod_{i=l+1}^{k}\|\xi_i\|_{C^1}|x-y|\right)$$

follows, since E is the polylinear mapping by arguments $\xi_{l+1}(z),\dots,\xi_k(z)\in X$, also the inequalities $|x_{s+1}-y_{s+1}|\leq|x-y|$ and $|x_{n+1}-x_n|\leq|x-y|$ and $|y_{n+1}-y_n|\leq|x-y|$ are fulfilled for each $n>s$, where $\rho^{s+1}\leq|x-y|<\rho^s$, $w=(\xi_{l+1},\dots,\xi_k)$.

4. Note. In particular, when $X=\mathbf{K}$, $l=0$, $k=1$, $A_1=1$ and $\xi(x)=x$ Lemma 3 provides the usual formula $d[\hat{P}_sG(s)](x)/dx=G(x)$.

5. Remark. Let X and Y be two Banach spaces over a (complete relative to its uniformity) local field $\mathbf{K}$ of zero characteristic. Let X and Y be isomorphic with the Banach spaces $c_0(\alpha,\mathbf{K})$ and $c_0(\beta,\mathbf{K})$ and there are given the standard orthonormal bases $\{e_j: j\in\alpha\}$ in X and $\{q_j: j\in\beta\}$ in Y respectively, where α and β are ordinals. Then each linear continuous operator $E\in L(X,Y)$ has its matrix realization $E_{j,k}:=q_k^*Ee_j$, where $q_k^*\in Y^*$ is a continuous $\mathbf{K}$-linear functional $q_k^*: Y\to\mathbf{K}$ corresponding to q_k under the natural embedding $Y\hookrightarrow Y^*$ associated with the chosen basis, Y^* is a topologically conjugated or dual space of $\mathbf{K}$-linear functionals on Y.

6. Notation. Suppose that A is a commutative Banach algebra over a field $\mathbf{K}$ and A^+ denotes the Gelfand space of A. This means that $A^+=Sp(A)$, where $Sp(A)$ is in another words spectrum of A, which was defined in Chapter 6 [99]. Let $C_\infty(A^+,\mathbf{K})$ be the same space as in [83, 99]. We remind their definitions.

The spectrum $Sp(A)$ is the set of all nonzero algebra homomorphisms $\phi: A\to\mathbf{K}$ topologized as the subset of the product space $\mathbf{K}^A$ supplied with the product (Tychonoff) topology. Every element $x\in A$ induces the function $G_x: Sp(A)\to\mathbf{K}$ given by the formula $G_x(\phi):=\phi(x)$, where $\phi\in Sp(A)$. This mapping G_x is called the Gelfand transform of x. The operator G is called the Gelfand transformation.

The spectral norm is defined by the formula:

$$\|x\|_{sp}:=\sup_{\phi\in Sp(A)}|G_x(\phi)|$$

for each element $x\in A$. If $Sp(A)=\emptyset$, then $\|x\|_{sp}:=0$ for each $x\in A$.

For a locally compact topological space E the $\mathbf{K}$ linear space $C_\infty(E,\mathbf{K})$ is defined as a subspace of the space $BUC(E,\mathbf{K})$ of bounded uniformly continuous functions $f: E\to\mathbf{K}$ such that for each $\varepsilon>0$ there exists a compact subset $V\subset E$ for which $|f(x)|<\varepsilon$ for each $x\in E\setminus V$.

When E is not locally compact and has an embedding into the product of balls $B(\mathbf{K},0,1)^\gamma$ such that $E\cup\{x_0\}=cl(E)$ we put

$$C_\infty(E,\mathbf{K}):=\{f\in C(E,\mathbf{K}):\ \lim_{x\to x_0}f(x)=0\},$$

where $B(X,x,r):=\{y\in X:\ d(x,y)\leq r\}$ denotes the ball in a metric space (X,d). The closure $cl(E)$ is taken in $B(\mathbf{K},0,1)^\gamma$, γ is an ordinal, $x_0\in B(\mathbf{K},0,1)^\gamma$.

7. Definition. A commutative Banach algebra A is called a C-algebra if it is isomorphic with $C_\infty(X,\mathbf{K})$ for a locally compact Hausdorff totally disconnected topological space X, where $f+g$ and fg are defined point-wise for each functions $f,g\in C_\infty(X,\mathbf{K})$.

8. Remark. We fix a Banach space H over a non-archimedean complete field $\mathbf{F}$, as above the Banach algebra of all bounded $\mathbf{F}$-linear operators on H is denoted by $L(H)$. If $b\in L(H)$ we write shortly $Sp(b)$ instead of $Sp_{L(H)}(b):=cl(Sp(span_{\mathbf{F}}\{b^n:n=1,2,3,\dots\}))$ (see also [99]).

It was proved in Theorem 2 of the article [100] in the case of the field $\mathbf{F}$ with the discrete normalization group, that each continuous $\mathbf{F}$-linear operator $A:E\to H$ with $\|A\|\le 1$ from one Banach space E into another H has the form

$$A=U\sum_{n=0}^{\infty}\pi^nP_{n,A},$$

where $P_n:=P_{n,A}$, $\{P_n:n\ge 0\}$ is a family of projections and $P_nP_m=0$ for each $n\ne m$, $\|P_n\|\le 1$ and $P_n^2=P_n$ for each n, U is a partially isometric operator, that is, $U|_{cl(\sum_n P_n(E))}$ is isometric, $U|_{E\ominus cl(\sum_n P_n(E))}=0$, $ker(U)\supset ker(A)$, $Im(U)=cl(Im(A))$, $\pi\in\mathbf{F}$, $|\pi|<1$ and π is the generator of the normalization group of $\mathbf{F}$.

We restrict our attention to the case of the local field $\mathbf{F}$ of zero characteristic, consequently, $\mathbf{F}$ has the discrete normalization group. If $\|A\|>0$ we get the series decomposition

$$(i)\quad A=\lambda_AU\sum_{n=0}^{\infty}\pi^nP_{n,A},$$

where $\lambda_A\in\mathbf{F}$ and $|\lambda_A|=\|A\|$. In view of [83] this is the particular case of the spectral integration on the discrete topological space X. Evidently, for each $1\le r<\infty$ there exists a linear operator $J\in L(H)$ for which

$$(ii)\quad \left\{\sum_{n\ge 0}s_n^r dim_{\mathbf{F}}P_{n,J}(H)\right\}^{1/r}<\infty$$

for $1\le r<\infty$, where J has the spectral decomposition given by Formula (i), $s_n:=|\lambda_J||\pi|^n\|P_n\|$. Using this result it is possible to give the following definition.

9. Definition. Let E and H be two normed $\mathbf{F}$-linear spaces, where $\mathbf{F}$ is an infinite spherically complete field with a nontrivial non-archimedean normalization. The continuous $\mathbf{F}$-linear operator $A\in L(E,H)$ is called of class $L_q(E,H)$ if sequences $a_n\in E^*$ and $y_n\in H$ for each $n\in\mathbf{N}$ exist such that

$$(i)\quad \left(\sum_{n=1}^{\infty}\|a_n\|_{E^*}^q\|y_n\|_H^q\right)<\infty$$

and A has the form

$$(ii)\quad Ax=\sum_{n=1}^{\infty}a_n(x)y_n$$

for each vector $x\in E$, where $0<q<\infty$. For each such linear continuous operator A we define the mapping

$$(iii)\quad \nu_q(A)=\inf\left\{\sum_{n=1}^{\infty}\|a_n\|_{E^*}^q\|y_n\|_H^q\right\}^{1/q},$$

where the infimum is taken by all such representations (ii) of A,

$$(iv) \quad \nu_\infty(A) := \|A\|$$

and $L_\infty(E,H) := L(E,H)$.

10. Proposition. *The space $L_q(E,H)$ is the normed* **F***-linear space with the norm ν_q, when $1 \le q$; it is the metric space, when $0 < q < 1$.*

Proof. Let us consider an operator $A \in L_q(E,H)$ and a positive number $1 \le q < \infty$, since the case $q = \infty$ follows from its definition. Then A has the representation $9(ii)$. From the ultrametric inequality

$$\|Ax\|_H \le \|x\|_E \sup_{n\in\mathbf{N}}(\|a_n\|_{E^*}\|y_n\|_H) \le \|x\|_E \left(\sum_{n=1}^{\infty} \|a_n\|^q_{E^*}\|y_n\|^q_H\right)^{1/q},$$

the estimate

$$\sup_{x\ne 0} \|Ax\|_H/\|x\|_E =: \|A\| \le \nu_q(A)$$

follows.

Let now $A,S \in L_q(E,H)$, then a positive number $0 < \delta < \infty$ exists and two representations

$$Ax = \sum_{n=1}^{\infty} a_n(x)y_n \quad \text{and}$$

$$Sx = \sum_{m=1}^{\infty} b_m(x)z_m$$

are valid for which the inequalities are satisfied:

$$\left(\sum_{n=1}^{\infty} \|a_n\|^q_{E^*}\|y_n\|^q_H\right)^{1/q} \le \nu_q(A) + \delta \quad \text{and}$$

$$\left(\sum_{n=1}^{\infty} \|b_n\|^q_{E^*}\|z_n\|^q_H\right)^{1/q} \le \nu_q(S) + \delta,$$

consequently,

$$(A+S)x = \sum_{n=1}^{\infty} (a_n(x)y_n + b_n(x)z_n) \quad \text{and}$$

$$\nu_q(A+S) \le \left(\sum_{n=1}^{\infty} \|a_n\|^q\|y_n\|^q\right)^{1/q} + \left(\sum_{n=1}^{\infty} \|b_n\|^q\|z_n\|^q\right)^{1/q} \le \nu_q(A) + \nu_q(S) + 2\delta$$

due to the Hölder inequality. The case $0 < q < 1$ is analogous to the classical one given in [95] using results described above.

11. Proposition. *If $J \in L_q(H)$, $S \in L_r(H)$ are commuting operators, the field* **F** *is with the discrete normalization group and $1/q + 1/r = 1/v$, then $JS \in L_v(H)$, where $1 \le q,r,v \le \infty$.*

Proof. Since $\mathbf{F}$ is with the discrete normalization, the linear continuous operators J and S have the Decompositions $8(i)$. Certainly each projector $P_{n,J}$ and $P_{m,S}$ belongs to $L_1(H)$ and has the Decomposition $9(ii)$. The $\mathbf{F}$-linear span of the set $\bigcup_{n,m} range(P_{n,J}P_{m,S})$ is dense in H. In particular, for each vector $x \in range(P_{n,J}P_{m,S})$ the equality

$$J^k S^l x = \lambda_J^k \lambda_S^l \pi^{nk+ml} P_{n,J}P_{m,S}x$$

is valid. Applying §2.8 to commuting operators J^k and S^l for each $k,l \in \mathbf{N}$ and using the base of the normed space H we get projectors $P_{n,J}$ and $P_{m,S}$ which commute for each n and m, consequently,

$$JS = U_J U_S \lambda_J \lambda_S \sum_{n\geq 0, m\geq 0} \pi^{n+m} P_{n,J}P_{m,S}.$$

Thus $U_{JS} = U_J U_S$, $\lambda_{JS} = \lambda_J \lambda_S$ and

$$P_{l,JS} = \sum_{n+m=l} P_{n,J}P_{m,S}.$$

The Hölder inequality implies the following estimate

$$\nu_v(JS) = \inf\left(\sum_{n=0}^{\infty} s_{n,JS}^v dim_{\mathbf{F}} P_{n,JS}(H)\right)^{1/v} \leq \nu_q(J)\nu_r(S).$$

12.1. Proposition. *If E is the normed space and H is the Banach space over the field $\mathbf{F}$ complete relative to its uniformity, then $L_r(E,H)$ is the Banach space such that if $J,S \in L_r(E,H)$, then*

$$\|J+S\|_r \leq \|J\|_r + \|S\|_r; \quad \|bJ\|_r = |b|\ \|J\|_r \textit{ for each } b \in \mathbf{K};$$

$\|J\|_r = 0$ *if and only if* $J = 0$, *where* $1 \leq r \leq \infty$, $\| * \|_q := \nu_q(*)$.

Proof. In view of Proposition 10 it remains to prove that $L_r(E,H)$ is complete, when H is complete. Let $\{T_\alpha\}$ be a Cauchy net in the space $L_r(E,H)$, then there exists an operator $T \in L(E,H)$ such that

$$\lim_\alpha T_\alpha x = Tx \quad \text{for each } \ x \in E,$$

since $L_r(E,H) \subset L(E,H)$ and $L(E,H)$ is complete. We demonstrate that $T \in L_r(E,H)$ and T_α converges to T relative to ν_r for $1 \leq r < \infty$. Let α_k be a monotone subsequence in $\{\alpha\}$ such that $\nu_r^r(T_\alpha - T_\beta) < 2^{-k-2}$ for each $\alpha, \beta \geq \alpha_k$, where $k \in \mathbf{N}$. Since the difference $T_{\alpha_{k+1}} - T_{\alpha_k}$ belongs to the space $L_r(E,H)$, we infer the decomposition

$$(T_{\alpha_{k+1}} - T_{\alpha_k})x = \sum_{n=1}^{\infty} a_{n,k}(x) y_{n,k}$$

with

$$\sum_{n=1}^{\infty} \|a_{n,k}\|^r \|y_{n,k}\|^r < 2^{-k-2}.$$

Therefore,

$$(T_{\alpha_{k+p}} - T_{\alpha_k})x = \sum_{h=k}^{k+p-1} \sum_{n=1}^{\infty} a_{n,h}(x) y_{n,h}$$

for each $p \in \mathbf{N}$. Therefore, using convergence while p tends to ∞ we get the converging series

$$(T - T_{\alpha_k})x = \sum_{h=k}^{\infty}\sum_{n=1}^{\infty} a_{n,h}(x)y_{n,h}.$$

Then the estimate

$$\nu_r^r(T - T_{\alpha_k}) \leq \sum_{h=k}^{\infty}\sum_{n=1}^{\infty} \|a_{n,h}\|^r \|y_{n,h}\|^r \leq 2^{-k-1}$$

is satisfied, consequently, $T - T_{\alpha_k} \in L_r(E,H)$ and inevitably $T \in L_r(E,H)$. Moreover, the inequality

$$\nu_r(T - T_\alpha) \leq \nu_r(T - T_{\alpha_k}) + \nu_r(T_{\alpha_k} - T_\alpha) \leq 2^{-(k-1)/r}2$$

is fulfilled for each $\alpha \geq \alpha_k$.

12.2. Proposition. *Let E,H,G be normed spaces over a spherically complete field $\mathbf{F}$. If $T \in L(E,H)$ and $S \in L_r(H,G)$ are two linear continuous operators, then their product ST belongs to the space $L_r(E,G)$ and $\nu_r(ST) \leq \nu_r(S)\|T\|$. If $T \in L_r(E,H)$ and $S \in L(H,G)$, then $ST \in L_r(E,G)$ and $\nu_r(ST) \leq \|S\|\nu_r(T)$.*

Proof. For each $\delta > 0$ there are continuous linear functionals $b_n \in H^*$ and vectors $z_n \in G$ such that

$$Sy = \sum_{n=1}^{\infty} b_n(y)z_n$$

for each vector $y \in H$ and the inequality

$$\sum_{n=1}^{\infty} \|b_n\|^r \|z_n\|^r \leq \nu_r^r(S) + \delta$$

is fulfilled. Therefore, we infer that

$$STx = \sum_{n=1}^{\infty} T^* b_n(x)z_n$$

for each $x \in E$, consequently, the inequality

$$\nu_r(ST) \leq \sum_{n=1}^{\infty} \|T^* b_n\|^r \|z_n\|^r \leq \|T\|[\nu_r^r(S) + \delta]$$

is satisfied, since

$$\|T^* b_n(x)\| = |b_n(Tx)| \leq \|b_n\|\|Tx\| \leq \|b_n\|\|T\|\|x\|,$$

where $T^* \in L(H^*,E^*)$ denotes the adjoint operator such that $b(Tx) =: (T^*b)(x)$ for each continuous linear functional $b \in H^*$ and every vector $x \in E$. The operator T^* exists due to the Hahn-Banach theorem for normed spaces over the spherically complete field $\mathbf{F}$ [99].

12.3. Proposition. *Suppose that $T \in L_r(E,H)$, where E and H are normed spaces over the spherically complete field $\mathbf{F}$. Then the adjoint operator T^* belongs to the space $L_r(H^*,E^*)$ and their norms are subordinated to the inequality $\nu_r(T^*) \leq \nu_r(T)$.*

Proof. For each $\delta > 0$ there are continuous linear functionals $a_n \in E^*$ and vectors $y_n \in H$ such that

$$Tx = \sum_{n=1}^{\infty} a_n(x) y_n$$

for each $x \in E$ and

$$\sum_{n=1}^{\infty} \|a_n\|^r \|y_n\|^r \leq \nu_r^r(T) + \delta.$$

Since

$$(T^* b)(x) = b(Tx) = \sum_{n=1}^{\infty} a_n(x) b(y_n)$$

for each continuous linear functional $b \in H^*$ and every vector $x \in E$, the adjoint operator has the decomposition

$$T^* b = \sum_{n=1}^{\infty} y_n^*(b) a_n,$$

where $y_n^*(b) := b(y_n)$. This is correct due to the Hahn-Banach theorem for normed spaces E and H over the spherically complete field $\mathbf{F}$ [99]. Therefore, the inequality

$$\nu_r^r(T^*) \leq \sum_{n=1}^{\infty} \|y_n\|^r \|a_n\|^r \leq \nu_r^r(T) + \delta$$

is satisfied, since $\|y^*\|_{H^*} = \|y\|_H$ for each $y \in H$.

13. For a space $L_k(H_1, \dots, H_k; H)$ of k-linear mappings of $H_1 \otimes \dots \otimes H_k$ into H we have its embedding into the space $L(E, H)$, where E is a normed space $H_1 \otimes \dots \otimes H_k$ in its maximum norm topology for normed spaces $H_1, \dots, H_k, H$ over $\mathbf{F}$ (see §§1, 9, 10). Therefore, we can define the following normed space $L_{k,r}(H_1, \dots, H_k; H) := L_k(H_1, \dots, H_k; H) \cap L_r(E; H)$ in particular $L_{k,\infty}(H_1, \dots, H_k; H) := L_k(H_1, \dots, H_k; H)$ supplied with the norm $\nu_r(J) =: \|J\|_r$, where $1 \leq r \leq \infty$. Certainly, $L_{k,r} \subset L_{k,q}$ for each $1 \leq r < q \leq \infty$.

Suppose that $(\Omega, \mathsf{B}, \lambda)$ is a probability space equipped with a non-negative probability measure λ, where B is a σ-algebra of subsets of Ω. We define a $\mathbf{K}$-linear Banach space $L^q(\Omega, \mathsf{B}, \lambda; L_{k,r}(H_1, \dots, H_k; H))$ and $L^q(\Omega, \mathsf{B}, \lambda; L_k(H_1, \dots, H_k; H))$ as a completion of a family of mappings $\sum_{j=1}^n A_j Ch_{W_j}$ with $A_j \in L_{k,r}(H_1, \dots, H_k; H)$ or $A_j \in L_k(H_1, \dots, H_k; H)$ respectively and $W_j \in \mathsf{B}$ and $n \in \mathbf{N}$. That is, as consisting of those mappings $\Omega \ni \nu \mapsto A(\nu) \in L_{k,r}(H_1, \dots, H_k; H)$ for which $\|A(\nu)\|_r$ is λ-measurable and with the finite norm:

$$\|A\|_{L^q} := \left\{ \int_\Omega \|A(\nu)\|_r^q \lambda(d\nu) \right\}^{1/q} < \infty, \text{ where } 1 \leq q < \infty;$$

$$\|A\|_{L^\infty} := ess - \sup_\lambda \|A(\nu)\|_r.$$

14. We consider a C^∞-manifold X with an atlas $At(X) = \{(U_j, \phi_j) : j \in \Lambda_X\}$, where $\bigcup_j U_j = X$, $\phi_j(U_j)$ are open in $c_0(\alpha, \mathbf{K})$ and U_j are open in X, $\phi_j : U_j \to \phi_j(U_j)$ are homeomorphisms, $\phi_i \circ \phi_j^{-1} \in C^\infty$ for each $U_i \cap U_j \neq \emptyset$ and of the finite norm

$$\|\phi_i \circ \phi_j^{-1}\|_{C^m} < \infty$$

for each $m \in \mathbf{N}$. We also suppose that each set $\phi_j(U_j)$ is bounded in the Banach space $c_0(\alpha, \mathbf{K})$ for each $j \in \Lambda_X$, Λ_X is a set, $C^n_b(X,H)$ denotes the completion of the set of all functions $f: X \to H$ such that $f \circ \phi_j^{-1} \in C^n(\phi_j(U_j),H)$ for each $j \in \Lambda_X$ and satisfying the condition $\sup_j \|f \circ \phi_j^{-1}\|_{C^n} =: \|f\|_{C^n(X,H)} < \infty$, where H is a Banach space over $\mathbf{K}$.

Then by our definition a linear space $C^n(X,H)$ is the set of all functions $f: X \to H$ such that for each $x \in X$ there exists a neighborhood $x \in U \subset X$ for which $f|_U \in C^n_b(U,H)$.

By $L^s(\Omega, \mathsf{B}, \lambda; C^n(X,H))$ we denote a completion of the space of all simple functions of the form $\sum_{j=1}^n \xi_j(x) Ch_{W_j}(\nu)$ with $\xi_j(x) \in C^n(X,H)$, $W_j \in \mathsf{B}$ and $n \in \mathbf{N}$, relative to the following norm

$$\|\xi\|_{L^s} := \left\{ \int_\Omega \|\xi(x,\nu)\|^s_{C^n(X,H)} \lambda(d\nu) \right\}^{1/s} < \infty$$

for each $1 \le s < \infty$ or

$$\|\xi\|_{L^\infty} := ess - \sup_\lambda \|\xi(x,\nu)\|_{C^n(X,H)} < \infty,$$

where X is the C^∞ Banach manifold on $c_0(\alpha, \mathbf{K})$. The norm $\|\xi(x,\nu)\|_{C^n(X,H)}$ is attached to ξ as a function by $x \in X$. This norm may depend on the parameter $\nu \in \Omega$ such that $\|\xi(x,\nu)\|_{C^n(X,H)}$ is a measurable function by ν. This means that $\|\xi(x,\nu)\|_{C^n(X,H)}$ is the random function of ν.

Theorem. *Let a mapping G be in the space*

$$L^r(\Omega, \mathsf{B}, \lambda; C^0(B_R \times L^q(\Omega, \mathsf{B}, \lambda; C^0(B_R, H))^{\otimes l}, L_{k-l}(H^{\otimes(k-l)}; H)),$$

$$\xi_1, \dots, \xi_k \in L^q(\Omega, \mathsf{B}, \lambda; C^0(B_R, H)),$$

let also a mapping A_{l+i} be in the space $C^0(B_R, L(H))$ for each $i = 1, \dots, k-l$ (see §1), where $B_R = B(\mathbf{K}, 0, R)$, $G = G(x; \xi_1, \dots, \xi_l; \nu)$, $\xi_i = \xi_i(x,\nu)$ with $x \in B_R$, $\nu \in \Omega$, $1/r + (k-l)/q = 1/s$ with $1 \le r,q,s \le \infty$. Then the anti-derivative $(\hat{P}_{(\xi_{l+1},\dots,\xi_k)} G \circ (A_{l+1} \otimes \dots \otimes A_k))$ is in the space $L^s(\Omega, \mathsf{B}, \lambda; C^0(B_R, H))$.

Proof. By the conditions of this theorem $\lambda(\Omega) = 1$ and λ is nonnegative [10]. Therefore, in the space $L^q(\Omega, \mathsf{F}, \lambda; C^0(B_R \times V, W))$ the family of all step functions

$$f(t,x,\omega) = \sum_{j=1}^n Ch_{U_j}(\omega) f_j(t,x)$$

is dense, where $f_j \in C^0(B_R \times V, W)$, as above Ch_U denotes the characteristic function of a set $U \in \mathsf{F}$, $n \in \mathbf{N}$ is a natural number, V and W are Banach spaces over $\mathbf{K}$, $t \in B_R$ is a variable, $x \in V$ is a vector, $\omega \in \Omega$ is an elementary event. Each matrix element $F_{h,b}(x,\nu)$ is in the space $L^r(\Omega, \mathsf{B}, \lambda; C^0(B_R, \mathbf{K}))$ and $\xi_j \in L^q(\Omega, \mathsf{B}, \lambda; C^0(B_R, \mathbf{K}))$, where

$$F(x,\nu) := G(x; a_1, \dots, a_l; \nu).(A_{l+1} a_{l+1}(x), \dots, A_k a_k(x)),$$

$h \in H^*$ is a continuous linear functional, $b \in H$ is a vector, $F_{h,b} := h(Fb)$, $a_i \in C^0(B_R, H)$ is a continuous function for each $i = 1, \dots, k$. Since the random norm mapping $\|\xi_j(x,\nu)\|_{C^0(X,H)}$ belongs to the space $L^q(\lambda)$, $\|F_{a,b}(x,\nu)\|_{C^n(X,H)} \in L^r(\lambda)$, the mapping $F(x,\nu).w(x,\nu)$ is in the space $L^s(\Omega,\mathsf{B},\lambda;C^0(B_R,H))$, where $w=(\xi_1,\dots,\xi_k)$. The operator $\hat{P}_wF$ is linear by w and F, hence it is defined on simple functions. In view of Lemma 2 the inequality

$$\|\hat{P}_wF(x,\nu)\|_H \le \|F(x,\nu)\|_{C^0(B_R\times H^{\otimes l},L_{k-l}(H^{\otimes(k-l)};H))}$$

$$\prod_{i=l+1}^{k} [\|A_i\|_{C^0(B_R,L(H))}\|\xi_i(x,\nu)\|_{C^0(B_R,H)}]$$

is satisfied for λ-a.e. $\nu\in\Omega$. Hence the estimate

$$\|(\hat{P}_wF)(x,\nu)\|_{L^s} \le \|G\|_{L^r}\prod_{i=l+1}^{k} [\|A_i\|_{C^0}\|\xi_i\|_{L^q}]$$

is valid.

15. Corollary. *If in suppositions of Theorem* 14 $\xi_i\in L^q(\Omega,\mathsf{B},\lambda;C^1(B_R,H))$ *for each* $i=1,\dots,k$, *then the anti-derivative* $(\hat{P}_wF)$ *of* F *is in the space* $L^s(\Omega,\mathsf{B},\lambda;$ $C^1(B_R,H))$ *and its norm satisfies the inequality:*

$$(i)\ \|(\hat{P}_wG.(A_{l+1}\otimes\cdots\otimes A_k))\|_{L^s(\lambda;C^1(B_R,H))} \le \|G\|_{L^r(\lambda;C^0(B_R\times H^{\otimes l},L_{k-l}(H^{\otimes(k-l)};H)))}$$

$$\prod_{i=l+1}^{k} [\|A_i\|_{C^0(B_R,L(H))}\|\xi_i\|_{L^q(\lambda;C^1(B_R,H))}.$$

Proof. In view of Lemma 3 and Theorem 14 the inequality

$$\|(\hat{P}_wF)(x,\nu)\|_{C^1(B_R,H)} \le \|G(x;\xi_1,\dots,\xi_l;\nu)\|_{C^0(B_R\times H^{\otimes l},L_{k-l}(H^{\otimes(k-l)},H))}$$

$$\prod_{i=l+1}^{k} [\|A_i\|_{C^0(B_R,L(H))}\|\xi_i(x,\nu)\|_{C^1(B_R,H)}]$$

is valid for λ-almost each $\nu\in\Omega$. From this Formula (i) follows.

3.6. Non-Archimedean Stochastic Processes

1. Remark and definition. Let $(\Omega,\mathsf{F},\lambda)$ be a probability space. Points $\omega\in\Omega$ are called elementary events and values $\lambda(S)$ probabilities of events $S\in\mathsf{F}$. A measurable map $\xi:(\Omega,\mathsf{F})\to(X,\mathsf{B})$ is called a random variable with values in X, where B is the σ-algebra of a locally $\mathbf{K}$-convex space X. The random variable ξ induces a normalized measure

$$\nu_\xi(A) := \lambda(\xi^{-1}(A))$$

in X and a new probability space (X,B,ν_ξ). We take $X=C^0(T,H)$ (see §3.1) and the σ-algebra B which is the subalgebra of the Borel σ-algebra $Bf(X)$ of X, where H is a Banach space over $\mathbf{K}$, $T=B(\mathbf{K},t_0,R)=:B_R$, $0<R<\infty$, $\mathbf{K}$ is the local field of zero characteristic or of the positive characteristic.

A random variable $\xi:\omega\mapsto\xi(t,\omega)$ with values in (X,B) is called a (non-archimedean) stochastic process on T with values in H.

Events $S_1, \ldots, S_n$ are called independent in total if

$$P(S_{k_1} \cdots S_{k_m}) = P(S_{k_1}) \cdots P(S_{k_m})$$

for any $1 \le k_1 < \cdots < k_m \le n$. σ-Subalgebras $\mathsf{F}_k \subset \mathsf{F}$ are said to be independent if all collections of events $S_k \in \mathsf{F}_k$ are independent in total, where $k = 1, \ldots, n$, $n \in \mathbf{N}$ ia natural number.

To each collection of random variables ξ_γ on (Ω, F) with $\gamma \in \Upsilon$ is related the minimal σ-algebra $\mathsf{F}_\Upsilon \subset \mathsf{F}$ with respect to which all ξ_γ are measurable, where Υ is a set.

Collections of random variables $\{\xi_\gamma : \gamma \in \Upsilon_j\}$ are called independent if such are F_{Υ_j}, where $\Upsilon_j \subset \Upsilon$ for each $j = 1, \ldots, n$, $n \in \mathbf{N}$.

Besides $X = C^0(T, H)$ it is possible to consider the product locally $\mathbf{K}$-convex spaces $X = H^T$.

2. We consider stochastic processes $E \in L^r(\Omega, \mathsf{F}, \lambda; C^0(T, L_v(H)))$ such that $E = E(t, \omega)$, where $1 \le v \le \infty$, $1 \le r \le \infty$, $t \in T = B(\mathbf{K}, t_0, R)$ and $\omega \in \Omega$ (see §5.14 and §3).

Definition. On the space $L^r(\Omega, \mathsf{F}, \lambda; C^0(T, L_v(H)))$ the non-archimedean stochastic integral is defined by the following equation:

$$(i)\ \mathsf{I}(E)(t, \omega) := (\hat{P}_w E)(t, \omega) = \sum_{j=0}^{\infty} E(t_j, \omega)[w(t_{j+1}, \omega) - w(t_j, \omega)],$$

where $w = w(t, \omega)$, $t_j = \sigma_j(t)$ (see §3.10).

3. Proposition. *The non-archimedean stochastic integral is the continuous* $\mathbf{K}$*-bilinear operator from the space* $L^r(\Omega, \mathsf{F}, \lambda; C^0(T, L_v(H))) \otimes L^q(\Omega, \mathsf{F}, \lambda; C^0_0(T, H))$ *into the space* $L^s(\Omega, \mathsf{F}, \lambda; C^0(T, H))$*, where* $1/q + 1/r = 1/s$ *and* $1 \le r, q, s \le \infty$.

Proof. It follows from Theorem 5.14, since $(\hat{P}_{aw+by} E) = (a\hat{P}_w E) + (b\hat{P}_y E)$ and $(\hat{P}_w (aE + bV)) = (a\hat{P}_w E) + b(\hat{P}_w V)$ for each $a, b \in \mathbf{K}$, each $w, y \in L^q(\Omega, \mathsf{F}, \lambda; C^0_0(T, H))$ and each $E, V \in L^r(\Omega, \mathsf{F}, \lambda; C^0(T, L_v(H)))$.

4. Consider a function f from $T \times H$ into the Banach space $Y = c_0(\beta, \mathbf{K})$ satisfying conditions $(a-d)$:

(a) $f \in C^1(T \times H, Y)$;

(b) $(\bar{\Phi}^n f)(t, x; h_1, \ldots, h_n; \zeta_1, \ldots, \zeta_n) \in C^0(T \times H^{n+1} \times \mathbf{K^n}, Y)$ for each $n \le m$,

(c) $(\bar{\Phi}^n f)(t, x; h_1, \ldots, h_n; \zeta_1, \ldots, \zeta_n) = 0$ for $n = m+1$,

(d) $f(t, x) - f(0, x) = (\hat{P}_t g)(t, x)$ with $g \in C^0(T \times H, Y)$, where $2 \le m \in \mathbf{N}$, $f = f(t, x)$ is a function, $t \in T$ is a time variable, $x \in H$ is a vector; also $h_1, \ldots, h_n \in H$ are vectors, $\zeta_1, \ldots, \zeta_n \in \mathbf{K}$ are numbers. Here $\hat{P}_u$ is the anti-derivation operator on the space $C^0(T, Y)$. The mapping $(\hat{P}_t g)(t, x)$ is defined for each fixed $x \in H$ by $t \in T$ such that $(\hat{P}_t g)(t, x) = \hat{P}_u g(u, x)|_{u=t}$ with $u \in T$ (see also §3.10 and Appendices A, B).

Suppose $a \in L^s(\Omega, \mathsf{F}, \lambda; C^0(T, H))$, $w \in L^q(\Omega, \mathsf{F}, \lambda; C^0_0(T, H))$ and $E \in L^r(\Omega, \mathsf{F}, \lambda; C^0(T, L(H)))$, where $1/r + 1/q = 1/s$, $1 \le r, q, s \le \infty$, $a = a(t, \omega)$, $E = E(t, \omega)$, $t \in T$, $\omega \in \Omega$. A stochastic process of the type

$$(i)\quad \xi(t, \omega) = \xi_0(\omega) + (\hat{P}_u a)(u, \omega)|_{u=t} + (\hat{P}_{w(u,\omega)} E)(u, \omega)|_{u=t}$$

is said to have a stochastic differential

$$(ii)\quad d\xi(t, \omega) = a(t, \omega)dt + E(t, \omega)dw(t, \omega),$$

because $(\hat{P}_t g)'(t) = g(t)$ for each $g \in C^0(T,H)$, where $\xi_0 \in L^s(\Omega,\mathsf{F},\lambda;H)$, $t_0, t \in T$, $w(t_0,\omega) = 0$. In view of Lemma 5.3, Theorem 5.14 and Proposition 3 the random function ξ belongs to the space $L^s(\Omega,\mathsf{F},\lambda;C^0(T,H))$.

Let $\hat{P}_{u^b,w^h}$ denote the anti-derivation operator $\hat{P}_{(\xi_1,\dots,\xi_{b+h})}$ given by Formula 5.1(4), where $\xi_1 = u,\dots,\xi_b = u$, $\xi_{b+1} = w,\dots,\xi_{b+h} = w$. Henceforth, it is used the notation

$$(iii)\ \tilde{P}^n_{a,Ew} f(u,\xi(u,\omega)) := \sum_{k=1}^{n} (k!)^{-1} \sum_{l=0}^{k} \binom{k}{l} (\hat{P}_{u^{k-l},w(u,\omega)^l}$$

$$[(\partial^k f/\partial x^k)(u,\xi(u,\omega)) \circ (a^{\otimes(k-l)} \otimes E^{\otimes l})])$$

for such operator, whenever it exists, where $n \in \mathbf{N}$ or $n = \infty$. For $char(\mathbf{K}) = 0$ there is not any restriction on n, for $char(\mathbf{K}) = p > 0$ we suppose that $n < p$ in Formula (iii).

Theorem. *Let Conditions* $4(a-d)$ *and* (i, ii) *be satisfied, either* $m \in \mathbf{N}$ *for* $char(\mathbf{K}) = 0$ *or* $m < p$ *for* $char(\mathbf{K}) = p > 0$. *Then the function* $f(t,\xi)$ *has the decomposition:*

$$(iv)\ f(t,\xi(t,\omega)) = f(t_0,\xi_0) + (\hat{P}_u f'_t(u,\xi(u,\omega)))|_{u=t} + \tilde{P}^m_{a,Ew} f(u,\xi(u,\omega))|_{u=t}.$$

Proof. Let $\{u_k : k = 0,1,\dots,n\}$ be a finite $|\pi|^l$ net in T, that is, for each $t \in T$ there exists k such that $|u_k - t| \le |\pi|^l$, where $n = n(k) \in \mathbf{N}$, $\pi \in \mathbf{K}$, $p^{-1} \le |\pi| < 1$ and $|\pi|$ is the generator of the normalization group of $\mathbf{K}$. Such net exists, since the ball T is compact. We choose $t = u_n$ and $t_0 = u_0$. Denote by $\eta(t)$ the random function $f(t,\xi(t,\omega))$. Then by the Taylor formula (see Theorem 29.4 [103] and the theorem in Appendix C) the increment of f is given by the equality:

$$\begin{gathered}(v)\quad f(t,\xi(t)) - f(u,\xi(u)) = f'_t(u,\xi(u))(t-u) + f'_x(u,\xi(u)).(\Delta\xi) \\ +(1/2)f''_{t,t}(u,\xi(u))(t-u)^2 + f''_{t,x}(u,\xi(u)).((t-u),\Delta\xi) \\ +(1/2)f''_{x,x}(u,\xi(u)).(\Delta\xi,\Delta\xi) + \{(\bar{\Phi}^2 f)(u,\xi(u);(t-u),(t-u);1,1) \\ -(1/2)f''_{t,t}(u,\xi(u))(t-u)^2\} \\ +\{(\bar{\Phi}^2 f)(u,\xi(u);(t-u),\Delta\xi;1,1) + (\bar{\Phi}^2 f)(u,\xi(u);\Delta\xi,(t-u);1,1) \\ -f''_{t,x}(u,\xi(u)).(t-u,\Delta\xi)\} \\ +\{(\bar{\Phi}^2 f)(u,\xi(u);\Delta\xi,\Delta\xi;1,1) - (1/2)f''_{x,x}(u,\xi(u)).(\Delta\xi,\Delta\xi)\},\end{gathered}$$

where $\Delta\xi = \xi(t) - \xi(u)$ denotes the increment of the stochastic process ξ. For a brevity we denote $\xi(t) = \xi(t,\omega)$ and $w(t) := w(t,\omega)$ for a chosen ω. If $t_n = \sigma_n(t)$ for each $n = 0,1,2,\dots$, then from Formulas (i) and 5.1(4) we deduce, that:

$$(vi)\ \xi(t_{n+1},\omega) - \xi(t_n,\omega) = a(t_n,\omega)(t_{n+1} - t_n) + E(t_n,\omega)(w(t_{n+1},\omega) - w(t_n,\omega)),$$

where $\{\sigma_n : n = 0,1,2,\dots\}$ is the approximation of the identity in T.

From Condition (d) it follows that

$$(\partial f(t,x)/\partial t) = g(t,x) = (\hat{P}_t g)'_t \quad \text{and}$$

$$\hat{P}_t(f'_t)(t,x) = f(t,x) - f(0,x),$$

which also leads to disappearance of terms $\partial^{m+b} f(t,x)/\partial t^b \partial x^m$ from Formula (iv) for each b,m such that $1 \le b$ and $2 \le m+b$.

Now we approximate $f(t,x)$ by functions of the form $\sum_j \phi_j(t)\psi_j(x)$, so the problem reduces to the consideration of functions $f(x)$ which are independent of t. Due to Conditions (i,ii) it is possible to put

$$\xi(t,\omega) = \xi_0(\omega) + a(\omega)(t-t_0) + E(\omega)[w(t) - w(t_0)].$$

By the Taylor formula we have the decomposition:

$$(vii)\ f(x) = f(x_0) + \sum_{n=1}^{m} (n!)^{-1} f^{(n)}(x_0).(x-x_0)^{\otimes n}$$

for each $x, x_0 \in H$, since $\bar{\Phi}^{m+1} f = 0$, either $char(\mathbf{K}) = 0$ and a number m may be any natural or $char(\mathbf{K}) = p > 0$ and $m < p$. We take the points $t_k = \sigma_k(t)$ for each $k = 0,1,2,\dots$, then the difference of the random function takes the form:

$$\eta(t) - \eta(t_0) = \sum_{j=0}^{\infty} \{f(\xi_{j+1}) - f(\xi_j)\},$$

where $\xi_j := \xi(t_j)$, since

$$\lim_{j\to\infty} \xi_j = \xi.$$

Then each term $f(\xi_{j+1}) - f(\xi_j)$ can be expressed by Formula (vii) due to Condition (b). On the other hand,

$$(\xi_{j+1} - \xi_j) = a(\omega)(t_{j+1} - t_j) + E(\omega)[w(t_{j+1}) - w(t_j)]$$

as the particular case of Formula (vi). From Formulas 5.1(4), $(v-vii)$ and Theorem 5.14 we get the statement of this theorem.

5. Corollary. *If Conditions* $4(a,d,i,ii)$ *are satisfied,* $4(b)$ *is accomplished for each* $n \in \mathbf{N}$ *with* $char(\mathbf{K}) = 0$ *and*

$$(c')\ \lim_{n\to\infty} \|(\bar{\Phi}^n_x f)(t,x;h_1,\dots,h_n;\zeta_1,\dots,\zeta_n)\|_{C^0(T\times(B(H,0,R_1))^{n+1}\times B(\mathbf{K}^{n+1},0,R_1),Y)} = 0$$

for each $0 < R_1 < \infty$*, then the random function* f *has the decomposition:*

$$(i)\ f(t,\xi(t,\omega)) = f(t_0,\xi_0) + (\hat{P}_u f'_t(u,\xi(u,\omega))|_{u=t} + (\tilde{P}^{\infty}_{a,Ew} f(u,x))|_{u=t}.$$

Proof. From the proof of Theorem 4 we get a function $f(x)$ for which

$$(ii)\ f(x) = f(x_0) + \sum_{n=1}^{\infty} (n!)^{-1} f^{(n)}(x).(x-x_0)^{\otimes n}$$

due to Condition (c') and convergence of the Taylor series when n tends to the infinity (see Appendix C). In view of Theorem 5.14 we infer the expression:

$$\lim_{m\to\infty} \|(m!)^{-1} \sum_{l=0}^{m} \binom{m}{l} (\hat{P}_{u^{m-l},w(u,\omega)^l}[(\partial^m f/\partial x^m)(u,\xi(u,\omega))$$

$$\circ(a^{\otimes(m-l)}\otimes E^{\otimes l})]])|_{u=t}\|_{L^s(\Omega,\mathsf{F},\lambda;C^0(T,Y))}=0.$$

Then for each chosen elementary event $\omega\in\Omega$ the functions $a(t,\omega)$ and $w(t,\omega)$ are bounded on the compact ball T in accordance with the imposed above conditions. Therefore, approximating the function $f(x)$ by the Taylor formula up to terms $\bar{\Phi}^m f$ by finite sums and taking the limit while m tends to the infinity one deduces Formula (i) of this Corollary from Formula $4(iv)$.

6. Theorem. *Let a function $f(u,x)$ belong to the space $C^\infty(T\times H,Y)$, where the field $\mathbf{K}$ is of zero characteristic, and let also the limit*

$$(i)\ \lim_{n\to\infty}\max_{0\le l\le n}\|(\bar{\Phi}^n f)(t,x;h_1,\dots,h_n;$$

$$\zeta_1,\dots,\zeta_n)\|_{C^0(T\times B(\mathbf{K},0,r)^l\times B(H,0,1)^{n-l}\times B(\mathbf{K},0,R_1)^{n-l},Y)}=0$$

be zero for each $0<R_1<\infty$, where $h_j=e_1$ and $\zeta_j\in B(\mathbf{K},0,r)$ for variables corresponding to $t\in T=B(\mathbf{K},t_0,r)$ and $h_j\in B(H,0,1)$, $\zeta_j\in B(\mathbf{K},0,R_1)$ for variables corresponding to $x\in H$. Then

$$(ii)\quad f(t,\xi(t,\omega))=f(t_0,\xi_0)+\sum_{m+b\ge1,0\le m\in\mathbf{Z},0\le b\in\mathbf{Z}}((m+b)!)^{-1}\sum_{l=0}^{m}\binom{m+b}{m}\binom{m}{l}$$

$$(\hat{P}_{u^{b+m-l},w(u,\omega)^l}[(\partial^{(m+b)}f/\partial u^b\partial x^m)(u,\xi(u,\omega))\circ(I^{\otimes b}\otimes a^{\otimes(m-l)}\otimes E^{\otimes l})]])|_{u=t}.$$

Proof. In view of the Taylor formula we have (see Appendix C and Theorem A5 [81]) we have the convergent series:

$$(iii)\quad f(t,x)=f(t_0,x_0)+\sum_{m+b=1}^{k}((m+b)!)^{-1}\binom{m+b}{m}(\partial^{(m+b)}f/\partial u^b\partial x^m)(t_0,x_0)$$

$$(t-t_0)^b.(x-x_0)^{\otimes m}+\sum_{m+b=k+1}\binom{k+1}{m}[(\bar{\Phi}^{k+1}f)(t_0,x_0;(t-t_0)^{\otimes b},(x-x_0)^{\otimes m};1^{\otimes(k+1)})$$

$$-((k+1)!)^{-1}(\partial^{(k+1)}f/\partial u^b\partial x^m)(t_0,x_0)(t-t_0)^b.(x-x_0)^{\otimes m}]$$

for each $k\in\mathbf{N}$. In view of Condition (i), Formulas (iii), $5.1(4)$, $4(vi)$ we get Formula (ii) (see the proof of Theorem 4).

Stochastic processes in Banach spaces can be used for studying stochastic processes in complete locally $\mathbf{K}$-convex spaces and complete ultra-metric and ultra-uniform spaces due to the following two theorems.

7. Theorem. *Let X be a complete ultra-uniform space and $\mathbf{K}$ be a local field. Then there exists an irreducible normal expansion of X into the limit of the inverse system $S=\{P_n,f_n^m,E\}$ of uniform polyhedra over $\mathbf{K}$, moreover, $\lim S$ is uniformly isomorphic with X, where E is an ordered set, $f_n^m:P_m\to P_n$ is a continuous mapping for each $m\ge n$; particularly for the ultra-metric space (X,d) with the ultra-metric d the inverse system S is the inverse sequence.*

8. Theorem. *Let X be a complete separable ultra-uniform space and let $\mathbf{K}$ be a local field. Then for each marked $b\in\mathbf{C_s}$ or $b\in\mathbf{C}$ there exists a nontrivial $\mathbf{F}$-valued* *measure μ on X which is a restriction of a measure ν in a measure space $(Y,\mathcal{B}(Y),\nu) = \lim\{(Y_m,\mathcal{B}(Y_m),\nu_m),\bar{f}_n^m,E\}$ on X and each ν_m is quasi-invariant and pseudo-differentiable for b relative to a dense subspace Y'_m, where $Y_n := c_0(\mathbf{K},\alpha_n)$, $\bar{f}_n^m : Y_m \to Y_n$ is a normal (that is, $\mathbf{K}$-simplicial non-expanding) mapping for each $m \geq n \in E$, $\bar{f}_n^m|_{P_m} = f_n^m$, $\mathbf{F}$ is either a non-archimedean field containing a local field or $\mathbf{F} = \mathbf{R}$, $\mathcal{B}(Y)$ is either the algebra of clopen subsets or the Borel σ-algebra in Y respectively. Moreover, if X is not locally compact, then the family $\mathcal{F}$ of all such μ contains a subfamily $\mathcal{G}$ of pairwise orthogonal measures with the cardinality $card(\mathcal{G}) = card(\mathbf{F})^{\mathsf{c}}$, $\mathsf{c} := card(\mathbf{Q_p})$.*

9. Remark. The proofs of these two preceding statements are given in [81] (see §§1.6 and 2.6 and the appendix there). Using projective limits $g = pr - \lim\{g_m, f_n^m, E\}$ of random functions g_m in Banach spaces Y_m and polyhedra P_m in them one can analyze random functions g in the complete locally $\mathbf{K}$-convex spaces, ultra-metric and ultra-uniform spaces.

3.6.1. Brownian Motion Controlled by Non-archimedean Valued Measures

We remind the following definitions.

1. Definitions and Notes. We consider the function

$$g(x,y,b) := s^{(-1-b)\times ord_p(x-y)}$$

with the corresponding Haar measure v with values in $\mathbf{K_s}$, where $\mathbf{K_s}$ is a local field of zero characteristic containing the field $\mathbf{Q_s}$, s is a prime number, $b \in \mathbf{C_s}$ and $|x|_\mathbf{K} = p^{-ord_p(x)}$, $\mathbf{C_s}$ denotes the field of complex numbers with the non-archimedean normalization extending that of $\mathbf{Q_s}$. Then $\mathbf{U_s}$ is a spherically complete field with a normalization group

$$\Gamma_{\mathbf{U_s}} := \{|x| : 0 \neq x \in \mathbf{U_s}\} = (0,\infty) \subset \mathbf{R}$$

such that $\mathbf{C_s} \subset \mathbf{U_s}$, $0 < s$ is a prime number [17,99,103,111].

A function $f : \mathbf{K} \to \mathbf{U_s}$ is called pseudo-differentiable of order b, if there exists the following integral:

$$PD(b,f(x)) := \int_\mathbf{K} [(f(x)-f(y)) \times g(x,y,b)]dv(y).$$

We introduce the following notation $PD_c(b,f(x))$ for such integral by $B(\mathbf{K},0,1)$ instead of the entire $\mathbf{K}$.

For each $\gamma \in (0,\infty)$ there exists $\alpha = log_s(\gamma) \in \mathbf{R}$, $\Gamma_{\mathbf{U_s}} = (0,\infty)$, hence $s^\alpha \in \mathbf{U_s}$ is defined for each $\alpha \in \mathbf{R}$, where $log_s(\gamma) = ln(\gamma)/ln(s)$, $ln : (0,\infty) \to \mathbf{R}$ is the natural logarithmic function such that $ln(e) = 1$. The function $s^{\alpha+i\beta} =: \xi(\alpha,\beta)$ with α and $\beta \in \mathbf{R}$ is defined (see [48]) in the following manner. We put

$$s^{\alpha+i\beta} := s^\alpha (s^i)^\beta$$

and choose as s^i a marked number in the field $\mathbf{U_s}$ such that

$$s^i := (EXP_s(i))^{ln\ s},$$

where $EXP_s : \mathbf{C_s} \to \mathbf{C_s^+}$ is the exponential function, $\mathbf{C_s^+} := \{x \in \mathbf{C_s} : |x-1|_s < 1\}$ (see Proposition 45.6 [103] or above). Therefore, the inequality

$$|EXP_s(i) - 1|_s < 1,$$

is satisfied, consequently, $|EXP_s(i)|_s = 1$ and inevitably $|s^i|_s = 1$. Therefore, the norm is:

$$|s^{\alpha+i\beta}|_s = s^{-\alpha}$$

for each α and $\beta \in \mathbf{R}$, where $|*|_s$ is the extension of the normalization from the s-adic field $\mathbf{Q_s}$ onto the field $\mathbf{U_s}$, consequently, $s^x \in \mathbf{U_s}$ is defined for each $x \in \mathbf{C_s}$.

A quasi-invariant measure μ on X is called pseudo-differentiable for $b \in \mathbf{C_s}$, if the pseudo-differential $PD(b, g(x))$ exists for the mapping $g(x) := \mu(-xz+S)$ for each clopen subset $S \in Bco(X)$ with $\|S\|_\mu < \infty$ and each $z \in J_\mu^b$, where J_μ^b is a $\mathbf{K}$-linear subspace dense in X. For a fixed $z \in X$ such measure is called pseudo-differentiable along z.

2. Definitions and Remarks. Let X be a locally $\mathbf{K}$-convex space equal to a projective limit

$$\lim\{X_j, \phi_l^j, \Upsilon\}$$

of Banach spaces over a local field $\mathbf{K}$ such that $X_j = c_0(\alpha_j, \mathbf{K})$, where Υ is an ordered set, $\phi_l^j : X_j \to X_l$ is a $\mathbf{K}$-linear continuous mapping for each $j \geq l \in \Upsilon$, $\phi_j : X \to X_j$ is a projection on X_j, $\phi_l \circ \phi_l^j = \phi_j$ for each $j \geq l \in \Upsilon$, $\phi_k^l \circ \phi_l^j = \phi_k^j$ for each $j \geq l \geq k$ in Υ.

Parallel we consider also a locally $\mathbf{R}$-convex space which is presentable as a projective limit

$$Y = \lim\{l_2(\alpha_j, \mathbf{R}), \psi_l^j, \Upsilon\},$$

where $l_2(\alpha_j, \mathbf{R})$ is the real Hilbert space of the topological weight $w(l_2(\alpha_j, \mathbf{R})) = card(\alpha_j)\aleph_0$. Suppose B is a symmetric non-negative definite (bilinear) nonzero functional $B : Y^2 \to \mathbf{R}$.

We take any non-archimedean field $\mathbf{F}$ such that $\mathbf{K_s} \subset \mathbf{F}$ and with the normalization group $\Gamma_\mathbf{F} = (0, \infty) \subset \mathbf{R}$ and $\mathbf{F}$ is complete relative to its uniformity (see [17, 20]). Suppose that B is a nonnegative definite bilinear $\mathbf{R}$-valued symmetric functional on a dense $\mathbf{R}$-linear subspace $\mathsf{D}_{B,Y}$ in Y^*, $B : \mathsf{D}_{B,Y}^2 \to \mathbf{R}$, $j \in \Upsilon$ may depend on z, $z_j : X_j \to \mathbf{K}$ is a continuous $\mathbf{K}$-linear functional having the decomposition:

$$z_j = \sum_{k \in \alpha_j} e_j^k z_{k,j},$$

where the latter series is countable and convergent such that $z_{k,j} \in \mathbf{K}$, e_j^k is a continuous $\mathbf{K}$-linear functional on X_j such that $e_j^k(e_{l,j}) = \delta_l^k$ is the Kroneker delta symbol. Here $e_{l,j}$ denotes the standard orthonormal (in the non-archimedean sense) basis in the Banach space $c_0(\alpha_j, \mathbf{K})$. The mappings

$$v_q^s(z) = v_q^s(z_j) := \{|s^{q\ ord_p(z_{k,j})/2}|_s : k \in \alpha_j\}$$

are used further. It is supposed that z is such that $v_q^s(z) \in l_2(\alpha_j, \mathbf{R})$, where q is a positive constant.

The group of all roots of unity in the complex field **C** is denoted by **T** and it is supplied with the discrete topology. We consider its subgroup

$$\mathbf{T_s} := \{t : t \in \mathbf{T},\ t^n = 1, n = s^k, k \in \mathbf{N}\}.$$

As usually

$$\chi_\gamma(z) : X \to \mathbf{T_s}$$

denotes a continuous character such that

$$\chi_\gamma(z) = \chi(z(\gamma)), \quad \gamma \in X,$$

$$\chi : \mathbf{K} \to \mathbf{T_s}$$

is a nontrivial character of **K** as an additive group (see [99] and §2.2.5 in [81]).

A measure $\mu = \mu_{q,B,\gamma}$ on X with values in the field $\mathbf{K_s}$ is called a q-Gaussian measure, if its characteristic functional $\hat{\mu}$ with values in **F** has the form

$$\hat{\mu}(z) = s^{[B(v^s_q(z), v^s_q(z))]}\chi_\gamma(z)$$

on a dense **K**-linear subspace $D_{q,B,X}$ in X^* of all continuous **K**-linear functionals $z : X \to \mathbf{K}$ of the form $z(x) = z_j(\phi_j(x))$ for each $x \in X$ with $v^s_q(z) \in D_{B,Y}$.

3. Proposition. *A q-Gaussian quasi-measure on an algebra of cylindrical subsets $\bigcup_j \pi_j^{-1}(\mathcal{R}_j)$, where X_j are finite-dimensional over* **K** *subspaces in X, is a measure on a covering separating ring $\mathcal{R}$ of subsets of X. Moreover, a correlation operator B is of class L_1, that is, $Tr(B) < \infty$, if and only if each finite dimensional over* **K** *projection of μ is a q-Gaussian measure (see §11).*

4. Corollary. *Let X be a complete locally* **K**-*convex space of separable type over a local field* **K** *of zero characteristic, then for each constant $q > 0$ there exists a nondegenerate symmetric positive definite operator $B \in L_1$ such that a q-Gaussian quasi-measure is a measure on $Bco(X)$ and each its one dimensional over* **K** *projection is absolutely continuous relative to the nonnegative Haar measure on* **K**.

5. Proposition. *Let $\mu_{q,B,\gamma}$ and $\mu_{q,E,\delta}$ be two q-Gaussian measures with correlation operators B and E of class L_1, then there exists a convolution of these measures $\mu_{q,B,\gamma} * \mu_{q,E,\delta}$, which is a q-Gaussian measure $\mu_{q,B+E,\gamma+\delta}$.*

Three previous statements are proved in Section 2.6 [81].

6. Definition. Let B and q be as in §2. We denote by $\mu_{q,B,\gamma}$ the corresponding q-Gaussian $\mathbf{K_s}$-valued measure on H. Let ξ be a stochastic process with a real time $t \in T \subset \mathbf{R}$. It is called a non-archimedean q-Brownian motion process with real time (and controlled by $\mathbf{K_s}$-valued measure), if the following condition

(i) the random variable $\xi(t,\omega) - \xi(u,\omega)$ has a distribution $\mu_{q,(t-u)B,\gamma}$ for each $t \neq u \in T$ is fulfilled.

In the particular case of $B = I$ and $\gamma = 0$ it is called q-Wiener's process.

Let ξ be a stochastic process with a non-archimedean time $t \in T \subset \mathbf{F}$, where **F** is a local field of zero characteristic, then ξ is called a non-archimedean q-Brownian motion process with **F**-time (and controlled by $\mathbf{K_s}$-valued measure), if

(ii) the random variable $\xi(t,\omega)-\xi(u,\omega)$ has a distribution $\mu_{q,ln[\chi_{\mathbf{F}}(t-u)]B,\gamma}$ for each $t\neq u\in T$, where $\chi_{\mathbf{F}}:\mathbf{F}\to\mathbf{T}$ is a continuous character of $\mathbf{F}$ as the additive group (see §2.5 [81]).

7. Proposition. *For each given q-Gaussian measure a non-archimedean q-Brownian motion process with real* (**F** *respectively*) *time exists.*

Proof. In view of Proposition 5 for each $t>u>b$ a random variable $\xi(t,\omega)-\xi(b,\omega)$ has a distribution $\mu_{q,(t-b)B,\gamma}$ for real time parameter. If t, u, b are pairwise different points in **F**, then $\xi(t,\omega)-\xi(b,\omega)$ has a distribution $\mu_{q,ln[\chi_{\mathbf{F}}(t-b)]B,\gamma}$, since

$$ln[\chi_{\mathbf{F}}(t-u)]+ln[\chi_{\mathbf{F}}(u-b)]=ln[\chi_{\mathbf{F}}(t-b)].$$

This induces the Markov quasi-measure $\mu^{(q)}_{x_0,\tau}$ on $(\prod_{t\in T}(H_t,\mathsf{U}_t))$, where $H_t=H$ and $\mathsf{U}_t=Bco(H)$ for each $t\in T$. In view of Theorem 2.2.39 [81] there exists an abstract probability space $(\Omega,\mathsf{F},\lambda)$, consequently, the corresponding space $L(\Omega,\mathsf{F},\lambda,\mathbf{K_s})$ exists.

Above non-archimedean analogs of Gaussian measures with specific properties were defined. Nevertheless, there do not exist any usual Gaussian $\mathbf{K_s}$-valued measures on non-archimedean Banach spaces (see Theorem 19 in [81]).

8. Theorem. *Let $\mu_{q,B,\gamma}$ and $\mu_{q,B,\delta}$ be two q-Gaussian $\mathbf{K_s}$-valued measures. Then $\mu_{q,B,\gamma}$ is equivalent to $\mu_{q,B,\delta}$ or $\mu_{q,B,\gamma}\perp\mu_{q,B,\delta}$ according to $v^s_q(\gamma-\delta)\in B^{1/2}(\mathsf{D}_{B,Y})$ or not. The measure $\mu_{q,B,\gamma}$ is orthogonal to $\mu_{g,B,\delta}$, when $q\neq g$. Two measures $\mu_{q,B,\gamma}$ and $\mu_{g,A,\delta}$ with positive definite nondegenerate A and B are either equivalent or orthogonal.*

9. Theorem. *The measures $\mu_{q,B,\gamma}$ and $\mu_{q,A,\gamma}$ are equivalent if and only if there exists a positive definite bounded invertible operator T such that $A=B^{1/2}TB^{1/2}$ and $T-I\in L_2(Y^*)$.*

The proof of the latter two theorems is given in Section 2.6 [81]. Other results including Feynman's quasi-measures ana applications to partial differential equations also are described there.

3.6.2. Brownian Motion Controlled by Real Valued Measures

1. Remark. Let $H=c_0(\alpha,\mathbf{K})$ be a Banach space over a local field $\mathbf{K}$ of zero or of positive characteristic. We take U^{P} be a cylindrical algebra generated by projections on finite-dimensional over $\mathbf{K}$ subspaces F in H and with bases of cylinders in Borel σ-algebras $Bf(F)$. We denote by U the minimal σ-algebra $\sigma(\mathsf{U}^{\mathsf{P}})$ generated by U^{P}.

Let us consider continuous real or complex-valued functions on the field $\mathbf{K}$, whose Fourier transform has the form:

$$\hat{f}(x)=\hat{f}_{\beta,\gamma,q}(x):=exp(-\beta|x|^q)\chi_\gamma(x), \tag{1}$$

where $\gamma\in\mathbf{K}$, $0<\beta<\infty$, $0<q<\infty$. Here $\chi_\gamma:\mathbf{K}\to S^1:=\{z\in\mathbf{C}:\ |z|=1\}$ is a continuous character of the field $\mathbf{K}$ as the additive group $\chi_\gamma(x)=\chi_1(\gamma x)$ (see above).

We recall that the Fourier transform is defined in accordance with the formula:

$$\hat{f}(x):=F[f](x):=\int_{\mathbf{K}}\chi(xy)f(y)\lambda(dy), \tag{2}$$

where λ denotes the non-negative Haar measure on $\mathbf{K}$ as the additive group, $\lambda(B(\mathbf{K}, 0,1)) = 1$, $\chi = \chi_1$ (see above or §7 [110] and [99]). For any $f \in L^1(\mathbf{K},\lambda,\mathbf{C})$ the integral in Formula (2) converges. On the Hilbert space the Fourier transform is defined with the help of the limit

$$\hat{f}(x) := \lim_{R\to\infty} \int_{B(\mathbf{K},0,R)} \chi(xy) f(y) \lambda(dy). \tag{3}$$

The Fourier transform is the linear isomorphism of the Hilbert space $L^2(\mathbf{K},\lambda,\mathbf{C})$ onto itself and the Parseval-Steklov formula

$$\int_{\mathbf{K}} \hat{f}(y)\bar{\hat{g}}(y)\lambda(dy) = \int_{\mathbf{K}} f(x)\bar{g}(x)\lambda(dx) \tag{4}$$

is fulfilled for all functions $f,g \in L^2(\mathbf{K},\lambda,\mathbf{C})$, where $\bar{g}$ denotes the complex conjugate function.

Let X be a locally $\mathbf{K}$-convex space equal to a projective limit $\lim\{X_j, \phi^j_l, \Upsilon\}$ of Banach spaces over a local field $\mathbf{K}$ such that $X_j = c_0(\alpha_j, \mathbf{K})$, where Υ is an ordered set, $\phi^j_l : X_j \to X_l$ is a $\mathbf{K}$-linear continuous mapping for each $j \ge l \in \Upsilon$, $\phi_j : X \to X_j$ is a projection on X_j, $\phi_l \circ \phi^j_l = \phi_j$ for each $j \ge l \in \Upsilon$, $\phi^l_k \circ \phi^j_l = \phi^j_k$ for each $j \ge l \ge k$ in Υ. Consider also a locally $\mathbf{R}$-convex space, that is a projective limit $Y = \lim\{l_2(\alpha_j,\mathbf{R}), \psi^j_l, \Upsilon\}$, where $l_2(\alpha_j,\mathbf{R})$ is the real Hilbert space of the topological weight $w(l_2(\alpha_j,\mathbf{R})) = card(\alpha_j)\aleph_0$. We take a symmetric non-negative definite (bilinear) nonzero functional $B : Y^2 \to \mathbf{R}$.

2. Definitions and Notes. A measure $\mu = \mu_{q,B,\gamma}$ on X with values in $\mathbf{R}$ is called a q-Gaussian measure, if its characteristic functional $\hat{\mu}$ has the form

$$\hat{\mu}(z) = \exp[-B(v_q(z), v_q(z))]\chi_\gamma(z)$$

on a dense $\mathbf{K}$-linear subspace $\mathsf{D}_{q,B,X}$ in X^* of all continuous $\mathbf{K}$-linear functionals $z : X \to \mathbf{K}$ of the form $z(x) = z_j(\phi_j(x))$ for each $x \in X$ with $v_q(z) \in \mathsf{D}_{B,Y}$, where B is a nonnegative definite bilinear $\mathbf{R}$-valued symmetric functional on a dense $\mathbf{R}$-linear subspace $\mathsf{D}_{B,Y}$ in Y^*, $B : \mathsf{D}^2_{B,Y} \to \mathbf{R}$, $j \in \Upsilon$ may depend on z, $z_j : X_j \to \mathbf{K}$ is a continuous $\mathbf{K}$-linear functional such that $z_j = \sum_{k\in\alpha_j} e^k_j z_{k,j}$ is a countable convergent series such that $z_{k,j} \in \mathbf{K}$, e^k_j is a continuous $\mathbf{K}$-linear functional on X_j. Moreover, they satisfy the condition

$$e^k_j(e_{l,j}) = \delta^k_l,$$

where δ^k_l denotes the Kroneker delta symbol, $e_{l,j}$ is the standard orthonormal (in the non-archimedean sense) basis in the Banach space $c_0(\alpha_j,\mathbf{K})$,

$$v_q(z) = v_q(z_j) := \{|z_{k,j}|^{q/2}_{\mathbf{K}} : k \in \alpha_j\}.$$

It is supposed that z is such that $v_q(z) \in l_2(\alpha_j,\mathbf{R})$, where q is a positive constant. Here $\chi_\gamma(z) = \chi_1(z(\gamma))$, $\gamma \in X$, $\chi_\gamma : X \to S^1$ is a character of X as the additive group.

If Y is a Hilbert space with a scalar product $(*,*)$, then due to the Riesz theorem there exists $E \in L(Y)$ such that $B(y_1,y_2) = (Ey_1,y_2)$ for each y_1, $y_2 \in Y$. A symmetric non-negative definite operator E (or sometimes the corresponding B) is called a correlation operator of a measure μ.

3. Proposition. *A q-Gaussian measure on X is σ-additive on some σ-algebra* A *of subsets of X. Moreover, a correlation operator B is of class* L_1*, that is,* $Tr(B) < \infty$*, if and only if each finite dimensional over* $\mathbf{K}$ *projection of* μ *is a σ-additive q-Gaussian Borel measure.*

4. Corollary. *A q-Gaussian measure* μ *from Proposition 6 with* $Tr(B) < \infty$ *is quasi-invariant and pseudo-differentiable for some* $b \in \mathbf{C}$ *relative to a dense subspace* $J_\mu \subset M_\mu = \{a \in X : v_q(a) \in E^{1/2}(Y)\}$*. Moreover, if B is diagonal, then each one-dimensional projection* μ^g *has the following characteristic functional:*

$$(i) \quad \hat{\mu}^g(h) = \exp\Big(- \Big(\sum_j \beta_j |g_j|^q\Big) |h|^q \Big) \chi_{g(\gamma)}(h),$$

where $g = (g_j : j \in \omega_0) \in c_0(\omega_0, \mathbf{K})^*$.

5. Corollary. *Let X be a complete locally* $\mathbf{K}$*-convex space of separable type over a local field* $\mathbf{K}$*, then for each constant* $q > 0$ *there exists a nondegenerate symmetric positive definite operator* $B \in L_1$ *such that a q-Gaussian measure is σ-additive on* $Bf(X)$ *and each its one dimensional over* $\mathbf{K}$ *projection is absolutely continuous relative to the nonnegative Haar measure on* $\mathbf{K}$.

6. Proposition. *Let* $\mu_{q,B,\gamma}$ *and* $\mu_{q,E,\delta}$ *be two q-Gaussian measures with correlation operators B and E of class* L_1*, then there exists a convolution of these measures* $\mu_{q,B,\gamma} * \mu_{q,E,\delta}$*, which is a q-Gaussian measure* $\mu_{q,B+E,\gamma+\delta}$.

The proofs of the latter propositions and corollaries are given in Section 1.6 [81].

7. Remark. Let Z be a compact subset without isolated points in a local field $\mathbf{K}$, for example, $Z = B(\mathbf{K}, t_0, 1)$. Then the Banach space $C^0(Z, \mathbf{K})$ has the Amice polynomial orthonormal base $Q_m(x)$, where $x \in Z$, $m \in \mathbf{N_o} := \{0, 1, 2, \dots\}$ [3]. Suppose $\tilde{P}^{n-1} : C^{n-1}(Z, \mathbf{K}) \to C^n(Z, \mathbf{K})$ are anti-derivations (see above or §80 [103]), where $n \in \mathbf{N}$. Each continuous function $f \in C^0$ has the decomposition

$$f(x) = \sum_m a_m(f) Q_m(x),$$

where $a_m \in \mathbf{K}$. This decomposition establish the isometric isomorphism $\theta : C^0(Z, \mathbf{K}) \to c_0(\omega_0, \mathbf{K})$ such that

$$\|f\|_{C^0} = \max_m |a_m(f)| = \|\theta(f)\|_{c_0}.$$

Since Z is homeomorphic with $\mathbf{Z_p}$, the operator $\tilde{P}^1\tilde{P}^0 : C^0(Z, \mathbf{K}) \to C^2(Z, \mathbf{K})$ is the linear, injective and compact operator so that $\tilde{P}^1\tilde{P}^0 \in L_1$, where $\tilde{P}^j$ here corresponds to $\tilde{P}_{j+1} : C^j \to C^{j+1}$ anti-derivation operator by Schikhof (see above or §§54, 80 [103]). The Banach space $C^2(Z, \mathbf{K})$ is dense in $C^0(Z, \mathbf{K})$. Using Proposition 3 and Corollaries 4, 5 above for $q \geq 1$ we get a q-Gaussian measure on the space $C^0(Z, \mathbf{K})$, where

$$\tilde{P}^1\tilde{P}^0 f = \sum_j \lambda_j P_j f$$

and

$$Jf = \sum_j \zeta_j P_j f$$

for each continuous function $f \in C^0$. We choose

$$|\lambda_j||\pi|^q \leq |\zeta_j|^q \leq |\lambda_j|$$

for each $j \in \mathbf{N}$. Here P_j are projectors, $\lambda_j, \zeta_j \in \mathbf{K}$ are numbers, $p^{-1} \leq |\pi| < 1$, $\pi \in \mathbf{K}$ and $|\pi|$ is the generator of the normalization group of $\mathbf{K}$.

If $H = c_0(\omega_0, \mathbf{K})$, then the Banach space $C^0(Z,H)$ is isomorphic with the tensor product $C^0(Z,\mathbf{K}) \otimes H$ (see §4.R [99]). Therefore, the anti-derivation operator $\tilde{P}^n$ on $C^n(Z,\mathbf{K})$ induces the anti-derivation operator $\tilde{P}^n$ on the Banach space $C^n(Z,H)$. If $J_i \in L_q(Y_i)$, then $J := J_1 \otimes J_2 \in L_q(Y_1 \otimes Y_2)$ (see also Theorem 4.33 [99]). Put $Y_1 = C^0(Z,\mathbf{K})$ and $Y_2 = H$, then each tensor product of operators of the form $J := J_1 \otimes J_2 \in L_q(Y_1 \otimes Y_2)$ induces the q-Gaussian measure μ on $C^0(Z,H)$ such that $\mu = \mu_1 \otimes \mu_2$, where μ_i are q-Gaussian measures on Y_i induced by J_i as above. In particular for $q = 1$ we also can take $J_1 = \tilde{P}^1\tilde{P}^0$.

The 1-Gaussian measure on $C^0(Z,H)$ induced by $J = J_1 \otimes J_2 \in L_1$ with $J_1 = \tilde{P}^1\tilde{P}^0$ we call standard. Analogously considering the following Banach subspace $C^0_0(Z,H) := \{f \in C^0(Z,H) : f(t_0) = 0\}$ and operators $J := J_1 \otimes J_2 \in L_1(C^0_0(Z,\mathbf{K}) \otimes H)$ we get the 1-Gaussian measures μ on it also, where $t_0 \in Z$ is a marked point. Certainly, we can take others operators $J_1 \in L_q(Y_1)$ not related with the anti-derivation as above.

8. We define a (non-archimedean) Brownian motion $w(t,\omega)$ with values in H as a stochastic process such that:

(i) the random variable $w(t,\omega) - w(u,\omega)$ has a distribution $\mu^{F_{t,u}}$, where μ is a probability Gaussian measure on $C^0(T,H)$ described in §§1,2 above.

9. If μ is the standard Gaussian measure on the Banach space $C^0_0(T,H)$, then Brownian motion is called standard (see also Chapter 1 Theorem 3.23, Lemmas 2.3, 2.5, 2.8 and §3.30 in [81]). For $B = I$ and $\gamma = 0$ and $q = 2$, the Brownian motion w is called Wiener's process. Sometimes Brownian motion is called also a Gaussian or Wiener or diffusion process.

10. Proposition. *For each given q-Gaussian measure non-archimedean q-Brownian motion process with real* (**F** *respectively*) *time exists.*

This proposition is proved in Section 1.6 [81] together with other results including Feynman's quasi-measures ana applications to partial differential equations.

3.7. Non-Archimedean Stochastic Anti-Derivational Equations

1. Remark. In Section 6 of this chapter the non-archimedean analogs of Itô's formula were proved. In the particular case $H = \mathbf{K}$ we have $a \in L^s(\Omega, \mathsf{F}, \lambda; C^0(T,\mathbf{K}))$, $E \in L^r(\Omega, \mathsf{F}, \lambda; C^0(T,\mathbf{K}))$, $f \in C^n(T \times \mathbf{K}, Y)$ and $w \in L^q(\Omega, \mathsf{F}, \lambda; C^0_0(T,\mathbf{K}))$ are functions (see Section 6), so that

$$\hat{P}_{u^{b+m-l}, w(u,\omega)^l}[(\partial^{m+b} f/\partial u^b \partial x^m)(u, \xi(u,\omega)) \circ (I^{\otimes b} \otimes a^{\otimes(m-l)} \otimes E^{\otimes l})]|_{u=t}$$

$$= \sum_j (\partial^{m+b} f/\partial u^b \partial x^m)(t_j, \xi(t_j,\omega))[t_{j+1} - t_j]^{b+m-l} a(t_j,\omega)^{k-l}$$

$$\times [E(t_j,\omega)(w(t_{j+1},\omega) - w(t_j,\omega))]^l$$

for each $m+b \leq n$, where $t_j = \sigma_j(t)$, $a(t,\omega)$, $E(t,\omega)$ and $w(t,\omega) \in \mathbf{K}$, that is a,E,w commute. In particular we get

$$\tilde{P}^m_{u,0} f(u) = \sum_{k=1}^{m} (k!)^{-1} \hat{P}_{u^k} f^{(k)}(u),$$

that is $\tilde{P}^m_{u,0} f(u)|_{u=t} = \tilde{P}_{m+1} f'(t)$, where $\tilde{P}_{m+1} : C^m(T,\mathbf{K}) \to C^{m+1}(T,\mathbf{K})$ is the Schikhof linear continuous anti-derivation operator, $m \in \mathbf{N}$ for $char(\mathbf{K}) = 0$, $m < p$ for $char(\mathbf{K}) = p > 0$.

In the non-archimedean case the mean value formula $M[(\int_S^T \phi(t,\omega) dB_t(\omega))^2] = M[\int_S^T \phi(t,\omega)^2 dt]$ (see Lemma 3.5 [89]) is not valid, but there exists its another analog. Let X be a locally compact Hausdorff space and let $BC_c(X,H)$ denote a subspace of the space $C^0(X,H)$ consisting of bounded continuous functions f such that for each $\varepsilon > 0$ there exists a compact subset $V \subset X$ for which

$$\|f(u)\|_H < \varepsilon$$

for each $u \in X \setminus V$. In particular for $X \subset \mathbf{K}$, $e^* \in H^*$ and a fixed $t \in X$ in accordance with Theorem 7.22 [99] there exists a $\mathbf{K}$-valued tight measure $\mu_{t,\omega,e^*,b,k}$ on the σ-algebra $Bco(X)$ of clopen subsets in X so that

$$e^* \hat{P}_{u^b,w^k} \psi(u,x,\omega) \circ (I^{\otimes b} \otimes E^{\otimes k})|_{u=t} = \int_X \psi(u, E(u,\omega) w(u,\omega), \omega) \mu_{t,\omega,e^*,b,k}(du)$$

for each $\psi \in L^r(\Omega, \mathsf{F}, \lambda; BC_c(X, L_k(H^{\otimes k}, H)))$ and $E \in L^q(\Omega, \mathsf{F}, \lambda; BC_c(X, L(H)))$, where H^* is a topologically adjoint space, $1 \leq r,q \leq \infty$, $1/r + 1/q \geq 1$.

If $\chi_\gamma : \mathbf{K} \to S^1 := \{z \in \mathbf{C} : |z| = 1\}$ is a continuous character of $\mathbf{K}$ as the additive group, then we infer the formula for the mean value:

$$M\chi_\gamma((e^* \hat{P}_{u^b,w^k} \psi(u,x,\omega) \circ (I^{\otimes b} \otimes E^{\otimes k})|_{u=t})^l)$$

$$= \prod_j M\chi_\gamma((e^* \psi(t_j,x,\omega)[t_{j+1} - t_j]^b \circ (1^{\otimes b} \otimes (E(t_j,\omega)[w(t_{j+1},\omega) - w(t_j,\omega)])^{\otimes k})^l)$$

due to Condition 6.6(i). For ψ independent of x, $l = 1$, $k = 2$, $b = 0$, $E = 1$ and $H = \mathbf{K}$ (so that $e^* = 1$) the latter formula takes a simpler form, which can be considered as another analog of the classical Itô's formula. For the evaluation of appearing integrals tables from §1.5.5 [110] can be used. Another important result in this area is the following theorem.

2. Theorem. *Let $\psi \in L^2(\Omega, \mathsf{F}, \lambda; C^0(T, L(H)))$, $w \in L^2(\Omega, \mathsf{F}, \lambda; C^0_0(T,H))$ be the random function with values in the Banach space H over a local field $\mathbf{K}$. Then there exists a function $\phi \in C^0(T,H)$ such that the mean value is:*

$$M\chi_\gamma(g\hat{P}_{w(u,\omega)} \psi(u,\omega) \circ I|_{u=t}) = \hat{\mu}(\gamma g \hat{P}_u \phi(u)|_{u=t})$$

for each $\gamma \in \mathbf{K}$ and each $t \in T$ and for each $g \in H^*$.

Proof. Take a point $t \in T$ and calculate values $t_j = \sigma_j(t)$, where σ_j is the approximation of the identity in T, $F_{a,b}(w) := w(a,\omega) - w(b,\omega)$ for $a,b \in T$ (see above). In view of

Conditions 6.6(i,ii) and the Hahn-Banach theorem (since the field is locally compact, see [99]) there exists a projection operator Pr_g so that

$$\hat{\mu}^{(F_{a,b}gE)}(h) = \hat{\mu}^{(F_{a,b}Pr_g)}(Pr_gEh).$$

Indeed,

$$F_{a,b}ghEw = ghE(w(a,\omega) - w(b,\omega)) = hgEF_{a,b}w$$

for each $a,b \in T$ and for each $h \in \mathbf{K}$, where $\hat{\mu}$ denotes the characteristic functional of the measure μ corresponding to w. That is,

$$\hat{\mu}(g) := \int_{C^0_0(T,H)} \chi_g(y)\mu(dy),$$

where $g \in C^0_0(T,H)^*$, $\chi_g : C^0_0(T,H) \to \mathbf{C}$ is the character of $C^0_0(T,H)$ as the additive group, $E \in L(H)$, $y \in C^0_0(T,H)$, μ is the Borel measure on $C^0_0(T,H)$. The random variable $E(w(a,\omega) - w(b,\omega))$ has the distribution $\mu^{F_{a,b}E}$ for each $a \neq b \in T$ and the linear operator $E \in L(H)$. On the other hand, the projection operator Pr_e commutes with the anti-derivation operator $\hat{P}_u$ on $C^0(T,H)$, where $(Pr_ef)(t) := Pr_ef(t)$ is defined point-wise for each continuous function $f \in C^0(T,H)$. In the Hilbert space $L^2(\Omega,\mathsf{F},\lambda;C^0(T,H))$ the family of step functions

$$f(t,\omega) = \sum_{j=1}^{n} Ch_{U_j}(\omega)f_j(t)$$

is dense, where $f_j \in C^0(T,H)$, Ch_U denotes the characteristic function of $U \in \mathsf{F}$, $n \in \mathbf{N}$, since $\lambda(\Omega) = 1$ and λ is nonnegative.

For each $t \in T$ the limit $\lim_{j\to\infty} \psi(t_j,\omega).(w(t_{j+1},\omega) - w(t_j,\omega))$ exists in the Hilbert space $L^2(\Omega,\mathsf{F},\lambda;H)$ (see Theorem 5.14 of this chapter).

If $A \in L(H)$ is a continuous linear operator, then

$$(i) \quad \chi_\gamma((g_1+g_2)Az) = \chi_\gamma(g_1Az)\chi_\gamma(g_2Az)$$

for each $g_1,g_2 \in H^*$ and $z \in H$,

$$(ii) \quad \chi_\gamma(gA(z_1+z_2)) = \chi_\gamma(gAz_1)\chi_\gamma(gAz_2)$$

for each $g \in H^*$ and $z_1,z_2 \in H$,

$$(iii) \quad \chi_\gamma(agAz) = [\chi_\gamma(gAz)]^{\zeta(a)}$$

for each $\{(e,\gamma gAz)\}_p \neq 0$ and $a \in \mathbf{K}$, where $\zeta(a) := \{(e,\gamma agAz)\}_p/\{(e,\gamma gAz)\}_p$. On the other hand the linear continuous operator A is completely defined by the family of values $\{e^*_iAe_j :\ i,j \in \alpha\}$, where $H = c_0(\alpha,\mathbf{K})$, $e^*_i(e_j) = \delta_{i,j}$, $e^*_i \in H^*$, $\{e_j : j \in \alpha\}$ is the standard orthonormal base of H. Hence the family $\{\chi_\gamma(ae^*_iAe_j) :\ i,j \in \alpha; a \in \mathbf{K}\}$ completely characterizes the linear continuous operator $A \in L(H)$ due to Equations $(i-iii)$, when $\gamma \neq 0$.

For each vector $y \in H$ and each number $\gamma \in \mathbf{K}$ the mean value function $M\chi_\gamma(g\psi(t,\omega)y)$ is continuous by the time variable $t \in T$, consequently, there exists a continuous function $\phi : T \to H$ such that the mean value satisfies the equality:

$$M\chi_\gamma(g\psi(t,\omega)y) = \chi_\gamma(g\phi(t)y)$$

for each vector $y \in H$ and $t \in T$. Indeed, the characters χ_γ are continuous from the field $\mathbf{K}$ as the additive group into the complex field $\mathbf{C}$ and $\chi_\gamma(h) = \chi_1(\gamma h)$ for each $0 \neq \gamma \in \mathbf{K}$ and $h \in \mathbf{K}$. Moreover, the $\mathbf{C}$-linear span of the family $\{\chi_\gamma : \gamma \in \mathbf{K}\}$ of characters is dense in the space $C^0(\mathbf{K}, \mathbf{C})$ by the Stone-Weierstrass theorem [26]. On the other hand, the limit

$$\lim_{j\to\infty} \chi_\gamma\left(\sum_{i=0}^{j} a_j\right) = \prod_{i=1}^{\infty} \chi_\gamma(a_i)$$

exists, when $\lim_j a_j = 0$ for a sequence a_j in $\mathbf{K}$. Therefore, the mean value can be calculated by the formula:

$$M\chi_\gamma\left(g\sum_{j=0}^{\infty} \psi(t_j,\omega).[w(t_{j+1},\omega) - w(t_j,\omega)]\right) = \prod_{j=0}^{\infty} \hat{\mu}(\gamma g \phi(t_j)(t_{j+1} - t_j))$$

$$= \hat{\mu}(\gamma g \hat{P}_u \phi(u)|_{u=t}) \quad \text{for each } t \in T \text{ and each } g \in H^*.$$

From the equality $\chi_{a+b}(c) = \chi_a(c)\chi_b(c)$ for each a, b and $c \in \mathbf{K}$ the statement of this theorem follows for each number $\gamma \in \mathbf{K}$.

3. Theorem. *Let $a \in L^q(\Omega, \mathsf{F}, \lambda; C^0(B_R, L^q(\Omega, \mathsf{F}, \lambda; C^0(B_R, H))))$ and $E \in L^r(\Omega, \mathsf{F}, \lambda; C^0(B_R, L(L^q(\Omega, \mathsf{F}, \lambda; C^0(B_R, H)))))$, $a = a(t,\omega,\xi)$, $E = E(t,\omega,\xi)$, $t \in B_R$, $\omega \in \Omega$, $\xi \in L^q(\Omega, \mathsf{F}, \lambda; C^0(B_R, H))$ and $\xi_0 \in L^q(\Omega, \mathsf{F}, \lambda; H)$, and $w \in L^s(\Omega, \mathsf{F}, \lambda; C^0_0(B_R, H))$ with parameters $1/r + 1/s = 1/q$, $1 \le r, s, q \le \infty$, where H is over a local field $\mathbf{K}$ of zero char$(\mathbf{K}) = 0$ or of the positive characteristic char$(\mathbf{K}) = p \ge 2$, the vector a and operator E random fields satisfy the local Lipschitz condition:*

(LLC) for each $0 < r < \infty$ there exists $K_r > 0$ such that

$$\max(\|a(t,\omega,x) - a(t,\omega,y)\|, \|E(t,\omega,x) - E(t,\omega,y)\|) \le K_r \|x - y\|$$

for each $x, y \in B(C^0(B_R, H), 0, r)$ and $t \in B_R$, $\omega \in \Omega$. Then the stochastic equation of the following type:

$$(i) \quad \xi(t,\omega) = \xi_0(\omega) + (\hat{P}_u a)(u,\omega,\xi)|_{u=t} + (\hat{P}_{w(u,\omega)} E)(u,\omega,\xi)|_{u=t}$$

has the unique solution.

4. Theorem. *Let $a \in L^\infty(\Omega, \mathsf{F}, \lambda; C^0(B_R, L^q(\Omega, \mathsf{F}, \lambda; C^0(B_R, H))))$ and $E \in L^\infty(\Omega, \mathsf{F}, \lambda; C^0(B_R, L(L^q(\Omega, \mathsf{F}, \lambda; C^0(B_R, H)))))$, $a = a(t,\omega,\xi)$, $E = E(t,\omega,\xi)$, $t \in B_R$, $\omega \in \Omega$, $\xi \in L^q(\Omega, \mathsf{F}, \lambda; C^0(B_R, H))$ and $\xi_0 \in L^q(\Omega, \mathsf{F}, \lambda; H)$, $w \in L^\infty(\Omega, \mathsf{F}, \lambda; C^0_0(B_R, H))$, $1 \le q \le \infty$, where a local field $\mathbf{K}$ is of zero characteristic char$(\mathbf{K}) = 0$, the vector a and operator E random fields satisfy the local Lipschitz condition (see 3(LLC)). The stochastic equation of the type*

$$(i) \quad \xi(t,\omega) = \xi_0(\omega)$$

$$+ \sum_{m+b=1}^{\infty} \sum_{l=0}^{m} (\hat{P}_{u^{b+m-l}, w(u,\omega)^l}[a_{m-l+b,l}(u,\xi(u,\omega)) \circ (I^{\otimes b} \otimes a^{\otimes(m-l)} \otimes E^{\otimes l})])|_{u=t}$$

such that $a_{m-l,l} \in C^0(B_{R_1} \times B(L^q(\Omega, \mathsf{F}, \lambda; C^0(B_R, H)), 0, R_2), L_m(H^{\otimes m}; H))$ *(continuous and bounded on its domain) for each* n, l, $0 < R_2 < \infty$ *and*

$$(ii) \quad \lim_{n\to\infty} \sup_{0\le l\le n} \|a_{n-l,l}\|_{C^0(B_{R_1}\times B(L^q(\Omega,\mathsf{F},\lambda;C^0(B_R,H)),0,R_2),L_n(H^{\otimes n},H))} = 0$$

for each $0 < R_1 \le R$ *when* $0 < R < \infty$, *or each* $0 < R_1 < R$ *when* $R = \infty$, *for each* $0 < R_2 < \infty$. *Then the stochastic equation* (i) *has the unique solution in* B_R.

Proof of Theorem 4. We have the estimate

$$\max(\|a(x) - a(y)\|^g, \|E(x) - E(y)\|^g) \le K\|x - y\|^g,$$

consequently,

$$\max(\|a(x)\|^g, \|E(x)\|^g) \le K_1(\|x\|^g + 1)$$

for each vectors $x, y \in H$ and for each number $1 \le g < \infty$ and each time variable $t \in B_R$ and each $\omega \in \Omega$, where K and K_1 are positive constants, $a(x)$ and $E(x)$ are short notations of the vector $a(t, \omega, x)$ and operator $E(t, \omega, x)$ random fields for $x = \xi(t, \omega)$ respectively. To solve the problem we organize the iteration procedure:

$$X_0(t) = x, \ \dots,$$

$$X_n(t) = x + \sum_{m+b=1}^{\infty} \sum_{l=0}^{m} (\hat{P}_{u^{b+m-l}, w(u,\omega)^l}[a_{m-l+b,l}(u, X_{n-1}(u,\omega))$$

$$\circ (I^{\otimes b} \otimes a^{\otimes(m-l)} \otimes E^{\otimes l})])|_{u=t},$$

consequently,

$$X_{n+1} - X_n(t) = \sum_{m+b=1}^{\infty} \sum_{l=0}^{m} (\hat{P}_{u^{b+m-l}, w(u,\omega)^l}[a_{m-l+b,l}(u, X_n(u)) - a_{m-l+b,l}(u, X_{n-1}(u))]$$

$$\circ (I^{\otimes b} \otimes a^{\otimes(m-l)} \otimes E^{\otimes l})])|_{u=t},$$

where in general

$$\hat{P}_{a(u,\xi)} 1|_{u=t} = a(t, \xi(t,\omega)) - a(t_0, \xi(t_0,\omega)) \ne \hat{P}_u a(u,\xi) = \sum_j a(t_j, \xi(t_j,\omega))[t_{j+1} - t_j],$$

$t_j = \sigma_j(t)$ for each $j = 0, 1, 2, \dots$. Let $M\eta$ be a mean value of a real-valued distribution $\eta(\omega)$ by $\omega \in \Omega$. Then we deduce the formula:

$$M\|\hat{P}_{u^{b+m-l}, w(u,\omega)^l}[a_{m-l+b,l}(u, X_n(u)) - a_{m-l+b,l}(u, X_{n-1}(u))]|_{(B_{R_1}\times B(L^q,0,R_2))}$$

$$\circ (I^{\otimes b} \otimes a^{\otimes(m-l)} \otimes E^{\otimes l})])|_{u=t}\|^g \le K(M\|\hat{P}_{u^{b+m-l}, w(u,\omega)^l}\|^g)\|a_{m-l+b,l}|_{(B_{R_1}\times B(L^q,0,R_2))}\|^g$$

$$(M \sup_u \|X_n(u) - X_{n-1}(u)\|^g)(M \sup_u \|a\|^{m-l})(M \sup_u \|E\|^l),$$

where $X_n \in C^0_0(B_R, H)$ for each n, $1 \le g < \infty$. On the other hand,

$$X_1(t) = x(t) + \sum_{m+b=1}^{\infty} \sum_{l=0}^{m} (\hat{P}_{u^{b+m-l}, w(u,\omega)^l}[a_{m-l+b,l}(u, x(u)) \circ (I^{\otimes b} \otimes a^{\otimes(m-l)} \otimes E^{\otimes l})])|_{u=t},$$

consequently,

$$\|X_1(t)-X_0(t)\|^g \le \sup_{m,l,b}(\|\hat{P}_{u^{b+m-l},w(u,\omega)^l}[a_{m-l+b,l}(u,x(u))\circ(I^{\otimes b}\otimes a^{\otimes(m-l)}\otimes E^{\otimes l})])|_{u=t}\|^g.$$

Due to Condition (ii) for each $\varepsilon>0$ and $0<R_2<\infty$ there exists a ball $B_\varepsilon\subset B_R$ such that the inequality

$$K\sup_{m,l,b}(\|\hat{P}_{u^{b+m-l},w(u,\omega)^l}|_{B_\varepsilon}[a_{m-l+b,l}(u,*)|_{(B_\varepsilon\times B(L^q,0,R_2))}$$

$$\circ(I^{\otimes b}\otimes a^{\otimes(m-l)}\otimes E^{\otimes l})])\|^g =: c<1$$

is satisfied. Therefore, the unique solution on each ball B_ε exists, since

$$\sup_u \|X_1(u)-X_0(u)\|<\infty \quad \text{and}$$

$$\lim_{l\to\infty} c^l C=0 \quad \text{for each} \quad C>0,$$

consequently, the limit

$$\lim_{n\to\infty} X_n(t)=X(t)=\xi(t,\omega)|_{B_\varepsilon}$$

exists, where

$$C:=M\sup_{u\in B_\varepsilon}\|X_1(u)-X_0(u)\|^g\le(c+1)K<\infty.$$

Here B_ε is an arbitrary ball of radius $\varepsilon>0$ in B_R, $t\in B_\varepsilon$.

If X^1 and X^2 are two solutions, then

$$X^1-X^2=:\psi=\sum_{j=1}^n C_j Ch_{B(\mathbf{K},x_j,r_j)},$$

where $n\in\mathbf{N}$ is a natural number, $C_j\in\mathbf{K}$ is a constant, $T=B_R$, since B_R has a disjoint covering by balls $B(\mathbf{K},x_j,r_j)$. While on each such ball as it was demonstrated above the unique solution exists with a given initial condition on it. That is, in a chosen point x_j such that C_j and $B(\mathbf{K},x_j,r_j)$ are independent of ω. If S is a poly-homogeneous function, then a natural number $n=deg(S)<\infty$ exists so that the differentials $D^mS=0$ are zero for each $m>n$, but its antiderivative $\hat{P}$ has $D^{n+1}\hat{P}S\ne 0$. If $\|S_1\|>\|S_2\|$, then $\|\hat{P}S_1\|>\|\hat{P}S_2\|$, which we can apply to a convergent series considering terms $\|D^m\hat{P}S\|(mod\ p^k)$ for each $k\in\mathbf{N}$. Therefore, the series decomposition

$$\psi=\sum_{m+b=1}^{\infty}\sum_{l=0}^{m}(\hat{P}_{u^{b+m-l},w(u,\omega)^l}[a_{m-l+b,l}(u,X^2)-a_{m-l+b,l}(u,X^1)]$$

$$\circ(I^{\otimes b}\otimes a^{\otimes(m-l)}\otimes E^{\otimes l})])|_{u=t}$$

converges, where the function ψ is locally constant by t and independent of ω. The term $(\Phi^1w)(t_i;1;t_{i+1}-t_i)=[w(t_{i+1})-w(t_i)]/(t_{i+1}-t_i)$ has the infinite-dimensional over $\mathbf{K}$ range in $C^0(B_R^2\setminus\Delta,H)$ for λ-a.e. $\omega\in\Omega$, where $\Delta:=\{(u,u):u\in B_R\}$. In view of Lemmas 5.2 and 5.3 we have $\psi=0$. Indeed, it is evident for $a(u,X)$ and $E(u,X)$ and $a_{k-l,l}(u,X)$ depending on X locally polynomially or poly-homogeneously for each u, but such locally polynomial or poly-homogeneous functions by X are dense in the Banach spaces

$$L^q(\Omega,\mathsf{F},\lambda;C^0(B_R,L^q(\Omega,\mathsf{F},\lambda;C^0(B_R,H)))) \quad \text{and}$$

$$L^q(\Omega,\mathsf{F},\lambda;C^0(B_R,L(L^q(\Omega,\mathsf{F},\lambda;C^0(B_R,H))))) \quad \text{and}$$

$$C^0(B_{R_1}\times B(L^q(\Omega,\mathsf{F},\lambda;C^0(B_R,H)),0,R_2),L_k(H^{\otimes k};H))$$

respectively.

The proof of Theorem 3 is the particular case of the latter proof, since the Banach space H is over a local field $\mathbf{K}$ of zero $char(\mathbf{K})=0$ or of positive characteristic $char(\mathbf{K})=p>2$ so that $m<p$ for $m=1$ in the considered case of the first order anti-derivation stochastic equation.

4.1. Remark. Evidently Theorem 4 also is valid in the case when random operator mappings $a_{m,l}$ are zero for all $m+l\geq p$ when $char(\mathbf{K})=p>0$, since in this case the series terminates by the finite sum of anti-derivation operators of order less than $p-1$ which exist in this case.

5. Proposition. *Let ξ be the Brownian motion process given by Equation 3(i) with the 1-Gaussian measure associated with the operator $\tilde{P}^1\tilde{P}^0$ as in §6.2.7 and let also*

$$\max(\|a(t,\omega,x)-a(v,\omega,x)\|,\|E(t,\omega,x)-E(v,\omega,x)\|)\leq|t-v|(C_1+C_2\|x\|^b)$$

for each t and $v\in B(\mathbf{K},t_0,R)$ λ-almost everywhere by $\omega\in\Omega$, where b, C_1 and C_2 are non-negative constants. Then ξ with probability 1 has a C^2-modification and

$$q(t)\leq\max\{M\|\xi_0\|^s,|t-t_0|(C_1+C_2q(t))\}$$

for each $t\in B(\mathbf{K},t_0,R)$, where

$$q(t):=\sup_{|u-t_0|\leq|t-t_0|} M\|\xi(u,\omega)\|^s$$

and $\mathbf{N}\ni s\geq b\geq 0$.

Proof. For the following function $f(t,x)=x^s$ in accordance with Theorem 6.6 we have the decomposition:

$$f(t,\xi(t,\omega))=f(t_0,\xi_0)$$

$$+\sum_{k=1}^{s}\sum_{l=0}^{k}\binom{k}{l}\left(\hat{P}_{u^{k-l},w(u,\omega)^l}\left[\binom{s}{k}\xi(t,\omega)^{s-k}(u,\xi(u,\omega))\circ(a^{\otimes(k-l)}\otimes E^{\otimes l})\right]\right)|_{u=t}.$$

For this function f we have

$$d(\hat{P}^s_*):=\sup_{a\neq 0,E\neq 0,f\neq 0}\max_{s\geq k\geq l\geq 0}\left\|(k!)^{-1}\binom{k}{l}\hat{P}_{u^{k-l},w^l}(\partial^k f/\partial^k x)\right.$$

$$\left.\circ(a^{\otimes(k-l)}\otimes E^{\otimes l})\right\|/(\|a\|^{k-l}_{C^0(B_R,H)}\|E\|^l_{C^0(B_R,L(H))}\|f\|_{C^s(B_R,H)}),$$

hence $d(\hat{P}^s_*) \le 1$, since $f \in C^s$ as a function by x and $(\bar{\Phi}^s g)(x;h_1,\dots,h_s;0,\dots,0) = D^s_x g(x).(h_1,\dots,h_s)/s!$ for each $g \in C^s$ and due to the definition of the norm $\|g\|_{C^s}$. Therefore, we deduce the inequality

$$M\|\xi(t,\omega)-\xi(v,\omega)\|^s \le |t-v|(1+C_1+C_2 d(\hat{P}^s_*) \sup_{|u-t_0|\le \max(|t-t_0|,|v-t_0|)} M\|\xi(u,\omega)\|^s),$$

since $|t_j-v_j| \le |t-v|+\rho^j$ for each $j \in \mathbf{N}$, where $0<\rho<1$. Hence the mean value can be estimated from above as:

$$M\|\xi(t,\omega)\|^s \le \max(M\|\xi_0\|^s, |t-t_0| d(\hat{P}^s_*)(C_1+C_2 \sup_{|u-t_0|\le|t-t_0|} M\|\xi(u,\omega)\|^s),$$

since $|t_j-t_0| \le |t-t_0|$ for each $j \in \mathbf{N}$. Considering in particular poly-homogeneous g on which $d(\hat{P}^s_*)$ takes its maximum value we get $d(\hat{P}^s_*) = 1$. Since $P(C^2)=1$ for the Markov measure P induced by the transition measures $P(v,x,t,S) := \mu^{F_{t,v}}(S|\xi(v)=x)$ for $t \ne v$ of the non-archimedean Brownian motion process, the random function ξ has with the probability 1 the C^2-modification.

6. Note. If consider a general stochastic process as in §3.10, then from the proof of Theorem 3.11 and Proposition 5 it follows, that the random function ξ with the probability 1 has a modification in the space $J(C^0_0(T,H))$, where J is a nondegenerate correlation operator of the product measure μ on $C^0_0(T,H)$.

7. Proposition. *Let ξ be a stochastic process given by Equation* $3(i)$ *and*

$$\max(\|a(t,\omega,x_1)-a(v,\omega,x_2)\|, \|E(t,\omega,x_1)-E(v,\omega,x_2)\|) \le |t-v|(C_1+C_2\|x_1-x_2\|^b)$$

for each t and $v \in B(\mathbf{K},t_0,R)$ λ-almost everywhere by $\omega \in \Omega$, where b, C_1 and C_2 are non-negative constants. Then two solutions ξ_1 and ξ_2 with initial conditions $\xi_{1,0}$ and $\xi_{2,0}$ satisfy the following inequality:

$$y(t) \le \max\{M\|\xi_{1,0}-\xi_{2,0}\|^s, |t-t_0|(C_1+C_2 y(t))\}$$

for each $t \in B(\mathbf{K},t_0,R)$, where

$$y(t) := \sup_{|u-t_0|\le|t-t_0|} M\|\xi_1(u,\omega)-\xi_2(u,\omega)\|^s$$

and $\mathbf{N} \ni s \ge b \ge 0$.

Proof. From §5 it follows, that the mean value satisfies the inequality:

$$M\|\xi_1(t,\omega)-\xi_2(t,\omega)\|^s \le |t-t_0|(C_1+C_2 \sup_{|u-t_0|\le|t-t_0|} M\|\xi_1(u,\omega)-\xi_2(u,\omega)\|^s),$$

since $d(\hat{P}^s_*) \le 1$.

8. Remark. Let us consider two random functions of the form

$$X_t = X_0 + \hat{P}_t a + \hat{P}_w v$$

and $Y_t = Y_0 + \hat{P}_t q + \hat{P}_w s$ corresponding to the unit operator field $E = I$ and a Banach algebra H over $\mathbf{K}$ (see Section 6). Then we get

$$X_u Y_u - X_t Y_t = (X_u - X_t)(Y_u - Y_t) + X_t(Y_u - Y_t) + (X_u - X_t)Y_t,$$

where $u, t \in T$. Hence the differential of their product is given by the equality:

$$d(X_t Y_t) = X_t dY_t + (dX_t)Y_t + (dX_t)(dY_t).$$

Therefore,

$$\hat{P}_{X_t} Y_t = X_t Y_t - X_0 Y_0 - \hat{P}_{Y_t} X_t - \hat{P}_{(X_t, Y_t)} 1,$$

which is the non-archimedean analog of the integration by parts formula, where in all terms X_t is displayed on the left from Y_t. For two C^1 functions f and g we have

$$(fg)' = f'g + fg' \quad \text{or} \quad d(fg) = gdf + fdg,$$

that is terms with $(dt)(dt)$ are absent, consequently, $(dt)(dt) = 0$. In the particular case

$$\hat{P}_{w_t} t = w_t t - \hat{P}_t w_t - \hat{P}_{(t, w_t)} 1.$$

Moreover, we have that

$$\hat{P}_{(t,w_t)} 1 = \sum_j t_j [w_{t_{j+1}} - w_{t_j}] - w_t t + \sum_j w_{t_j} [t_{j+1} - t_j] \neq 0,$$

for example, for $t = 1$, $w \in C^0_0(T, H)$, $T = \mathbf{Z_p}$ and $t_0 = 0$ this gives

$$\hat{P}_{(t,w_t)} 1 = w_1 - w_0 = w_1.$$

Therefore, $(dt)(dw_t) \neq 0$. The latter inequality provides the important difference of the non-archimedean and classical cases (see for comparison Exer. 4.3 and Theorem 4.5 [89]).

If H is a Banach space over the local field $\mathbf{K}$ and $f(x,y) = x^* y$ is a $\mathbf{K}$-bilinear functional on it, where x^* is an image of $x \in H$ under an embedding $H \hookrightarrow H^*$ associated with the standard orthonormal base $\{e_j\}$ in H, then

$$\hat{P}_{X_t^*} Y_t = X_t^* Y_t - X_0^* Y_0 - \hat{P}_{Y_t^*} X_t - \hat{P}_{(X_t^*, Y_t)} 1,$$

hence

$$d(X_t^* Y_t) = X_t^* dY_t + (dX_t^*)Y_t + (dX_t^*)(dY_t) \quad \text{and}$$
$$d(w^* w) = w^* dw + (dw^*)w + (dw^*)(dw).$$

9. Definition. If $\xi(t,\omega) \in L^q(\Omega, \mathsf{F}, \lambda; C^0(B_R, H)) =: Z$ is a stochastic process and $T(t,s)$ is a family of bounded linear operators satisfying the following Conditions $(i - iv)$:

(i) $T(t,s) : H_s \to H_t$, where $H_s := L^q(\Omega, \mathsf{F}, \lambda; C^0(B(\mathbf{K}, 0, |s|), H))$,

(ii) $T(t,t) = I$,

(iii) $T(t,s)T(s,v) = T(t,v)$ for each $t, s, v \in B_R$,

(iv) $M_s\{\|T(t,s)\eta\|_H^q\} \leq C\|\eta\|_H^q$ for each $\eta \in H_s$,

where C is a positive non-random constant, $1 \leq q \leq \infty$, then $T(t,s)$ is called a multiplicative operator functional of the stochastic process ξ.

If $T(t,s;\omega)$ is a system of random variables on Ω with values in the linear space $L(H)$ of all continuous linear operators from H into H, satisfying almost surely Conditions $(i-iii)$ and uniformly by $t,s \in B_R$ Condition (iv) so that

(v) $(T(t,s)\eta)(\omega) = T(t,s;\omega)\eta(\omega)$, then such multiplicative operator functional is called homogeneous. An operator

$$(vi) \quad A(t) = \lim_{s\to 0}[T(t,t+s)-I]/s$$

is called the generating operator of the evolution family $T(t,v)$. If $T(t,v) = T(t,v;\omega)$ depends on ω, then $A(t) = A(t;\omega)$ is also considered as the random variable on Ω (depending on the parameter ω) with values in the linear space $L(H)$.

10. Remark. Let $A(t)$ be a linear continuous operator on a Banach space Y over the field $\mathbf{K}$ such that it depends strongly continuously on the variable $t \in B(\mathbf{K},0,R)$, that is $A(t)y$ is continuous by t for each chosen $y \in Y$ and $A(t) \in L(Y)$. Then the solution of the differential equation

$$dx(t)/dt = A(t)x(t),\ x(s) = x_0 \text{ has a solution} \tag{1}$$

$$x(t) = U(t,s)x(s), \text{ where } U(t,s) \text{ is a generating operator such that} \tag{2}$$

$$U(t,s) = I + \hat{P}_u A(u)U(u,s)|_{u=s}^{u=t}, \tag{3}$$

though $x(t)$ may be non-unique, where

$$x(s) = x_0 \quad \text{is the initial condition,}$$

$x,t \in B(\mathbf{K},0,R)$. The solution of Equation (3) exists using the method of iterations (see §4 above in this section). Indeed, in view of 6.3 $U(s,s) = I$ and

$$dx(t)/dt = \partial U(t,s)x(s)/\partial t = A(t)U(t,s)x(s) = A(t)x(t). \tag{4}$$

If consider a solution of the anti-derivational equation

$$V(t,s) = I + \hat{P}_u V(t,u)A(u)|_{u=s}^{u=t}, \tag{5}$$

then it is a solution of the Cauchy problem

$$\partial V(t,s)/\partial s = -V(t,s)A(s), \quad V(t,t) = I. \tag{6}$$

Therefore, we have

$$\partial[V(t,s)U(s,v)]/\partial s = -V(t,s)A(s)U(s,v) + V(t,s)A(s)U(s,v) = 0,$$

consequently, $V(t,s)U(s,v)$ is not dependent from s. Thus operators U and V exist such that

$$V(t,s) = U(t,s) \quad \text{for each} \quad t,s \in B(\mathbf{K},0,R). \tag{7}$$

The latter implies, that

$$U(t,s)U(s,u) = U(t,u) \quad \text{for each} \quad s,u,t \in B(\mathbf{K},0,R). \tag{8}$$

In particular, if $A(t) = A$ is a constant operator, then there exists the solution $U(t,s) = EXP((t-s)A)$ (see about the exponential function EXP above or Proposition 45.6 [103]). Equation (3) has a solution under milder conditions, for example, $A(t)$ is weakly continuous, that is $e^*A(t)\eta$ is continuous for each $e^* \in Y^*$ and $\eta \in Y$, then $e^*U(t,s)\eta$ is differentiable by t and $U(t,s)$ satisfies Equation (4) in the weak sense and there exists a weak solution of (5) coinciding with $U(t,s)$. If substitute $A(t)$ on another operator $\tilde{A}(t)$, then for the corresponding evolution operator $\tilde{U}(t,s)$ the following inequality

$$\|\tilde{U}(t,s) - U(t,s)\| \le M\tilde{M} \sup_{u \in B(\mathbf{K},0,R)} \|\tilde{A}(u) - A(u)\| R \tag{9}$$

is fulfilled, where

$$M := 1 + \sup_{s,t \in B(\mathbf{K},0,R)} \|U(t,s)\|$$

and $\tilde{M}$ corresponds to $\tilde{U}$.

11. Proposition. *Let $B(t)$ and two sequences $A_n(t)$ and $B_n(t)$ be given of strongly continuous on $B(\mathbf{K},0,R)$ bounded linear operators and let $\tilde{U}_n(t,s)$ also be evolution operators corresponding to the sums*

$$\tilde{A}_n(t) := A_n(t) + B_n(t), \tag{1}$$

where

$$\sup_{n \in \mathbf{N}, u \in B(\mathbf{K},0,R)} \|B_n(u)\| \le \sup_{u \in B(\mathbf{K},0,R)} \|B(u)\| = C < \infty. \tag{2}$$

If $MCR < 1$, then there exists a sequence $\tilde{U}_n(t,s)$ which is also uniformly bounded. If there exists $U_n(t,s)$ strongly and uniformly converging to $U(t,s)$ in $B(\mathbf{K},0,R)$, then $\tilde{U}_n(t,s)$ also can be chosen strongly and uniformly convergent.

Proof. From the use of Equations 10(3,8) iteratively for $U_n(\sigma_{j+1}(t),\sigma_j(t))$ and $U_n(\sigma_j(t),s)$ and also for $\tilde{U}_n$ and taking the difference of operators $\tilde{U}_n - U_n$ it follows, that

$$\tilde{U}_n(t,s) = U_n(t,s) + \hat{P}_v U_n(t,v) B_n(v) \tilde{U}_n(v,s)|_{v=s}^{v=t} \tag{3}$$

for each $n \in \mathbf{N}$. Therefore,

$$\|\tilde{U}_n(t,s)\| \le M + MC \sup_v \|\tilde{U}_n(v,s)\| R,$$

hence $\|\tilde{U}_n(t,s)\| \le M/[1 - MCR]$, since $MCR < 1$. If $\lim_n x_n = x$ in Y and $U_n(t,s)x$ is uniformly convergent to $U(t,s)x$, then for each positive number $\varepsilon > 0$ there exist a positive number $\delta > 0$ and a natural number $m \in \mathbf{N}$ such that

$$\sup_{t,s \in B(\mathbf{K},0,R)} \|U_n(t+h,s+v)x_n - U_n(t,s)x_n\| < \varepsilon$$

for each $n > m$ and $\max(|h|,|v|) < \delta$ due to Equality (3).

12. Proposition. *Let a, $a_{m-l+b,l}$ and E be the same as in §4 and 4.1. Then Equation 4(i) has the unique solution ξ in B_R for each initial value $\xi(t_0,\omega) \in L^q(\Omega,\mathsf{F},\lambda;H)$ and it can be represented in the following form:*

$$\xi(t,\omega) = T(t,t_0;\omega)\xi(t_0;\omega), \tag{2}$$

where $T(t,v;\omega)$ is the multiplicative operator functional.

Proof. In view of Theorem 4, Definition 9, Remark 10 and Proposition 11 with the use of a parameter $\omega \in \Omega$ the statement of Proposition 12 follows.

13. Let us now consider the case $J(C_0^0(T,H)) \subset C^1(T,H)$ (see §5), for example, the standard Brownian motion process. For a function $f(t,x)$ of two variables t and x we consider partial difference quotient operators by the first $\Phi_t^m f$ and

14. Corollary. *Let a function $f(t,x)$ satisfies conditions of Note 6, where the local field **K** is of zero characteristic. Then a generating operator of an evolution family $T(t,v)$ of the random function $\eta = f(t,\xi(t,\omega))$ is given by the following equation:*

$$A(t)\eta(t) = f_t'(t,\xi(t,\omega)) + f_x'(t,\xi(t,\omega)) \circ a(t,\omega)$$

$$+f_x'(t,\xi(t,\omega)) \circ E(t,\omega)w'_t(t,\omega)$$

$$+ \sum_{m+b\geq 2, 0\leq m\in \mathbf{Z}, 0\leq b\in \mathbf{Z}} ((m+b)!)^{-1} \sum_{l=0}^{m} \binom{m+b}{m}\binom{m}{l}$$

$$\{b(\hat{P}_{u^{b+m-l-1},w(u,\omega)^l}[(\partial^{(m+b)} f/\partial u^b \partial x^m)(u,\xi(u,\omega)) \circ (I^{\otimes(b-1)} \otimes a^{\otimes(m-l)} \otimes E^{\otimes l})])|_{u=t}$$

$$+\{(m-l)(\hat{P}_{u^{b+m-l-1},w(u,\omega)^l}[(\partial^{(m+b)} f/\partial u^b \partial x^m)(u,\xi(u,\omega))$$

$$\circ(I^{\otimes b} \otimes a^{\otimes(m-l-1)} \otimes E^{\otimes l})]a)|_{u=t} + l(\hat{P}_{u^{b+m-l},w(u,\omega)^{l-1}}[(\partial^{(m+b)} f/\partial u^b \partial x^m)(u,\xi(u,\omega))$$

$$\circ(I^{\otimes b} \otimes a^{\otimes(m-l)} \otimes E^{\otimes(l-1)})]Ew'_u(u,\omega))|_{u=t}\}. \qquad (1)$$

Proof. In view of Theorem 6.6 and Proposition 12 a generating operator of an evolution family exists. From Lemma 5.3 and Formula 6.6(ii) the statement of this corollary follows using the partial difference operators over the field **K** (see also Appendices A and C).

15. Remark. If $f(t,x)$ satisfies conditions either of §6.4 or of §6.5, then Formula 14(1) takes simpler forms, since the corresponding terms vanish.

Chapter 4

Random Functions in Manifolds, Their Transition Probabilities

4.1. Introduction

Stochastic processes and stochastic differential equations in real Banach spaces and manifolds on them were intensively studied (see, for example, [6, 15, 16, 32, 33, 38, 39, 85, 86, 89] and references therein). On the other hand, the development of the non-archimedean functional analysis and non-archimedean quantum physical theories and quantum mechanics demands to possess measure theory and stochastic processes on non-archimedean Banach spaces and manifolds on them [40, 42, 43, 99, 103, 110].

It is necessary to mention that the property of having a compact support is unacceptable for quasi-invariance of transition measures on a Banach space either infinite dimensional over a field **K** or when **K** is non locally compact. But the quasi-invariance is one of the main properties to satisfy in many situations, moreover, for wider classes of measures. Then each field that is an infinite extension of a local field and which is complete relative to its uniformity has structure of a locally convex space over a local field, hence it is encompassed in the present investigation.

This chapter presents results on stochastic processes with non-archimedean values and having quasi-invariant transition measures. Moreover, the considered here variant permits to have an analog of Itô's bundle on a non-archimedean manifold. On the other hand, Itô's formula is very useful for random functions in manifolds.

In this chapter non-archimedean stochastic processes and stochastic anti-derivational equations on manifolds modelled on Banach spaces over non-archimedean fields are investigated. Moreover, wider classes of random functions are considered below, for example, of Lèvy's type. Then it is found and proved a non-archimedean analog of Itô's formula.

It is necessary to note that here not only manifolds treated by the rigid geometry, but much wider classes are considered. For them the existence of an exponential mapping is proved. A rigid non-archimedean geometry serves mainly for needs of the cohomology theory on such manifolds, but it is too restrictive and operates with narrow classes of analytic functions [29]. The rigid geometry was introduced at the beginning of sixties of the 20-th century. Few years later wider classes of functions were investigated by Schikhof [103]. In this chapter classes of functions and anti-derivation operators from the preceding chapter are used.

The results of this chapter permit to consider random functions having values in non-archimedean manifolds.

4.2. Stochastic Processes for Non-Archimedean Locally K-convex Spaces

1. Definitions and Notes. A measurable space (Ω, F) with a probability real-valued σ-additive measure λ on a covering σ-algebra F of a set Ω is called a probability space. It is denoted by the triple $(\Omega, \mathsf{F}, \lambda)$. In the case of a complex-valued σ-additive measure λ we suppose, that its variation $|\lambda|$ is a probability real-valued σ-additive measure. The latter is natural, since the variation $|\lambda|$ is the non-negative σ-additive measure.

We recall that each complex (σ-additive) measure λ can be written in the form

$$\lambda = \lambda_0^+ - \lambda_0^- + i(\lambda_1^+ - \lambda_1^-),$$

where $i = \sqrt{-1}$, while λ_0^+, λ_0^-, λ_1^+ and λ_1^- are non-negative (σ-additive) measures so that

$$|\lambda|(A) = \lambda_0^+(A) + \lambda_0^-(A) + \lambda_1^+(A) + \lambda_1^-(A)$$

for any $A \in \mathsf{F}$.

Let T be a set with a covering σ-algebra $\mathcal{R}$ and a σ-additive measure $\eta : \mathcal{R} \to \mathbf{C}$. We consider the following Banach space $L^q(T, \mathcal{R}, \eta, H)$ as the completion of the set of all $\mathcal{R}$-step functions $f : T \to H$ relative to the following norm:

$$\|f\|_{\eta,q} := \left[\int_T \|f(t)\|_H^q |\eta|(dt)\right]^{1/q} \tag{1}$$

for $1 \le q < \infty$ and

$$\|f\|_{\eta,\infty} := ess - \sup_{t \in T, \eta} \|f(t)\|_H, \tag{2}$$

where H is a Banach space over $\mathbf{K}$, $|\eta|$ is a variation of η. For $0 < q < 1$ this is the metric space with the metric

$$\rho_q(f,g) := \left[\int_T \|f(t) - g(t)\|_H^q |\eta|(dt)\right]^{1/q}. \tag{3}$$

More generally one can consider now a complete locally $\mathbf{K}$-convex space H. This space H is linearly topologically isomorphic with a projective limit of Banach spaces

$$H = \lim\{H_\alpha, \pi_\beta^\alpha, \Upsilon\},$$

where Υ is a directed set, $\pi_\beta^\alpha : H_\alpha \to H_\beta$ is a $\mathbf{K}$-linear continuous mapping for each $\alpha \ge \beta$, $\pi_\alpha : H \to H_\alpha$ is a $\mathbf{K}$-linear continuous mapping such that $\pi_\beta^\alpha \circ \pi_\alpha = \pi_\beta$ for each $\alpha \ge \beta$. Each norm p_α on H_α induces a semi-norm $\tilde{p}_\alpha$ on H. If $f : T \to H$, then $\pi_\alpha \circ f =: f_\alpha : T \to H_\alpha$.

In the considered case a space $L^q(T,\mathcal{R},\eta,H)$ is defined as the completion of the family of all step functions $f:T\to H$ relative to the family of semi-norms

$$\|f\|_{\eta,q,\alpha} := \left[\int_T \tilde{p}_\alpha(f(t))^q|\eta|(dt)\right]^{1/q}, \tag{1'}$$

$\alpha\in\Upsilon$, for $1\le q<\infty$ and

$$\|f\|_{\eta,\infty,\alpha} := ess-\sup_{t\in T,\eta} \tilde{p}_\alpha(f(t)), \tag{2'}$$

$\alpha\in\Upsilon$, or pseudo-metrics

$$\rho_{q,\alpha}(f,g) := \left[\int_T \tilde{p}_\alpha((f(t)-g(t))^q|\eta|(dt)\right]^{1/q}, \tag{3'}$$

$\alpha\in\Upsilon$, for $0<q<1$. Thus this locally convex space can be written is the projective limit

$$L^q(T,\mathcal{R},\eta,H) = \lim\{L^q(T,\mathcal{R},\eta,H_\alpha),\pi^\alpha_\beta,\Upsilon\}.$$

For example, T may be a subset of $\mathbf{F}$ or of $\mathbf{R}$, where $\mathbf{F}$ denotes a non-archimedean field.

If T is a zero-dimensional T_1-space, then denote by $C^0_b(T,H)$ the Banach space of all continuous bounded functions $f:T\to H$ supplied with the norm:

$$\|f\|_{C^0} := \sup_{t\in T}\|f(t)\|_H<\infty. \tag{4}$$

Let a set T be such that $card(T)>n$. For $X=C^0_b(T,H)$ or $X=H^T$ we define $X(T,H;(t_1,\dots,t_n);(z_1,\dots,z_n))$ as a closed sub-manifold in X of all functions $f:T\to H$, $f\in X$ such that $f(t_1)=z_1,\dots,f(t_n)=z_n$, where $t_1,\dots,t_n$ are pairwise distinct points in T and $z_1,\dots,z_n$ are points in H. For $X=L^q(T,\mathcal{R},\eta,H)$ and pairwise distinct points $t_1,\dots,t_n$ in $T\cap\operatorname{supp}(|\eta|)$ one defines a set $X(T,H;(t_1,\dots,t_n);\,(z_1,\dots,z_n))$ as a closed sub-manifold which is the completion relative to the metric ρ_q (or a family of pseudo-metrics $\{\rho_{q,\alpha}:\alpha\}$ respectively), where $0<q\le\infty$, of a family of $\mathcal{R}$-step functions $f:T\to H$ such that $f(t_1)=z_1,\dots,f(t_n)=z_n$. In these cases $X(T,H;(t_1,\dots,t_n);(0,\dots,0))$ is the proper $\mathbf{K}$-linear sub-space of $X(T,H)$ such that $X(T,H)$ is isomorphic with $X(T,H;(t_1,\dots,t_n);(0,\dots,0))\oplus H^n$, since if $f\in X$, then $f(t)-f(t_1)=:g(t)\in X(T,H;t_1;0)$. In the third case we use that $T\in\mathcal{R}$ and hence there exists the embedding $H\hookrightarrow X$. For $n=1$ and $t_0\in T$ and $z_1=0$ we use the notation $X_0:=X_0(T,H):=X(T,H;t_0;0)$.

2. Definition. The Borel σ-algebra $Bf(X)$ of a topological space X is the minimal σ-algebra generated by the family of all open subsets in X.

Let B be a σ-algebra on $X(T,H)$ containing the Borel σ-algebra $Bf(X(T,H))$ of $X(T,H)$. Let also a σ-algebra $\mathcal{R}_H$ on H contain the Borel σ-algebra $Bf(H)$ of H.

We define a stochastic process $\xi(t,\omega)$ with values in H as a random variable such that:

(i) the differences $\xi(t_4,\omega)-\xi(t_3,\omega)$ and $\xi(t_2,\omega)-\xi(t_1,\omega)$ are independent for each chosen (t_1,t_2) and (t_3,t_4) with $t_1\ne t_2$, $t_3\ne t_4$, such that neither t_1 nor t_2 is in the two-element set $\{t_3,t_4\}$, when T is a subset in $\mathbf{R}$ we suppose additionally, that $t_1<t_2\le t_3<t_4$, where $\omega\in\Omega$;

(ii) the random variable $\xi(t,\omega)-\xi(u,\omega)$ has a distribution $\mu^{F_{t,u}}$, where μ is a probability complex-valued measure on the measurable space $(X(T,H),\mathsf{B})$ from §1, $\mu^g(A) := \mu(g^{-1}(A))$ for $g: X \to H$ such that $g^{-1}(\mathcal{R}_H) \subset \mathsf{B}$ and for each $A \in \mathcal{R}_H$, a continuous linear operator $F_{t,u}: X \to H$ is given by the formula $F_{t,u}(\xi) := \xi(t,\omega)-\xi(u,\omega)$ for each $\xi \in L^r(\Omega,\mathsf{F},\lambda;X)$, where $0<r\le\infty$, $\mathcal{R}_H$ is a separating covering σ-algebra of H such that $F^{-1}_{t,u}(\mathcal{R}_H) \subset \mathsf{B}$ for each $t \ne u$ in T;

(iii) we also put $\xi(0,\omega)=0$, that is, we consider a **K**-linear subspace $L^r(\Omega,\mathsf{F},\lambda;X_0)$ of $L^r(\Omega,\mathsf{F},\lambda;X)$, where $\Omega \ne \emptyset$, X_0 is the closed subspace of X as in §1.

Therefore, $\xi(t,\omega)$ is a Markov process with the transition measure $P(u,x,t,A) = \mu^{F_{t,u}}(A-x)$.

3. Remark. In Section 4 of Chapter 3 Poisson random functions controlled by non-archimedean $\mathbf{K_s}$ valued measures were considered. In this section we describe them also with values in non-archimedean spaces, but controlled by real-valued transition measures.

4. Definition. Let T be an additive group contained in **R**. Consider a stochastic process $\xi \in L^r(\Omega,\mathsf{F},\lambda,X_0(T,H))$ such that a transition measure has the form

$$P(t_1,x,t_2,A) := P(t_2-t_1,x,A) := \exp(-\rho(t_2-t_1))P(A-x)$$

(see §2 above) for each $x \in H$ and $A \in \mathcal{R}_H$ and t_1 and t_2 in T, where $\rho>0$ is a positive constant. Then such random function is called the Poisson process.

5. Proposition. *Let*

$$P(A-x) = \int_H P(-x+dy)P(A-y)$$

for each $x \in H$ and $A \in \mathcal{R}_H$, where P is a probability measure on H and T is an interval in **R**, *then there exists a measure μ on $X_0(T,H)$ for which the Poisson process exists.*

Proof. Consider separating covering σ-algebras $\mathcal{R}_t = \mathcal{R}_H$ on $t \times H$ for each $t \in T$. We take measures

$$\mu_{t_1,\dots,t_n} := P(t_2-t_1,0,*)\dots P(t_n-t_{n-1},0,*)$$

on the σ-algebras $\mathcal{R}_{t_1,\dots,t_n} := \mathcal{R}_{t_1}\times\dots\times\mathcal{R}_{t_n}$ for each pairwise distinct points $t_1,\dots,t_n \in T$, where $\mu_{t_1,\dots,t_n} = \pi_{t_1,\dots,t_n}(\mu)$ and $\pi_{t_1,\dots,t_n}: X_0(T,H) \to H_{t_1}\times\dots\times H_{t_n}$ is the natural linear projection operator, $H_t = H$ for each $t \in T$. There is a family Λ of all finite subsets of T directed by inclusion.

If H is not a separable metrizable complete space, we take a locally compact subfield **S** in the normed field **K** complete relative to its norm, since H is the Banach space. Then we can embed the finite dimensional space $\mathbf{S}^n$ over **S** into H.

The normed field **K** can be considered as the complete linear space over **S**. The field **S** is locally compact, consequently, **K** as the Banach space over **S** is isomorphic with $c_0(\alpha,\mathbf{S})$ (see Theorem 5.13 [99]). Therefore, there are continuous **S**-linear projection operators $\kappa: \mathbf{K} \to \mathbf{S}$. The space H is either Banach or the projective limit of Banach spaces over **K**, consequently, over the subfield **S** as well. Therefore, we can take $\kappa_n: H \to \mathbf{S}^n$ as the **S**-linear continuous projection operator. Thus $\kappa_n^{-1}(A)$ is open in H for each A open in $\mathbf{S}^n$. If construct a Poisson process ψ on $X_0(T,\mathbf{S}^n)$, then it will induce the Poisson process on ξ on

$X_0(T,H)$ also so that $\kappa_n \circ \xi = \psi$. Thus we can reduce the proof to the case, when H is the complete separable normed space, since the locally compact normed field $\mathbf{S}$ is separable.

A class $\mathcal{K}$ of subsets in a set C is called compact if for each sequence K_n in $\mathcal{K}$ from $\bigcap_{n=1}^{\infty} K_n = \emptyset$ it follows that a natural number m exists for which the finite intersection $\bigcap_{n=1}^{m} K_n = \emptyset$ is void.

We recall that a measurable space $(H,\mathcal{U})$ is called Radon if it contains a compact subclass $\mathcal{K}$ of subsets approximating from below every real measure on $(H,\mathcal{U})$. Then $(H,\mathcal{U},\mathcal{K})$ is called a (measurable) Radon space. In accordance with Theorem I.1.2 [16] each separable complete metric space is a topological Radon space.

Let $\mathcal{U}_j$ be a family of algebras of subsets of a topological space X such that for each pair j_1, j_2 in a given set Λ there exists $j \in \Lambda$ so that $\mathcal{U}_{j_1} \cup \mathcal{U}_{j_2} \subset \mathcal{U}_j$. Then the family $\{\mathcal{U}_j : j \in \Lambda\}$ is called directed and the algebra $\mathcal{U} = \bigcup_{j\in\Lambda} \mathcal{U}_j = \lim_{\Lambda} \mathcal{U}_j$ is said to be its limit.

Theorem I.1.3 [16] states that if $(X,\mathcal{U}) = \lim_{\Lambda}(X,\mathcal{U}_j)$, a compact class $\mathcal{K} \subset \mathcal{U}$ exists such that each $(X,\mathcal{U}_j,\mathcal{K}_j)$ is a measurable Radon space with $\mathcal{K}_j = \mathcal{K} \cap \mathcal{U}_j$, then every bounded non-negative quasi-measure on $(X,\mathcal{U})$ is a measure.

Thus the cylindrical distribution μ described above and generated by the family $P(t_2 - t_1, x, A)$ has an extension to a measure on $X_0(T,H)$ with $\mathcal{U} \subset Bf(X_0(T,H))$. All others conditions are satisfied in accordance with §2 and §4. A construction of an elementary events space Ω and a random function is analogous to that of §3.11 in Chapter 3, but here separating covering σ-algebras on H are considered instead of separating covering rings.

6. Definitions and Remarks. Let the configuration space Γ_X be the same as it was described in Sections 4.3 and 4.4 of Chapter 3 above.

Let $K \in \{K_n : n \in \mathbf{N}\}$, then m_K denotes the restriction $m|_K$, where $m : \mathcal{R} \to \mathbf{C}$ is a σ-additive measure on a covering σ-algebra $\mathcal{R}_m$ of X, $K_n \in \mathcal{R}_m$ for each $n \in \mathbf{N}$. Suppose that $K_l^n \in \mathcal{R}_{m^n}$ for each n and l in $\mathbf{N}$, where $\mathcal{R}_{m^n}$ is the completion of the covering σ-algebra $\mathcal{R}^n$ of X^n relative to the product measure $m^n = \bigotimes_{j=1}^n m_j$, $m_j = m$ for each j, where $\mathcal{R} \supset Bf(X)$. Then

$$m_K^n := \bigotimes_{j=1}^{n} (m_K)_j$$

is the product measure on K^n and hence on $\tilde{K}^n$, when m is such that

$$|m|(K^n \setminus \tilde{K}^n) = 0,$$

where $(m_K)_j = m_K$ for each j. Let its variation $|m|(X) < \infty$ be bounded. Then

$$(i) \quad P_{K,m} := \exp(-m(K)) \sum_{n=0}^{\infty} m_{K,n}/n!$$

is the measure on $\mathcal{R}(B_K)$, where

$$\mathcal{R}(B_K) = B_K \cap \left(\bigoplus_{n=0}^{\infty} \mathcal{R}_{m^n} \right),$$

$m_{K,0}$ is a probability measure on the singleton B_K^0, and $m_{K,n}$ are images of m_K^n under the following mappings:

$$p_K^n : \tilde{K}^n \ni (x_1,\dots,x_n) \mapsto \{x_1,\dots,x_n\} \in B_K^n.$$

Such system of measures $P_{K,n}$ is consistent, that is,

$$\pi_l^n(P_{K_l,m}) = P_{K_n,m}$$

for each $n \le l$. This defines the unique measure P_m on $\mathcal{R}(\Gamma_X)$, which is called the Poisson measure, where $\pi_n : Y \to B_{K_n}$ is the natural projection for each $n \in \mathbf{N}$. For each $n_1,\dots,n_l \in \mathbf{N_o}$ and disjoint subsets $B_1,\dots,B_l$ in X belonging to $\mathcal{R}_m$ there is the following equality:

$$(ii) \quad P_m\Big(\bigcap_{j=1}^{l}\{\gamma : card(\gamma\cap B_i) = n_i\}\Big) = \prod_{i=1}^{l} m(B_i)^{n_i}\exp(-m(B_i))/n_i!.$$

There exists the following embedding $\Gamma_X \hookrightarrow S_X$, where $S_X := \lim\{E_{K_n}, \pi_m^n, \mathbf{N}\}$ is the limit of an inverse mapping sequence,

$$E_K := \bigoplus_{l=0}^{\infty} K^l \quad \text{for each} \quad K \in \{K_n : n = 0,1,2,\dots\}.$$

The Poisson measure P_m on $\mathcal{R}(\Gamma_X)$ considered above has an extension on $\mathcal{R}(S_X)$ such that $|P_m|(S_X \setminus \Gamma_X) = 0$. Here $\mathcal{R}(S_X)$ contains the Borel σ-algebra $Bf(S_X)$, since $\mathcal{R}(\Gamma_X) \supset Bf(\Gamma_X)$. If each K_n is a complete $\mathbf{K}$-linear space (not open in K), then E_K and S_X are complete $\mathbf{K}$-linear spaces, since

$$S_X \subset \Big(\prod_{n=1}^{\infty} E_{K_n}\Big).$$

Then on $\mathcal{R}(S_X)$ there exists the Poisson measure P_m, but without the restriction $|m_K^n|(K^n \setminus \tilde{K}^n) = 0$, where

$$(iii) \quad P_{K,m} := \exp(-m(K)) \sum_{n=0}^{\infty} m_K^n/n!,$$

$\pi_l^n(P_{K_l,m}) = P_{K_n,m}$ for each $n \le l$.

7. Corollary. *Let suppositions of Proposition 5 be satisfied with $H = S_X$ for a complete* $\mathbf{K}$*-linear space X and $P(A) = P_m(A)$ for each $A \in \mathcal{R}(S_X)$ (see §6), then there exists a measure μ on $X_0(T,H)$ for which the Poisson process exists.*

8. Definition. The stochastic process of Corollary 7 is called the Poisson process with values in X.

9. Note. If $\xi \in L^r(\Omega, \mathsf{F}, \lambda; X_0(T,Y))$ is a stochastic process with values in a Hilbert space Y over $\mathbf{C}$, then its mean value for $t_1, t_2 \in T$ is defined by the following formula:

$$(i) \quad M_{t_1,t_2}(\eta) := \int_Y yP(t_1,0,t_2,dy),$$

where $P(t_1,y_1,t_2,A)$ is a transition probability of ξ corresponding to $\xi(t_1,\omega) = y_1$, $\xi(t_2,\omega) \in A$, $A \in Bf(Y)$. For $t_1 = t_0$ it is simpler to write M_{t_2}. If t_1 and t_2 are definite moments, then they may be omitted and we may write M instead of M_{t_1,t_2}.

Let $H = \mathbf{K}$ be a field, where $\mathbf{Q_p} \subset \mathbf{K} \subset \mathbf{C_p}$, let also λ be a probability real-valued measure. We suppose that T is an interval $[t_0, R]$ in $\mathbf{R}$, where $R > t_0$.

Consider a multiplicative character for a field $\mathbf{K}$, $\pi : \mathbf{K} \setminus \{0\} \to \mathbf{C}$, $\pi = \pi_a$ for some $a \in \mathbf{C}$ such that

$$\pi_a(x) := |x|_{\mathbf{K}}^{a-1} \pi_0(x|x|_{\mathbf{K}}),$$

where $\pi_0 : S_1 \to S^1$ is the multiplicative character on the multiplicative commutative group $S_1 := \{x \in \mathbf{K} : |x|_{\mathbf{K}} = 1\}$, the field $\mathbf{K}$ is so that $\mathbf{C_p} \supset \mathbf{K} \supset \mathbf{Q_p}$, $S^1 := \{z \in \mathbf{C} : |z| = 1\}$ denotes the unit circle in the complex plane. Characters are described in the literature, for example, in §§VI.23-25 [37], [99] and §III.2 [110]. We take the parameter a with $Re(a) > 1$ and consider an extension of π_a as a continuous function such that $\pi_a(0) = 0$.

The character π_0 on S_1 is continuous, but the group S_1 is totally disconnected and S^1 is connected, consequently, the character π_0 is locally constant and $\pi_0^{-1}(1) =: S_0$ is the clopen subgroup in S_1. Thus any character $\phi : S_1/S_0 \to S^1$ of the discrete multiplicative commutative quotient group $G := S_1/S_0$ generates the character π_0 of S_1 so that $\pi_0 = \phi \circ \theta$, where $\theta : S_1 \to S_1/S_0$ is the quotient mapping.

A subgroup G^* of the direct product group $G = \prod_{j \in \Lambda} G_j$ consisting of all elements $g = \{g_j : g_j \in G_j\}$ so that $g_j \neq e_j$ only for a finite number of indices j is called the weak direct product of groups, where e_j is the unit element in G_j, Λ is a non-void set. The multiplication of elements $g, h \in G$ is given by the rule $gh = \{g_j h_j : j \in \Lambda\}$.

Theorems A15 and A14 [37] imply that each commutative group G has a group embedding into the group equal to the weak direct product $\mathbf{Q}^{r_0 *} \times \prod^*_{p \in \mathbf{P}} (Z(p^\infty))^{r_p *}$, where r_0 and r_p are cardinals, $\mathbf{P}$ denotes the set of all prime numbers, the star $*$ in the product indicates that the product is weak. Here $Z(p^\infty)$ denotes the multiplicative group of all numbers $\exp(2\pi i k/p^n)$ with $k \in \mathbf{Z}$, $n \in \mathbf{N}$, $i = \sqrt{-1}$, p is any marked prime number. As usually $\mathbf{Z}$ denotes the ring of all integers, $\mathbf{N} = \{1, 2, 3, \dots\}$ denotes the set of all natural numbers.

For a group G isomorphic to the weak direct product $\prod^*_j G_j$ of discrete commutative groups G_j its character group X is given by Theorem 23.22 [37] so that when G is discrete and $\prod^*_j G_j$ is supplied with the discrete topology, then X is compact and isomorphic with the direct product $\prod_j X_j$ of compact character groups X_j of G_j.

In particular, if the field $\mathbf{K}$ is locally compact, then S_1 is compact and the quotient group S_1/S_0 is finite. A group G is called a periodic (or torsion) group if each element of G has a finite order. A periodic group is called p-prime, where p is a prime number, if an order of each its element is the power of p. In accordance with Theorem A3 [37] each torsion group is the weak direct product of its p-prime subgroups. For each finite commutative group G the group X of all its characters is isomorphic with G (see §25.24 [37]).

10. Theorem. *Let ψ be a continuously differentiable function, from an interval $T \subset \mathbf{R}$ into $\mathbf{R}$ and $\psi(0) = 0$. Then a stochastic process $\xi(t, \omega)$ exists such that*

$$M_t(\exp(-\rho\pi[\xi(t, \omega)])) = \exp(-t\psi(\rho))$$

for each t in T and each constant $\rho > 0$, where $\pi : \mathbf{K} \setminus \{0\} \to \mathbf{C}$ is a multiplicative continuous character as in §9.

Proof. We consider solution of the following equation

$$M_t[\exp(-\rho\pi[\xi(t, \omega)])] = \exp(-t\psi(\rho))$$

taking $t_0 = 0$ without loss of generality. Then $e(t) = e(t-s)e(s)$ for each t and s in T and each $\rho > 0$, where

$$e_\rho(t) := e(t) := M_t(\exp(-\rho\pi[\xi(t,\omega)])).$$

Hence

$$\partial e_\rho(t)/\partial\rho = -t\psi'(\rho)\exp(-t\psi(\rho)),$$

consequently,

$$\psi'(\rho) = t^{-1}\Big(\int_{\mathbf{K}} \pi(l)\exp(-\rho\pi(l))\Big)P(\{\omega : \xi(t,\omega) \in dl\})$$

for each $t \neq 0$. In particular, we get the equation:

$$\psi'(\rho) = \lim_{t\to 0, t\neq 0} t^{-1}\Big(\int_{\mathbf{K}} \pi(l)\exp(-\rho\pi(l))\Big)P(\{\omega : \xi(t,\omega) \in dl\}).$$

By the conditions of this theorem we have the equality:

$$\psi(\alpha) = \int_0^\alpha \psi'(\beta)d\beta.$$

Let us consider a σ-additive measure m on a separating covering ring $\mathcal{R}(\mathbf{K})$ such that $\mathcal{R}(\mathbf{K}) \supset Bf(\mathbf{K}) \cup \{0\}$ with values in $\mathbf{C}$ given by the formula

$$m(dl) := \lim_{t\to 0, t\neq 0} \pi(l)P(\{\omega : \xi(t,\omega) \in dl\})/t$$

on $\mathbf{K}\setminus\{0\}$, $m(\{0\}) = m_0$ and consider a σ-additive measure n satisfying the relation:

$$m(dl) = \pi(l)n(dl) \text{ for } l \neq 0.$$

Therefore, the function ψ has the form:

$$\psi(\rho) = \int_0^\rho \Big(\int_{\mathbf{K}} \exp(-\beta\pi(l))m(dl)\Big)d\beta.$$

The character π is continuous and multiplicative, that is,

$$\pi(ab) = \pi(a)\pi(b)$$

for each a and $b \in \mathbf{K}\setminus\{0\}$. From $\psi(0) = 0$ we have $e_0(t) = 1$ for each $t \geq 0$, consequently,

$$\psi(\rho) = \rho m_0 + \int_{\mathbf{K}\setminus\{0\}} [1 - \exp(-\rho\pi(l))]n(dl),$$

since π as the continuous function has the extension on $\mathbf{K}$ with $\pi(0) = 0$ and the limits

$$\lim_{l\to 0, l\neq 0} [1 - \exp(-\rho\pi(l))]/\pi(l) = \rho$$

and

$$\lim_{\rho\to 0, \rho\neq 0} \int_{B(\mathbf{K},0,k)} [1 - \exp(-\rho\pi(l))]n(dl) = 0$$

for each $k > 0$ exist. Then the expression for the function ψ simplifies:

$$\psi(\rho) = \int_{\mathbf{K}} [1 - \exp(-\rho\pi(l))] n(dl).$$

We search a solution of this problem in the form

$$\pi(\xi(t,\omega)) = tm_0 + \int_{\mathbf{K}} \pi(l)\eta([0,t],dl,\omega).$$

Here we take $\eta(dt,dl,\omega)$ as the real-valued σ-additive measure on $Bf(T) \times \mathcal{R}(\mathbf{K})$ for each $\omega \in \Omega$ such that its moments satisfy the Poisson distribution with the Poisson measure P_{tn}, that is,

$$M_t[\eta^k([0,t],dl,\omega)] = \sum_{s \le k} a_{s,k}(tn)^s(dl)/s!$$

for each $0 < t \in T$, where

$$a_{k,j} = \sum_{s_1+\dots+s_k=j, s_1 \ge 1,\dots,s_k \ge 1} j!/(s_1!\dots s_k!)$$

for each $k \le j$. Using the fact that the set of step functions is dense in the linear space $L^r(\mathbf{K}, \mathcal{R}(\mathbf{K}), n, \mathbf{C_p})$ we get the equalities:

$$M_t\left[\exp\left(-\rho\int_{\mathbf{K}} \pi(l)\eta([0,t],dl,\omega)\right)\right] = \lim_{\mathcal{Z}} M_t\left[\prod_j \exp(-\rho\pi(l_j)\eta([0,t],\delta_j,\omega))\right]$$

$$= \lim_{\mathcal{Z}} \prod_j M_t[\exp(-\rho\pi(l_j)\eta([0,t],\delta_j,\omega))] = \lim_{\mathcal{Z}} \exp\left(-\rho t \sum_j (1 - \exp(-\rho\pi(l_j)))n(\delta_j)\right)$$

$$= \exp\left[-\rho t \int_{\mathbf{K}} \{1 - \exp(-\rho\pi(l))\} n(dl)\right],$$

where $\mathcal{Z}$ is an ordered family of partitions $\mathcal{U}$ of the field $\mathbf{K}$ into disjoint union of elements of $\mathcal{R}(\mathbf{K})$. We adopt the ordering $\mathcal{U} \le \mathcal{V}$ in $\mathcal{Z}$ if and only if each element of the disjoint covering $\mathcal{U}$ is a union of elements of $\mathcal{V}$, $l_j \in \delta_j \in \mathcal{U} \in \mathcal{Z}$. Then we deduce the equation

$$M_t[\exp(-\rho\pi[\xi(t,\omega)])] = M_t\left[\exp\left(-\rho\int_{\mathbf{K}} \pi(l)\eta([0,t],dl,\omega)\right)\right].$$

This defines the stochastic process $\pi[\xi(t,\omega)]$ with the probability space $(\Omega, \mathsf{F}, \lambda)$. If $f \in L^r(\Omega, \mathsf{F}, \lambda; X_0(T,\mathbf{C}))$ and $f(\Omega \times T) \subset \pi(\mathbf{K}) \subset \mathbf{C}$, then there exists $h \in L^r(\Omega, \mathsf{F}, \lambda; X_0(T,\mathbf{K}))$ such that $f = \pi[h]$. Since the character $\pi = \pi_a$ is continuous on $\mathbf{K}$ and locally constant on $\mathbf{K} \setminus \{0\}$, due to §3 and Corollary 7 above there exists a $\mathbf{K}$-valued stochastic process ξ for a given complex-valued stochastic process $\pi[\xi]$ with the measure space $(\Omega, \mathsf{F}, \lambda)$.

11. Notes. For a continuous function $c(t)$ and a nonnegative measure n on the product $T \times \mathbf{K}$ such that

$$\int_{\mathbf{K}\setminus\{0\}} (1 - \exp(-\pi(l))n((0,t] \times dl) < \infty$$

for each $t > 0$ there exists a $\mathbf{K}$-valued stochastic process $\xi(t,\omega)$ satisfying the equation:

$$M_{t_1,t_2}[\exp\{-\rho(\pi(\xi(t_2,\omega)) - \pi(\xi(t_1,\omega)))\}] = \exp[-\rho(c(t_2) - c(t_1))]$$

$$-\int_{\mathbf{K}\setminus\{0\}}[1-\exp(-\rho\pi(l))]n((t_1,t_2]\times dl)]$$

for each $\rho > 0$ and $t_1 < t_2$ with a Poisson measure η having the mean $n(dt \times dl)$. This can be proved analogously to §4.10 [39] using §10 above and with the help of disjoint pavings of $\mathbf{K}$ by clopen balls instead of intervals $(l_1, l_2]$ in the real case. For this we put

$$\pi(\xi(t,\omega)) = c(t) + \int_{\mathbf{K}\setminus\{0\}} \pi(l)\eta([0,t]\times dl)$$

for each $t \geq 0$, where $\eta(dt \times dl)$ is a number of jumps of magnitude $s \in dl$ in time dt.

4.3. Stochastic Processes on Non-archimedean Manifolds

To avoid misunderstanding at first definitions and notations are given.

1. Definitions and Notes. Let M be a C^n-manifold on a Banach space X over a non-archimedean field $\mathbf{K}$ complete relative to its norm. We take an atlas $At(M) := \{(U_j, \phi_j) : j \in \Lambda_M\}$ of the manifold M so that the family $\{U_j\}$ is an open covering of M and $\phi_j : U_j \to \phi_j(U_j)$ is a homeomorphism for each j. Here each set $\phi_j(U_j)$ is open in X, each mapping $\phi_l \circ \phi_j^{-1} : \phi_j(U_j \cap U_l) \to \phi_l(U_j \cap U_l)$ is a diffeomorphism of class C^n for each $U_j \cap U_l \neq \emptyset$ (see also Appendix A).

Since the derivative $(\phi_l \circ \phi_j^{-1})'(x)$ is a linear continuous operator on the open set $\phi_j(U_j \cap U_l) \times X$ of class C^{n-1} for each $n \geq 1$ and the derivative of an inverse operator $(\phi_j \circ \phi_l^{-1})'(y)$ exists on the set $\phi_l(U_j \cap U_l) \times X$, the derivative $(\phi_l \circ \phi_j^{-1})'(x)$ belongs to the group $GL(X)$ for each $x \in \phi_j(U_j \cap U_l)$, where $GL(X)$ is the group of invertible $\mathbf{K}$-linear bounded operators of X onto X. Therefore, for each $n \geq 1$ a functor T such that $T(\phi_l \circ \phi_j^{-1})(x) := (\phi_{l,j}(x), \phi'_{l,j}(x))$ for each point $x \in (\phi_l \circ \phi_j^{-1})(U_j \cap U_l)$ exists, also $T(\phi_j(U_j)) := \phi_j(U_j) \times X$, where $\phi_{l,j} := (\phi_l \circ \phi_j^{-1})$.

For $n \geq 1$ we define

$$TM = \bigcup_{j\in\Lambda_M} TU_j$$

with the atlas

$$At(TM) := \{(U_j \times X, T\phi_j) : j \in \Lambda_M\}$$

such that $T\phi_j : TU_j \to \phi_j(U_j) \times X$ is a homeomorphism, $T\phi_j|_{\{x\}\times X} =: T_x\phi_j$ is a bounded continuous operator on X by the second argument for each $x \in U_j$. Thus we get

$$TM = \bigcup_{x\in M} T_x M,$$

where $T_x\phi_j : T_xU_j \to T_{\phi_j(x)}\phi_j(U_j)$ is a $\mathbf{K}$-linear isomorphism for each $j \in \Upsilon_M$, where

$$T_{\phi_j(x)}\phi_j(U_j) = \{\phi_j(x)\} \times X.$$

TM is called the total tangent space of M, T_xM is called the tangent space of M at x. The projection $\tau := \tau_M : TM \to M$ is given by

$$\tau_M(s) = x$$

for each vector $s \in T_xM$, τ_M is called the tangent bundle.

1.1 If M and N are two C^l-manifolds on Banach spaces X and S over the field $\mathbf{K}$ with $l \geq n$, where an atlas of N is $At(N) := \{(V_j, \psi_j) : j \in \Lambda_N\}$ and $f : M \to N$ is a continuous mapping, then by the definition $f \in C^n(M,N)$, if $\psi_l \circ f \circ \phi_j^{-1} \in C^n(W_{l,j}, Y)$ for each $W_{l,j} := \phi_j(f^{-1}(V_l) \cap U_j) \neq \emptyset$. A norm in $C^n(X,S)$ induces a complete uniformity in $C^n(M,N)$. If $n \geq 1$ and $f \in C^n(M,N)$, then the tangent mappings $Tf : TM \to TN$ and $Tf \in C^{n-1}(TM,TN)$ exist.

1.2. Let H and X be two Banach spaces over the non-archimedean field $\mathbf{K}$. Let M be a C^l-manifold on the Banach space X and let P be a manifold with a continuous mapping $\pi : P \to M$ such that π is surjective and

$$\pi^{-1}(x) =: P_x =: H_x$$

is a Banach space over $\mathbf{K}$ linearly topologically isomorphic to H for each $x \in M$, π is called a projection, $\pi^{-1}(x)$ is called a fibre of π over x. Suppose that P is supplied with an atlas

$$At(P) = \{(U_j, \phi_j, P\phi_j) : \ j \in \Lambda_M\}$$

consistent with $At(M)$ such that

$$pr_1 \circ P\phi_j = \phi_j \circ \pi|_{PU_j}$$

on $\pi^{-1}(U_j)$ for each j, where

$$pr_1 : \ U_j \times H \to U_j \quad \text{and} \quad pr_2 : \ U_j \times H \to H$$

are projections, $P\phi_j$ is bijective. Here a mapping $P_x\phi_j$ is a Banach space isomorphism

$$P_x\phi_j = P\phi_j|_{H_x} : H_x \to \{\phi_j(x)\} \times H,$$

moreover the mapping

$$P\phi_l \circ (P\phi_j|_{P(U_l \cap U_j)})^{-1} : \phi_j(U_j \cap U_l) \times H \to \phi_l(U_j \cap U_l) \times H$$

is the C^n-diffeomorphism, where $l \geq n$.

Two atlases are called equivalent, if their union is an atlas. The triple (P, M, π) (or shortly also denotes by π) with the described above restrictions is called a vector bundle over M with a fibre on H. The object P is called the total space of the vector bundle π and the manifold M is known as the base space of π.

Let (P_1, M_1, π_1) and (P_2, M_2, π_2) be two vector bundles with spaces H_1 and H_2 for the fibres of π_1 and π_2 respectively. Suppose there are two C^n-mappings $F : M_1 \to M_2$ and $PF : P_1 \to P_2$ so that

$$\pi_2 \circ PF = F \circ \pi$$

on P_1 and the restriction

$$P_xF := PF|_{H_{1,x}} : H_{1,x} \to H_{2,F(x)}$$

is a $\mathbf{K}$-linear mapping. Then (F, PF) is called a morphism from the vector bundle (P_1, M_1, π_1) into the vector bundle (P_2, M_2, π_2).

1.3. A C^m-vector field on M is a C^m-mapping $\Psi : M \to TM$ satisfying the equality $\tau_M \circ \Psi = id$, where id is the identity function.

If $F : M \to N$ is a C^m-morphism and a mapping $\Psi : M \to TN$ fulfilling the equality $\tau_N \circ \Psi = F$, then Ψ is called a vector field along F.

Suppose that a field $\mathbf{K}$ is spherically complete, then a topologically adjoint space H^* of all continuous $\mathbf{K}$-linear functionals on a Banach space H over $\mathbf{K}$ separates points of H, $H^* \neq \emptyset$ (see Lemma 4.3.5 [99]).

The bundle of r-fold contra-variant and s-fold covariant tensors over M is defined by

$$L(\tau^*,\dots,\tau^*,\tau,\dots,\tau;\rho) : L(T^*M,\dots,T^*M,TM,\dots,TM;\mathbf{K}M) \to M$$

or shortly

$$\tau^r_s : T^r_s M \to M,$$

where τ^* and T^*M are repeated r times, τ and TM are repeated s times,

$$\rho : \mathbf{K}M = M \times \mathbf{K} \to M$$

is the trivial bundle over M. A mapping $L(\alpha_1,\dots,\alpha_r;\beta) : L(A_1,\dots,A_r;B) \to M$ denotes a vector bundle over M, where (A_j, M, α_j) and (B, M, β) are vector bundles,

$$\alpha_k^{-1}(x) = H_{k,x}, \quad k = 1,\dots,r, \quad \beta_H^{-1}(x) = Y_x,$$

$$L(A_1,\dots,A_r;B) := \bigcup_{x\in M} L(H_{1,x},\dots,H_{r,x};Y_x),$$

$H_{k,x}$ and Y_x are isomorphic to Banach spaces H_k and Y respectively over the field $\mathbf{K}$. For each chart (U_j,ϕ_j) of the manifold M the bundle chart $(U_j,\phi_j,L(A_1,\dots,$ $A_r;B)\phi_j)$ is prescribed by the mapping

$$L(A_1\phi_j(x),\dots,A_r\phi_j(x);B\phi_j(x)) : L(H_{1,x},\dots,H_{r,x};Y_x) \to L(H_1,\dots,H_r;Y)$$

such that for $\Psi_x \in L(H_{1,x},\dots,H_{r,x};Y_x)$ its image is

$$L(A_1\phi_j(x),\dots;B\phi_j(x))\Psi_x = B\phi_j(x) \circ \Psi_x \circ (A_1\phi_j(x)^{-1} \times \dots \times A_r\phi_j(x)^{-1}),$$

$$A_k\phi_j(x)^{-1} : H_{k,x} \to H_k$$

is the $\mathbf{K}$-linear isomorphism of Banach spaces. As usually $L(H_1,\dots,H_r;Y)$ denotes the Banach space of all continuous mappings $f : H_1 \times \dots \times H_r \to Y$ such that f is $\mathbf{K}$-linear by each variable $z_k \in H_k$, $k = 1,\dots,r$, that is f is r-linear over the field $\mathbf{K}$.

If $\Psi : M \to \kappa TM$ is a C^m-mapping for which the equality

$$\kappa \circ \Psi = id$$

is satisfied, then Ψ is called a tensor field (of type κ), where $(\kappa TM, M, \kappa(\tau))$ is a tensor bundle over M.

If (P,N,π) is a vector bundle and $F : M \to N$ is a morphism, then a morphism $\theta : M \to P$ so that

$$\pi \circ \theta = F$$

is called a section along F.

1.4. Let M be a C^n-manifold on a Banach space X over a spherically complete non-archimedean field $\mathbf{K}$. Let also $\mathcal{B}_n M$ denote the set of all C^n-vector fields on M, where $n \geq 2$. Suppose that

$$\Gamma = {}_j\Gamma : \phi_j(U_j) \ni y_j \mapsto \Gamma(y_j) \in L(X,X;X)$$

is a C^{n-2}-mapping so that

$$(i) \quad \phi'_{l,j} \cdot {}_j\Gamma(y_j) = \phi_{l,j}" + {}_l\Gamma(y_l) \circ (\phi'_{l,j} \times \phi'_{l,j})$$

for each two charts with $U_j \cap U_l \neq \emptyset$.

This set $\{ {}_j\Gamma \}$ is called the family of Christoffel symbols ${}_j\Gamma$ on M.

A covariant derivation $\mathcal{B}_{n-1} M^2 \ni (\Psi, \Phi) \mapsto \nabla_\Psi \Phi \in \mathcal{B}_{n-2} M$ is given by the equation

$$(ii) \quad \nabla_\Psi \Phi(y_j) = \Phi'(y_j).\Psi(y_j) + \Gamma(y_j)(\Psi(y_j), \Phi(y_j)),$$

where $\Psi(y_j)$ and $\Phi(y_j)$ are principal parts of vector fields Ψ and Φ on (U_j, ϕ_j). If the manifold M with the atlas $At(M)$ is supplied with the family of Christoffel symbols Γ, then M possesses a covariant derivation.

1.5. For a C^n-vector bundle (P, M, π) on $X \times H$ with $n \geq 2$ a $\mathbf{K}$-(linear) connection is defined as a bundle morphism $K : TP \to P$ such that

$$\pi \circ K = \pi \circ \tau_P.$$

This mapping K in its local representation

$${}_jK = P\phi_j \circ K \circ TP\phi_j^{-1}$$

for bundle charts $(U_j, \phi_j, P\phi_j)$ of (P, M, π) and $(TU_j, T\phi_j, TP\phi_j)$ of (TP, P, τ_P) is given by the formula:

$$\{U_j, \Xi\} \times (X \times H) \ni (x, \Psi, \Phi, z) \mapsto (x, z + {}_j\Gamma(x)(\Phi, \Psi)) \in \{x\} \times H.$$

The Christoffel symbol ${}_j\Gamma(x) : U_j \to L(X, H; H)$ is of class of smoothness C^{n-2}. For it the horizontal space $T_{\Psi h}$ is defined as the kernel of the mapping $K|_{T_\Psi P} : T_\Psi P \to H_q$, $q = \pi(\Psi)$.

For a section $\Psi : M \to P$ in the bundle (P, M, π) the covariant derivation of Ψ in the direction $\Phi \in T_x M$ is defined by the expression

$$(i) \quad \nabla_\Phi \Psi(x) = K \circ T_x \Psi.\Phi.$$

2. Let X be a $\mathbf{K}$-linear either finite dimensional over a local field $\mathbf{K}$ or of countable type space supplied with a sequence of subspaces S_n so that $S_n \subset S_{n+1}$ and $S_n \neq S_{n+1}$ for each $n \in \mathbf{N}$, $cl(\bigcup_n S_n) = X$, a dimension $dim_{\mathbf{K}} S_n =: m(n)$ of S_n over $\mathbf{K}$ is finite. Let U be a clopen bounded subset in S_n.

We consider an anti-derivation operator $P(l,s)$ on the Banach space $C^{(t,s-1)}(U, \mathbf{K})$ of functions $f : U \to \mathbf{K}$ with definite partial difference quotients having continuous extensions and denote $P(l,s)$ on U by $P_U(l,s)$, where $t \in [0, \infty)$, $1 \leq s \in \mathbf{Z}$,

$l=[t]+1$, $[t]$ is an integer part of t. In particular, $C^{(t,0)}(U,\mathbf{K})$ is denoted here by $C^t(U,\mathbf{K})$ (see Appendix D).

3. Definition and Note. Let now U be a clopen bounded subset in X with $dim_{\mathbf{K}}X=\infty$. For each $f\in C_0((t,s-1),U\to\mathbf{K})$ by the definition a sequence of cylindrical functions f_n exists so that $f=\sum_n f_n$ and the limit

$$\lim_n \|\hat{f}_n\|_{C^{(t,s-1)}(U_n,\mathbf{K})}=0$$

is zero, where f_n is a cylindrical function on U such that $f_n(x)=\hat{f}_n(\pi_n x)$, $\hat{f}_n$ is a function on $U_n:=S_n\cap U$, $\pi_n:X\to S_n$ is a projection on S_n. For each $0\le t<\infty$ there exists U of sufficiently small diameter δ such that the norm of the anti-derivation operator satisfies the inequality:

$$\|P_{U_n}(l,s)\|\le 1$$

for each n, since it is sufficient to take

$$\delta^{|j|+n}/|(j+\bar{u})!|\le 1$$

for each j with $|j'|=0,\dots,l-1$, $j=j'+s'\bar{u}$, $s'\in\{0,1,\dots,s-1\}$ (see Appendix D). For the chart U of the diameter $diam(U)$ satisfying such condition we define the anti-derivation operator

$$P_U(l,s)f:=\sum_n P_{U_n}(l,s)f_n.$$

For U as above the space

$${}_PC_0((t,s),U\to Y):=P_U(l,s)C_0((t,s-1),U\to Y)$$

is defined, where Y is a Banach space over the field $\mathbf{K}$.

4. Lemma. *An image $P_U(t,s)(C^{(t,s-1)}(U,Y))$ denoted by ${}_PC^{(t,s)}(U,Y)$ is contained in the space $C^{(t,s)}(U,Y)$ and does not coincide with the latter space. The space ${}_PC^{(t,s)}(U,Y)$ can be supplied with a norm denoted by $\|*\|_{U,(t,s),P}$ relative to which it is complete and the anti-derivation operator*

$$P_U(l,s):(C^{(t,s-1)}(U,Y),\|*\|_{C^{(t,s-1)}(U,Y)})\to({}_PC^{(t,s)}(U,Y),\|*\|_{U,(t,s),P})$$

is continuous.

Proof. We consider at first the space X of a finite dimension $dim_{\mathbf{K}}X=k<\infty$ over the field $\mathbf{K}$. If $f\in{}_PC^{(t,s)}(U,Y)$, then

$$\partial^{\bar{u}}(P(t,s)f)(x)=f(x)$$

for each $x\in U$ (Appendix D). On the other hand, there are functions $g\in C^{(t,s)}(U,Y)$ for which

$$\partial^{e_j}g(x)=0,$$

where $\partial^{e_j}g(x):=\partial g(x)/\partial x_j$, $x=(x_1,\dots,x_k)$, $x_j\in\mathbf{K}$ for each j. This demand is satisfied not only for any locally constant function g, but also for a wider class of functions (see [103,104]).

Let now X may be infinite dimensional, then from taking the limit of f_n this statement follows in the general case.

We consider the image $P_U(l,s)(B(C^{(t,s-1)}(U,Y),0,1))) =: V$ of the closed ball in the space $C^{(t,s)}(U,Y)$ containing 0 and of the unit radius. Let $f \in {}_PC^{(t,s)}(U,Y)$, then there exists a function $g \in C^{(t,s-1)}(U,Y)$ so that

$$P_U(l,s)g = f.$$

On the other hand, its norm is finite

$$\|g\|_{C^{(t,s-1)}(U,Y)} < \infty$$

and there a constant $0 \neq c \in \mathbf{K}$ exists such that $cg \in B(C^{(t,s-1)}(U,Y),0,1))$. Therefore, $cf \in V$, since $P_U(l,s)$ is the **K**-linear operator. This means that V is the absorbing subset. Since the ball $B(C^{(t,s-1)}(U,Y),0,1)$ is **K**-convex, the set V also is **K**-convex. Evidently, $0 \in V$.

We now take a weak topology on $C^{(t,s)}(U,Y)$, then it induces a weak topology on its **K**-linear subspace ${}_PC^{(t,s)}(U,Y)$. In particular, each evaluation functional $h_x(f) := f(x)$ is **K**-linear and continuous on the latter space, where $x \in U$. In view of the theorem from Appendix D the anti-derivation operator $P_U(l,s)$ is continuous from the space $C^{(t,s-1)}(U,Y)$ into $C^{(t,s)}(U,Y)$. Therefore, V is bounded relative to the weak topology, since U is compact and V is bounded relative to a weaker topology generated by evaluation functionals.

Let η be a Minkowski functional on the space ${}_PC^{(t,s)}(U,Y)$ generated by V (see [87]). It generates a norm in ${}_PC^{(t,s)}(U,Y)$ relative to which it is complete. Since V is the unit ball relative to this norm and $P_U(l,s)^{-1}(V)$ is the unit ball in $C^{(t,s-1)}(U,Y)$, we deduce that the anti-derivation operator $P_U(l,s)$ is continuous relative to this topology.

5. Note. In view of Lemma 4 Definitions 1.1-1.5 can be spread on $C_0((t,s))$ and ${}_PC_0((t,s))$-manifolds. This means that

$$(\phi_{l,j} - id) \in C_0((t,s), W_{l,j} \to X)$$

and

$$(\phi_{l,j} - id) \in {}_PC_0((t,s), W_{l,j} \to X)$$

respectively for all charts U_l and U_j with the non-void intersection $U_l \cap U_j \neq \emptyset$, where each $\phi_j(U_j)$ is the bounded clopen subset in X of a sufficiently small diameter as in §3 if X is infinite dimensional over **K**.

6. Note. We now consider the space of functions

$$\mathcal{F}_{(t,s)}M = C_0((t,s), M \to \mathbf{K}).$$

In this space the equalities

$$\nabla_S(aV + bW) = a\nabla_S V + b\nabla_S W \quad \text{and} \nabla_S(fV) = S(f)V + f\nabla_S V$$

are satisfied, where S, V, $W \in \mathcal{B}_{(t,s)}M$, $\mathcal{B}_{(t,s)}M$ denotes the set of all $C_0((t,s))$-vector fields on M. Considering the foliation of M and taking the limit we get for each given chart (U_j, ϕ_j) the relation:

$$\nabla_S V(\phi_j) = \sum_k \Big\{ \sum_i S^i(\phi_j)(\partial V^k / \partial \phi_j^i)(\phi_j) + \sum_{i,l} S^i(\phi_j) V^l(\phi_j) \Gamma^k_{i,l}(\phi_j) \Big\} e_k,$$

where (ϕ_j, e_i) are basic vector fields on $\phi_j(U_j)$,

$$S(\phi_j) = \sum_i S^i(\phi_j)e_i,$$

$$\Gamma(\phi_j) = \sum_{i,l,k} \Gamma^k_{i,l}(\phi_j)e^i \otimes e^j \otimes e_k,$$

$e^i(e_j) = \delta^i_j$ for each i and $j \in \alpha$. Therefore, there exists a torsion tensor

$$T(S,V) = \nabla_S V - \nabla_V S - [S,V]$$

and a curvature tensor

$$R(S,V)W = \nabla_S \nabla_V W - \nabla_V \nabla_S W - \nabla_{[S,V]} W$$

for each vector fields S, V and $W \in \mathcal{B}_{(t,s)}M$. In the standard way this implies the relations

$$T(S,V) = -T(V,S), \quad R(S,V)W = -R(V,S)W \quad \text{and}$$

$$T(\phi_j)(S,V) = \Gamma(\phi_j)(S,V) - \Gamma(\phi_j)(V,S) \in L(X,X;X),$$
$$R(\phi_j)(S,V)W = D\Gamma(\phi_j).S(V,W) - D\Gamma(\phi_j).V(S,W)$$
$$+\Gamma(\phi_j)(S,\Gamma(\phi_j)(V,W)) - \Gamma(\phi_j)(V,\Gamma(\phi_j)(S,W)) \in L(X,X,X;X).$$

They are satisfied analogously to Lemma 1.5.3 [47].

7. Theorem. *Let M be a ${}_PC_0((t,s))$-manifold with $s \geq 2$, then there exists a clopen neighborhood $\tilde{T}M$ of M in TM and an exponential $C_0((t,s))$-mapping*

$$\exp : \tilde{T}M \to M$$

of $\tilde{T}M$ on M.

Proof. Let the manifold M be embedded into TM as the zero section of the bundle τ_M. We consider the non-archimedean geodesic equation

$$\nabla_{\dot{c}}\dot{c} = 0$$

with initial conditions

$$c(0) = x_0, \;\; \dot{c}(0) = y_0, \;\; x_0 \in M, \;\; y_0 \in T_{x_0}M,$$

where $c(b)$ is a ${}_PC_0((t,s))$-curve on the manifold M, $c : B(\mathbf{K},0,1) \to M$. For a chart (U_j, ϕ_j) containing a point x of M let us consider the composition

$$\phi_j \circ c(b) := \psi_j(b).$$

Therefore,

$$(i) \quad \psi_j\text{''}(b) + \Gamma(\psi_j(b))(\dot{\psi}_j(b), \dot{\psi}_j(b)) = 0.$$

Since $\psi_j \in {}_PC_0((t,s))$, certainly a function $f \in C^{(t,s-2)}(B,X)$ exists such that $\psi_j = P_B(l,s)P_B(l,s-1)f$, where $B := B(\mathbf{K},0,1)$. Therefore, the equations

$$\dot{\psi}_j = P_B(l,s-1)f$$

and $\psi_j" = f$ are fulfilled, consequently, the function f satisfies the equation

$$(ii) \quad f(b) + \Gamma(P^2 f|_b)(P^1 f|_b, P^1 f|_b) = 0,$$

where we have put $P^2 := P_B(l,s)P_B(l,s-1)$ and $P^1 := P_B(l,s-1)$.

Consider a marked point $b_0 \in B$. At first a positive number $r > 0$ exists so that Equation (ii) and hence (i) has a unique solution in the ball $B(\mathbf{K},b_0,r)$. For this we consider the iteration equation:

$$(iii) \quad f_{m+1}(b) + \Gamma(P^2 f_m|_b)(P^1 f_m|_b, P^1 f_m|_b) = 0,$$

where f_m is a sequence of functions. From the inclusion $\Gamma \in {}_PC_0^{(t,s-2)}$, since M is the ${}_PC_0^{(t,s)}$-manifold, it follows, that $f_{m+1} \in {}_PC^{(t,s-2)}$ for each $f_m \in {}_PC^{(t,s-2)}$. Then we infer that

$$f_{m+1}(t) - f_m(t) = -\Gamma(P^2 f_m|_t)(P^1 f_m|_t, P^1 f_m|_t)$$
$$+\Gamma(P^2 f_m|_t)(P^1 f_{m-1}|_t, P^1 f_{m-1}|_t) - \Gamma(P^2 f_m|_t)(P^1 f_{m-1}|_t, P^1 f_{m-1}|_t)$$
$$+\Gamma(P^2 f_{m-1}|_t)(P^1 f_{m-1}|_t, P^1 f_{m-1}|_t).$$

In view of the ultra-metric inequality, the bi-linearity of the mapping $\Gamma(x)(a,b)$ by arguments a, b and continuity by x, and the continuity of the anti-derivation operators P^1 and P^2 for each $x_0 \in M$ and each $t_0 \in B(\mathbf{K},0,1)$ positive numbers $r > 0$ and $\varepsilon > 0$ exist so that

$$(iv) \quad \|f_{m+1} - f_m\| \le C\varepsilon^2 \|\Gamma\| \|f_m - f_{m-1}\|$$

for each $t \in B(\mathbf{K},t_0,r)$ and each $\|y_0\| < \varepsilon$, where $C > 0$ is a constant related with the anti-derivation operators P^1 and P^2. Then a positive number $0 < r < \infty$ exists such that the norms

$$\|P^1\| \le 1 \quad \text{and} \quad \|P^2\| \le 1$$

satisfy these restrictions and $P^2 f \in G_{j,k} \subset U_j$ for each $f \in G_{j,k}$, since t and s are finite (see above). Here $G_{j,k}$ is a clopen subset in the chart U_j, $\|\Gamma\|$ is a norm of Γ on $G_{j,k} \times X^2$ as a bi-linear operator on the space X for each $x \in G_{j,k}$. In view of continuity of Γ and boundedness of the set $\phi_j(U_j)$ for each j it is possible to choose a locally finite covering $G_{j,k}$ subordinated to U_j such that $\|\Gamma\|$ is finite on $G_{j,k}$, $k \in \mathbf{N}$. Therefore, choosing a constant C so that $C\varepsilon^2\|\Gamma\| < 1$ we get a convergent sequence on $B(\mathbf{K},t_0,r) \times G_{j,k} \times B(X,0,\delta)$. Thus due to the fixed point theorem there exists a unique solution in the ball $B(\mathbf{K},t_0,r)$. In view of compactness of the ball $B(\mathbf{K},0,1)$ there exists a solution on it. Let f and g be two functions providing solutions

$$\psi^f = P^2 f \quad \text{and} \quad \psi^g = P^2 g$$

of the problem on the unit ball $B(\mathbf{K},0,1)$, then the equalities

$$P^2 f(t_l) = P^2 g(t_l), \quad P^1 f(t_l) = P^1 g(t_l)$$

are satisfied for a finite number of points $t_0 = 0, t_1, \dots, t_k \in B(\mathbf{K}, 0, 1)$ such that on each ball $B(\mathbf{K}, t_j, r_j)$ a solution is unique for given initial conditions. Moreover, we have $0 < r_j \le 1$ for each j and

$$\bigcup_j B(\mathbf{K}, t_j, r_j) = B(\mathbf{K}, 0, 1).$$

This imply that

$$(v) \quad \Gamma(P^2 f|_t)(P^1 f|_t, P^1 f|_t) - \Gamma(P^2(f + c_{1,l})|_t)(P^1(f + c_{2,l})|_t, P^1(f + c_{2,l})|_t) = c_{1,l}$$

for each natural number l and each point $t \in B(\mathbf{K}, t_l, r_l)$. On the other hand, the functions $P^1 c$ and $P^2 c$ are not locally constant for a constant $c \ne 0$, while $\Gamma(\phi_j)(a, b)$ is bi-linear by $(a, b) \in X^2$ and satisfies Equation 1.4(i), consequently, Equation (v) may be satisfied only for the data $c_{1,l} = c_{2,l} = 0$ for each l, consequently, a solution is unique.

Since $f \in {}_P C_0((t, s-2))$, we get the inclusion

$$\psi_j \in {}_P C_0((t, s))$$

for each j. Moreover,

$$c_{aS}(t) = c_S(at)$$

for each $a \in B(\mathbf{K}, 0, 1)$ such that the inequality

$$|aS(\phi_j(q))| < \varepsilon$$

is satisfied, since

$$dc_S(at)/dt = a(dc_S(z)/dz)|_{z=at}.$$

In view of continuity of the anti-derivation operators P^2 and P^1 and the Γ operator, for each point $x_0 \in M$ a chart (U_j, ϕ_j) and clopen neighborhoods V_1 and V_2 with $\phi_j(x_0) \in V_1 \subset V_2 \subset \phi_j(U_j)$ and $\delta > 0$ exist such that from the inclusion $S \in TM$ with $\tau_M S = q \in \phi_j^{-1}(V_1)$ and the inequality

$$|S(\phi_j(q))| < \delta$$

it follows, that the geodesic c_S with $c_S(0) = S$ is defined for each $t \in B(\mathbf{K}, 0, 1)$ and $c_S(t) \in \phi_j^{-1}(V_2)$. Due to the para-compactness of the manifolds TM and M this covering can be chosen locally finite [19].

This means that a clopen neighborhood $\tilde{T}M$ of the manifold M in TM exist such that a geodesic $c_S(t)$ is defined for each $S \in \tilde{T}M$ and each $t \in B(\mathbf{K}, 0, 1)$. Therefore, we can define the so called exponential mapping

$$\exp : \tilde{T}M \to M$$

by the rule

$$S \mapsto c_S(1).$$

Customary $\exp_x := \exp|_{\tilde{T}M \cap T_x M}$ denotes a restriction to a fibre. Then the exponential mapping exp has a local representation $(x_0, y_0) \in V_1 \times B(X, 0, \delta) \mapsto \psi_j(1; x_0, y_0) \in V_2 \subset \phi_j(U_j)$. From Equations (i, ii) it follows that the exponential mapping exp is of the $C_0((t, s))$-class of smoothness from $\tilde{T}M$ onto M.

8. Corollary. *If M is a ${}_P C_0((t,s)) \cap C^\infty$-manifold with $s \geq 2$, then* $\exp \in C^\infty(\tilde{T}M, M)$.

9. Note. If M is an analytic manifold, then $\exp : \tilde{T}M \to M$ is a locally analytic mapping. Thus as it was observed above Theorem 7 gives an exponential manifold mapping for wider class of manifolds, than treated by the rigid geometry.

10. Note and Definitions. Let M be a C^∞-manifold and let $\tau_M : TM \to M$ be the tangent bundle, $\theta : M \times H \to M$ be a trivial bundle over M with a Banach space fibre H over $\mathbf{K}$. There exists the bundle $L_{r,1}(\theta, \tau_M)$ over M with the fibre $L_{r,1}(H,X)$, where $r \geq 1$ and spaces $L_{r,n}(H,X)$ were defined in Section 5.13 of Chapter 3.

Let M be a C^∞-manifold with functions $\phi_{l,j}$ satisfying Conditions 6.6(i) of Chapter 3. Suppose that w is a stochastic process with values in the Banach space H and ξ is a stochastic process with values in the Banach space X such that $\lambda\{\omega : w(t,\omega) \in C^0 \setminus C^1\} = 0$, where H and X are over a local field $\mathbf{K}$.

Let the following conditions

$$a \in L^q(\Omega, \mathsf{F}, \lambda; C^0(B_R, L^q(\Omega, \mathsf{F}, \lambda; C^0(B_R, X))))$$

and

$$E \in L^r(\Omega, \mathsf{F}, \lambda; C^0(B_R, L(L^q(\Omega, \mathsf{F}, \lambda; C^0(B_R, H)), L^q(\Omega, \mathsf{F}, \lambda; C^0(B_R, X))))),$$

$$(i) \quad \xi(t,\omega) = \xi_0(\omega) + (\hat{P}_u a)(u,\omega,\xi)|_{u=t} + (\hat{P}_{w(u,\omega)} E)(u,\omega,\xi)|_{u=t}$$

be satisfied, where $1 \leq r,s,q \leq \infty$, $1/r + 1/s = 1/q$, $w \in L^s(\Omega, \mathsf{F}, \lambda; C^0_0(B_R, H))$, $\xi \in L^q(\Omega, \mathsf{F}, \lambda; C^0(B_R, X))$. Since H and X are isomorphic with $c_0(\alpha_H, \mathbf{K})$ and $c_0(\alpha_X, \mathbf{K})$, the space $L_{r,n}(X,H)$ has the embedding into the space $L_{r,n}(H,H)$ for $\alpha_X \subset \alpha_H$ and the space $L_{r,n}(H,H)$ has an embedding into the space $L_{r,n}(X,X)$ for $\alpha_H \subset \alpha_X$. Inclusions $Range(E) \subset X$, $Range(w) \subset H$ and $Range(\xi) \subset X$ reduce this case to Theorems 7.3 or 7.4 of Chapter 3. In view of Lemma 5.3 and Formula 6.6(ii) of Chapter 3 we get the equality:

$$(ii) \quad d\phi(\xi(t,\omega)) = J(\phi,a,E)a\,dt + J(\phi,a,E)E\,dw,$$

where

$$(iii) \quad J(\phi,a,E) := \sum_{m=0}^{\infty} [m!]^{-1} \sum_{l=0}^{m} \binom{m}{l} \hat{P}_{u^l, w^{m-l}} \phi^{(m+1)} \circ (a^{\otimes l} \otimes E^{\otimes(m-l)})|_{u=t},$$

when the field $\mathbf{K}$ is of zero characteristic.

We consider the following generalization of Theorems 6.6 and 7.4 of Chapter 3.

11. Note. We take any random vector field

$$a \in L^\infty(\Omega, \mathsf{F}, \lambda; C^0(B_R, L^q(\Omega, \mathsf{F}, \lambda; C^0(B_R, X))))$$

and a random operator field

$$E \in L^\infty(\Omega, \mathsf{F}, \lambda; C^0(B_R, L(L^q(\Omega, \mathsf{F}, \lambda; C^0(B_R, H)), L^q(\Omega, \mathsf{F}, \lambda; C^0(B_R, X))))),$$

so that $a = a(t,\omega,\xi)$, $E = E(t,\omega,\xi)$, $t \in B_R$, where $\omega \in \Omega$, $\xi \in L^q(\Omega, \mathsf{F}, \lambda; C^0(B_R, X))$ and $\xi_0 \in L^q(\Omega, \mathsf{F}, \lambda; X)$, $w \in L^\infty(\Omega, \mathsf{F}, \lambda; C^0_0(B_R, H))$, $1 \leq q \leq \infty$. It is supposed that the vector a

and the operator E vector fields satisfy the local Lipschitz condition 7.3(LLC) of Chapter 3. Let us suppose that ξ is a stochastic process of the type

$$(i) \quad \xi(t,\omega) = \xi_0(\omega)$$

$$+ \sum_{m+b=1}^{\infty} \sum_{l=0}^{m} (\hat{P}_{u^{b+m-l},w(u,\omega)^l}[a_{m-l+b,l}(u,\xi(u,\omega)) \circ (I^{\otimes b} \otimes a^{\otimes(m-l)} \otimes E^{\otimes l})])|_{u=t}$$

such that a random operator

$$a_{m-l,l} \in C^0(B_{R_1} \times B(L^q(\Omega,\mathsf{F},\lambda;C^0(B_R,X)),0,R_2),L_m(X^{\otimes m};X))$$

is continuous and bounded on its domain for each n,l, $0 < R_2 < \infty$ and the limit

$$(ii) \quad \lim_{n\to\infty} \sup_{0\le l\le n} \|a_{n-l,l}\|_{C^0(B_{R_1}\times B(L^q(\Omega,\mathsf{F},\lambda;C^0(B_R,X)),0,R_2),L_n(X^{\otimes n},X))} = 0$$

is zero for each $0 < R_1 \le R$ when $0 < R < \infty$, or each $0 < R_1 < R$ when $R = \infty$, for each $0 < R_2 < \infty$.

Naturally as above in (i) it is supposed that either the field $\mathbf{K}$ is of zero characteristic $char(\mathbf{K}) = 0$, or the field $\mathbf{K}$ is of the positive characteristic $char(\mathbf{K}) = p > 0$ and $a_{m,l} = 0$ for all m and l satisfying the condition $m + l \ge p$.

Moreover, we suppose that a function f satisfies the conditions:

$$(iii) \quad f(u,x) \in C^\infty(T \times H,X)$$

and

$$(iv) \quad \lim_{n\to\infty} \max_{0\le l\le n} \|(\bar{\Phi}^n f)(t,x;h_1,\dots,h_n;$$

$$\zeta_1,\dots,\zeta_n)\|_{C^0(T\times B(\mathbf{K},0,r)^l\times B(H,0,1)^{n-l}\times B(\mathbf{K},0,R_1)^{n-l},X)} = 0$$

for each $0 < R_1 < \infty$, where $h_j = e_1$ and $\zeta_j \in B(\mathbf{K},0,r)$ for variables corresponding to $t \in T = B(\mathbf{K},t_0,r)$ and $h_j \in B(H,0,1)$, $\zeta_j \in B(\mathbf{K},0,R_1)$ for variables corresponding to $x \in H$.

Analogously vector a and operator E, $a_{l,m}$ random fields for ξ with values in M are considered substituting the Banach space $C^0(B_R,H)$ on the uniform space $C^0(B_R,M)$.

12. Theorem. *If Conditions* 11(ii) *are satisfied, then Equation* (i) *has the unique solution in the bounded clopen ball* B_R. *If in addition Conditions* 11(iii,iv) *are satisfied, then*

$$(i) \quad f(t,\xi(t,\omega)) = f(t_0,\xi_0) + \sum_{m+b\ge 1, 0\le m\in\mathbf{Z}, 0\le b\in\mathbf{Z}} ((m+b)!)^{-1}$$

$$\times \sum_{l_1,\dots,l_m} \binom{m+b}{m} (\hat{P}_{u^{b+m-l},w(u,\omega)^l}[(\partial^{(m+b)} f/\partial u^b \partial x^m)(u,\xi(u,\omega))$$

$$\circ(a_{l_1,n_1} \otimes \dots \otimes a_{l_m,n_m}) \circ (I^{\otimes b} \otimes a^{\otimes(m-l)} \otimes E^{\otimes l})])|_{u=t},$$

where $l_1 + \dots + l_m = m + b - l$, $n_1 + \dots + n_m = l$, $l_1,\dots,l_m,n_1,\dots,n_m$ *are nonnegative integers.*

Proof. The first part of the theorem follows from Theorem 7.4 of Chapter 3 and embeddings of §10. Since the decomposition of unity satisfies the equality

$$\sigma_n \circ \sigma_m(t) = \sigma_n \circ \sigma_{m+j}(t)$$

for each $n \geq m$, $j > 0$ and

$$\sigma_0(t) = t_0,$$

there follows from Formula 5.1(4) of Chapter 3, that

$$\hat{P}_{u^{l+b},w^m} a_{l+b,m} \circ (I^{\otimes b} \otimes a^{\otimes l} \otimes E^{\otimes m})|_{u=t_n}^{u=t_{n+1}}$$

$$= a_{l+b,m}(t_n) \circ ((t_{n+1} - t_n)^{\otimes b} \otimes (a(t_n)(t_{n+1} - t_n))^{\otimes l} \otimes (E(t_n)(w(t_{n+1} - w(t_n)))^{\otimes m}),$$

where other arguments are omitted for shortening the notation. Therefore, the second part of this theorem follows from Formulas 6.6(iii) of Chapter 3 and 11(i).

13. Note. Let Conditions 11$(i-iv)$ be satisfied and $\phi = f$ be independent of the variable t. Then due to Lemma 6.3 of Chapter 3 and Theorem 12 above Formula 10(ii) is valid with the new operator J:

$$(i) \quad J(\phi, a, E) := \sum_{m=0}^{\infty} [m!]^{-1} \sum_{l_1,\dots,l_m} \hat{P}_{u^l,w^{m-l}} \phi^{(m+1)} \circ (a_{l_1,n_1} \dots a_{l_m,n_m})$$

$$\circ (a^{\otimes l} \otimes E^{\otimes (m-l)}),$$

where $l_1 + \dots + l_m = l$, $n_1 + \dots + n_m = m - l$.

14. Definition. Let (Π, M, π) be a bundle on a manifold M with fibres $X \oplus L(H,X)$ for each $x \in M$ and with transition functions

$$J(\phi, a, E) : (a, E) \mapsto (J(\phi, a, E)a, J(\phi, a, E)E),$$

where $\phi = \phi_{j,l}$ for each pair of charts (U_j, ϕ_j) and (U_l, ϕ_l) with $U_j \cap U_l \neq \emptyset$, a vector a is in X, an operator E is in $L(H,X)$, the operator $J(\phi, a, E)$ is given either by Equation 10(iii) or by 13(i).

15. Definition and Note. Let $t \in T \subset \mathbf{K}$, where $\mathbf{K}$ is a local field, let also T be a clopen subset in $\mathbf{K}$. Let also (U_j, ϕ_j) be a chart of a manifold M on a Banach space X over $\mathbf{K}$, $x \in U_j \subset M$, $(a, E) \in \pi^{-1}(x)$ (see §14). A collection of all M-valued random functions ξ such that $\xi \in U_j$ with probability 1, for which $\phi_j \circ \xi$ is a solution of Equation either 10(i) or 11(i) for each j, is denoted by $\mathcal{G}_x(a, E)$.

Then $\mathcal{G}_x(a, E)$ is called the germ of the diffusion process at the point x defined by a pair (a, E).

This germ is supplied in addition with a given family of sections $a_{l,m}$ of bundles $(\Pi_{l+m}, M, \pi_{l+m})$ with fibres $L_{m+l}(X^{\otimes m+l}; X)$ such that $a_{l,m,x} \in \pi_{l+m}^{-1}(x)$ in the case 11. Therefore, §10 is the particular case of §11.

A section $\mathcal{U}$ of the vector bundle (Π, M, π) is the non-archimedean analog of Itô's field over M.

16. Theorem. *Let ϕ and ψ be two functions satisfying conditions either of §10 or §11 such that $Dom(\phi) \supset Range(\psi)$, where $Dom(\phi)$ denotes a domain of definition of ϕ. Then the operator field J_x satisfies the equalities:*

$$(i) \quad J_{\psi(x)}(\phi, a, E) \circ J_x(\psi, a, E) = J_x(\phi \circ \psi, a, E),$$

$$(ii) \quad J_x(id, a, E) = id.$$

Proof. Since

$$a_{l,m,x} \in L_{l+m}(X^{\otimes l+m}; X),$$

the identity

$$J_x(\phi, a, E) a_{l,m,x} \circ (a^{\otimes l} \otimes E^{\otimes m}) = a_{l,m,x} \circ ((J_x(\phi, a, E)a)^{\otimes l} \otimes (J_x(\phi, a, E)E)^{\otimes m})$$

is accomplished for each $0 \le l, m \in \mathbf{Z}$ and $x \in M$, where $a = a_x$, $E = E_x$, $(a_x, E_x) \in \pi^{-1}(x)$. Each derivative $\phi^{(m)}$ and $\psi^{(m)}$ is a m-polylinear operator on X. Therefore, the m-th derivative operator is given by the equality:

$$(\phi \circ \psi)^{(m)}(x) = \sum_{l_1 + \dots + l_b \ge m, 1 \le b \le m} R_b \circ (Q_{l_1} \otimes \dots \otimes Q_{l_b}),$$

where R_b and Q_l are the b-linear and l-linear operators corresponding up to constant multipliers to

$$\phi^{(b)}(z)|_{z=\psi(x)} \quad \text{and} \quad \psi^{(l)}(x).$$

Then we deduce that

$$\sum_k Q_{l_j}(\Delta_k \xi_1, \dots, \Delta_k \xi_{l_j}) = \hat{P}_{u^{l_{j,1}}, w^{l_{j,2}}} Q_{l_j}(a^{\otimes l_{j,1}} \otimes E^{\otimes l_{j,2}})$$

for non-negative $l_{j,1}$ and $l_{j,2}$ with $l_{j,1} + l_{j,2} = l_j$ and

$$\xi_i(t, \omega) = \hat{P}_u a|_{u=t}$$

for $i = 1, \dots, l_{j,1}$,

$$\xi_i(t, \omega) = \hat{P}_w E|_{u=t} \text{ for } i = l_{j,1} + 1, \dots, l_j.$$

Moreover, one gets

$$\sum_k Q_{l_j}(\Delta_k \xi_1, \dots, \Delta_k \xi_{l_j - 1}, \xi_{l_j}) = \hat{P}_{u^{l_{j,1}}, w^{l_{j,2}}} Q_{l_j}(a^{\otimes l_{j,1}} \otimes E^{\otimes l_{j,2}}) v$$

for non-negative integers $l_{j,1}$ and $l_{j,2}$ $l_{j,1} + l_{j,2} = l_j - 1$ and the equality

$$\xi_i(t, \omega) = \hat{P}_u a|_{u=t}$$

is fulfilled for each $i = 1, \dots, l_{j,1}$. Moreover,

$$\xi_i(t, \omega) = \hat{P}_w E|_{u=t}$$

for all $i = l_{j,1}+1,\dots,l_j-1$, $\xi_{l_j} = v$, where either $v = a$ or $v = E$. Therefore, the mapping ϕ is so that

$$\phi : \mathcal{G}_x(a,E) \to \mathcal{G}_{\phi(x)}(Ja,JE),$$

where $J = J(\phi,a,E)$ is the short notation. In view of Theorems 6.6 of Chapter 3 and 12 in this section, Formulas $10(iii)$ and $13(i)$ the equality

$$(iii)\quad J_x(\phi,a,E) = \phi'(\xi^0_x)$$

is satisfied, where ξ^0 is a stochastic process being the solution either of Equation $10(i)$ or $11(i)$, $x \in M$, $\xi^0_x \in T_xM$. On the other hand, the derivative of the composition is:

$$(\phi(\psi)(x))' = \phi'(\psi(x)).\psi'(x)$$

for each $x \in Dom(\psi)$. Therefore, from $\xi \in Dom(\psi)$ and $Range(\psi) \subset Dom(\phi)$, Formula $16(i)$ follows. Evidently, $id' = I$, where I is a unit operator, and this implies Formula $16(ii)$.

17. Remark and Definition. The bundle associated with the operator $J(\phi,a,E)$ in general is not quadratic. This bundle may be polynomial only in a particular case, when $a_{l,m,x} = 0$ for all $l+m > q$ and each $x \in M$, where q is some marked natural number.

Let us consider the function $f = \exp$, where $\exp := \exp^M$ is the exponential mapping for M. Take $\mathcal{G}_{(x,0)}(a,E)$ a stochastic processes germ at a point $y = 0$ in the tangent space T_xM. Then certainly the equality

$$\exp^*_x \mathcal{G}_{(x,0)}(a,E) = \mathcal{G}_x(J(\exp_x,a,E))(a,E)$$

defines a stochastic processes germ at a point $x \in M$. Therefore, we get that

$$\phi_j \circ \exp^*_x \mathcal{G}_{(x,0)}(a,E) = \mathcal{G}_{\phi(x)}(J(\phi_j \circ \exp_x,a,E))(a,E)$$

for each chart (U_j,ϕ_j) of the manifold M.

The germ $\exp^*_x \mathcal{G}_{(x,0)}(a,E)$ is called the stochastic differential bundle.

18. Corollary. *To a functor J a bundle (J^M,M,π_J) corresponds and a fibre*

$$J^M_x := \pi_J^{-1}(x)$$

may be identified with the space $\mathcal{G}_x(J^M)$ of stochastic processes germ. To a morphism $f : M \to N$ of manifolds a bundle morphism $\mathcal{G}(f) = f * f^*$ corresponds, where $f^*\xi := f(\xi)$.

Proof. If $f : M \to N$ is a manifold morphism, then $\mathcal{U}$ is transformed in accordance with the formula:

$$(a_x,E_x) \mapsto (a^f_{f(x)},E^f_{f(x)}),$$

where $a^f_{f(x)} = J(f,a,E)a_x$ and $E^f_{f(x)} = J(f,a,E)E_x$, $a^f_{l,m,f(x)}(t,f^*\xi) = a_{l,m,x}(t,\xi)$ for each $x \in M$. The stochastic process ξ^0_x satisfies the anti-derivational equation

$$(i)\quad \xi^0_x = \sum_{l,m} \hat{P}_{u^l,w^m} a_{l,m,x} \circ (a_x^{\otimes l} \otimes E_x^{\otimes m})$$

and its differential has the form:

$$(ii)\quad d\xi_x^0 = \sum_{l,m} l\hat{P}_{u^{l-1},w^m} a_{l,m,x} \circ (a_x^{\otimes(l-1)} \otimes E_x^{\otimes m}) a_x dt$$

$$+ \sum_{l,m} m\hat{P}_{u^l,w^{m-1}} a_{l,m,x} \circ (a_x^{\otimes l} \otimes E_x^{\otimes(m-1)}) E_x dw_x.$$

Hence

$$(iii)\quad f(\xi_x^0(t,\omega)) = \sum_{l,m} \hat{P}_{u^l,w^m} a^f_{l,m,f(x)}(u, f(\xi_x^0(u,\omega))) \circ ({a^f_{f(x)}}^{\otimes l} \otimes {E^f_{f(x)}}^{\otimes m})|_{u=t}.$$

Therefore, we get the mapping

$$f^* : \exp_x^{M*}(d\xi_x^0) \mapsto \exp_{f(x)}^{N*}(f^* d\xi_x^0),$$

where

$$(iv)\quad f^* d\xi_x^0 = \sum_{l,m} l\hat{P}_{u^{l-1},w^m} a^f_{l,m,f(x)} \circ \left({a^f_{f(x)}}^{\otimes(l-1)} \otimes {E^f_{f(x)}}^{\otimes m}\right) a^f_{f(x)} dt$$

$$+ \sum_{l,m} m\hat{P}_{u^l,w^{m-1}} a^f_{l,m,f(x)} \circ \left({a^f_{f(x)}}^{\otimes l} \otimes {E^f_{f(x)}}^{\otimes(m-1)}\right) E^f_{f(x)} df(w_x)$$

for f-related mappings $\exp^M$ and $\exp^N$.

19. Theorem. *Let* $\exp : \tilde{T}M \to M$ *be the exponential mapping of a manifold* M*. Then the mapping*

$$J(\exp, a, E) : J^{\tilde{T}M} \to J^M$$

is a bundle morphism. If (U,ϕ) *is a chart of* M*, then the formula*

$$(i)\quad J(\exp^\phi_{\phi(x)}, a^\phi_x, E^\phi_x)(a^\phi_x, E^\phi_x) = (Sa^\phi_x, SE^\phi_x)$$

is accomplished, where

$$S := \left(d[\phi \circ \exp_x \circ [\phi'(x)]^{-1}](z)/dz\right)\Big|_{z=\xi_x^0}.$$

Proof. The first statement of the theorem follows from Theorem 16 and Corollary 18 above. Next we consider the mapping

$$F(z) := [\phi \circ \exp_x \circ [\phi'(x)]^{-1}](z)$$

for a chart $(TU_j, T\phi_j)$ of TM, where $\phi = \phi_j$. The mapping F is the local representation of the exponential mapping exp in terms of coordinate mappings. Hence

$$J(\exp^\phi_{\phi(x)}, a^\phi_x, E^\phi_x) = [dF(z)/dz]|_{z=\xi_x^0},$$

where ξ_x^0 is a solution of Equation either $10(i)$ or $11(i)$ in T_xM. In particular, one has

$$F'(0) = id \quad \text{and} \quad F''(0).(v,v) = -\Gamma(x)(v,v),$$

but generally ξ^0_x may be nonzero.

20. Definition. Let $\mathcal{U}$ be a section of the bundle (Π, M, π). We consider the differential

$$(i) \quad d\xi(t,\omega) = \exp^*_{\xi(t,\omega)} \mathcal{G}\left(a_{\xi(t,\omega)}, E_{\xi(t,\omega)}\right)$$

and the corresponding anti-derivational equation:

$$(ii) \quad \xi(t,\omega) = \exp_{\xi(t,\omega)} \Big\{ \sum_{l,b,m} \hat{P}_{u^{l+b},w^m} a_{b+l,m,\xi(t,\omega)}(u, \xi(u,\omega)))$$

$$\circ (I^{\otimes b} \otimes a^{\otimes l}_{\xi(t,\omega)} \otimes E^{\otimes m}_{\xi(t,\omega)})|_{u=t}\Big\}.$$

Suppose that there exists a neighborhood $V_x \ni x$ and a stochastic process belonging to the germ

$$\exp_x(\mathcal{G}(a_x, E_x)) = \mathcal{G}_x(J(\exp_x, a_x, E_x))(a_x, E_x)$$

such that $P_{s,x}\{\omega : \xi_x(t,\omega) \in V_x, t \neq s\} = 1$, where

$$P_{s,x}(W) := P(W : \xi(s,\omega) = x), \quad W \in \mathsf{F}.$$

If this is satisfied for $\nu_{\xi(s)}$-a.e. $x \in M$, then it is said, that $\xi(t,\omega)$ possesses a stochastic differential governed by the field $\mathcal{U}$, where

$$\nu_{\xi(s)}(*) := P \circ \xi^{-1}(s, *).$$

An M-valued ξ satisfying Formula (ii) is called an integral process of the field $\mathcal{U}(t)$.

21. Definition. An atlas $At(M) = \{(U_j, \phi_j) : j\}$ of a manifold M on a Banach space X over $\mathbf{K}$ is called uniform, if its charts satisfy the following conditions:

$(U1)$ for each $x \in M$ there exist neighborhoods $U^2_x \subset U^1_x \subset U_j$ such that for each $y \in U^2_x$ the inclusion $U^2_x \subset U^1_y$ is fulfilled;

$(U2)$ the image $\phi_j(U^2_x) \subset X$ contains a ball of the fixed positive radius $\phi_j(U^2_x) \supset B(X,0,r) := \{y : y \in X, \|y\| \leq r\}$; $(U3)$ for each pair of intersecting charts (U_1, ϕ_1) and (U_2, ϕ_2) transition mappings $\phi_{l,j} = \phi_l \circ \phi_j^{-1}$ are such that the suprema

$$\sup_x \|\phi'_{l,j}\| \leq C \quad \text{and} \quad \sup_x \|\phi_{l,j}(x)\| \leq C$$

are finite, where $C = const > 0$ does not depend on ϕ_l and ϕ_j.

22. Remark. Take a measurable space $(M, \mathcal{L})$, where $\mathcal{L}$ is a σ-algebra on M, define a random mapping $S(t,\tau;\omega) : M \to M$ for each $t, \tau \in T$ by the formula

$$x \mapsto S(t,\tau;\omega;x) = S(t,\tau;\omega) \circ x.$$

Let

(1) the mapping $x \times \omega \mapsto S(t,\tau;\omega;x)$ be $\mathcal{L} \times \mathsf{F}$-measurable for each $t, \tau \in T$;

(2) the random variable $S(t,\tau;\omega;x)$ be F-measurable and does not depend on F for each t, τ, x. Suppose that all others conditions of §7.9 in Chapter 3 also are satisfied with the notation $S(t,\tau;\omega)$ here instead of $T(t,s;\omega)$ there.

23. Proposition. *Let ξ be a stochastic process given by Equation* $11(i)$ *and let also the estimate*

$$\max(\|a(t,\omega,x)-a(v,\omega,x)\|,\|E(t,\omega,x)-E(v,\omega,x)\|)\le|t-v|(C_1+C_2\|x\|^b)$$

be satisfied for each t and $v\in B(\mathbf{K},t_0,R)$ λ-almost everywhere by $\omega\in\Omega$, where b, C_1 and C_2 are non-negative constants. Then ξ with the probability 1 *has a C^0-modification and the inequality*

$$q(t)\le\max\{M\|\xi_0\|^s,|t-t_0|(C_1+C_2q(t))\}$$

is satisfied for each $t\in B(\mathbf{K},t_0,R)$, where

$$q(t):=\sup_{|u-t_0|\le|t-t_0|}M\|\xi(u,\omega)\|^s$$

and $\mathbf{N}\ni s\ge b\ge 0$. Moreover, if

$$\lambda\{\omega:w(t,\omega)\in C^0\setminus C^1\}=0,$$

then for λ-a.e. elementary events ω the derivative ξ' exists and $\lambda\{\omega:\xi(t,\omega)\in C^0\setminus C^1\}=0$.

Proof. In view of Theorem 12 applied to the function $f(t,x)=x^s$ the equality

$$f(t,\xi(t,\omega))=f(t_0,\xi_0)+\sum_{k=1}^{s}\sum_{l_1,\dots,l_k}\left(\hat{P}_{u^{k-l},w(u,\omega)^l}\left[\left(\binom{s}{k}\xi(t,\omega)^{s-k}(u,\xi(u,\omega))\right.\right.\right.$$

$$\left.\left.\left.\circ(a_{l_1,n_1}\otimes\dots\otimes a_{l_k,n_k})\circ(a^{\otimes(k-l)}\otimes E^{\otimes l})\right)\right]\right)\Big|_{u=t}$$

is satisfied, where $l_1+\dots+l_k=k-l$, $n_1+\dots+n_k=l$. From Conditions of §11 and in particular $11(ii)$ the inequality

$$M\|\xi(t,\omega)\|^s\le\max(M\|\xi_0\|^s,|t-t_0|d(\hat{P}^s_*)(C_1+C_2\sup_{|u-t_0|\le|t-t_0|}M\|\xi(u,\omega)\|^s)$$

follows, since $|t_j-t_0|\le|t-t_0|$ for each $j\in\mathbf{N}$ and

$$M\|\xi(t,\omega)-\xi(v,\omega)\|^s\le|t-v|(1+C_1+C_2d(\hat{P}^s_*)\sup_{|u-t_0|\le\max(|t-t_0|,|v-t_0|)}M\|\xi(u,\omega)\|^s),$$

since $|t_j-v_j|\le|t-v|+\rho^j$ for each $j\in\mathbf{N}$, where $0<\rho<1$,

$$d(\hat{P}^s_*):=\sup_{(a\ne0,E\ne0,f\ne0,a_{l_j,n_j}\ne0,j=1,\dots,k)}\max_{s\ge k\ge l\ge0}\Big\|(k!)^{-1}\hat{P}_{u^{k-l},w^l}(\partial^kf/\partial^kx)$$

$$\circ(a_{l_1,n_1}\otimes\dots\otimes a_{l_k,n_k})$$

$$\circ(a^{\otimes(k-l)}\otimes E^{\otimes l})\Big\|/\left(\|a\|^{k-l}_{C^0(B_R,H)}\|E\|^l_{C^0(B_R,L(H))}\|f\|_{C^s(B_R,H)}\prod_{j=1}^k\|a_{l_j,n_j}\|\right),$$

consequently, $d(\hat{P}^s_*) \le 1$, since $f \in C^s$ as a function by x and $s!(\bar{\Phi}^s g)(x;h_1,\dots,h_s;0,\dots,0) = D^s_x g(x).(h_1,\dots,h_s)$ for each $g \in C^s$ and due to the definition of $\|g\|_{C^s}$. Considering in particular any poly-homogeneous function g on which $d(\hat{P}^s_*)$ takes its maximum value we get

$$d(\hat{P}^s_*) = 1.$$

From conditions on w, $a_{l,k}$, a and E it follows, that $\xi(t,\omega)$ with the probability 1 has a C^0-modification (see Theorem 7.4 in Chapter 3), $\xi \in L^q(\Omega, \mathsf{F}, \lambda; C^0(B_R, H))$.

The last statement of this proposition follows from Lemma 6.3 of Chapter 3.

24. Theorem. *Suppose that M is a manifold either satisfying conditions of Corollary 8 or M is analytic, its atlas $At(M)$ is uniform (see §21). Let a, E, $a_{m,l}$ and w corresponding to a section $\mathcal{U}$ satisfy conditions of §11 with $\lambda\{\omega : w(t,\omega) \in C^0 \setminus C^1\} = 0$. Then a unique up to stochastic equivalence random evolution family $S(t,\tau;\omega)$ exists for a solution $\xi(t,\omega)$ of Equation 20(ii).*

Proof. Let us consider a solution of the stochastic anti-derivation equation:

$$(i) \quad \xi(t,\omega) = \exp_{\xi(t,\omega)}\Big\{ \sum_{m,b,l} \hat{P}_{u^{m+b},w^l} a_{m+b,l,\xi(t,\omega)}(u,\xi(u,\omega))$$

$$\circ (I^{\otimes b} \otimes a^{\otimes m}_{\xi(t,\omega)} \otimes E^{\otimes l}_{\xi(t,\omega)})|_{u=t}\Big\}$$

corresponding to 20(i). On each chart of the uniform atlas $At(M)$ of the manifold M random fields $\{a_{m,l} : m,l\}$, a, E and w are λ-a.e. bounded due to conditions of §11. For each two charts (U_j,ϕ_j) and (U_l,ϕ_l) with the non-void intersection $U_j \cap U_l \ne \emptyset$ a transition mapping $\phi := \phi_{j,l}$ is bounded together with its derivatives, hence the Christoffel symbol operator field Γ is bounded on each chart U_j, since the covering $\{U_j : j\}$ of M can be chosen locally finite due to para-compactness of M [19].

In view of Theorem 7.4 of Chapter 3, Corollary 18 and Theorem 19 Equation (i) has a unique solution on the manifold M. Let (a,E) be a section of the bundle (Π, M, π) and $a_{l,m}$ be sections of the bundles $(\Pi_{l+m}, M, \pi_{l+m})$ (see §§14 and 15). Take a family $\zeta_y(x)$ of functions on M of the class $C^1(M,\mathbf{K})$ such that $\zeta_y(x) = 0$ if $x \notin U^1_y$, $\zeta_y(x) = 1$ if $x \in U^2_y$ of the uniform atlas (see §21), then we put

$$a^y_x := \zeta_y(x) a_x, \; E^y_x := \zeta_y(x) E_x, \; a^y_{l,m,x} := \zeta_y(x) a_{l,m}$$

for such local function fields. Then there exists the local evolution family $S_y(t,\tau;\omega)$ for each local solution (that is, with local coefficients):

$$(ii) \quad \xi^y(t,\omega) = \exp_{\xi^y(t,\omega)}\Big\{ \sum_{m,b,l} \hat{P}_{u^{m+b},w^l} a^y_{m+b,l,\xi^y(t,\omega)}$$

$$(u,\xi(u,\omega)) \circ (I^{\otimes b} \otimes (a^y_{\xi^y(t,\omega)})^{\otimes m} \otimes (E^y_{\xi^y(t,\omega)})^{\otimes l})|_{u=t}\Big\}$$

due to Theorem 7.4 of Chapter 3 and Theorem 12 above. Therefore, we have $S_y(t,\tau;\omega) \circ x \in U^1_y$ for each $x \in U^2_y$. Gluing together local solutions with the help of transition functions $\phi_{l,j}$ of charts with non-void intersections $U_l \cap U_j$ leads to the conclusion that a stochastic process ξ is a solution of the stochastic anti-derivational equation (i) on the manifold M if and only if for each point $t \in T$ for $\nu_{\xi(t)}$-a.e. points $x \in M$ it coincides $P_{t,x}$-a.e. with some local solution of this equation inside U_x^2, since $\{U_x^2 : x \in M\}$ is the covering of M.

Consider a local representation $\xi^\phi := \phi(\xi)$. For it the evolution family S^ϕ corresponds generated by the differential $d\xi^\phi$ such that

$$S^\phi(t,\tau;\omega)\circ\xi^\phi(\tau,\omega) = \phi(S(t,\tau;\omega)\circ\xi(\tau,\omega))$$

for each t and $\tau \in T \subset \mathbf{K}$, $\phi \in \{\phi_j : j\}$.

In view of Proposition 23 a positive number $\delta > 0$ exists such that

$$P\{\omega : S_x(t,\tau;\omega)\circ x \notin U_x^2\} \le P\{\sup \|\phi(S_x(t,\tau;\omega)\circ x)\| > 1\} \le C|t-\tau|$$

for each $t,\tau \in T$ such that $|t-\tau| < \delta$, where $C > 0$ is a constant. We consider a family Υ of all finite partitions q of T into disjoint unions of balls $B(\mathbf{K},t_k^q,r_k^q)$, where $t_k^q \in T$, $0 < r_k^q \le \varepsilon_q$, $0 < \varepsilon_q < \delta$ for each $q \in \Upsilon$. Let $q \le v$ if and only if $q \subset v$, then Υ is ordered by this relation. We consider a linearly ordered subsequence

$$\Upsilon_0 := \{q_k : \ k \in \mathbf{N}\}$$

in Υ with the zero limit

$$\lim_{k\to\infty} \sup\{r_j^{q_k} : j \in q_k\} = 0$$

and for it we define the random function

$$\xi_k(t,\omega) := S_{\xi_{k-1}(t_l,\omega)}(t,t_l;\omega)\circ\xi_{k-1}(t_l,\omega)$$

for each natural number k and each point $t \in B(\mathbf{K},t_l^v,r_l^v)$ for each $t_l \in q_{k-1}$ and each $k \ge 1$, where $v = q_{k-1}$, $\xi(t_0,\omega) = x$, $\xi_1(t_l,\omega) := \xi(t_l,\omega)$ for each $t_l \in q_1$. Thus the evolution family

$$S^k(t,t_0;\omega)\circ x = S_{\xi_{k-1}(t_l,\omega)}(t,t_l;\omega)\circ\xi_{k-1}(t_l)$$

is also defined. Consider the random function

$$z(s,\omega) := S_y(s,t_k^q;\omega)\circ y \in U_y^2$$

for each point $s \in B(\mathbf{K},t_k^q,r_k^q)$. For each point $t \in B(\mathbf{K},t_k^q,r_k^q)$ the equality

$$S_y(t,t_k^q;\omega)\circ y = S_{z(s,\omega)}(t,s;\omega)\circ z(s,\omega)$$

is fulfilled, since

$$S_y(t,t_k^q;\omega)\circ y = S_y(t,s;\omega)\circ S_y(s,t_k^q;\omega)\circ y$$

due to the existence of a local solution.

Next we put

$$\Omega^{\Upsilon_0} := \bigcup_{k\in\Upsilon_0} \Omega_k,$$

where $\Omega_k := \bigcap_{l\in q_k} \Omega_{k,l}$, while

$$\Omega_{k,l} := \{\omega : S_{\xi_k(t_l,\omega)}(s,t_l)\circ\xi_k(t_l,\omega) \in U^2_{\xi_k(t_l,\omega)}, s \in B(\mathbf{K},t_l^v,r_l^v)\}.$$

From the existence of a local solution the equality

$$S^k(t,t_0)\circ x = S^l(t,t_0)\circ x$$

follows for each $k \geq l$ and each $\omega \in \Omega_l$. In view of Theorem 19 the limit

$$\lim_{q_k \in \Upsilon_0} S^k(t,t_0;\omega) = S(t,t_0;\omega)$$

exists. For each two linearly ordered subsets Λ_1 and Λ_2 in Υ a linearly ordered subset Λ in Υ exists such that

$$\Lambda \supset \Lambda_1 \cup \Lambda_2,$$

consequently, this limit does not depend on the choice of Υ_0. Events $\Omega_{k,l}$ and $\Omega_{k,j}$ are independent in total for each $l \neq j$:

$$P(\Omega_{k,l} \cap \Omega_{k,j}) = P(\Omega_{k,l})P(\Omega_{k,j}).$$

Since $\mathbf{K}$ is the local field, its residue class field k is finite and of positive characteristic $p = char(k)$ [111]. On can choose Υ_0 such that for each $q_k \in \Upsilon_0$ the supremum

$$\sup_{l \in q_k} r_l^{q_k} =: \delta_k \leq p^{-k}$$

and

$$card(t_l : t_l \in q_k \cap B(\mathbf{K}, t_0, p^s)) =: m_{k,s} \leq p^{snk}.$$

In view of the ultra-metric inequality from the inequality

$$\|\alpha(\omega) + \beta(\omega)\| \geq \delta$$

it follows, that

$$\max(|\alpha(\omega)|, |\beta(\omega)|) \geq \delta$$

for each two random variables α and β. Therefore, from Proposition 23 applied to the difference $\phi_j(\xi) - \phi_j(\xi_0)$, and the inclusion

$$\xi(t,\omega) \in L^q(\Omega, \mathsf{F}, \lambda; C^0(T,M))$$

the inequality

$$P\{\Omega_k : t \in T \cap B(\mathbf{K}, t_0, p^s)\} \geq (1 - C_k p^{-k})^{p^{snk}}$$

follows, where

$$\lim_k C_k = 0,$$

since $\xi(t,\omega)$ is uniformly continuous by the variable t on $T \cap B(\mathbf{K}, t_0, p^s)$ for λ-a.e. elementary events ω. Therefore,

$$P\{\Omega^{\Upsilon_0} : t \in T \cap B(\mathbf{K}, t_0, p^s)\} \geq \lim_k \exp(-C_k sn) = 1$$

for each given natural number $s \in \mathbf{N}$.

From the equality

$$S^k(t,t_0;\omega)\circ x = S^k(t,s;\omega)\circ S^k(s,t_0;\omega)\circ x$$

and taking the limit by $q \in \Upsilon$ it follows, that S satisfies the evolution property

$$S(t,t_0;\omega)\circ x = S(t,s;\omega)\circ S(s,t_0;\omega)\circ x.$$

Then $S(t,t_0;\omega)\circ x$ is a measurable function of x, since it is the following superposition

$$S(t,t_0;\omega)\circ x = S^k(t,t_0;\omega)\circ x$$

of locally measurable functions.

25. Corollary. *Let conditions of Theorem* 24 *be satisfied, let also f be a function on $T\times M$ such that each composition $f\circ\phi_j^{-1}$ satisfies Conditions* $11(iii,iv)$ *on its domain. Then a generating operator of an evolution family $S(t,\tau;\omega)$ of a stochastic process*

$$\eta(t,\omega) = f(t,\xi(t,\omega))$$

is given by the equation:

$$(i)\quad A(t;\omega)\eta(t,\omega) = \sum_{m+b\ge 1, 0\le m\in\mathbf{Z}, 0\le b\in\mathbf{Z}} ((m+b)!)^{-1}$$

$$\times \sum_{l=0}^{m} \binom{m+b}{m} \sum_{l_1+\dots+l_m=m-l, n_1+\dots+n_m=l} \{b(\hat{P}_{u^{b+m-l-1},w(u,\omega)^l}$$

$$\times[(\partial^b\nabla^m f/\partial u^b\partial x^m)(u,\xi(u,\omega))\circ(a_{l_1,n_1}\otimes\dots\otimes a_{l_m,n_m})\circ(I^{\otimes(b-1)}\otimes a^{\otimes(m-l)}\otimes E^{\otimes l})])|_{u=t}$$

$$+(m-l)(\hat{P}_{u^{b+m-l-1},w(u,\omega)^l}[(\partial^b\nabla^m f/\partial u^b\partial x^m)(u,\xi(u,\omega))\circ(a_{l_1,n_1}\otimes\dots\otimes a_{l_m,n_m})$$

$$\circ(I^{\otimes b}\otimes a^{\otimes(m-l-1)}\otimes E^{\otimes l})]a)|_{u=t}$$

$$+l(\hat{P}_{u^{b+m-l},w(u,\omega)^{l-1}}[(\partial^b\nabla^m f/\partial u^b\partial x^m)(u,\xi(u,\omega))\circ(a_{l_1,n_1}\otimes\dots\otimes a_{l_m,n_m})$$

$$\circ(I^{\otimes b}\otimes a^{\otimes(m-l)}\otimes E^{\otimes(l-1)})]Ew'_u(u,\omega))|_{u=t}\}.$$

Proof. In view of Theorem 24 a generating operator $S(t,\tau;\omega)$ of an evolution family exists. For each chart (U_j,ϕ_j) the stochastic process $f\circ\phi_j^{-1}(\xi)$ is given by Equation $12(i)$. We consider the covariant differentiation

$$(\nabla f/\partial x).h = \nabla_h f$$

on the manifold M, where $h\in T_xM$. For a random variable belonging to $L^q(\Omega,\mathsf{F},\lambda;C^1(M,X))$ its derivative and partial difference quotients $\bar{\Phi}^1 f\circ\phi_j^{-1}(x;h;b)$ are naturally understood as elements of the corresponding spaces $L^q(\Omega,\mathsf{F},\lambda;$ $C^0(W_j,X))$ such that each limit $\lim_{x\to x_0} g(x,\omega) = c(\omega)$ is taken in the space

$L^q(\Omega,\mathsf{F},\lambda;C^0(M,X))$, where $W_j := \{(x,h,b)\in U_j\times X\times \mathbf{K} : x+bh\in U_j\}$. In another words it exists if and only if the limit

$$\lim_{x\to x_0}\|g(x,\omega)-c(\omega)\|_{L^q}=0$$

is zero, where $c\in L^q(\Omega,\mathsf{F},\lambda;X)$. Then it implies:

$$f(t,\xi(t,\omega))=\lim_k f(t,S^k(t,t_0;\omega)\circ x).$$

For each chart we put

$$f_j(t,*) := f(t,\phi_j^{-1}(*)\phi_j\circ S_y),$$

where $S_y(t,\tau;\omega)y$ does not leave for λ-a.e. $\omega\in\Omega$ a clopen subset U_j in the manifold M for each t and $\tau\in T_j$, $T_j\subset T$, where $\bigcup_j T_j=T$, T_j is a clopen subset in T. Then we define the generating operators

$$(f(t,S^k(t,\tau;\omega)x)'_x.h=f'_x(t,S^k(t,\tau)\circ x)A^k(t,\tau;\omega)h$$

and take their limit while k tends to the infinity. From Lemma 6.3 of Chapter 3 the statement of this corollary follows.

26. Remarks. In the particular case of §10 Formula $25(i)$ simplifies. When the family of Γ together with all its covariant derivatives along a and Ew' is equi-uniformly bounded on each chart U_j, then Formula $25(i)$ can be written in another form using the identity

$$\nabla^{m+1}f.(h_1,\dots,h_m)=\nabla_{h_{m+1}}(\nabla^m f.(h_1,\dots,h_m))$$

$$-\sum_{l=1}^{m}(\nabla^m f).(h_1,\dots,h_{l-1},\nabla_{h_{m+1}}h_l,h_{l+1},\dots,h_m),$$

in particular with $h_l\in\{a,Ew'\}$.

Take on the locally convex space $C^0(T,c_0(\alpha,\mathbf{K}))$ a weaker topology making it linearly topologically isomorphic with $c_0(\alpha,\mathbf{K})^T$ supplied with the product (Tychonoff) topology. The product topology is weaker than the box topology (see [87]). Let $\theta:\mathbf{K}\to\mathbf{R}$ be a continuous surjective quotient mapping such that

$$\theta(B(\mathbf{K},0,1))=[0,1]$$

(see, for example, [19]). Then for each random function

$$\xi\in L^q(\Omega,\mathsf{F},\lambda;c_0(\alpha,\mathbf{K})^T)$$

the random function

$$\theta(\xi)\in L^q(\Omega,\mathsf{F},\lambda;c_0(\alpha,\mathbf{R})^{\theta(T)})$$

exists, that induces a surjective mapping θ^* from the space $L^q(\Omega,\mathsf{F},\lambda;c_0(\alpha,\mathbf{K})^T)$ onto the space $L^q(\Omega,\mathsf{F},\lambda;c_0(\alpha,\mathbf{R})^{\theta(T)})$. Therefore, for each random function η in the space $L^q(\Omega,\mathsf{F},\lambda;c_0(\alpha,\mathbf{R})^{\theta(T)})$ a random function ξ in the space $L^q(\Omega,\mathsf{F},\lambda;c_0(\alpha,\mathbf{K})^T)$ can be counterposed such that

$$\theta(\xi)=\eta.$$

A non-archimedean normed field $\mathbf{K}$ can be considered as the module over the ring $B(\mathbf{K},0,1)$. We take rings B_{π^n} isomorphic with the quotient $B(\mathbf{K},0,1)/\Xi_n$ of the unit ball $B(\mathbf{K},0,1)$ in the normed locally compact field $\mathbf{K}$ by the equivalence relation Ξ_n associated with the disjoint subsets $x_j+\pi^n B(\mathbf{K},0,1)$ in $B(\mathbf{K},0,1)$, $j=0,1,2,\dots,x_0:=0$, where $\pi\in\mathbf{K}$ with the norm $|\pi|=\max\{|x|:|x|<1,x\in\mathbf{K}\}<1$, $n=1,2,3,...$, since $\pi^n B(\mathbf{K},0,1)$ is the ideal in the commutative ring $B(\mathbf{K},0,1)$ (see also [5, 111]). On the other hand, the set $\mathbf{K}$ we can topologize as a projective limit of an inverse sequence of the discrete topological spaces $\mathbf{S}_{\pi^n}=\mathbf{K}/\Xi_n$, where $n\in\mathbf{N}$. When the field $\mathbf{K}$ is locally compact, each set $\mathbf{S}_{\pi^n}$ is countable. The projective limit topology is inherited from the product topology on the countable product of these discrete topological spaces $\mathbf{S}_{\pi^n}$, which is metrizable (see §§2.4, 2.5 and 7.3.15 [19]).

Therefore, the topological space $L^q(\Omega,\mathsf{F},\lambda;c_0(\alpha,\mathbf{K})^{\mathbf{K}})$ is isomorphic to the projective limit of the topological spaces $L^q(\Omega,\mathsf{F},\lambda;c_0(\alpha,\mathbf{S}_{\pi^n})^{\mathbf{S}_{\pi^n}})$, since simple functions are dense in L^q. Thus ξ is equal to the projective limit of stochastic processes with values in discrete topological spaces $c_0(\alpha,\mathbf{S}_{\pi^n})$.

This opens a possibility of approximation of stochastic processes by stochastic processes with values in discrete modules. Certainly there is not any simple relation between classical and non-archimedean stochastic equations, so if ξ satisfies definite stochastic antiderivational equation relative to w it is a problem to find a classical stochastic equation to which $\theta(\xi)$ satisfies relative to either $\theta(w)$ or a standard stochastic process (Brownian motion, Lèvy) and vice versa.

Particularly, one can consider a sequence of Markov's stochastic processes with a finite number of states to approximate processes with a continuum of states over $\mathbf{K}$ or $\mathbf{R}$.

Theorem 24 and Corollary 25 are applicable in particular to totally disconnected Lie groups over non-archimedean fields.

Chapter 5

Random Functions in Topological Groups

5.1. Introduction

This chapter is devoted to stochastic processes on a totally disconnected topological group which is complete, separable and ultrametrizable. In particular stochastic processes on diffeomorphism groups and wrap groups of manifolds on Banach spaces over a local field are considered. These groups were defined and investigated in previous articles of the author (see [55, 56, 62, 66, 82]). These groups are non-locally compact and for them the Campbell-Hausdorff formula is not valid in an open local subgroup. In this chapter topological groups satisfying locally the Campbell-Hausdorff formula also are considered.

As it is well-known finite-dimensional Lie groups satisfy locally the Campbell-Hausdorff formula. This is guarantied, if impose on a locally compact topological Hausdorff group G two conditions: it is a C^∞-manifold and the following mapping $(f,g) \mapsto f \circ g^{-1}$ from $G \times G$ into G is of class C^∞. But for the infinite-dimensional group G the Campbell-Hausdorff formula does not follow from these conditions. Frequently topological Hausdorff groups satisfying these two conditions also are called Lie groups, though they can not have all properties of finite-dimensional Lie groups, so that the Lie algebras for them do not play the same role as in the finite-dimensional case. Therefore Lie algebras for them are not so helpful.

If G is a Lie group and its tangent space T_eG is a Banach space, then it is called a Banach-Lie group. Sometimes for a Lie group it is undermined, that it satisfies the Campbell-Hausdorff formula at least locally for a Banach-Lie algebra T_eG. In some papers the Lie group terminology undermines, that it is finite-dimensional.

In view of this it is worthwhile to call the Lie group satisfying the Campbell-Hausdorff formula locally in an open local subgroup by the Lie group in the narrow sense. In the contrary case one can call them by Lie groups in the broad sense.

In this chapter also theorems about a quasi-invariance and a pseudo-differentiability of transition measures of random functions on the totally disconnected topological group G relative to the dense subgroup G' are proved. In each concrete case of G it its necessary to construct a stochastic process and G'. Below path spaces, wrap spaces, wrap monoids, wrap groups and diffeomorphism groups are considered not only for finite-dimensional, but also for infinite-dimensional over a field **K** manifolds.

It is worthwhile to mention that wrap and diffeomorphism groups are important for the development of the representation theory of non-locally compact groups. Their theory has many principal differences with the traditional representation theory of locally compact groups and finite-dimensional Lie groups. This is caused by the fact that non-locally compact groups have not any non-trivial C^*-algebras associated with a quasi-invariant measure. Frequently such groups have not underlying Lie algebras and as the consequence any relations between representations of groups and underlying algebras.

In accordance with the A. Weil theorem if a topological Hausdorff group G has a non-trivial non-negative quasi-invariant Borel σ-additive measure relative to the entire G, then G is locally compact. Thus each non locally compact group can not have a non-trivial quasi-invariant measure relative to the entire group, but only relative to proper subgroups G'.

In this chapter notations and definitions from the preceding chapters and the cited above works are used.

5.2. Stochastic Anti-derivational Equations and Measures on Totally Disconnected Topological Groups

1. Note. Let X be a Banach space over a local field $\mathbf{K}$. Suppose M is an analytic manifold modelled on the Banach space X with an atlas $At(M)$ consisting of disjoint clopen charts (U_j,ϕ_j), $j\in\Lambda_M$, $\Lambda_M\subset\mathbf{N}$. That is, U_j and $\phi_j(U_j)$ are clopen in M and X respectively, $\phi_j : U_j\to\phi_j(U_j)$ are homeomorphisms, each set $\phi_j(U_j)$ is bounded in the Banach space X.

2. Note. Let $\Omega_\xi^{\{k\}}(M,N)$ be the wrap submonoid as in Appendix E below such that $c>0$ and $c'>0$. Then it generates the wrap group $G' := L_\xi^{\{k\}}(M,N)$ as in Appendix E such that G' is the dense subgroup in $G = L_\xi(M,N)$.

3. Remarks. Let M be a manifold on the Banach space X with an atlas $At(M)$ consisting of disjunctive charts (U_j,ϕ_j), $j\in\Lambda$, $\Lambda\subset\mathbf{N}$, where U_j and $\phi_j(U_j)$, are clopen subsets in M and X respectively, $\phi_j : U_j\to\phi_j(U_j)$ is a homeomorphism, also $\phi_j(U_j) = B(X,x_j,r_j)$ is a ball in X of a radius $0<r_j<\infty$ for each j.

For $\Lambda=\omega_0$ we define a Banach space

$$\tilde{C}_*(t,M\to X) := \{f|_{U_j}\in C_*(t,U_j\to X), \|f\|_{C_*(t,M\to X)}$$

$$:= \sup_{j\in\Lambda}(\|f|_{U_j}\|_{C_*(t,U_j\to X)}/\min(1,r_j)) < \infty \quad \text{and}$$

$$(\|f|_{U_j}\|_{C_*(t,U_j\to X)})/\min(1,r_j)) \to 0 \text{ while } j\to\infty\},$$

where ω_0 denotes the first countable infinite ordinal, $card(\omega_0)=\aleph_0$, $0\le t<\infty$. Here the following notation is used: $*=0$ for spaces $C_0(t,U\to X)$, $*=\emptyset$ or simply is omitted for $C(t,U\to X)$. For the finite atlas $At(M)$ the spaces $\tilde{C}_*(t,U\to X)$ and $C_*(t,U\to X)$ are linearly topologically isomorphic. By $C_*^\theta(t,M\to M)$ for $0\le t\le\infty$ is denoted the following space of functions $f : M \to M$ such that $(f_i - \theta_i) \in C_*(t, M \to X)$ for each $i \in \Lambda$ and $f_i = \psi_i \circ f$, $\theta_i = \psi_i \circ \theta$. We introduce the following group

$$G(t,M) := \tilde{C}_0^{id}(t, M \to M) \cap Hom(M),$$

which is called the diffeomorphism group for $t \geq 1$ and the homeomorphism group for $0 \leq t < 1$, where $Hom(M)$ is the group of continuous homeomorphisms of the manifold M.

Each continuous function $f \in C_0(t, M \to X)$ has the following decomposition:

$$f(x)|_{U_j} = \sum_{(i\in\mathbf{N}, n\in\mathbf{N_o})} f^i(n;x)|_{U_j} e_i \tilde{z}(n), \text{ and } \{e_i \tilde{z}(n)(\bar{Q}_m(x)|_{U_j}) : i, n, Ord(m) = n, j\}$$

is the orthogonal basis, moreover,

$$f_n(x)|_{U_j} := \sum_i f^i(n;x)|_{U_j} e_i \in C_0(t, U_j \to X),$$

where

$$X_{\tilde{z}(n)} := \{f_n(x) : f_n|_{U_j} \in C_0(t, U_j \to X)\}$$

is the Banach space with the norm induced from $C_0(t, M \to X)$ such that

$$f^i(n;x)|_{U_j} := \sum_{(Ord\ m=n, m=(m(1),\dots,m(n)), m(j)\in\mathbf{N_o})} a(m, f^i|_{U_j}) \bar{Q}_m(x)|_{U_j},$$

where $\bar{Q}_m(x)|_{U_j} = 0$ for $x \in M \setminus U_j$.

For the manifold M we fix a subsequence $\{M_n : n \in \mathbf{N_o}\}$ of sub-manifolds in M such that $M_n \hookrightarrow M_{n+1} \hookrightarrow \dots M$ for each n, $dim_{\mathbf{K}} M_n = \beta(n) \in \mathbf{N}$ for each $n \in \mathbf{N_o}$, $\bigcup_n M_n$ is dense in M, where $\beta(n) < \beta(n+1)$ for each n and there exists $n_0 \in \mathbf{N}$ with $\beta(n) = n$ for each $n > n_0$.

We take the following subgroup

$$G' := \{f \in G(t,M) : (f^i(n;x) - id^i(n;x)) =: g^i(n;x) \in C_0(t_n, M_n \to \mathbf{K})$$

and

$$|\,a(m; g^i(n;x)|_{U_j})\,|\,J_j(t_n, m) \leq c(f) p^{v'(m,j,i)}\},$$

where $c(f) > 0$ is a constant, $v'(m,j,i) = -c'i - c'n - c"j$, $n = Ord(m)$, $c' = const > 0$ and $c" = const \geq 0$, $c" > 0$ for $\Lambda = \omega_0$, $t_n = t + s(n)$ for $0 \leq t < \infty$, $s(n) > n$ for each n and $\liminf_{n\to\infty} s(n)/n =: \zeta > 1$. Then there exists the following ultra-metric in G':

$$d(f, id) = \sup_{m,n,j} \{|\,a(m; g^i(n;x)|_{U_j})\,|\,J_j(t_n, m) p^{-v'(m,j,i)}\}.$$

4. Note. At first it is necessary to prove theorems about the quasi-invariance and the pseudo-differentiability of transition measures of stochastic processes on Banach spaces over local fields. We consider two types of measures on the Banach space $c_0(\omega_0, \mathbf{K})$. The first is the q-Gaussian measure (see Section 6.2 of Chapter 3 above). The characteristic functional of the q-Gaussian measure is positive definite, hence μ is non-negative. The second is specified below and is the particular case of measures considered in §6.6 of Chapter 3.

Let w be the real-valued non-negative Haar measure on the local field $\mathbf{K}$ with $w(B(\mathbf{K},0,1)) = 1$. We consider the following measure μ on the Banach space $c_0(\omega_0,\mathbf{K})$:

$$(i) \quad \mu(dx) = \bigotimes_{j=1}^{\infty} \mu_j(dx^j).$$

This Borel measure is described in details below and it is the extension of a quasi-measure induced by the cylindrical distribution on the cylinder algebra $\mathcal{U}$ of cylinder subsets with Borel bases in finite-dimensional over the field $\mathbf{K}$ subspaces $span_{\mathbf{K}}\{e_1,\dots,e_j\}$, $j \in \mathbf{N}$, where $\{e_j : j\}$ denotes the standard orthonormal base in the Banach space $c_0(\omega_0,\mathbf{K})$. Here x is a point in $c_0(\omega_0,\mathbf{K})$, $x = (x^j : j \in \omega_0)$, $x^j \in \mathbf{K}$, $x = \sum_j x^j e_j$.

Let now on the Banach space $c_0 := c_0(\omega_0,\mathbf{K})$ an operator $J \in L_1(c_0)$ be given so that it is diagonal $Je_i = v_i e_i$ relative to the standard orthonormal base with $v_i \neq 0$ for each i. We consider a measure

$$\nu_i(dx) := f_i(x)w(dx)$$

on the field $\mathbf{K}$, where $f_i : \mathbf{K} \to [0,1]$ is a function belonging to the space $L^1(\mathbf{K},w,\mathbf{R})$ such that

$$f_i(x) = f(x/v_i) + h_i(x/v_i),$$

where f is a locally constant positive function,

$$f(x) = \sum_{j=1}^{\infty} C_j Ch_{B_j}(x),$$

$B_j := B(\mathbf{K},x_j,r_j)$ is a ball in $\mathbf{K}$, Ch_V is the characteristic function of a subset V in $\mathbf{K}$. That is $Ch_V(x) = 1$ for each $x \in V$, while $Ch_V(x) = 0$ for each $x \in \mathbf{K} \setminus V$. Here $x_1 := 0$, $r_1 := 1$, $\inf_j r_j = 1$, $\{B_j : j\}$ is the disjoint clopen covering of the field $\mathbf{K}$, $1 \geq C_j > 0$, the limit

$$\lim_{|x|\to\infty} f(x) = 0$$

is zero, $h_i \in L^1(\mathbf{K},w,\mathbf{R})$ such that

$$ess_w - \sup_{x\in\mathbf{K}} |h_i(x)/f(x)| = \delta_i < 1,$$

$$\sum_i \delta_i < \infty \quad \text{and} \quad \nu_i(\mathbf{K}) = 1.$$

Then $\nu_i(S) > 0$ for each open subset S in the local field $\mathbf{K}$. There exists a σ-additive product measure

$$(ii) \quad \mu_J(dx) := \prod_{i=1}^{\infty} \mu_i(dx^i)$$

on the σ-algebra of Borel subsets of c_0, since the Borel σ-algebras defined for the weak topology of c_0 and for the norm topology of c_0 coincide, where

$$\mu_i(dx^i) := \nu(dx^i/v_i).$$

Let $A : c_0 \to c_0$ be a linear topological isomorphism, that is, A and $A^{-1} \in L(c_0)$, then for a measure μ on c_0 there exists its image $\mu_A(S) := \mu(A^{-1}S)$ for each Borel subset S in c_0.

In view of Proposition 5.12.2 of Chapter 3 $L_q(c_0)$ is the ideal in $L(c_0)$. This produces new q-Gaussian measures

$$(\mu_{J,\gamma,q})_A =: \mu_{AJ,A^*\gamma,q}$$

and measures of the second type

$$(\mu_J)_A =: \mu_{AJ}.$$

In view of Remark §5.8 of Chapter 3 each injective linear operator $S \in L_q(c_0)$ with the range $E(c_0)$ dense in the Banach space c_0 can be presented in as the product $S = AJ$. Hence for each such S the σ-additive measures $\mu_{S,S^*\gamma,q}$ and μ_S exist. Here the algebra U of cylindrical subsets is generated by subsets $\pi_V^{-1}(A)$, where A is a Borel subset in $\mathbf{K^n}$, $card(V) = n < \aleph_0$, $V \subset \mathbf{N}$,

$$\pi_V : \mathbf{K}^{\aleph_0} \to \prod_{i \in V} \mathbf{K}_i$$

denotes the natural projection.

On the space $C_0^0(T,H) = C_0^0(T,\mathbf{K}) \otimes H$ let an operator S be written as the tensor product $S = S_1 \otimes S_2$ and a vector γ as the tensor product of two vectors $\gamma = \gamma^1 \otimes \gamma^2$, where S_1 is a linear operator on $C_0^0(T,\mathbf{K})$ and S_2 is a linear operator on H, $\gamma^1 \in C_0^0(T,\mathbf{K})$, $\gamma^2 \in H$. This makes the measure $\mu_{S,\gamma,q}$ to be the product of measures $\mu_{S_1,\gamma^1,q}$ on $C_0^0(T,\mathbf{K})$ and $\mu_{S_2,\gamma^2,q}$ on H. Analogously μ_S is the product of measures μ_{S_1} on $C_0^0(T,\mathbf{K})$ and μ_{S_2} on H.

With the help of such measures on the space $C_0^0(T,H)$ the stochastic process $w(t,\omega)$ is defined as in §§3.10 and 6.2.8 of Chapter 3.

5. Let Y be a Banach space over the local field $\mathbf{K}$ and V be a neighborhood of zero in Y. Consider either the measure $\mu_{S,\gamma,\psi}$ or μ_S outlined in §4 above. Suppose that in stochastic anti-derivational equations 6.7.3(i) and 6.7.4(i) in Chapter 3 mappings a and E may be dependent on the parameter $y \in V$, that is, $a = a(t,\omega,\xi,y)$ and $E = E(t,\omega,\xi,y)$; moreover, $a_{k,l} = a_{k,l}(t,\xi,y)$ for each k and l in the latter equation. Let Condition 6.7.3(LLC) in Chapter 3 be satisfied for each $0 < r < \infty$ with the constant K_r independent of $y \in V$ for each $y \in V$. Evidently, Equation 6.7.3(i) in Chapter 3 is the particular case of 6.7.4(i) in Chapter 3, when in the latter equation the corresponding terms $a_{0,1}$ and $a_{1,0}$ are chosen with all others operators $a_{k,l} = 0$ when $k + l \ne 1$. Let also

(i) a, E and $a_{k,l}$ be of class C^1 by $y \in V$ such that

$$a \in C^1(V, L^q(\Omega,\mathsf{F},\lambda; C^0(B_R, L^q(\Omega,\mathsf{F},\lambda; C^0(B_R,H)))))$$

and

$$E \in C^1(V, L^r(\Omega,\mathsf{F},\lambda; C^0(B_R, L(L^q(\Omega,\mathsf{F},\lambda; C^0(B_R,H)))))),$$

$$a_{m-l,l} \in C^1(V, C^0(B_{R_1} \times B(L^q(\Omega,\mathsf{F},\lambda; C^0(B_R,H)),0,R_2), L_m(H^{\otimes m};H)))$$

(continuous and bounded on its domain) for each n,l, $0 < R_2 < \infty$ and

$$\lim_{n\to\infty} \sup_{0 \le l \le n} \|a_{n-l,l}\|_{C^1(V,C^0(B_{R_1}\times B(L^q(\Omega,\mathsf{F},\lambda;C^0(B_R,H)),0,R_2),L_n(H^{\otimes n},H)))} = 0$$

for each $0 < R_1 \le R$ when $0 < R < \infty$, or each $0 < R_1 < R$ when $R = \infty$, for each $0 < R_2 < \infty$; the kernel of the operator E is zero

$$(ii) \quad ker(E(t,\omega,\xi,y)) = 0$$

for each t, ξ and y, also for λ-almost every ω;

$$(iii) \quad a_y(t,\omega,\xi,y)$$

and $\partial a(t,\omega,\xi,y)/\partial y \in X_{0,d}(H) := \{z : S^{-1}z \in H_d\}$ and $\partial E(t,\omega,\xi,y)/\partial y \in L_b(H)$ for λ-almost all elementary events ω and each t, ξ, y, where

$$H_d := \left\{z : z \in H; \sum_{j=1}^{\infty} |z_j|^d < \infty\right\}$$

for any $0 < d < \infty$, while

$$H_\infty := H,$$

with $d = b = \psi$ for the measure $\mu_{S,\gamma,\psi}$; $d = \infty$ and $b = 0$ for the measure of the second type μ_S. Here z_j are the coordinates of the vector z in the standard orthonormal base in H. In addition for Equation 6.7.4(i) we suppose that

$$(iv) \quad \partial a_{l,k}(t,\omega,\xi,y)/\partial y \in L_{k+l,b}(H^{\otimes(k+l)};H)$$

for each l and each k with either $b = \psi$ or $b = 0$ correspondingly, where parameters r, s, q are the same as in Theorems 6.7.3 and 6.7.4 of Chapter 3 respectively. In addition in the case of a local field $\mathbf{K}$ of a positive characteristic $char(\mathbf{K}) = p > 0$ we suppose that $a_{l,k} = 0$ for all $k + l \geq p$. The following theorem states the quasi-invariance of the transition measure

$$\mu^{F_{t,t_0}}(\{\omega : \xi(t_0,\omega,y) = 0, \xi(t,\omega,y) \in A\}) =: P_y(A),$$

where $F_{t,u}(\xi) := \xi(t,\omega,y) - \xi(u,\omega,y)$ denotes the difference operator.

5.1. Theorem. *Let either Conditions $(i - iii)$ or $(i - iv)$ be satisfied, then the transition measure $P_y(A)$ of the stochastic process $\xi(t,\omega,y)$ being the solution of Equation either 6.7.3(i) or 6.7.4(i) in Chapter 3 and depending on the parameter $y \in V$ is quasi-invariant relative to each mapping*

$$U(y_2, y; \xi(t,\omega,y)) := \xi(t,\omega,y_2)$$

for each y and $y_2 \in V$.

Proof. The Banach spaces on which the measures μ were constructed above are linearly topologically isomorphic with the Banach space $c_0(\omega_0,\mathbf{K})$. The continuous natural embedding

$$\theta : c_0(\omega_0,\mathbf{K}) \to \mathbf{K}^{\omega_0}$$

associated with the decomposition of vectors x in the Banach space $c_0(\omega_0,\mathbf{K})$ by the standard orthonormal base $\{e_j : j\}$ exists, where $x = \sum_j x_j e_j$, $x_j \in \mathbf{K}$ denotes the j-th coordinate of the vector x. Here the space $\mathbf{K}^{\omega_0}$ is supplied with the product (Tychonoff or weak) topology. The base of the latter topology form open sets of the form $\pi_V^{-1}(A)$, where A is open in $span_{\mathbf{K}}\{e_j : j \in V\}$, V is a finite subset in the set of natural numbers $\mathbf{N}$, $\pi_V : c_0 \to span_{\mathbf{K}}\{e_j : j \in V\}$ denotes the $\mathbf{K}$-linear projection operator.

The field $\mathbf{K}$ is locally compact, consequently, it is separable relative to its non-trivial norm topology. We remind that in accordance with Theorem I.1.2 [16] each separable complete metric space is a topological Radon space. By the constriction each measure μ on the Banach space $c_0(\omega_0, \mathbf{K})$ is σ-additive and Borel, while $(c_0(\omega_0, \mathbf{K}), \| * \|)$ is the Radon space relative to the norm topology (see also Section 2.5 in Chapter 4). That is μ is approximated from below by the class of compact subsets. Therefore, μ can be extended on the Borel σ-algebra $Bf(\mathbf{K}^{\omega_0})$, since the Borel σ-algebras of c_0 generated by the norm topology and the metrizable topology inherited from $\mathbf{K}^{\omega_0}$ coincide. By the construction of the non-negative σ-additive bounded measure μ for each Borel subset $A \in Bf(\mathbf{K}^{\omega_0})$ and every positive number $\varepsilon > 0$ a sequence $\{K_n : n \in \mathbf{N}\}$ of compact subsets K_n in $(c_0, \| * \|)$ exists so that $K_n \subset A$ and $\mu(A \setminus K_n) < \varepsilon/2^n$. Therefore, $\mu(\mathbf{K}^{\omega_0} \setminus \theta(c_0)) = 0$. Thus we can consider the product measure μ on $Bf(\mathbf{K}^{\omega_0})$ as well.

The Kakutani theorem (see §II.4.1 in [16]) states, whether $\prod_{k=1}^{\infty} \alpha_k$ converges to a positive number or diverges to zero, the measure μ is absolutely continuous or orthogonal with respect to ν, correspondingly, where

$$\alpha_k := \int_{X_k} (p_k(x_k))^{1/2} \nu_k(dx_k),$$

each measure μ_k is absolutely continuous relative to ν_k, $\mu = \otimes_k \mu_k$, $\nu = \otimes_k \nu_k$, μ_k and ν_k are probability measures on measurable spaces X_k for each $k \in \mathbf{N}$,

$$p_k(x) := \mu_k(dx)/\nu_k(dx)$$

denotes the Radon-Nykodim derivative. In the first case $\prod_k p_k(x_k)$ converges in the mean to $\mu(dx)/\nu(dx)$. In the considered here case let $X_k = \mathbf{K}$ for each $k \in \mathbf{N}$. Let us consider the measures

$$\mu_k(dx) = Cf(x-y)v(dx), \quad \nu_k(dx) = Cf(x)v(dx),$$

where v is the non-negative Haar measure on the field $\mathbf{K}$, f is a positive function such that $f \in L^1(\mathbf{K}, v, \mathbf{R})$, $C = const > 0$ is a positive constant so that $\nu(\mathbf{K}) = 1$. Then we get:

$$p_k(x) = f(x-y)/f(x) \quad \text{and} \quad \alpha_k = \int_{\mathbf{K}} (f(x-y)f(x))^{1/2} v(dx).$$

For the ψ-Gaussian measure using the inverse Fourier transform we infer that

$$f(x) = \int_{\mathbf{K}} [\exp(-\beta|x|^{\Psi})\chi_\gamma(x)]\chi_1(-zx)v(dx)$$

(see §6.2 of Chapter 3). If the inequality $|yx| \le 1$ is satisfied, then by the definition of the character of $\mathbf{K}$ as the additive group we have $\chi_1(yx) = 1$. Therefore, a positive constant $C_1 > 0$ independent of β and γ exists such that the inequality

$$|f(z-y) - f(z)| \le |f(z)|(1 + C_1 \exp(-\beta r^{-\Psi}))$$

is satisfied for each y with $|y| < r$, where $\beta r^{-\Psi} > 1$, since due to Cauchy-Schwarz-Bunyakovskii inequality the estimate

$$\left| \int_{|x|>1/r} \exp(-\beta|x|^{\Psi})\chi_\gamma(x)\chi_1(-(z-y)x)v(dx) \right|$$

$$\leq \left| \int_{|x|>1/r} \exp(-\beta|x|^{\Psi})\chi_{\gamma}(x)\chi_1(-zx)v(dx) \right| g(y,z) \leq |f(z)|g(y,z)$$

is valid, where

$$g(y,z) := \left| \int_{|x|>1/r} \exp(-\beta|x|^{\Psi})\chi_{\gamma}(x)\chi_1(-zx)\chi_1(2yx)v(dx) \right|.$$

Let $|y_j/v_j| =: r_j < 1$ for each $j > j_0$, then this implies the estimate

$$|\alpha_j - 1| \leq C\exp(-\beta_j r_j^{-\Psi})$$

for each $j > j_0$, where $C = const > 0$. In view of Proposition 6.12.2 in Chapter 3 and the Kakutani theorem the measure $\mu^z_{S,\gamma,\psi}$ is equivalent to the measure $\mu_{S,\gamma,\psi}$ for each vector $z \in X_{0,\psi}(C^0_0(T,H))$, where $\mu^z(A) := \mu(A-z)$ for each Borel subset A in the Banach space $C^0_0(T,H)$, that is, $\mu_{S,\gamma,\psi}$ is quasi-invariant relative to shifts on vectors z belonging to the dense proper subspace $X_{0,\psi}(C^0_0(T,H))$.

For the measure μ_J and $|y| < 1/|v|$ the equality $f((x-y)/v) = f(x/v)$ is fulfilled for each numbers $x \in \mathbf{K}$ and $0 \neq v \in \mathbf{K}$. In view of the definition of the function f_k the equality

$$p_k(x) = f_k(x-y_k)/f_k(x) = [f((x-y_k)/v_k)/f(x/v_k)]$$

$$\times[1 + h_k((x-y_k)/v_k)/f((x-y_k)/v_k)]/[1 + h_k(x/v_k)/f(x/v_k)]$$

is valid. If $|y_k/v_k| \leq 1$, then

$$f((x-y_k)/v_k)/f(x/v_k) = 1$$

for each number $x \in \mathbf{K}$. From the conditions imposed on h_k and f and the Kakutani theorem and Proposition 6.12.2 of Chapter 3 it follows, that the measure μ_S is quasi-invariant relative to shifts on vectors z from the dense proper subspace $X_{0,\infty}(C^0_0(T,H))$.

The Lebesgue majorized convergence theorem states the following (see also §2.4.9 [25]). If ϕ is a non-negative σ-additive measure on X and $h : X \to \mathbf{R}$ is a ϕ-integrable function; $f_1, f_2, f_3, \dots$ and g are ϕ-measurable functions such that $|f_n(x)| \leq h(x)$ for each $n = 1, 2, 3, \dots$ and

$$\lim_{n\to\infty} f_n(x) = g(x)$$

for each point $x \in X$, then

$$\lim_{n\to\infty} \int_X |f_n - g| d\phi = 0$$

and

$$\lim_{n\to\infty} \int_X f_n d\phi = \int_X g d\phi.$$

The quasi-invariance factor

$$\rho(z,x) := \mu^z(dx)/\mu(dx)$$

is Borel measurable as follows from the construction of the measure μ and the Kakutani theorem and the Lebesgue theorem about majorized convergence, since this is true for each its one-dimensional projection.

The Banach theorem states: if G is a topological group and $A \subset G$ is a Borel measurable set of second category, then $A \circ A^{-1}$ is a neighborhood of the unit (see §5.5 [14]).

The quasi-invariance factor satisfies the co-cycle condition:

$$\rho(z+h,x) = \rho(z,x-h)\rho(h,x)$$

for each vectors z and $h \in X_{0,d}(C^0_0(T,H))$ and each $x \in C^0_0(T,H)$.

We shall use Lusin's theorem stating the following (see also §2.3.5 [25]). If ϕ is a Borel regular non-negative σ-additive measure on a metric space X (or a Radon measure on a locally compact Hausdorff space X), f is a ϕ-measurable function with values in a separable metric space Y, A is a ϕ-measurable set, $\phi(A) < \infty$, $\varepsilon > 0$ is a positive number, then the set A contains a closed (compact correspondingly) subset E such that $\phi(A \setminus E) < \varepsilon$ and the restriction $f|_E$ of the function f is continuous.

Therefore, in view of the Lusin theorem the quasi-invariance factor

$$\rho(z,x) := \mu^z(dx)/\mu(dx)$$

is such that $\mu(W_L) = 1$ for each finite-dimensional subspace L in $X_{0,d}(C^0_0(T,H))$, where either $\mu = \mu_{S,\gamma,\psi}$ or $\mu = \mu_S$,

$$W_L := \{x : \rho(z,x) \text{ is defined and continuous by } z \in L\}.$$

In view of the preceding consideration

$$\lim_{n\to\infty} \rho(\hat{P}_n z,x) = \rho(z,x)$$

for μ-almost all $x \in C^0_0(T,H)$, moreover, this convergence is uniform by the variable z in each ball $B(L,0,c)$ for each finite-dimensional subspace L in the linear space $X_{0,s}(C^0_0(T,H))$, where $\hat{P}_n$ denotes a projection linear operator on a linear span subspace $span_{\mathbf{K}}\{e_1,\dots,e_n\}$ isomorphic with $\mathbf{K}^n$, where $\{e_j : j\}$ is the orthonormal base in the space $X_{0,s}(C^0_0(T,H))$. Evidently, the space $X_{0,s}(C^0_0(T,H))$ is dense in the space $C^0_0(T,H)$.

Stochastic anti-derivational Equation 6.3(i) in Chapter 3 is the particular case of 6.4(i). Therefore, it is sufficient to consider the latter equation. It is shown below, that the one-parameter family of solutions $\xi(t,\omega,y)$ is almost everywhere of class C^1 by $y \in V$. Let us construct the iteration process

$$X_0(t,y) = x(y), \dots, X_n(t,y) = x(y)$$

$$+\sum_{m+b=1}^{\infty}\sum_{l=0}^{m}(\hat{P}_{u^{b+m-l},w(u,\omega)^l}[a_{m-l+b,l}(u,X_{n-1}(u,\omega,y),y)\circ(I^{\otimes b}\otimes a^{\otimes(m-l)}\otimes E^{\otimes l})])|_{u=t},$$

consequently,

$$X_{n+1}(t,y) - X_n(t,y)$$

$$= \sum_{m+b=1}^{\infty}\sum_{l=0}^{m}(\hat{P}_{u^{b+m-l},w(u,\omega)^l}[a_{m-l+b,l}(u,X_n(u,y),y) - a_{m-l+b,l}(u,X_{n-1}(u,y),y)]$$

$$\circ(I^{\otimes b}\otimes a^{\otimes(m-l)}\otimes E^{\otimes l})])|_{u=t},$$

where $t_j = \sigma_j(t)$ for each $j = 0,1,2,\dots$, for the shortening of the notation X_n, x and $a_{l,k}$ are written without the argument ω, a and E are written without their variables. Then we infer

$$M\sup_y \|\hat{P}_{u^{b+m-l},w(u,\omega)^l}[a_{m-l+b,l}(u,X_n(u,y),y)$$

$$a_{m-l+b,l}(u,X_{n-1}(u,y),y)]|_{(B_{R_1}\times B(L^q,0,R_2)\times V)} \circ (I^{\otimes b}\otimes a^{\otimes(m-l)}\otimes E^{\otimes l})])|_{u=t}\|^g$$

$$\leq K(M\|\hat{P}_{u^{b+m-l},w(u,\omega)^l}\|^g)\|a_{m-l+b,l}|_{(B_{R_1}\times B(L^q,0,R_2)\times V)}\|^g$$

$$(M\sup_{u,y}\|X_n(u,y)-X_{n-1}(u,y)\|^g)(M\sup_{u,y}\|a\|^{m-l})(M\sup_{u,y}\|E\|^l),$$

where $X_n \in C^0_0(B_R,H)$ for each ω, $y \in V$ and for each n, K is the same constant as in §6.4 of Chapter 3, $1 \leq g < \infty$. On the other hand, we have

$$X_1(t,y) = x(t,y) + \sum_{m+b=1}^{\infty}\sum_{l=0}^{m}(\hat{P}_{u^{b+m-l},w(u,\omega)^l}[a_{m-l+b,l}(u,x(u,y),y)\circ$$

$$(I^{\otimes b}\otimes a^{\otimes(m-l)}\otimes E^{\otimes l})])|_{u=t},$$

consequently,

$$\|X_1(t,y)-X_0(t,y)\|^g$$

$$\leq \sup_{m,l,b}(\|\hat{P}_{u^{b+m-l},w(u,\omega)^l}[a_{m-l+b,l}(u,x(u,y),y)\circ(I^{\otimes b}\otimes a^{\otimes(m-l)}\otimes E^{\otimes l})])|_{u=t}\|^g.$$

Due to Condition (ii) for each $\varepsilon > 0$ and $0 < R_2 < \infty$ there exists a ball $B_\varepsilon \subset B_R$ such that

$$K\sup_{m,l,b}(\|\hat{P}_{u^{b+m-l},w(u,\omega)^l}|_{B_\varepsilon}[a_{m-l+b,l}(u,*,y)|_{(B_\varepsilon\times B(L^q,0,R_2)\times V)}$$

$$\circ(I^{\otimes b}\otimes a^{\otimes(m-l)}\otimes E^{\otimes l})])\|^g =: c < 1.$$

On the other hand, the partial difference quotient has the continuous extension $\bar{\Phi}^1(X_{n+1} - X_n)(y;h;\zeta)$, that is expressible through $\bar{\Phi}^1$ of $a_{l,k}$, a and E, and also through $a_{l,k}$, a and E themselves, where $y \in V$, $h \in Y$, $\zeta \in \mathbf{K}$ such that $y+\zeta h \in V$, since analogous to $(X_{n+1}-X_n)$ estimates are true for $\bar{\Phi}^1(X_{n+1}-X_n)$. Therefore, a unique solution on each ball B_ε exists and it is of class C^1 by $y \in V$, since

$$\sup_{u,y}\max(\|X_1(u,y)-X_0(u,y)\|_{L^q(\Omega,H)}, \|\bar{\Phi}^1(X_1(u,y)-X_0(u,y))\|_{L^q(\Omega,H)} < \infty$$

and

$$\lim_{l\to\infty} c^l C = 0$$

for each positive constant $C > 0$, hence the limit

$$\lim_{n\to\infty} X_n(t,y) = X(t,y) = \xi(t,\omega,y)|_{B_\varepsilon}$$

exists, where

$$C := M\sup_{u\in B_\varepsilon, y\in V}\max(\|X_1(u,y)-X_0(u,y)\|^q_{L^q(\Omega,H)}, \|\bar{\Phi}^1(X_1(u,y)-X_0(u,y))\|^q_{L^q(\Omega,H)}$$

$$\leq (c+1)K < \infty,$$

here B_ε is an arbitrary ball of radius ε in B_R so that $t \in B_\varepsilon$. Therefore, we get the solution $\xi(t,\omega,y) \in C^1(V, L^q(\Omega,\mathsf{F},\lambda;C^0(B_R,H))$.

From Proposition 7.11 of Chapter 3 it follows, that the multiplicative operator functional $T(t,v;\omega;y)$ is of class C^1 by the parameter $y \in V$ such that

$$\xi(t,\omega,y) = T(t,v;\omega;y)\xi(v,\omega,y)$$

for each t and $v \in T$.

Due to the existence and uniqueness of the solution $\xi(t,\omega,y)$ for each $y \in V$, the operator

$$U(y_2,y;\xi(t,\omega,y)) := \xi(t,\omega,y_2)$$

exists, which may be nonlinear by the argument ξ. The variation of the family of solutions $\{\xi(t,\omega,y) : y\}$ corresponds to the differential $D_y\xi(t,\omega,y)$. Since the random function $\xi(t,\omega,y)$ is of class C^1 by y, the mapping $U(y_2,y;\xi(t,\omega,y)$ is of class C^1 by the arguments y and y_2. The operator $U(y_2,y;*)$ has the inverse, since

$$U(y,y_2;U(y_2,y;\xi(t,\omega,y))) = \xi(t,\omega,y)$$

for each y_2 and $y \in V$, $t \in T$ and $\omega \in \Omega$. Therefore, the mapping $U^{-1}(y_2,y;*)$ is also of class C^1 by y_2 and y. In view of Conditions (iii,iv) and the inclusion

$$\xi(t,\omega,y_2) - \xi(t,\omega,y) \in X_{0,d}(H)$$

is valid. On the other hand, the measure either $\mu_{S,\gamma,\psi}$ or μ_S is quasi-invariant relative to shifts on vectors z from the dense proper subspace $X_{0,d}(C^0_0(T,H))$ and the operator S is the tensor product $S = S_1 \otimes S_2$, consequently, the transition measure P_y is quasi-invariant relative to shifts $z \in X_{0,d}(H)$. In view of Conditions $(ii-iv)$ the inclusion

$$\partial U(y_2,y;\eta)/\partial\eta - I \in L_b(H)$$

is fulfilled for each variables y_2 and $y \in V$, where $\eta \in \{\xi(t,\omega,y) : y\}$, either $b = \psi$ or $b = 0$ respectively. The equality

$$\mu_S(C^0_0(T,H)) = 1,$$

implies that $P_y(H) = 1$, consequently, the mapping $U(y_2,y;*)$ is defined P_y-almost everywhere on H for each y_2 and $y \in V$. Therefore, there exists a natural number n such that for each $j > n$ the mappings

$$V(j;x) := x + P_j(U^{-1}(x) - x) \quad \text{and} \quad U(j;x) := x + P_j(U(x) - x)$$

are invertible and the limits

$$\lim_j |detU'x(j;x)| = |detU'_x(x)| \quad \text{and}$$

$$\lim_j |detV'_x(j;x)| = 1/|detU'_x(x)|$$

exist, where $U(x) := U(y_2, y; x)$, y_2 and $y \in V$.

In view of Theorem 3.28 [81] for each y_2 and $y \in V$ the transition measures P_{y_2} and P_y are equivalent.

6. Definition and notes. A function $f : \mathbf{K} \to \mathbf{R}$ is called pseudo-differentiable of order b, if there exists the following integral:

$$PD(b, f(x)) := \int_{\mathbf{K}} [(f(x) - f(y)) \times g(x, y, b)] dv(y). \qquad (1)$$

We introduce the following notation $PD_c(b, f(x))$ for such integral by $B(\mathbf{K}, 0, 1)$ instead of the entire $\mathbf{K}$. Where $g(x, y, b) :=| x - y |^{-1-b}$ with the corresponding Haar measure v with values in $\mathbf{R}$, where $b \in \mathbf{C}$ and $|x|_{\mathbf{K}} = p^{-ord_p(x)}$.

Obviously, the definitions of differentiability of measures can not be transferred from [16] onto the case considered here. This is the reason why the notion of pseudo-differentiability is introduced here. A quasi-invariant measure μ on X is called pseudo-differentiable for $b \in \mathbf{C}$, if there exists $PD(b, g(x))$ for $g(x) := \mu(-xz + S)$ for each $S \in Af(X, \mu)$ with $|m|(S) < \infty$ and each $z \in J^b_\mu$, where J^b_μ is a $\mathbf{K}$-linear subspace dense in X. For a fixed $z \in X$ such measure is called pseudo-differentiable along z.

For a one-parameter subfamily of operators $B(\mathbf{K}, 0, 1) \ni t \to U_t : X \to X$ quasi-invariant measure μ is called pseudo-differentiable for $b \in \mathbf{C}$, if for each S the same as above there exists $PD_c(b, g(t))$ for a function $g(t) := \mu(U_t^{-1}(S)$.

7. Theorem. *Let Conditions $5(i - iv)$ be satisfied and let ϕ be a C^1-diffeomorphism of a subset V clopen in the local field $\mathbf{K}$ onto the unit ball $B(\mathbf{K}, 0, 1)$. Then*

(1) the transition measure P_y corresponding to $\mu_{S,\gamma,d}$ is pseudo-differentiable by the parameter $y = \phi(z)$ of order $b \in \mathbf{C}$ for each $Re(b) \geq 0$, where $z \in V$;

(2) P_y corresponding to μ_S with h_k such that $\sum_k \delta_k < \infty$, where

$$\delta_k := \sup_{x \in B(\mathbf{K},0,1)} |PD_c(b, h_k(x))|,$$

is pseudo-differentiable by the parameter $y = \phi(z)$ of order b for each $b \geq 0$, moreover, P_y is pseudo-differentiable for each $b \in C$, when each f_k is locally constant, that is, $h_k = 0$ for each $k \in \mathbf{N}$.

Proof. If $\psi \in L^2(\mathbf{K}, w, \mathbf{C})$ and $b > 0$, then due to the Cauchy-Schwarz-Bunyakovskii inequality the integral

$$\int_{\mathbf{K} \setminus B(\mathbf{K},x,1)} [\psi(x) - \psi(y)] |x - y|^{-1-b} w(dy)$$

exists, where w denotes the Haar non-negative measure on the local field $\mathbf{K}$ considered as the additive group. Then

(3) $F[D^b(h(x))] = |x|^b F[h(x)]$, where

$$F(h)(x) := \int_{\mathbf{K}} h(y) \chi_1((x, y)) w(dx)$$

is the Fourier transform,

(4) $D^b h(x) = CPD(b, h(x))$, C is a fixed real constant independent of h. If the field $\mathbf{K}$ is of zero characteristic $C = (\Gamma_p(b))^{-m}$, where m is the dimension of $\mathbf{K}$ as the $\mathbf{Q}_p$ linear space, p is the corresponding characteristic $p = char(k)$ of the residue class field k of the field $\mathbf{K}$, $Re(b) \neq -1$,

$$\Gamma_p(b) := [1 - p^{b-1}]/[1 - p^b]$$

(see also [111] and §II.9 [110]). Analogously $C = 1/\Gamma_p(b)$ for the local field $\mathbf{F_p}(\theta)$ of the positive characteristic $p > 0$.

As it was mentioned above (see also [37]) the Fourier transform $f \mapsto F[f]$ is the bijective continuous isomorphism of the complex Hilbert space $L^2(\mathbf{K}, w, \mathbf{C})$ onto itself such that

$$f(x) = \lim_{r \to \infty} \int_{B(\mathbf{K},0,r)} F[f](y) \chi_1(-(y,x)) w(dy)$$

and $(f, g) = (F[f], F[g])$ for each $f, g \in L^2(\mathbf{K}, w, \mathbf{C})$. If

$$F[\psi](x) = C \exp(-\beta |x|^d) \chi_\gamma(x),$$

then there exists $D^b \psi(x)$ for each $b \geq 0$. We certainly have the equality:

(5) $\int_{\mathbf{K}} \chi_\gamma(x) w(dx) = 0$ for each $\gamma \neq 0$ (see also Example 4.3.9 [110] and §4 of Chapter 1 above). In view of Example 4.3.10 [110] and Lemma 4.1 of Chapter 1 one has

(6) $\int_{\mathbf{F}} |x|^{nd} \chi_1(yx) w(dx) = [1 - p^{nd}][1 - p^{-n(d+1)}]^{-1} |y|^{-n(d+1)}$ for each $d \in \mathbf{C}$ with $Re(d) > 0$ and $n \in \{1, 2, 3, \dots\}$, where either $\mathbf{F} = \mathbf{Q_p}$ or $\mathbf{F} = \mathbf{F_p}(\theta)$.

If f is a locally constant function as in §4, then $PD_c(b, f)$ exists for each $b \in \mathbf{C}$. On the other hand, the pseudo-differential operator is $\mathbf{R}$-linear:

$$PD_c(b, \alpha f + \beta h) = \alpha PD_c(b, f) + \beta PD_c(b, h)$$

for all pseudo-differentiable functions f and h of order b and all real numbers α and β.

Let g be a continuously differentiable bounded with its derivative function $g : \mathbf{R} \to \mathbf{R}$ such that

$$\|g\|_{C^1(\mathbf{R},\mathbf{R})} := \sup_x |g(x)| + \sup_x |g'(x)| < \infty,$$

that is $g \in C^1_b(\mathbf{R}, \mathbf{R})$. If for $f : \mathbf{K} \to \mathbf{R}$ and $x \in \mathbf{K}$ the function $[f(x) - f(y)]|x - y|^{-1-b}$ belongs to the space $L^1(\mathbf{K}, w, \mathbf{C})$ as the function by the variable $y \in \mathbf{K}$, then

$$\int_{\mathbf{K}} [g \circ f(x) - g \circ f(y)] |x - y|^{-1-b} w(dy)$$

$$= \int_{S(f,x)} [g \circ f(x) - g \circ f(y)][f(x) - f(y)]^{-1} [f(x) - f(y)] |x - y|^{-1-b} w(dy),$$

where

$$S(f, x) := \{y : y \in \mathbf{K}, f(x) \neq f(y)\},$$

consequently, the pseudo-differential $PD(b, g \circ f)(x)$ exists.

If instead of g a differentiable function $h \in C^1(\mathbf{K}, \mathbf{K})$ exists such that

$$\|h\|_{C^1(\mathbf{K},\mathbf{K})} := \max(\sup_x |h(x)|, \sup_{x,y} |\bar{\Phi}^1 h(x; 1; y)|) < \infty,$$

that is $h \in C^1_b(\mathbf{K},\mathbf{K})$, then the integral

$$\int_{\mathbf{K}} [f \circ h(x) - f \circ h(y)]|x-y|^{-1-b} w(dy)$$

$$= \int_{S(h,x)} [f \circ h(x) - f \circ h(y)]|h(x)-h(y)|^{-1-b}|h(x)-h(y)|^{1+b}|x-y|^{-1-b} w(dy)$$

converges, consequently, the pseudo-differential $PD(b, f \circ h)(x)$ exists. Analogous two statements are true for the operator PD_c instead of PD.

Let $\mathsf{D}' = \mathsf{D}^*$ denotes the topologically dual space to the space D of all locally constant functions $\phi : \mathbf{K} \to \mathbf{R}$. Then Formulas $(3,4)$ above imply that

$$D^{\alpha} D^{\beta} \psi = D^{\beta} D^{\alpha} \psi = D^{\alpha+\beta} \psi$$

for each $\alpha \neq -1$, $\beta \neq -1$ and $\alpha+\beta \neq -1$ for each $\psi \in \mathsf{D}'$ such that the pseudo-differentials $D^{\alpha}\psi$ and $D^{\beta}\psi$ and $D^{\alpha+\beta}\psi$ exist (see also Equation 9.(1.5) in [110]). On the other hand, the space D is dense in the space D' in the weak topology (see also §6 in [110]). Evidently, the linear space $L^2 \cap \mathsf{D}$ over the real field is (everywhere) dense in the $\mathbf{R}$-linear space $L^2(\mathbf{K}, w, \mathbf{R})$ as well. The characteristic functional of the Gaussian measure belongs to the space D' and is locally constant on the set $\mathbf{K} \setminus \{0\}$. The Fourier transform is the linear topological isomorphism of D on D and of D' on D' (see also [37] and §§7.2 and 7.3 [110]). This implies the inclusion

$$\mu^g_{S,\gamma,d}(dx)/w(dx) \in L^1(\mathbf{K}, w, \mathbf{R}) \cap \mathsf{D}'$$

for each $g \in C^0_0(T,H)^*$.

In view of Theorem 4.3 of Chapter 1 in [81] and using the Kakutani theorem as in §4 we get the statements of this theorem, since the quasi-invariance factor $P_y(dx)/P_u(dx)$ is pseudo-differentiable as the function by y of order b for each fixed $u \in B(\mathbf{K}, 0, 1)$.

8. Theorem. *Let G be either a diffeomorphism group or a wrap group defined as in Section 3 above and Appendix E below, then there exists a random function $\xi(t,\omega)$ with values in G which induces a quasi-invariant transition measure P on G relative to G' and P is pseudo-differentiable of order b for each $b \in \mathbf{C}$ such that $Re(b) \geq 0$ relative to G', where a dense subgroup G' is given in Section 3 and Appendix E.*

Proof. These topological groups also have structures of C^∞-manifolds, which are infinite-dimensional over the local field $\mathbf{K}$. But these groups do not satisfy the Campbell-Hausdorff formula in any open local subgroup [61,62]. Nevertheless, their manifold structures and actions of G' on G will be sufficient for the construction of desired measures. These separable Polish groups have embeddings as clopen subsets into the corresponding tangent Banach spaces Y' and Y in accordance with [57,62] and §2.3 above, where Y' is the dense subspace of Y. As usually $TG = \bigcup_{x \in G} T_xG$ and $T_xG = (x,Y)$.

Let the group G be the complete separable relative to its metric ρ C^∞-manifold on a Banach space Y over $\mathbf{K}$ such that it has an embedding into Y as the clopen subset. We consider the tangent bundle $\tau_G : TG \to G$ on the group G. It is trivial, since $TG = G \times Y$ for the considered here case.

Therefore, without loss of generality we take a trivial bundle $\theta : Z_G \to G$ on the group G with the fibre Z such that $Z_G = Z \times G$. Then $L_{2,1}(\theta, \tau_G)$ is an operator bundle with a fibre $L_{2,1}(Z,Y)$ (see §3.10 of Chapter 4).

In more detailed notation let $E(M,F,G,\pi,\Psi)$ be a C^t bundle with a bundle space E, a base space M, a fiber F, a structural group G, a projection $\pi : E \to M$ and an atlas Ψ, $F_x = \pi^{-1}(x)$ is diffeomorphic with F for each $x \in M$ (see also Appendix F). We remind that for two vector bundles $E^j(M,F^j,G^j,\pi^j,\Psi^j)$, where F^j are vector spaces over the field $\mathbf{K}$, $j = 1,2$, Witney's sum is defined as a vector bundle $E(M,F,G,\pi,\Psi)$ with the direct sum fiber $F = F^1 \oplus F^2$, the direct product structural group $G = G^1 \times G^2$ acting on the fiber by the formula $F^1 \oplus F^2 \ni (v_1,v_2) \mapsto (g_1,g_2)(v_1+v_2) = g_1v_1+g_2v_2 \in F$ for each element $g = (g_1,g_2)$ of the structural group G, where $v_j \in F^j$, $g_j \in G^j$. An action of the group G^j on the vector fiber F^j is denoted by $G^j \times F^j \ni (g_j,v_j) \mapsto g_jv_j \in F^j$. That is $E(M,F,G,\pi,\Psi) = E^1(M,F^1,G^1,\pi^1,\Psi^1) \oplus E^2(M,F^2,G^2,\pi^2,\Psi^2)$. A covering of M for Witney's sum may be defined common for E^1 and E^2 taking intersections of charts of two atlases in a case of necessity and constructing an atlas $At(M)$ refining two initial atlases $At^1(M)$ and $At^2(M)$. Thus $\pi^1 \oplus \pi^2 = \pi : E \to M$ is the projection of the vector bundle $E(M,F,G,\pi,\Psi)$.

Let $\Pi := \tau_G \oplus L_{2,1}(\theta,\tau_G)$ be a Whitney sum of bundles τ and $L_{2,1}(\theta,\tau_G)$.

Since G is clopen in Y, the normalization group of $\mathbf{K}$ is discrete in $(0,\infty)$, the set G has a clopen disjoint covering by balls $B(Y,x_j,r_j)$. That is, the atlas $At(G)$ of G has a refinement $At'(G)$ being a disjoint atlas.

On the Banach space Y we consider the measure either $\mu_{S,\gamma,d}$ or μ_S as in §4. Then in view of Theorem 3.11 and Proposition 6.2.10 in Chapter 3 the stochastic process $w(t,\omega)$ corresponding to either $\mu_{S,\gamma,d}$ or μ_S exists. We take functions f and h_k for each $k \in \mathbf{N}$ defining the measure μ_S and satisfying the Conditions of §4 and of Theorem 7.

Now let G be a wrap or a diffeomorphism group of the corresponding manifolds over the local field $\mathbf{K}$. We consider for the group G a field U with a principal part (a_η,E_η), where $a_\eta \in T_\eta G$ and $E_\eta \in L_{2,1}(H,T_\eta G)$ and $ker(E_\eta) = \{0\}$, where $\theta : H_G \to G$ is a trivial bundle with a Banach fiber H and $H_G := G \times H$. We remind that $L_{2,1}(\theta,\tau_\eta)$ denotes an operator bundle with a fibre $L_{2,1}(H,T_\eta G)$ such that (a_η,E_η) satisfies Conditions of Theorem 7.3 in Chapter 3. For Equation 3.7.4(i) we take additionally $(a_{l,k})_\eta$ for each l,k satisfying conditions of Theorem 7.4 in Chapter 3.

To satisfy conditions of quasi-invariance and pseudo-differentiability of transition measures theorems we choose a_η, E_η and $(a_{k,l})_\eta$ of class C^1 and fulfilling Conditions 5(iii,iv) by the variable $y := \eta \in G' =: V$ for each k,l.

For simplicity we can take initially a cylindrical distribution either $\mu_{I,\gamma,d}$ or μ_I on a Banach space X' such that $T_\eta G' \subset X' \subset T_\eta G$. If A_η is the L_d-operator or the L_1-operator with $ker(A_\eta) = \{0\}$, then the operator field A_η gives the σ-additive measure either $\mu_{A_\eta,A^*_\eta z,d}$ or μ_{A_η} in the completion $X'_{1,\eta}$ of the space X' with respect to the norm

$$\|x\|_1 := \|A_\eta x\|$$

(see also §4).

There exists the solution $\xi(t,\omega,\eta) = \xi_\eta(t,\omega)$ of stochastic anti-derivational Equation 7.3(i) or 7.4(i) of Chapter 3. When the embedding θ of the space $T_\eta G'$ into the space $T_\eta G$ is $\theta = \theta_1\theta_2$ with θ_1 and θ_2 of class L_d for either $\mu_{S,\gamma,d}$ or of class L_1 for μ_S, then the operator field A_η exists such that either $\mu_{A_\eta, A_\eta z, d}$ or μ_{A_η} is the quasi-invariant and pseudo-differentiable of order b measure on the space $T_\eta G$ relative to shifts on vectors from $T_\eta G'$ (see Theorems 5.1 and 7). Henceforth we impose such demand on the operator field A_η for each $\eta \in G'$.

Consider left shifts $L_h : G \to G$ on the group G given by the formula

$$L_h \eta := h \circ \eta.$$

Let us take a vector field $a_e \in T_e G'$, an operator field either $A_e \in L_{d,1}(T_e G', T_e G)$ or $A_e \in L_{1,1}(T_e G', T_e G)$ respectively, operator field $(a_{k,l})_\eta \in L_{k+l}((T_e G)^{\otimes(k+l)}; T_e G)$ for each k and each l. The linear spaces H, $T_e G'$ and $T_e G$ in their own norm uniformities are isomorphic with the Banach space $c_0(\omega_0, \mathbf{K})$. Then we put $a_x = (DL_x) a_e$ and $A_x = (DL_x) \circ A_e$ for each $x \in G$, consequently, $a_x \in T_e G$ and $A_x \in L_{\kappa,1}(H_x, (DL_x) T_e G)$, where

$$(DL_x) T_e G = T_x G \quad \text{and} \quad T_e G' \subset T_e G,$$

$$H_x := (DL_x) T_e G',$$

either $\kappa = d$ or $\kappa = 1$. Each operators L_h is (strongly) C^∞-differentiable diffeomorphisms of G such that its differential

$$D_h L_h : T_\eta G \to T_{h\eta} G$$

is correctly defined, since $D_h L_h = h_*$ is the differential of h for the considered group G. In view of the choice of G' in G each partial difference quotient $\bar{\Phi}^n L_h(X_1, \dots, X_n; \zeta_1, \dots, \zeta_n)$ is of class C^0, i.e. continuous, and the n-th differential $D^n L_h$ is of class $L_{\kappa,n+1}(TG'^{\otimes n} \times G', TG)$ for each vector fields $X_1, \dots, X_n$ on G', numbers $\zeta_1, \dots, \zeta_n \in \mathbf{K}$ with $\zeta_j p_2(X_j) + h \in G'$ and $h \in G'$. Indeed for each $0 \le l \in \mathbf{Z}$ the embedding of of the bundle $T^l G'$ into $T^l G$ is the product of two operators of the L_d-class or the embedding is of the L_1-class, where $T^0 G := G$, $X = (x, X_x) \in T_x G$, $x \in G'$, $X_x \in Y'$, $p_1(X) = x$, $p_2(X) = X_x$. We take a dense subgroup G' from Appendix E or §3 correspondingly and consider left shifts L_h for $h \in G'$.

Let B_n denotes the Borel σ-algebra of the vector space $\mathbf{K^n}$. The considered here groups G are separable, hence the minimal σ-algebra generated by cylindrical subalgebras $f^{-1}(\mathsf{B}_n)$, n=1,2,..., coincides with the σ-algebra B of Borel subsets of G, where $f : G \to \mathbf{K^n}$ are continuous functions. Moreover, G is metrizable, separable and complete, consequently, G is the topological Radon space (see above). We have got that each measure $\mu_{A_\eta, A^*_\eta z, q}$ is defined on the σ-algebra of Borel subsets of $T_\eta G$ (see above). So we can take the transition probability

$$P(t_0, \psi, t, W) := P(\{\omega : \xi(t_0, \omega) = \psi, \xi(t, \omega) \in W\})$$

of the stochastic process ξ for a marked value of the variable $t \in T$. The transition probability $P(t_0, \psi, t, W)$ is certainly the conditional probability and it is defined on a σ-algebra B of Borel subsets in G, $W \in \mathsf{B}$. On the other hand,

$$T(t, \tau, \omega) g x = g T(t, \tau, \omega) x$$

is the stochastic evolution family of operators for each $\tau \ne t \in T$. One can take, for example, $t_0 = 0$ and $\psi = e$ with the fixed $t_0 \in T$. Applying the construction above and Theorem 7 we infer, that the transition measure $P(t_0,\psi,t,W)$ is σ-additive and quasi-invariant and pseudo-differentiable of order b relative to the action of G' by the left shifts L_h on μ measure on G.

9. Note. In §8 the group G' as the manifold is modelled on the Banach space Y' and the group G as the manifold is modelled on the Banach space Y over the local field $\mathbf{K}$ such that G' and G are complete relative to their uniformities $\mathsf{U}_{G'}$ and U_G. There are inclusions $TG' = G' \times Y' \subset G \times Y' \subset G \times Y = TG$. The completion of TG' relative to the uniformity $\mathsf{U}_G \times \mathsf{U}_{Y'}$ produces the uniform space $G \times Y'$ up to a uniform isomorphism. Therefore, each $\mathsf{U}_G \times \mathsf{U}_{Y'}$-uniformly continuous vector field $X = (x, X_x)$ on G' has the unique extension on G such that $X_x \in Y'$ for each $x \in G$, where $\mathsf{U}_G|_{G'} \subset \mathsf{U}_{G'}$. Thus the $\mathsf{U}_G \times \mathsf{U}_{Y'}$-$C^\infty$-vector field X on G' has the $\mathsf{U}_G \times \mathsf{U}_{Y'}$-$C^\infty$-extension on G and it provides the 1-parameter group

$$\rho : \mathbf{K} \times G \to G$$

of C^∞-diffeomorphisms of G generated by a $\mathsf{U}_G \times \mathsf{U}_{Y'}$-$C^\infty$-vector field X_ρ on G', where the field $\mathbf{K}$ is considered as the additive group. This means that

$$(\partial\rho(v,x)/\partial v)|_{v=0} = X_{\rho(0,x)}$$

for each $x \in G$, where $v \in \mathbf{K}$, $X_\rho(x) \in G \times Y'$. From the theory of the differential equations over the field $\mathbf{K}$ we get that at least a local solution ρ of the differential equation

$$(\partial\rho(v,x)/\partial v) = X_{\rho(v,x)}$$

with the initial condition $\rho(0,x) = x$ for each x exists. This provides the local one-parameter transformation diffeomorphism group. A local partial solution can be obtained with the help of the anti-derivation operator which is continuous from C^0 into C^1.

In view of §8 the transition measure P is quasi-invariant and pseudo-differentiable of order b relative to the 1-parameter local group ρ.

This approach is also applicable to the case of two Polish manifolds G' and G of class C^∞ on Y' and Y over $\mathbf{K}$. The quasi-invariance and pseudo-differentiability of the measure P on G relative to the 1-parameter group ρ corresponding to X by our definition means such properties of P relative to the $\mathsf{U}_G \times \mathsf{U}_{Y'}$-$C^\infty$-vector field X on G'.

Evidently, considering different fields (a,E) and $\{a_{k,l} : k,l\}$ we see that a family of the cardinality $\mathsf{c} = card(\mathbf{R})$ exists, which consists of non-equivalent stochastic (in particular, Brownian motion) processes on G and c with orthogonal quasi-invariant pseudo-differentiable of order $b \in \mathbf{C}$ with $Re(b) > 0$ transition measures on G relative to G'.

If the manifold M is compact, then in the case of the diffeomorphism group its dense subgroup G' can be chosen such that $G' \supset Diff(t',M)$. The dimension of the compact manifold M over the field $\mathbf{K}$ is finite $dim_{\mathbf{K}} M = n \in \mathbf{N}$ and we can take $t' = t+s$ for $0 \le t \in \mathbf{R}$, where $s > nv$, $v = dim_{\mathbf{Q_p}}(\mathbf{K})$.

Analogously can be considered the manifold $M \subset B(\mathbf{K^n},0,r)$ and the group $G := Diff(an_r,M)$ of analytic diffeomorphisms $f : M \to M$ having analytic extensions on the clopen ball $B(\mathbf{K},0,r)$ with the corresponding norm topology, where a radius is finite $0 < r < \infty$. Then a random function ξ with values in T_eG exists such that it generates the transition measure P on the tangent space T_eG. In this analytic plane case its restriction on the clopen subset G embedded into the tangent space T_eG produces the quasi-invariant and pseudo-differentiable of each order $b \in \mathbf{C}$ with $Re(b) \geq 0$ measure $P|_G$ relative to the dense subgroup $G' := Diff(an_R, M)$ for $R > r > 0$, since the embedding T_eG' into T_eG is of class L_1.

10. Theorem. *Let G be a separable Banach-Lie group over a local field $\mathbf{K}$. Then there exists a probability quasi-invariant and pseudo-differentiable of each order $b \in \mathbf{C}$ with $Re(b) > 0$ transition measure P on G relative to a dense subgroup G' such that P is associated with a random function with values in G.*

Proof. We consider two cases:

(I) G satisfies locally the Campbell-Hausdorff formula and the characteristic of the local field is zero;

(II) G does not satisfy it in any neighborhood of the unit element e in G and the characteristic of the field $\mathbf{K}$ may be zero or positive.

Certainly, the first case permits to describe G' more concretely. In both cases an embedding of G into the Banach space T_eG as a clopen subset exists, since G is the Polish group.

The second case can be considered quite analogously to §8. In this variant the dense subgroup G' can be characterized by the condition that the embedding of the linear space T_eG' into the Banach space T_eG is $\theta = \theta_1\theta_2$ with θ_1 and θ_2 of class L_d or θ of class L_1, where $d = \psi$ or $d = 1$ for random functions associated with a transition measure equivalent either $\mu_{S,\gamma,\psi}$ or μ_S respectively.

Therefore, it remains to consider the first case. For the Banach-Lie group G a Banach-Lie algebra g exists and the exponential mapping $\exp : V \to U$ is provided, where V is a neighborhood of zero 0 in the algebra g and U is a neighborhood of the unit element e in the group G such that $\exp(V) = U$, where $\exp(X+Y) = \exp(X)\exp(Y)$ for commuting elements X and Y of g, that is, $[X,Y] = 0$, $\exp(X)Y\exp(-X) = \exp(ad\, X)Y$,

$$\exp(\lambda X) = \sum_{j=0}^{\infty} \lambda^j X^j / j!,$$

$V = B(\mathrm{g}, 0, r)$ is a ball of radius $0 < r < \infty$ in g, $\lambda \in \mathbf{K}$, $\lambda X \in V$, $\mathrm{g} = T_eG$. The radii of convergence of the exponential function and Hausdorff's series corresponding to $\ln(\exp(X).\exp(Y))$ are positive such that for each $0 < R < p^{1/(1-p)}$ to a ball $B(\mathrm{g}, 0, R)$ a clopen subgroup G_1 supplied with Hausdorff's function corresponds (see §II.6 and §II.8 [9]). Therefore, the exponential mapping exp supplies the group G with the structure of the analytic manifold over the local field $\mathbf{K}$.

By $At(G) = \{(U_j, \phi_j) : j \in \mathbf{N}\}$ is denoted the analytic atlas of the group G, that is $\phi_j : U_j \to V_j$ are diffeomorphisms of U_j onto V_j, where U_j and V_j are clopen in G and in g respectively. Here connecting mappings $\phi_j \circ \phi_i^{-1}$ are analytic on clopen non-void subsets $\phi_i(U_i \cap U_j) \subset \mathrm{g}$. Therefore, the exponential mapping provides the group G with the covariant derivation ∇ and a bilinear Christoffel tensor Γ such that

$$\nabla_X Y = {}_L\nabla_X Y - {}_LT(X,Y)/2$$

and

$$\nabla_X Y_u = DY_u.X_u + \Gamma_u(X_u, Y_u),$$

where the left-invariant derivation on the group G is defined by the formula

$$_L\nabla_X \tilde{Y} = 0$$

for an arbitrary left-invariant vector field $\tilde{Y}$ and all vector fields X on G.

A vector field $\tilde{Y}$ is called left-invariant if $TL_g\tilde{Y}(h) = \tilde{Y}(gh)$, where $L_gh := gh$ for each $g,h \in G$, TL_g denotes the tangent mapping of the left shift mapping L_g. For such covariant derivation ∇ the torsion tensor is zero (see above Chapter 4 and also §1.7 [47], [29]). It defines the rigid analytic geometry and the corresponding atlas on G.

The group G has a countable base of open neighborhoods of the unit element and is metrizable by a left-invariant metric in accordance with Theorem 8.3 [37]. But the Banach space T_eG is ultra-normed, consequently, G is ultra-metrizable. Therefore, the initial atlas $At(G)$ has the refinement $At'(G)$ such that charts of $At'(G)$ compose the disjoint covering of G.

Each local field is spherically complete. The Banach algebra g considered as the Banach space is of separable type over the local field $\mathbf{K}$ and hence isomorphic with the Banach space $c_0(\omega_0,\mathbf{K})$ (see Theorem 5.13 [99]). Let a_x be a an analytic vector field and A_x be an analytic operator field on the group G such that A_x is an injective compact operator of class L_d for each $x \in G$, where $d = \psi$ or $d = 1$. Let $w_x(t,\omega)$ be a random function with values in T_xG (or, in particular, Brownian motion) such that $a_xt + A_xw_x(t) \in T_xG$, since the Banach space $C^0_0(T,T_xG)$ is isomorphic with c_0 when T is a compact subset in the field $\mathbf{K}$. For a ball $B_R := B(\mathbf{K},0,R)$ in $\mathbf{K}$ for $0 < R < \infty$ let $\{B(\mathbf{K},t_j,r) : j\}$ be a disjoint paving of B_R of a sufficiently small radius $0 < r < \infty$ for which the random function

$$\xi^q_x(t) := \exp_{\xi^q_{x,k}}\{a_{\xi^q_{x,k}}(t-t_k) + A_{\xi^q_{x,k}}[w_{\xi^q_{x,k}}(t) - w_{\xi^q_{x,k}}(t_k)]\}$$

is defined, where $\xi^q_{x,k} = \xi^q_x(t_k)$ for $k = 0,1,\dots,n$, $\xi^q_x(0) = x$, q denotes the partition of B_R into balls $B(\mathbf{K},t_j,r)$. Then the limit random function

$$\xi = \lim_q \xi^q(t)$$

exists which is by our definition a solution of the following stochastic equation:

$$(i)\quad d\xi(t,\omega) = \exp_{\xi(t,\omega)}\{a_{\xi(t,\omega)}dt + A_{\xi(t,\omega)}dw(t,\omega)\}$$

for the variable $t \in B_R$. A function $f(t,x)$ such that

$$f(t,\xi) := \ln_{\xi(t,\omega)}\xi(t,\omega)$$

satisfies the conditions of Theorem 6.6 of Chapter 3 on the corresponding domain W, where $(t,x) \in W \subset T \times H$. In view of Theorem 6.6 of Chapter 3 after coordinate mapping of a chart (U,ϕ) this equation takes the following form on the algebra g:

$$(ii)\quad \phi(\xi(t,\omega)) = \phi(\xi(t_0,\omega)) + (\hat{P}_u a^\phi_{\xi(u)})|_{u=t} + (\hat{P}_{w^\phi_{\xi(u)}(u,\omega)}E)|_{u=t},$$

$$-\sum_{m=2}^{\infty}(m!)^{-1}\sum_{l=0}^{m}\binom{m}{l}(\hat{P}_{u^{m-l},w^\phi_{\xi(u)}(u,\omega)^l}[(\partial^{m-2}\Gamma^\phi_{\phi(\xi(u))}/\partial x^{m-2})$$

$$\circ(a^{\phi}_{\xi(u)}(u)^{\otimes(m-l)} \otimes E^{\otimes l})])|_{u=t},$$

where $E = A^{\phi}_{\xi(u,\omega)}$, $a^{\phi} = (\partial\phi_x/\partial x)a_x$, $A^{\phi}_x = (\partial\phi_x/\partial x)A_x(\partial\phi_x^{-1}/\partial x)$, since

$$h^{\phi} = (\partial g^{\phi}/\partial x)f^{\phi} + \Gamma^{\phi}_{\phi(x)}(f^{\phi}, g^{\phi})$$

for any $h = \nabla_f g$, $f^{\phi} = (\partial\phi/\partial x)f$, $g^{\phi} = (\partial\phi/\partial x)g$, $h^{\phi} = (\partial\phi/\partial x)h$, Γ^{ϕ} denotes the bi-linear Christoffel operator in g considered as the manifold. Christoffel's operator has the transformation property:

$$D(\psi\circ\phi^{-1}).\Gamma^{\phi}_{\phi(x)} = D^2(\psi\circ\phi^{-1}) + \Gamma^{\psi}_{\psi(x)} \circ (D(\psi\circ\phi^{-1}) \times D(\psi\circ\phi^{-1}))$$

such that

$$\nabla_X Y_{\phi} = DY_{\phi}.X_{\phi} + \Gamma^{\phi}_{\phi(x)}(X_{\phi}, Y_{\phi}),$$

where $\Gamma^{\phi}_{\phi(x)}$ denotes Γ for the chart (U,ϕ), ψ corresponds to another chart (V,ψ) such that the intersection $U \cap V \neq \emptyset$ is non-void, f, g and h are vector fields, since

$$[\partial(\psi\circ\phi^{-1})/\partial t] = 0.$$

Thus Corollary 6.5 of Chapter 3 is applicable instead of Theorem 3.6.6, because f corresponds to $(\psi\circ\phi^{-1})$ (see §1.5 [47] and [11]). Since the vector field a_x and the operator field A_x are analytic, these fields a and E satisfy the conditions of Theorem 7.3 of Chapter 3 as well.

Christoffel's mapping Γ is analytic on the corresponding domain. On the other hand, g is isomorphic with $c_0(\omega_0, \mathbf{K})$ as the Banach space. If Z is the center of the algebra g, then

$$ad:\ \mathrm{g}/Z \to gl(c_0(\omega_0, \mathbf{K}))$$

is the injective representation, where $gl(c_0)$ denotes the general linear algebra of all linear continuous operators on c_0 with values in c_0. As usually $ad(x)y := [x,y]$ denotes the adjoint linear mapping for each $x, y \in$ g. Since Z is commutative, it also has an injective representation in the general linear algebra $gl(c_0)$. Hence the Banach algebra g has an embedding into the general linear algebra $gl(c_0(\omega_0, \mathbf{K})$, since the direct sum $c_0 \oplus c_0$ is isomorphic with the Banach space c_0. Therefore, each element $x \in$ g can be written in the form

$$x = \sum_{i,j} x^{i,j} X_{i,j},$$

where $\{X_{i,j} : i, j \in \mathbf{N}\}$ is the orthonormal basis of g as the Banach space, $x^{i,j} \in \mathbf{K}$ are numbers, the limit

$$\lim_{i+j\to\infty} x^{i,j} = 0$$

is zero, consequently, g has an embedding into $L_0(c_0(\omega_0, \mathbf{K})$. Then Christoffel's mapping Γ can be written in local coordinates $x_{s(i,j)} := x^{i,j}$, where $s : \mathbf{N}^2 \to \mathbf{N}$ is a bijection for which

$$\lim_{i+j\to\infty} s(i,j) = \infty,$$

$X_{i,j} =: q_{s(i,j)}$, since
$$(x^n)^{j,k} = \sum_{l_1,\dots,l_{n-1}} x^{j,l_1} x^{l_1,l_2} \dots x^{l_{n-1},k} X_{j,k},$$
when $X_{j,k} = (\delta_{a,j}\delta_{b,k} : a,b \in \mathbf{N})$,
$$\psi \circ \phi^{-1}(x) = \sum_s q_s a^s_m x^m$$
with $a^s_m \in \mathbf{K}$ and
$$\lim_{s+|m|+Ord(m)\to\infty} a^s_m = 0,$$
since the exponential mapping $\exp(x)$ has the convergence radius $0 < \tilde{r} = p^{-1}$ for $char(\mathbf{K}) = 0$ (see above or Theorem 25.6 in [103]), where $m = (m_1,\dots,m_k)$, $k = Ord(m)$, $0 \le m_1 \in \mathbf{Z}$,..., $0 \le m_{k-1} \in \mathbf{Z}$, $0 < m_k \in \mathbf{Z}$, $0 \le k \in \mathbf{Z}$. Evidently, there exists a radius $0 < r < \infty$ such that the series for $\psi \circ \phi^{-1}$ converges on the ball $B(c_0,0,r)$ for the non-void intersection $V \cap U \neq \emptyset$. Hence each operator field of the form
$$a_{m-l,l} := (\partial^{m-2}\Gamma^{\phi}_{\phi(x)}/\partial^{m-2}x)/m! \text{ for } m \ge 2$$
and
$$a_{1,0} = a_{0,1} = (\partial\phi/\partial x)$$
satisfies Condition 3.7.4(ii). Due to Theorem 3.7.4 a unique solution of Equation 3.7.4(i) exists.

We consider a group G' corresponding to a dense subalgebra g' in g such that the embedding θ of g' into g is of class L_1 for the measure μ_S or $\theta = \theta_1\theta_2$ with θ_1 and θ_2 of class L_ψ for the measure $\mu_{S,\gamma,\psi}$.

Let us take an operator $T \in L_d(\mathrm{g})$ or $T = T_1T_2$ with T_1 and $T_2 \in L_d(\mathrm{g})$, where $d = 1$ or $d = \psi$. One can consider $\mathsf{h}_1 := T(\mathrm{g})$,
$$\mathsf{h}_2 := span_{\mathbf{K}}([\mathsf{h}_1,\mathrm{g}] \cup \mathsf{h}_1)$$
and by induction $\mathsf{h}_{n+1} := span_{\mathbf{K}}([\mathsf{h}_n,\mathrm{g}] \cup \mathsf{h}_n)$, then $\mathsf{h}_{n+1} \supset \mathsf{h}_n$ and h_n is the subalgebra in g for each $n \in \mathbf{N}$. In view of Proposition 5.12.2 of Chapter 3 the space $L_d(\mathrm{g})$ is the ideal in the algebra $L(\mathrm{g})$. Therefore, $\mathsf{h} := \bigcup_n \mathsf{h}_n$ is the ideal in the algebra g due to the anti-commutativity and the Jacobi identity. Since the field $\mathbf{K}$ is spherically complete, the spaces $\mathsf{h}_{n+1} \ominus \mathsf{h}_n =: \mathsf{t}_{n+1}$ for each $n \in \mathbf{N}$ and $\mathsf{t}_1 := \mathsf{h}_1$ exist such that t_n is the $\mathbf{K}$-linear subspace of g. We take $c_0(\mathrm{g},\{\mathsf{t}_n : n\}) =: \mathsf{y}$ as the completion in g of vectors z such that $z = \sum_n z_n$ with $z_n \in \mathsf{t}_n$ for each n and for which
$$\lim_{n\to\infty} z_n = 0.$$
Evidently y is the proper ideal in the algebra g such that $\mathsf{h} \subset \mathsf{y}$, since g is infinite dimensional over the field $\mathbf{K}$. Then the embedding θ of y into g is either of class L_1 or $\theta = \theta_1\theta_2$ such that θ_1 and θ_2 belong to L_d.

Due to this let the vector field a and the operator field A be such that
$$[a_x,\mathsf{y}] \subset \mathsf{y} \quad \text{and} \quad [\mathsf{A}_x, \mathsf{ad}(\mathsf{y})] \subset \mathsf{ad}(\mathsf{y})$$

for each $x \in G$, where $ad(x)g := [x,g]$ for each $x, g \in \mathrm{g}$, that is, $ad(x) \in L(\mathrm{g})$. If $g \in \mathrm{y} \cap V$, then $\exp(ad(g)) - I$ is either of the class L_1 or the product of two operators each of which is of the class L_d.

A a countable family $(g_j, W_j) : j \in \mathbf{N})$ exists consisting of elements $g_j \in G \setminus W$ for each $j > 1$ and clopen subsets W_j with $e \in W_j \subset W$ such that $g_1 = e$, $W_1 = W$ and $\{g_j W_j : j\}$ is a locally finite covering of the group G, since G is separable and ultra-metrizable (see above). If P is a quasi-invariant and pseudo-differentiable of order b measure on a clopen subgroup W relative to a dense subgroup W', then the measure

$$P(S) := \Big(\sum_j P((g_j^{-1}S) \cap W_j) 2^{-j}\Big) \Big(\sum_j P(W_j) 2^{-j}\Big)^{-1}$$

for each Borel subset S in G is quasi-invariant and pseudo-differentiable of order b on the group G relative to the dense subgroup

$$G' := \bigcup_j g_j (W_j \cap W').$$

The group G is totally disconnected and is left-invariantly ultra-metrizable (see §8 and Theorem 5.5 [37]), consequently, in each neighborhood of the unit element e a clopen subgroup in G exists. Then conditions of Theorems 5.1 and 7 are satisfied. Therefore, analogously to §8 there are S, γ and the random function corresponding to either $\mu_{S,\gamma,\psi}$ or μ_S such that the transition measure P is quasi-invariant and pseudo-differentiable relative to G'.

11. Theorem 8 gives the subgroup G' concretely for the given group G, but Theorem 10 describes concretely G' only for the case of G satisfying the Campbell-Hausdorff formula. For a Banach-Lie group not satisfying locally the Campbell-Hausdorff formula Theorem 10 gives only the existence of G'.

These transition measures $P =: \nu$ on the Borel σ-algebra $Bf(G)$ of the totally disconnected group G induce strongly continuous unitary regular representations of G' given by the following formula:

$$T_h^\nu f(g) := (\nu^h(dg)/\nu(dg))^{i\beta + 1/2} f(h^{-1}g) \text{ for } f \in H,$$

where $H := L^2(G, \nu, \mathbf{C}) =: H$, $T_h^\nu \in U(H)$, $U(H)$ denotes the unitary group of the Hilbert space H, $i = \sqrt{-1}$, β is a marked real number. For the strong continuity of T_h^ν the continuity of the mapping $G' \ni h \mapsto \rho_\nu(h,g) \in L^1(G, \nu, \mathbf{C})$ and that ν is the Borel measure are sufficient, where $g \in G$, since G is the complete metric (i.e. Polish) space and hence the Radon space (see above).

Then analogously to §8 random functions with values in the manifold M can be constructed with quasi-invariant and pseudo-differentiable transition measures P on the manifold M relative to the action of the diffeomorphism subgroup G', $G' \subset Diff^\infty(M)$. When the manifold M is compact and hence finite dimensional over a local field $\mathbf{K}$, the group G' can be taken as the entire group $Diff^\infty(M)$.

Random functions with values in the group G can be used for constructions of infinite dimensional topologically irreducible strongly continuous unitary representations of the dense subgroup G' for the pairs (G, G') considered above (see also [65, 66, 73]).

Analogously random functions with values in the semidirect product $G := L_\xi(M,N) \otimes^s Diff^\xi(N)$ with quasi-invariant and pseudo-differentiable transition measures P relative to a dense subgroup G' can be constructed. Another related example is the direct product group $G := L_\xi(M,N) \otimes N$, when the manifold N is supplied with the additive group structure due to the embedding into the corresponding Banach space Y over the local field $\mathbf{K}$, for example, $N = B(Y,0,R)$, $0 < R \leq \infty$.

12. Theorem. *On the wrap monoid $G = \Omega_\xi(M,N)$ (see Appendix E) for each $b_0 \in \mathbf{C}$ with $Re(b_0) \geq 0$ a random function $\eta(t,\omega)$ with values in G exists so that the transition measure P is quasi-invariant and pseudo-differentiable of each order $b \in \mathbf{C}$ with $Re(b) \geq Re(b_0)$ relative to the dense wrap submonoid*

$$G' := \Omega_\xi^{\{k\}}(M,N)$$

(from Appendix E with $c > 0$ and $c' > 0$).

Proof. In view of Lemma I.2.17 [62] it is sufficient to consider the case of the manifold M with the finite atlas $At'(M)$. The rest of the proof is quite analogous to that of Theorem 4.8 using the definitions of the quasi-invariance and the pseudo-differentiability of measures for semigroups instead of groups.

Chapter 6

Appendices

6.1. Appendix A. Spaces of Continuously Differentiable Functions

1. Definitions. Let $\mathbf{K}$ be an infinite field with a non trivial non archimedean multiplicative norm. Let also X and Y be topological vector spaces over the field $\mathbf{K}$ and U be an open subset in X. Partial difference quotients of two types and spaces C^n and $C^{[n]}$ are defined in Appendix B.

Subspaces of C^n or $C^{[n]}$ of all bounded uniformly continuous functions together with $\bar{\Phi}^k f$ or $\Upsilon^k f$ on bounded open subsets of U and $U^{(k)}$ or $U^{[k]}$ for $k = 1, \dots, n$ are denoted by $C^n_b(U,Y)$ or $C^{[n]}_b(U,Y)$ respectively.

2. Definitions. Let M be a manifold modelled on a topological vector space X over the field $\mathbf{K}$ such that its atlas $At(M) := \{(U_j,\ {}_M\phi_j) :\ j \in \Lambda_M\}$ is of class $C^{\alpha'}_{\beta'}$, that is the following four conditions are satisfied:

$(M1)$ $\{U_j : j \in \Lambda_M\}$ is an open covering of M, $U_j = {}_MU_j$,

$(M2)$ $\bigcup_{j\in\Lambda_M} U_j = M$,

$(M3)$ ${}_M\phi_j := \phi_j : U_j \to \phi_j(U_j)$ is a homeomorphism for each $j \in \Lambda_M$, $\phi_j(U_j) \subset X$,

$(M4)$ $\phi_j \circ \phi_i^{-1} \in C^{\alpha'}_{\beta'}$ on its domain for each $U_i \cap U_j \neq \emptyset$, where Λ_M is a set,

$$C^\infty := \bigcap_{l=1}^{\infty} C^l_\beta,\quad C^{[\infty]}_\beta := \bigcap_{l=1}^{\infty} C^{[l]},$$

$\alpha' \in \{n, [n] : 1 \le n \le \infty\}$, $\beta \in \{\emptyset, b\}$, $C^{\alpha'}_\emptyset := C^{\alpha'}$.

Let us supply the space $C^{\alpha}_{\beta}(U,Y)$ with the bounded-open C^{α}_{β} topology. It is denoted by $\tau_{\alpha,\beta}$ generally or τ_{α} for $\beta=\emptyset$ or for compact U. Its base is:

$$W(P,V)=\{f\in C^{\alpha}_{\beta}(X,Y): S^k f|_P\in V, k=0,\dots,n\}$$

of neighborhoods of zero, where P is bounded and open in $U\subset X$, $P\subset U$, V is open in Y, $0\in V$,

$S^k=\bar{\Phi}^k$ or $S^k=\Upsilon^k$ for $\alpha=n$ or $\alpha=[n]$ respectively, $v_1,\dots,v_n\in(P-y_0)$, $v_l^{[k]}\in(P-y_0)$ for each k,l for some marked $y_0\in P$ and $|t_j|\le 1$ for every j.

If M and N are $C^{\alpha'}_{\beta}$ manifolds on topological vector spaces X and Y over $\mathbf{K}$, then we consider the uniform space $C^{\alpha}_{\beta}(M,N)$ of all mappings $f:M\to N$ such that $f_{j,i}\in C^{\alpha}_{\beta}$ on its domain for each $j\in\Lambda_N$, $i\in\Lambda_M$, where

$$f_{j,i}:={}_N\phi_j\circ f\circ{}_M\phi_i^{-1}$$

is with values in Y, $\alpha\le\alpha'$. The uniformity in the space $C^{\alpha}_{\beta}(M,N)$ is inherited from the uniformity in $C^{\alpha}_{\beta}(X,Y)$ with the help of charts of atlases of M and N. If M is compact, then the uniform spaces $C^{\alpha}_{b}(M,N)$ and $C^{\alpha}(M,N)$ coincide.

The family of all homeomorphisms $f:M\to M$ of class $C^{\alpha,id}_{\beta}$, that is $f\in C^{\alpha}(M,M)$ and $(f_{j,i}-id_{j,i})\in C^{\alpha}_{\beta}$ for all j,i, we denote by $Diff^{\alpha}_{\beta}(M)$.

Henceforth, suppose that $X=c_0(\gamma_X,\mathbf{K})$ and $Y=c_0(\gamma_Y,\mathbf{K})$ are two normed spaces over the field $\mathbf{K}$. We remind, that each vector $x\in X$ is $x=(x_j:\ x_j\in\mathbf{K},\ j\in\gamma_X)$,

$$\|x\|:=\sup_j|x_j|<\infty,$$

γ_X is a set which can be considered as an ordinal due to the Kuratowski-Zorn lemma, for each $\varepsilon>0$ the set $\{j:|x_j|>\varepsilon\}$ is finite. If the field $\mathbf{K}$ is complete relative to its non-archimedean multiplicative norm, then these spaces X and Y are complete, i.e. the Banach spaces.

3. Theorem. *The uniform space $Diff^{\alpha}_{b}(M)$ (see §2) is the topological group relative to compositions of mappings. If $f\in C^{\alpha}(M,N)$ and $g\in C^{\alpha}(N,P)$, where M, N and P are $C^{\alpha'}$-manifolds over a field $\mathbf{K}$, $\alpha'\ge\alpha$, then $g\circ f\in C^{\alpha}(M,P)$.*

Proof. The group operation in the uniform space $Diff^{\alpha}_{\beta}(M)$ is $(f,g)\mapsto f\circ g$, where $f\circ g(x):=f(g(x))$ for each $x\in M$. Then $f=id$ is the unit element in $Diff^{\alpha}_{\beta}(M)$, where $id(x)=x$ for each $x\in M$. Since the composition of mappings is associative, the composition $f\circ(g\circ h)=(f\circ g)\circ h$ is associative as the group operation. For each mapping $f\in Diff^{\alpha}_{\beta}$ there exists its inverse mapping f^{-1} such that $f^{-1}(y)=x$ for each $y=f(x)$, $x\in M$, since $f:M\to M$ is the homeomorphism.

It remains to verify that $f^{-1}\in Diff^{\alpha}_{\beta}(M)$ for each $f\in Diff^{\alpha}_{\beta}(M)$ and the composition $(Diff^{\alpha}_{\beta})^2\ni(f,g)\mapsto f\circ g\in Diff^{\alpha}_{\beta}(M)$ and inversion $f\mapsto f^{-1}$ are continuous operations.

In the normed space $Y = c_0(\gamma_Y, \mathbf{K})$ a subspace $span_{\mathbf{K}}\{e_j : j \in \gamma_X\}$ consisting of all finite $\mathbf{K}$-linear combinations of vectors $e_j = (0,\dots,0,1,0,\dots)$ with 1 on the j-th place is everywhere dense. Therefore, each $f \in C^{\alpha}_b(U,Y)$ is the uniform limit of mappings $(f_1,\dots,f_j,0,\dots) \in C^{\alpha}_b(U,Y)$ together with $\bar{\Phi}^k f$ or $\Upsilon^k f$ on bounded subsets of U and $U^{(k)}$ or $U^{[k]}$ for $1 \le k \le n$, $n \in \mathbf{N}$, $n \le \alpha$. In particular, consider ${}_N\phi_l \circ f \circ {}_M\phi_i^{-1}$ for $f \in C^{\alpha}_b(M,N)$ taking U as a finite union of ${}_M\phi_i({}_MU_j)$. Consider all possible embeddings of $\mathbf{K}^v$ into X, particularly, containing $x^{(k)}$ or $x^{[k]}$ for each $0 \le k \le n$, where $n \in \mathbf{N}$, $n \le \alpha$. In view of Formulas 9(1) or 10(1) in Appendix B, using restrictions on different embedded subspaces $\mathbf{K}^{[k]}$ or $\mathbf{K}^{(k)}$ into $X^{[k]}$ or $X^{(k)}$ and uniform continuity of Υ^k and $\bar{\Phi}^k$ on bounded open subsets, $k = 0,1,2,\dots$, we get, that $f \circ g \in C^{\alpha}_b(M,M)$ for each $f,g \in Diff^{\alpha}_b(M)$, since $(f_{l,s} \circ g_{s,i} - id_{l,s,i}) \in C^{\alpha}_b(U_{l,s,i},X)$ on the corresponding domains $U_{l,s,i}$ in X.

From $f,g \in Hom(M,M)$ it follows, that $f \circ g \in Hom(M,M)$, hence $f \circ g \in Diff^{\alpha}_b(M)$. Applying to $id_{l,i} = f_{l,s}^{-1} \circ f_{s,i}$ on corresponding domains Formulas 9(1) or 10(1) and Lemma 21 of Appendix B and restricting on different embedded subspaces $\mathbf{K}^{[k]}$ or $\mathbf{K}^{(k)}$ in $X^{[k]}$ or $X^{(k)}$ and using uniform continuity on bounded open subsets for Υ^k or $\bar{\Phi}^k$, $k = 0,1,2,\dots$, to both sides of this equality gives that $f^{-1} \in Diff^{\alpha}_b(M)$ for each $f \in Diff^{\alpha}_b(M)$.

The space $C^{\phi}_b(U,Y)$ is normed for U bounded in X for $\phi \in \{n,[n]\}$ with $n \in \mathbf{N}$ such that

$$\|f\|_{C^n_b(U,Y)} := \sup_{0 \le k \le n; z \in V^{(k)}} \|\bar{\Phi}^k f(z)\|_Y \tag{1}$$

or

$$\|f\|_{C^{[n]}_b(U,Y)} := \sup_{0 \le k \le n; z \in V^{[k]}} \|\Upsilon^k f(z)\|_Y, \tag{2}$$

where $V^{(k)} := \{z \in U^{(k)} : z = (x; v_1, v_2, \dots; t_1, t_2, \dots), \|v_j\|_X = 1 \forall j\}$,

$$V^{[k]} := \{z = x^{[k]} \in U^{[k]} : \|v_1^{[k-1]}\|_X = 1, |\ {}_l v_2^{[q]} t_{q+1}| \le 1, |v_3^{[q]}| \le 1 \quad \forall l,q\}.$$

The uniformity of the linear space $C^{\infty}_b(U,Y)$ or $C^{[\infty]}_b(U,Y)$ is defined by the family of such norms. Then the uniformity in $C^{\alpha}_b(M,N)$ is induced by the uniformity in $C^{\alpha}_b(U,Y)$ by all bounded subsets U in finite unions of ${}_M\phi_i({}_MU_i)$, since to each $f \in C^{\alpha,id}_b(M,N)$ the functions $f_{j,i} = {}_N\phi_j \circ f \circ {}_M\phi_i^{-1}$ such that $(f_{j,i} - id_{j,i}) \in C^{\alpha}_b(U_{j,i},Y)$ for all j,i correspond on their domains $U_{j,i} \subset X$. Then application of Formulas 9(1) or 10(1) and Lemma 21 of Appendix B by induction on k and restricting on different embedded subspaces $\mathbf{K}^{[k]}$ or $\mathbf{K}^{(k)}$ in $X^{[k]}$ or $X^{(k)}$ and using uniform continuity on bounded open subsets gives that $(f,g) \mapsto f^{-1} \circ g$ is C^{α}_b uniformly continuous on bounded subsets of U and $U^{(k)}$ or $U^{[k]}$, where U is a finite union of charts U_j of M.

The second statement follows from the proof above for three manifolds M, N and P modelled on topological vector spaces X_M, X_N and X_P respectively over the field $\mathbf{K}$ instead of one M. For $\beta = \emptyset$ the inclusion $f_{l,s} \circ g_{s,i} \in C^{\alpha}(U_{l,s,i},X_P)$ follows for all l,s,i on the corresponding non-void domain $U_{l,s,i}$ due to Formulas 9(1) or 10(1) and Lemma 21 of Appendix B and using continuity on open subsets in $X_M^{(k)}$ and $X_N^{(k)}$ or $X_M^{[k]}$ and $X_N^{[k]}$ of $\bar{\Phi}^k f_{l,s}$ and $\bar{\Phi}^k g_{s,i}$ or $\Upsilon^k f_{l,s}$ and $\Upsilon^k g_{s,i}$ correspondingly.

6.2. Appendix B. Functions Differentiability over Non-archimedean Fields

6.2.1. Introduction

In this appendix basic facts about differentiability of functions are written.

In the non-archimedean analysis classes of smoothness are defined in another fashion as in the classical case over the real field **R**, since locally constant functions on fields **K** with non-archimedean norms are infinite differentiable. On the other hand, non trivial non locally constant functions infinite differentiable with identically zero derivatives exist [103, 104]. This is caused by the stronger ultra-metric inequality $|x+y| \le \max(|x|,|y|)$ in comparison with the usual triangle inequality, where $|x|$ is a multiplicative norm in **K** [99].

This appendix describes the smoothness of functions $f(x_1,\dots,x_m)$ of variables $x_1,\dots,x_m$ in infinite fields with non trivial non-archimedean norms, where $m \ge 2$. Here fields locally compact and as well as non locally compact are considered. Theorems about classes of smoothness C^n or C^n_b of functions with continuous or bounded uniformly continuous on bounded domains partial difference quotients up to the order n are investigated.

Moreover, classes of smoothness $C^{n,r}$ and $C^{n,r}_b$ and more general in the sense of Lipschitz for partial difference quotients are considered and theorems for them are proved. In the proof of Theorem 2.42 it was used specific feature of the non-archimedean analysis of analytic functions for which an analog of the Louiville theorem is not true (see also [103]).

In the third section the approximate limits and approximate differentiability in the sense of partial difference quotients are defined and investigated over locally compact fields relative to the Haar non-negative measures on fields. This is important for treatments of random functions. Non-archimedean analogs of classical theorems of Kirzsbraun, Rademacher, Stepanoff, Whitney are formulated and proved (see Theorems 3.8, 9, 17, 22 respectively). Their relations with the lipschitzian property and almost everywhere differentiability are studied (see Theorems 3.15, 19, 20, 23, 24).

6.2.2. Smoothness of Functions

1. Definitions. Let **K** be an infinite field with a non trivial non-archimedean multiplicative norm. Let also X and Y be topological vector spaces over the field **K** and U be an open subset in X. For a function $f: U \to Y$ consider the associated function

$$f^{[1]}(x,v,t) := [f(x+tv) - f(x)]/t$$

on a set $U^{[1]}$ at first for $t \neq 0$. The latter set is defined as:

$$U^{[1]} := \{(x,v,t) \in X^2 \times \mathbf{K}, x \in U, x+tv \in U\}.$$

If f is a continuous function on the set U and $f^{[1]}$ has a continuous extension on the set $U^{[1]}$, then we say, that f is continuously differentiable or belongs to the class C^1.

The **K**-linear space of all such continuously differentiable functions f on U is denoted by $C^{[1]}(U,Y)$.

By induction we define functions

$$f^{[n+1]} := (f^{[n]})^{[1]}$$

and spaces $C^{[n+1]}(U,Y)$ for $n = 1,2,3,\ldots$, where $f^{[0]} := f$, $f^{[n+1]} \in C^{[n+1]}(U,Y)$ has as the domain $U^{[n+1]} := (U^{[n]})^{[1]}$.

The differential $df(x) : X \to Y$ of the function f is defined as $df(x)v := f^{[1]}(x,v,0)$.

We define also partial difference quotient operators Φ^n by variables corresponding to x only such that

$$\Phi^1 f(x;v;t) = f^{[1]}(x,v,t)$$

at first for $t \neq 0$ and if $\Phi^1 f$ is continuous for $t \neq 0$ and has a continuous extension on $U^{[1]} =: U^{(1)}$, then we denote it by $\bar{\Phi}^1 f(x;v;t)$. Define by induction

$$\Phi^{n+1} f(x;v_1,\ldots,v_{n+1};t_1,\ldots,t_{n+1}) := \Phi^1(\Phi^n f(x;v_1,\ldots,v_n;t_1,\ldots,t_n))(x;v_{n+1};t_{n+1})$$

at first for $t_1 \neq 0,\ldots,t_{n+1} \neq 0$ on the domain

$$U^{(n+1)} := \{(x;v_1,\ldots,v_{n+1};t_1,\ldots,t_{n+1}) :$$

$$x \in U; v_1,\ldots,v_{n+1} \in X; t_1,\ldots,t_{n+1} \in \mathbf{K};$$

$$x + v_1t_1 \in U,\ldots,x + v_1t_1 + \cdots + v_{n+1}t_{n+1} \in U\}.$$

If f is continuous on the set U and the partial difference quotients $\Phi^1 f,\ldots,\Phi^{n+1} f$ has continuous extensions denoted by $\bar{\Phi}^1 f,\ldots,\bar{\Phi}^{n+1} f$ on the domains $U^{(1)},\ldots,U^{(n+1)}$ respectively, then we say that f is of class of smoothness C^{n+1}.

The **K** linear space of all C^{n+1} functions on U is denoted by $C^{n+1}(U,Y)$, where $\Phi^0 f := f$, $C^0(U,Y)$ is the space of all continuous functions $f : U \to Y$.

Then the differential of the function f is given by the equation

$$d^n f(x).(v_1,\ldots,v_n) := n!\bar{\Phi}^n f(x;v_1,\ldots,v_n;0,\ldots,0),$$

where $n \geq 1$, also one denotes $D^n f = d^n f$.

Shortly we shall write the argument of $f^{[n]}$ as $x^{[n]} \in U^{[n]}$ and of $\bar{\Phi}^n f$ as $x^{(n)} \in U^{(n)}$, where

$$x^{[0]} = x^{(0)} = x, \quad x^{[1]} = x^{(1)} = (x,v,t),$$

$$v^{[0]} = v^{(0)} = v, \quad t_1 = t, \quad x^{[k]} = (x^{[k-1]}, v^{[k-1]}, t_k)$$

for each $k \geq 1$, $x^{(k)} := (x;v_1,\ldots,v_k;t_1,\ldots,t_k)$.

Subspaces of uniformly C^n or $C^{[n]}$ bounded continuous functions together with $\bar{\Phi}^k f$ or $\Upsilon^k f$ on bounded open subsets of U and $U^{(k)}$ or $U^{[k]}$ for $k = 1,\ldots,n$ we denote by $C^n_b(U,Y)$ or $C^{[n]}_b(U,Y)$ respectively.

Consider partial difference quotients of products and compositions of functions and relations between partial difference quotients and differentiability of both types. One denotes by $L(X,Y)$ the space of all continuous **K**-linear mappings $A : X \to Y$. By $L_n(X^{\otimes n},Y)$ is denoted the space of all continuous **K** n-linear mappings $A : X^{\otimes n} \to Y$, particularly,

$L(X,Y) = L_1(X^{\otimes 1},Y)$. If X and Y are normed spaces, then $L_n(X^{\otimes n},Y)$ is supplied with the operator norm:

$$\|A\| := \sup_{h_1\neq 0,\dots,h_n\neq 0; h_1,\dots,h_n\in X} \|A.(h_1,\dots,h_n)\|_Y/(\|h_1\|_X\dots\|h_n\|_X).$$

2. Lemma. *The spaces $C^{[1]}(U,Y)$ and $C^1(U,Y)$ are linearly topologically isomorphic. If $f\in C^n(U,Y)$, then $\bar\Phi^n f(x;*;0,\dots,0): X^{\otimes n}\to Y$ is a $\mathbf{K}$ n-linear $C^0(U,L_n(X^{\otimes n},Y))$ symmetric map.*

Proof. From Definition 1 it follows, that $f^{[1]}(x,v,t) = \bar\Phi^1 f(x;v;t)$ on the domain $U^{[1]} = U^{(1)}$, so both $\mathbf{K}$-linear spaces are linearly topologically isomorphic.

On the other hand, due to its definition $\Phi^n f(x;*,0,\dots,0)$ is the $\mathbf{K}$ n-linear symmetric mapping for each $x\in U$ and it belongs to the space $C^0(U,L_n(X^{\otimes n},Y))$, since $\bar\Phi^n f(x;v_1,\dots,v_n;t_1,\dots,t_n)$ is continuous on the domain $U^{(n)}$ and for each point $x\in U$ and vectors $v_1,\dots,v_n\in X$ there exist neighborhoods V_i of v_i in X and W of zero in $\mathbf{K}$ such that the inclusion $x+WV_1+\dots+WV_n\subset U$ is fulfilled.

3. Lemma. *Operators $\Upsilon^n(f) := f^{[n]}$ from the space $C^{[n]}(U,Y)$ into the space $C^0(U^{[n]},Y)$ and $\bar\Phi^n: C^n(U,Y)\to C^0(U^{(n)},Y)$ are $\mathbf{K}$-linear and continuous.*

Proof. Since $[(af+bg)(x+vt)-(af+bg)(x)]/t = a(f(x+vt)-f(x))/t + b(g(x+vt)-g(x))/t$ for each $f,g\in C^1(U,Y)$ and each $a,b\in\mathbf{K}$, applying this formula by induction and using definitions of operators Υ^n and $\bar\Phi^n$ we get their $\mathbf{K}$-linearity. Indeed,

$$\Upsilon^n(af+bg)(x^{[n]}) = \Upsilon^1(\Upsilon^{n-1}(af+bg)(x^{[n-1]}))(x^{[n]}) = \Upsilon^1(af^{[n-1]}+bg^{[n-1]})(x^{[n]})$$
$$= af^{[n]}(x^{[n]}) + bg^{[n]}(x^{[n]})$$

and

$$\bar\Phi^n(af+bg)(x^{(n)}) = \bar\Phi^1(\bar\Phi^{n-1}(af+bg)(x^{(n-1)}))(x^{(n)}) = \bar\Phi^1(af^{(n-1)}+bg^{(n-1)})(x^{(n)})$$
$$= af^{(n)}(x^{(n)}) + bg^{(n)}(x^{(n)}).$$

The continuity of Υ^n and $\bar\Phi^n$ follows from definitions of spaces $C^{[n]}(U,Y)$ and $C^n(U,Y)$ respectively.

4. Lemma. *Let either $f,g\in C^{[n]}(U,Y)$, where U is an open subset in X, Y is an algebra over $\mathbf{K}$; or $f\in C^{[n]}(U,\mathbf{K})$ and $g\in C^{[n]}(U,Y)$, where Y is a topological vector space over $\mathbf{K}$. Then the partial difference quotient $(fg)^{[n]}$ of the product fg of two functions f and g is given by the equations:*

$$(fg)^{[n]}(x^{[n]}) = (\Upsilon\otimes\hat P + \hat\pi\otimes\Upsilon)^n.(f\otimes g)(x^{[n]}) \quad (1)$$

and $(fg)^{[n]}\in C^0(U^{[n]},Y)$, where

$$(\hat\pi^k g)(x^{[k]}) := g\circ\pi_1^0\circ\pi_2^1\circ\dots\circ\pi_k^{k-1}(x^{[k]}),$$

$$\hat P^n g := P_n P_{n-1}\dots P_1 g,$$

$$\pi_k^{k-1}(x^{[k]}) := x^{[k-1]},$$

$$(A \otimes B).(f \otimes g) := (Af)(Bg)$$

for $A, B \in L(C^n(U,Y), C^m(U,Y))$, $m \leq n$,

$$(A_1 \otimes B_1) \dots (A_k \otimes B_k).(f \otimes g):$$

$$= (A_1 \dots A_k \otimes B_1 \dots B_k).(f \otimes g) := (A_1 \dots A_k f)(B_1 \dots B_k g)$$

for the corresponding operators,

$$\Upsilon^n f := f^{[n]}, \quad (P_k g)(x^{[k]}) := g(x^{[k-1]} + v^{[k-1]} t_k),$$

$$\hat{P}^k \hat{\pi}^{a_1} \Upsilon^{b_1} \dots \hat{\pi}^{a_l} \Upsilon^{b_l} g = P_{k+s} \dots P_{s+1} \hat{\pi}^{a_1} \Upsilon^{b_1} \dots \hat{\pi}^{a_l} \Upsilon^{b_l} g$$

with $s = b_1 + \dots + b_l - a_1 - \dots - a_l \geq 0$, $a_1, \dots, a_l, b_1, \dots, b_l \in \{0, 1, 2, 3, \dots\}$.

Proof. Let at first $n = 1$, then

$$(fg)^{[1]}(x^{[1]}) = [(fg)(x+vt) - (fg)(x)]/t = [(f(x+vt) - f(x))g(x+vt)$$

$$+ f(x)(g(x+vt) - g(x))]/t = (\Upsilon^1 f)(x^{[1]})(P_1 g)(x^{[1]}) + (\hat{\pi}_1^0 f)(x^{[1]}) \Upsilon^1 g(x^{[1]}), \qquad (2)$$

since $\hat{\pi}_1^0(x^{[1]}) = x$ and P_1 is the composition of the projection $\hat{\pi}_1^0$ and the shift operator on vt.

Let now $n = 2$, then applying Formula (2) we get:

$$(fg)^{[2]}(x^{[2]}) = ((fg)^{[1]}(x^{[1]}))^{[1]}(x^{[2]})$$

$$= (\Upsilon^1(f^{[1]}(x^{[1]})(x^{[2]}))g(x + (v^{[0]} + v_2^{[1]} t_2)(t_1 + v_3^{[1]} t_2) + v_1^{[1]} t_2)$$

$$+ f^{[1]}(x^{[1]}) g^{[1]}(x + v^{[0]} t_1, v_1^{[1]} + v_2^{[1]}(t_1 + v_3^{[1]} t_2), t_2)$$

$$+ f^{[1]}(x, v_1^{[1]}, t_2) g^{[1]}(x^{[1]} + v_1^{[1]} t_2) + f(x) g^{[2]}(x^{[2]}), \qquad (3)$$

where $v^{[k]} = (v_1^{[k]}, v_2^{[k]}, v_3^{[k]})$ for each $k \geq 1$ and $v^{[0]} = v_1^{[0]}$ such that $x^{[k]} + v^{[k]} t_{k+1} = (x^{[k]} + v_1^{[k]} t_{k+1}, v^{[k-1]} + v_2^{[k]} t_{k+1}, t_k + v_3^{[k]} t_{k+1})$ for each $1 \leq k \in \mathbf{Z}$. For $n = 3$ we deduce the formula

$$(fg)^{[3]}(x^{[3]}) = [(\Upsilon^3 f)(\hat{P}^3 g) + (\hat{\pi}^1 \Upsilon^2 f)(\Upsilon^1 \hat{P}^2 g)$$

$$+ (\Upsilon^1(\hat{\pi}^1 \Upsilon^1 f))(\hat{P}^1 \Upsilon^1 \hat{P}^1 g) + (\hat{\pi}^2 \Upsilon^1 f)(\Upsilon^2 \hat{P}^1 g) + (\Upsilon^2 \hat{\pi}^1 f)(\hat{P}^2 \Upsilon^1 g)$$

$$+ (\hat{\pi}^1 \Upsilon^1 \hat{\pi}^1 f)(\Upsilon^1 \hat{P}^1 \Upsilon^1 g) + (\Upsilon^1(\hat{\pi}^2 f))(\hat{P}^1 \Upsilon^2 g) + (\hat{\pi}^3 f)(\Upsilon^3 g)](x^{[3]}), \qquad (4)$$

since by our definition

$$\hat{P}^k \hat{\pi}^{a_1} \Upsilon^{b_1} \dots \hat{\pi}^{a_l} \Upsilon^{b_l} g = P_{k+s} \dots P_{s+1} \hat{\pi}^{a_1} \Upsilon^{b_1} \dots \hat{\pi}^{a_l} \Upsilon^{b_l} g$$

with $s = b_1 + \dots + b_l - a_1 - \dots - a_l \geq 0$, $a_1, \dots, a_l, b_1, \dots, b_l \in \{0, 1, 2, 3, \dots\}$.

Therefore, Formula (1) for $n = 1$ and $n = 2$ and $n = 3$ is demonstrated by Formulas $(2-4)$. If $f, g \in C^0(U^{[k]}, Y)$, $a, b \in \mathbf{K}$ are numbers, then

$$(P_k(af + bg))(x^{[k]}) := (af + bg)(x^{[k-1]} + v^{[k-1]} t_k)$$

$$= af(x^{[k-1]} + v^{[k-1]}t_k) + bg(x^{[k-1]} + v^{[k-1]}t_k).$$

Moreover,

$$\hat{\pi}^k(af+bg)(x^{[k]}) = (af+bg)\circ\pi_1^0\circ\pi_2^1\circ\dots\circ\pi_k^{k-1}(x^{[k]}) = (af+bg)(x)$$

$$= af(x)+bg(x) = a\hat{\pi}^k f(x^{[k]}) + b\hat{\pi}^k g(x^{[k]})$$

for each $x^{[k]} \in U^{[k]}$. Thus, $\hat{\pi}^k$ and P_k and $\hat{P}^k$ are **K**-linear operators for each $k \in \mathbf{N}$. Suppose that Formula (1) is proved for $n = 1,\dots,m$. Then for $n = m+1$ by application of Formula (2) to both sides of Formula (1) for $n = m$ we deduce:

$$(fg)^{m+1}(x^{[m+1]}) = ((fg)^{[m]}(x^{[m]}))^{[1]}(x^{[m+1]})$$

$$= ((\Upsilon\otimes\hat{P}+\hat{\pi}\otimes\Upsilon)^m.(f\otimes g)(x^{[m]}))^{[1]}(x^{[m+1]})$$

$$= (\Upsilon\otimes\hat{P}+\hat{\pi}\otimes\Upsilon)^{m+1}.(f\otimes g)(x^{[m+1]}),$$

since $x^{[m+1]} = (x^{[m]})^{[1]}$ and more generally $x^{[m+k]} = (x^{[m]})^{[k]}$ for any non-negative integers m and k such that $\pi_k^{k-1}(x^{[m+k]}) = x^{[m+k-1]}$ for $k \geq 1$. Indeed, the operators Υ^k, $\hat{P}^k$ and $\hat{\pi}$ are **K**-linear on corresponding spaces of functions (see above and Lemma 3) and

$$(\Upsilon\otimes\hat{P}+\hat{\pi}\otimes\Upsilon)^{m+1}.(f\otimes g)(x^{[m+1]})$$

$$= \sum_{a_1+\dots+a_{m+1}+b_1+\dots+b_{m+1}=m+1} (\Upsilon^{a_1}\otimes\hat{P}^{a_1})$$

$$(\hat{\pi}^{b_1}\otimes\Upsilon^{b_1})\dots(\Upsilon^{a_{m+1}}\otimes\hat{P}^{a_{m+1}})(\hat{\pi}^{b_{m+1}}\otimes\Upsilon^{b_{m+1}}).(f\otimes g)(x^{[m+1]}),$$

where a_j and b_j are nonnegative integers for each $j = 1,\dots,m+1$,

$$(A_1\otimes B_1)\dots(A_k\otimes B_k).(f\otimes g) := (A_1\dots A_k\otimes B_1\dots B_k).(f\otimes g):$$

$$= (A_1\dots A_k f)(B_1\dots B_k g).$$

5. Note. One can consider the projection

$$\psi_n : X^{m(n)}\times\mathbf{K}^{s(n)} \to X^{l(n)}\times\mathbf{K}^n, \tag{1}$$

where $m(n) = 2m(n-1)$, $s(n) = 2s(n-1)+1$, $l(n) = n+1$ for each natural number $n \in \mathbf{N}$ so that $m(0) = 1$, $s(0) = 0$, $m(n) = 2^n$, $s(n) = 1+2+2^2+\dots+2^{n-1} = 2^n-1$. Then $m(n)$, $s(n)$, $l(n)$ and n correspond to number of variables in X, **K** for Υ^n, in X and **K** for Φ^n respectively. Therefore, $\psi(x^{[n]}) = x^{(n)}$ and $\psi_n(U^{[n]}) = U^{(n)}$ for each natural number $n \in \mathbf{N}$ for suitable ordering of variables. Thus we get the formulas

$$\bar{\Phi}^n f(x^{(n)}) = \hat{\psi}_n\Upsilon^n f(x^{[n]}) = f^{[n]}(x^{[n]})|_{W^{(n)}},$$

where $\hat{\psi}_n g(y) := g(\psi_n(y))$ for a function g on a subset V in $X^{l(n)}\times\mathbf{K}^n$ for each $y \in \psi_n^{-1}(V) \subset X^{m(n)}\times\mathbf{K}^{s(n)}$, $W^{(n)} = U^{(n)}\times 0$, $0 \in X^{m(n)-l(n)}\times\mathbf{K}^{s(n)-n}$ for the corresponding ordering of variables.

6. Corollary. *Let either $f,g \in C^n(U,Y)$, where U is an open subset in X, Y is an algebra over $\mathbf{K}$; or $f \in C^n(U,\mathbf{K})$ and $g \in C^n(U,Y)$, where Y is a topological vector space over $\mathbf{K}$. Then the partial difference quotient $\bar{\Phi}^n(fg)$ of the product fg of two functions f and g is the following:*

$$\bar{\Phi}^n(fg)(x^{(n)}) = (\bar{\Phi}\otimes\hat{P}+\hat{\pi}\otimes\bar{\Phi})^n.(f\otimes g)(x^{(n)}) \tag{1}$$

and $\bar{\Phi}^n(fg) \in C^0(U^{(n)},Y)$. In more details:

$$\bar{\Phi}^n(fg)(x^{(n)}) = \sum_{0\le a,0\le b,a+b=n}\ \sum_{j_1<\dots<j_a;s_1<\dots<s_b;\{j_1,\dots,j_a\}\cup\{s_1,\dots,s_b\}=\{1,\dots,n\}}$$
$$\bar{\Phi}^a f(x;v_{j_1},\dots,v_{j_a};t_{j_1},\dots,t_{j_a})$$
$$\bar{\Phi}^b g(x+v_{j_1}t_{j_1}+\dots+v_{j_a}t_{j_a};v_{s_1},\dots,v_{s_b};t_{s_1},\dots,t_{s_b}). \tag{2}$$

Proof. The operator $\hat{\psi}_n$ is $\mathbf{K}$-linear, since

$$\hat{\psi}_n(af+bg)(y) = (af+bg)(\psi_n(y)) = af(\psi_n(y))+bg(\psi_n(y))$$

for any numbers $a,b \in \mathbf{K}$ and functions f,g on a subset V in $X^{l(n)}\times\mathbf{K}^n$ and each point $y \in \psi_n^{-1}(V) \subset X^{m(n)}\times\mathbf{K}^{s(n)}$. Mention that the restrictions of $\hat{\pi}_k^{k-1}$ and P_k on the domain $W^{(k)}$ give $\pi_k^{k-1}(x^{(k)}) := x^{(k-1)}$ and $(P_k g)(x^{(k)}) := g(x^{(k-1)}+v_k t_k)$ in the notation of §1. The application of the operator $\hat{\psi}_n$ to both sides of Equation 4(1) gives Equation (1) of this corollary, since $\hat{\psi}_n\Upsilon^n = \bar{\Phi}^n$ for each nonnegative integer n, where $\Upsilon^0 = I$ and $\bar{\Phi}^0 = I$ and $\hat{\psi}_0 = I$ are the unit operators.

7. Lemma. *Let $f_1,\dots,f_k \in C^{[n]}(U,Y)$, where U is an open subset in X, either Y is an algebra over $\mathbf{K}$; or $f_1,\dots,f_{k-1} \in C^{[n]}(U,\mathbf{K})$ and $f_k \in C^{[n]}(U,Y)$, where Y is a topological vector space over $\mathbf{K}$. Then the partial difference quotient $(f_1\dots f_k)^{[n]}$ of the product of functions $f_1,\dots,f_k$ is given by the formula:*

$$(f_1\dots f_k)^{[n]}(x^{[n]}) = \left[\sum_{\alpha=0}^{k-1}\hat{\pi}^{\otimes\alpha}\otimes\Upsilon\otimes\hat{P}^{\otimes(k-\alpha-1)}\right]^n.(f_1\otimes\dots\otimes f_k)(x^{[n]}) \tag{1}$$

and $(f_1\dots f_k)^{[n]} \in C^0(U^{[n]},Y)$, where

$$\hat{\pi}^{\otimes\alpha}\otimes\Upsilon\otimes\hat{P}^{\otimes(k-\alpha-1)}.(f_1\otimes\dots\otimes f_k) := (\hat{\pi}(f_1\dots f_\alpha))(\Upsilon f_{\alpha+1})(\hat{P}(f_{\alpha+2}\dots f_k)),$$

particularly $\hat{\pi}^0 := I$, $\hat{P}^0 = I$, I denotes the unit operator, $\hat{\pi}f_0 := 1$, $\hat{P}f_{k+1} := 1$ (see Lemma 4).

Proof. We consider at first the case $n=1$ and apply Formula 4(1) by induction to appearing products of functions. Then we infer that

$$\Upsilon^1(f_1\dots f_k)(x^{[1]}) = [(\Upsilon^1(f_1\dots f_{k-1}))(P_1 f_k)+(\hat{\pi}^1(f_1\dots f_{k-1}))(\Upsilon^1 f_k)](x^{[1]})$$
$$= [(\Upsilon^1(f_1\dots f_{k-2}))(P_1 f_{k-1})(P_1 f_k)+(\hat{\pi}^1(f_1\dots f_{k-2}))(\Upsilon^1 f_{k-1})(P_1 f_k)$$

$$+(\hat{\pi}^1(f_1 \dots f_{k-1}))(\Upsilon^1 f_k)](x^{[1]}) = \dots$$

$$= \Big(\sum_{\alpha=0}^{k-1} (\hat{\pi}^1)^{\otimes \alpha} \otimes \Upsilon^1 \otimes P_1^{\otimes(k-\alpha-1)}\Big).(f_1 \otimes \dots \otimes f_k), \tag{2}$$

where

$$A^{\otimes \alpha} \otimes B \otimes C^{\otimes(k-\alpha-1)}.(f_1 \otimes \dots \otimes f_k) := (A(f_1 \dots f_\alpha))(B f_{\alpha+1})(C(f_{\alpha+2} \dots f_k))$$

for operators A, B and C and each nonnegative integer α, where $A^0 := I$ and $C^0 = I$ are the unit operators, $A f_0 := 1$, $C f_{k+1} := 1$. Particularly we have $A = \hat{\pi}^1$, $B = \Upsilon^1$, $C = P_1$. Thus, acting by induction on both sides by Υ^1 from Formula (2) we deduce Formula (1) of this lemma, since the product (i.e. composition) of n operators $\Upsilon^1 \dots \Upsilon^1$ is equal to Υ^n.

8. Corollary. *Let $f_1, \dots, f_k \in C^n(U,Y)$, where U is an open subset in X, either Y is an algebra over $\mathbf{K}$; or $f_1, \dots, f_{k-1} \in C^n(U, \mathbf{K})$ and $f_k \in C^n(U,Y)$, where Y is a topological vector space over $\mathbf{K}$. Then the partial difference quotient $\bar{\Phi}^n(f_1 \dots f_k)$ of the product of functions $f_1, \dots, f_k$ has the following expression:*

$$\bar{\Phi}^n(f_1 \dots f_k)(x^{(n)}) = \Big[\sum_{\alpha=0}^{k-1} \hat{\pi}^{\otimes \alpha} \otimes \bar{\Phi} \otimes \hat{P}^{\otimes(k-\alpha-1)}\Big]^n.(f_1 \otimes \dots \otimes f_k)(x^{(n)}) \tag{1}$$

and $\bar{\Phi}^n(f_1 \dots f_k) \in C^0(U^{(n)}, Y)$, where

$$\hat{\pi}^{\otimes \alpha} \otimes \bar{\Phi} \otimes \hat{P}^{\otimes(k-\alpha-1)}.(f_1 \otimes \dots \otimes f_k) := (\hat{\pi}(f_1 \dots f_\alpha))(\bar{\Phi} f_{\alpha+1})(\hat{P}(f_{\alpha+2} \dots f_k))$$

(see Lemma 7).

Proof. Applying operator $\hat{\psi}_n$ from Note 5 to both sides of Equation 7(1) we get Formula (1) of this Corollary.

9. Lemma. *Let two functions f and u be given so that $u \in C^{[n]}(\mathbf{K}^s, \mathbf{K}^m)$, $u(\mathbf{K}^s) \subset U$ and $f \in C^{[n]}(U,Y)$, where U is an open subset in $\mathbf{K}^m$, $s, m \in \mathbf{N}$, Y is a $\mathbf{K}$-linear space. Then their composition has the partial difference quotient in the form:*

$$(f \circ u)^{[n]}(x^{[n]}) = \Big[\sum_{j_1=1}^{m} \dots \sum_{j_n=1}^{m(n)} (A_{j_n, v^{[n-1]}, t_n} \dots A_{j_1, v^{[0]}, t_1} f \circ u)(\Upsilon^1 \circ p_{j_n} \hat{S}_{j_{n-1}+1, v^{[n-2]} t_{n-1}}$$

$$\dots \hat{S}_{j_1+1, v^{[0]} t_1} u^{n-1})(P_n \Upsilon^1 \circ p_{j_{n-1}} \hat{S}_{j_{n-2}+1, v^{[n-3]} t_{n-2}} \dots \hat{S}_{j_1+1, v^{[0]} t_1} u^{n-2})$$

$$\dots (P_n \dots P_2 \Upsilon^1 \circ p_{j_1} u) + \sum_{j_1=1}^{m} \dots \sum_{j_{n-1}=1}^{m(n-1)} (\hat{\pi}^1 (A_{j_{n-1}, v^{[n-2]}, t_{n-1}} \dots A_{j_1, v^{[0]}, t_1} f \circ u)$$

$$\Big[\sum_{\alpha=0}^{n-2} \hat{\pi}^{\otimes \alpha} \otimes \Upsilon \otimes \hat{P}^{\otimes(n-\alpha-2)}\Big]((\Upsilon^1 \circ p_{j_{n-1}} \hat{S}_{j_{n-2}+1, v^{[n-3]} t_{n-2}} \dots \hat{S}_{j_1+1, v^{[0]} t_1} u^{n-2})$$

$$\otimes \dots \otimes (P_{n-1} \dots P_2 \Upsilon^1 \circ p_{j_1} u)) + \Big[\sum_{\alpha=0}^{n-2} \hat{\pi}^{\otimes \alpha} \otimes \Upsilon \otimes \hat{P}^{\otimes(n-\alpha-2)}\Big]$$

$$\left(\sum_{j_1=1}^{m}\cdots\sum_{j_{n-2}=1}^{m(n-2)}(\hat{\pi}^1(A_{j_{n-2},v^{[n-3]},t_{n-2}}\dots A_{j_1,v^{[0]},t_1}f\circ u))\otimes\right.$$

$$\left[\sum_{\alpha=0}^{n-3}\hat{\pi}^{\otimes\alpha}\otimes\Upsilon\otimes\hat{P}^{\otimes(n-\alpha-3)}\right]((\Upsilon^1\circ p_{j_{n-2}}\hat{S}_{j_{n-3}+1,v^{[n-4]}t_{n-3}}\dots\hat{S}_{j_1+1,v^{[0]}t_1}u^{n-3})$$

$$\otimes\dots\otimes(P_{n-2}\dots P_2\Upsilon^1\circ p_{j_1}u))+\dots$$

$$+\left[\sum_{\alpha=0}^{2}\hat{\pi}^{\otimes\alpha}\otimes\Upsilon\otimes\hat{P}^{\otimes(2-\alpha)}\right]^{n-3}\left\{\sum_{j_1=1}^{m}\sum_{j_2=1}^{m(2)}(\hat{\pi}^1A_{j_2,v^{[1]},t_2}A_{j_1,v^{[0]},t_1}f\circ u)\right.$$

$$\left.(\Upsilon^1\otimes\hat{P}^1+\hat{\pi}^1\otimes\Upsilon^1)((\Upsilon^1\circ p_{j_2}\hat{S}_{j_1+1,v^{[0]}t_1}u)\otimes(P_2\Upsilon^1\circ p_{j_1}u))\right\}$$

$$\left.+(\Upsilon\otimes\hat{P}+\hat{\pi}\otimes\Upsilon)^{n-2}\left\{\sum_{j_1=1}^{m}(\hat{\pi}^1A_{j_1,v^{[0]},t_1}f\circ u)\otimes(\Upsilon^2\circ p_{j_1}u)\right\}\right](x^{[n]}) \qquad (1)$$

and $f\circ u\in C^0((\mathbf{K}^s)^{[n]},Y)$, *where*

$$S_{j,\tau}u(y):=(u_1(y),\dots,u_{j-1}(y),u_j(y+\tau_{(s)}),u_{j+1}(y+\tau_{(s)}),\dots,u_m(y+\tau_{(s)})),$$

$u=(u_1,\dots,u_m)$, $u_j\in\mathbf{K}$ *for each* $j=1,\dots,m$, $y\in\mathbf{K}^s$, $\tau=(\tau_1,\dots,\tau_k)\in\mathbf{K}^k$, $k\geq s$, $\tau_{(s)}:=(\tau_1,\dots,\tau_s)$, $p_j(x):=x_j$, $x=(x_1,\dots,x_m)$, $x_j\in\mathbf{K}$ *for each* $j=1,\dots,m$, $\hat{S}_{j+1,\tau}g(u(y),\beta):=g(S_{j+1,\tau}u(y),\beta)$, $y\in\mathbf{K}^s$, β *is some parameter,*

$$A_{j,v,t}:=(\hat{S}_{j+1,vt}\otimes t\Upsilon^1\circ p_j)^*\Upsilon^1_j,$$

where Υ^1 *is taken for variables* (x,v,t) *or corresponding to them after actions of preceding operations as* Υ^k,

$$\Upsilon^1_jf(x,v_j,t):=[f(x+e_jv_jt)-f(x)]/t,$$

$$(B\otimes A)^*\Upsilon^1f_i\circ u^i(x,v,t):=\Upsilon^1_jf_i(Bu^i,v,Au^i),$$

$B:\mathbf{K}^{m(i)}\to\mathbf{K}^{m(i)}$, $A:\mathbf{K}^{m(i)}\to\mathbf{K}$, *each vector* e_j *is so that* $e_j=(0,\dots,0,1,0,\dots,0)\in\mathbf{K}^{m(i)}$ *with* 1 *on j-th place;* $m(i)=m+i-1$, $j_i=1,\dots,m(i)$; $u^1:=u$, $u^2:=(u^1,t_1\Upsilon^1\circ p_{j_1}u^1),\dots,u^n=(u^{n-1},t_{n-1}\Upsilon^1\circ p_{j_{n-1}}u^{n-1})$, $A_{j_1,v^{[0]},t_1}f\circ u=:f_1\circ u^1$, $A_{j_n,v^{[n-1]},t_n}f_{n-1}\circ u^{n-1}=:f_n\circ u^n$, $\hat{S}_*\Upsilon^1f(z):=\Upsilon^1f(\hat{S}_*z)$.

Proof. At first consider the base of the induction with $n=1$. Then

$$(f\circ u)^{[1]}(t_0,v,t)=[f(u(t_0+vt))-f(u(t_0))]/t,$$

where $t_0\in\mathbf{K}^s$, $t\in\mathbf{K}$, $v\in\mathbf{K}^s$. Though we consider here the general case mention, that in the particular case $s=1$ one has $t_0\in\mathbf{K}$, $v\in\mathbf{K}$. The first difference quotient of the composition takes the form:

$$(f\circ u)^{[1]}(t_0,v,t)=[f(u(t_0+vt))-f(u_1(t_0),u_2(t_0+vt),\dots,u_m(t_0+vt))]/t$$

$$+[f(u_1(t_0),u_2(t_0+vt),u_3(t_0+vt),\dots,u_m(t_0+vt))$$

$$-f(u_1(t_0),u_2(t_0),u_3(t_0+vt),\dots,u_m(t_0+vt))]/t+\dots$$
$$+[f(u_1(t_0),\dots,u_{m-1}(t_0),u_m(t_0+vt))-f(u(t_0))]/t,$$

where $u=(u_1,\dots,u_m)$, $u_j\in\mathbf{K}$ for each $j=1,\dots,m$. Since $u_j(t_0+vt)-u_j(t_0)=tu_j^{[1]}(t_0,v,t)$, we infer the formula

$$(f\circ u)^{[1]}(t_0,v,t)=\Upsilon^1 f((u_1(t_0),u_2(t_0+vt),\dots,u_m(t_0+vt)),e_1,t\Upsilon^1 u_1(t_0,v,t))$$
$$\Upsilon^1 u_1(t_0,v,t)+\Upsilon^1 f((u_1(t_0),u_2(t_0),u_3(t_0+vt),\dots,u_m(t_0+vt)),e_2,t\Upsilon^1 u_2(t_0,v,t))$$
$$\Upsilon^1 u_2(t_0,v,t)+\dots+\Upsilon^1 f(u(t_0),e_m,t\Upsilon^1 u_m(t_0,v,t))\Upsilon^1 u_m(t_0,v,t),$$

since $u_j\in\mathbf{K}$ for each $j=1,\dots,m$ and $\mathbf{K}$ is the field, where $e_j=(0,\dots,0,1,0,\dots,0)\in\mathbf{K}^m$ with 1 on j-th place for each $j=1,\dots,m$. With the help of the shift operators it is possible to write the latter formula shorter:

$$\Upsilon^1(f\circ u)(y,v,t)=\sum_{j=1}^{m}\hat{S}_{j+1,vt}\Upsilon^1 f(u(y),e_j,t\Upsilon^1\circ p_ju(y,v,t))(\Upsilon^1\circ p_ju(y,v,t)), \qquad (2)$$

where $p_j(x):=x_j$, $x=(x_1,\dots,x_m)$, $x_j\in\mathbf{K}$ for each $j=1,\dots,m$,

$$\hat{S}_{j+1,\tau}g(u(y),\beta):=g(S_{j+1,\tau}u(y),\beta),$$

$y\in\mathbf{K}^s$ is a vector, $\tau\in\mathbf{K}^k$, $k\ge s$, β is some parameter. For our convenience we introduce the operators

$$A_{j,v,t}:=(\hat{S}_{j+1,vt}\otimes t\Upsilon^1\circ p_j)^*\Upsilon^1_j,$$

where Υ^1 is taken for variables (y,v,t) or corresponding to them after actions of preceding operators as Υ^k remembering that $y^{[k]},v^{[k]}\in(\mathbf{K}^s)^{[k]}$, $t\in\mathbf{K}$,

$$v^{[k]}=(v_1^{[k]},v_2^{[k]},v_3^{[k]})$$

with $v_1^{[k]},v_2^{[k]}\in(\mathbf{K}^s)^{[k-1]}$,

$$v_3^{[k]}\in\mathbf{K}^k \quad \text{for each} \quad k\ge 1,$$

in particular, $v^{[0]}=v_1^{[0]}$ for $k=0$. One certainly also has

$$\Upsilon^1_j f(x,v,t):=[f(x+e_jv_jt)-f(x)]/t,$$

$$(B\otimes A)^*\Upsilon^1 f_i\circ u^i(y,v,t):=\Upsilon^1_j f_i(Bu^i,v,Au^i),$$

where
$$B:\mathbf{K}^{m(i)}\to\mathbf{K}^{m(i)},\ A:\mathbf{K}^{m(i)}\to\mathbf{K}.$$

For example, in the particular case of $s=1$ we have $v^{[k]}\in(\mathbf{K})^{[k]}$. Therefore, in the general case Formula (2) takes the form:

$$\Upsilon^1 f\circ u(y,v,t)=\sum_{j=1}^{m}(A_{j,v,t}f\circ u)(\Upsilon^1\circ p_ju)(y,v,t). \qquad (3)$$

Taking now $n = 2$ one deduces that

$$\Upsilon^2 f \circ u(y^{[2]}) = \Upsilon^1 \sum_{j=1}^{m} [(A_{j,v,t} f \circ u)(\Upsilon^1 \circ p_j u)(y,v,t)](y^{[2]}).$$

Here in the square brackets the product is written. Hence from Formula 4(1) and Lemma 3 we infer:

$$\Upsilon^2 f \circ u(y^{[2]}) = \sum_{j=1}^{m} [(\Upsilon^1 A_{j,v^{[0]},t} f \circ u)(P_2 \Upsilon^1 \circ p_j u) + (\hat{\pi}^1 A_{j,v^{[0]},t} f \circ u)(\Upsilon^2 \circ p_j u)](y^{[2]}). \quad (4)$$

Then from Formula (3) applied to terms $A_{j,v,t} f \circ u$ it follows, that

$$\Upsilon^1 A_{j_1,v^{[0]},t_1} f \circ u(y^{[2]}) = \sum_{j_2=1}^{m(2)} (A_{j_2,v^{[1]},t_2} A_{j_1,v^{[0]},t_1} f \circ u)(\Upsilon^1 \circ p_{j_2} S_{j_1+1,v^{[0]}t_1} u)(y^{[2]}),$$

where $v^{[0]} = v$, $t_1 = t$ (see also Lemma 4). Therefore, the formula

$$\Upsilon^2 f \circ u(y^{[2]}) = \Bigg[\sum_{j_1=1}^{m} \sum_{j_2=1}^{m(2)} (A_{j_2,v^{[1]},t_2} A_{j_1,v^{[0]},t_1} f \circ u)(\Upsilon^1 \circ p_{j_2} \hat{S}_{j_1+1,v^{[0]}t_1} u)$$

$$\times (P_2 \Upsilon^1 \circ p_{j_1} u) + \sum_{j_1=1}^{m} (\hat{\pi}^1 A_{j_1,v^{[0]},t_1} f \circ u)(\Upsilon^2 \circ p_{j_1} u) \Bigg] (y^{[2]}) \quad (5)$$

is valid. Then for $n = 3$ applying Formulas (3) and 7(1) to (5) we deduce that

$$\Upsilon^3 f \circ u(y^{[3]}) = \Bigg[\sum_{j_1=1}^{m} \sum_{j_2=1}^{m(2)} \sum_{j_3=1}^{m(3)} (A_{j_3,v^{[2]},t_3} A_{j_2,v^{[1]},t_2} A_{j_1,v^{[0]},t_1} f \circ u)$$

$$(\Upsilon^1 \circ p_{j_3} \hat{S}_{j_2+1,v^{[1]}t_2} \hat{S}_{j_1+1,v^{[0]}t_1} u^2)(P_2 \Upsilon^1 \circ p_{j_2} \hat{S}_{j_1+1,v^{[0]}t_1} u)(P_3 P_2 \Upsilon^1 \circ p_{j_1} u)$$

$$+ \sum_{j_1=1}^{m} \sum_{j_2=1}^{m(2)} [(\hat{\pi}^1 (A_{j_2,v^{[1]},t_2} A_{j_1,v^{[0]},t_1} f \circ u))(\Upsilon^2 \circ p_{j_2} \hat{S}_{j_1+1,v^{[0]}t_1} u)(P_3 P_2 \Upsilon^1 \circ p_{j_1} u)$$

$$+ (\hat{\pi}^1 \{(A_{j_2,v^{[1]},t_2} A_{j_1,v^{[0]},t_1} f \circ u)(\Upsilon^1 \circ p_{j_2} \hat{S}_{j_1+1,v^{[0]}t_1} u)\}(\Upsilon^1 P_2 \Upsilon^1 \circ p_{j_1} u)]$$

$$+ \sum_{j_1=1}^{m} \sum_{j_3=1}^{m(3)} (A_{j_3,v^{[2]},t_3} \hat{\pi}^1 A_{j_1,v^{[0]},t_1} f \circ u)(\Upsilon^1 \circ p_{j_3} \hat{S}_{j_1+1,v^{[0]}t_1} u)(P_3 \Upsilon^2 \circ p_{j_1} u)$$

$$+ \sum_{j_1=1}^{m} (\hat{\pi}^2 A_{j_1,v^{[0]},t_1} f \circ u)(\Upsilon^3 \circ p_{j_1} u) \Bigg] (y^{[3]}). \quad (6)$$

Thus Formula (1) is proved for $n = 1,2,3$. We proceed by the mathematical induction and suppose that the needed formula is true for $k = 1,\dots,n$ and prove it for $k = n+1$. Applying Formula 7(1) to both sides of (1) we get:

$$\Upsilon^{n+1} f \circ u(y^{[n+1]}) = \Bigg[\sum_{j_1=1}^{m} \cdots \sum_{j_{n+1}=1}^{m(n+1)} (A_{j_{n+1},v^{[n]},t_{n+1}} \dots A_{j_1,v^{[0]},t_1} f \circ u)$$

$$(\Upsilon^1 \circ p_{j_{n+1}} \hat{S}_{j_n+1,v^{[n-1]}t_n} \dots \hat{S}_{j_1+1,v^{[0]}t_1} u^n)$$

$$(P_{n+1}\Upsilon^1 \circ p_{j_n} \hat{S}_{j_{n-1}+1,v^{[n-2]}t_{n-1}} \dots \hat{S}_{j_1+1,v^{[0]}t_1} u^{n-1}) \dots$$

$$(P_{n+1} \dots P_2 \Upsilon^1 \circ p_{j_1} u) + \sum_{j_1=1}^{m} \cdots \sum_{j_n=1}^{m(n)} (\hat{\pi}^1 (A_{j_n,v^{[n-1]},t_n} \dots A_{j_1,v^{[0]},t_1} f \circ u)$$

$$\Upsilon^1((\Upsilon^1 \circ p_{j_n} \hat{S}_{j_{n-1}+1,v^{[n-2]}t_{n-1}} \dots \hat{S}_{j_1+1,v^{[0]}t_1} u^{n-1}) \dots (P_n \dots P_2 \Upsilon^1 \circ p_{j_1} u))$$

$$+\Upsilon^1 \Bigg(\sum_{j_1=1}^{m} \cdots \sum_{j_{n-1}=1}^{m(n-1)} (\hat{\pi}^1 (A_{j_{n-1},v^{[n-2]},t_{n-1}} \dots A_{j_1,v^{[0]},t_1} f \circ u))$$

$$\Upsilon^1((\Upsilon^1 \circ p_{j_{n-1}} \hat{S}_{j_{n-2}+1,v^{[n-3]}t_{n-2}} \dots \hat{S}_{j_1+1,v^{[0]}t_1} u^{n-2})$$

$$\dots (P_{n-1} \dots P_2 \Upsilon^1 \circ p_{j_1} u)) + \dots + \Upsilon^{n-2} \Bigg\{ \sum_{j_1=1}^{m} \sum_{j_2=1}^{m(2)}$$

$$(\hat{\pi}^1 A_{j_2,v^{[1]},t_2} A_{j_1,v^{[0]},t_1} f \circ u) \Upsilon^1((\Upsilon^1 \circ p_{j_2} \hat{S}_{j_1+1,v^{[0]}t_1} u)(P_2 \Upsilon^1 \circ p_{j_1} u)) \Bigg\}$$

$$+\Upsilon^{n-1} \Bigg\{ \sum_{j_1=1}^{m} \hat{\pi}^1 A_{j_1,v^{[0]},t_1} f \circ u)(\Upsilon^2 \circ p_{j_1} u) \Bigg\} \Bigg] (y^{[n+1]} \Bigg\} \Bigg) =$$

$$= \Bigg[\sum_{j_1=1}^{m} \cdots \sum_{j_{n+1}=1}^{m(n+1)} (A_{j_{n+1},v^{[n]},t_{n+1}} \dots A_{j_1,v^{[0]},t_1} f \circ u)(\Upsilon^1 \circ p_{j_{n+1}} \hat{S}_{j_n+1,v^{[n-1]}t_n}$$

$$\dots \hat{S}_{j_1+1,v^{[0]}t_1} u^n)(P_{n+1}\Upsilon^1 \circ p_{j_n} \hat{S}_{j_{n-1}+1,v^{[n-2]}t_{n-1}} \dots \hat{S}_{j_1+1,v^{[0]}t_1} u^{n-1})$$

$$\dots (P_{n+1} \dots P_2 \Upsilon^1 \circ p_{j_1} u)$$

$$+ \sum_{j_1=1}^{m} \cdots \sum_{j_n=1}^{m(n)} (\hat{\pi}^1 (A_{j_n,v^{[n-1]},t_n} \dots A_{j_1,v^{[0]},t_1} f \circ u) \left[\sum_{\alpha=0}^{n-1} \hat{\pi}^{\otimes\alpha} \otimes \Upsilon \otimes \hat{P}^{\otimes(n-\alpha-1)} \right]$$

$$((\Upsilon^1 \circ p_{j_n} \hat{S}_{j_{n-1}+1,v^{[n-2]}t_{n-1}} \dots \hat{S}_{j_1+1,v^{[0]}t_1} u^{n-1}) \otimes \dots \otimes (P_n \dots P_2 \Upsilon^1 \circ p_{j_1} u))$$

$$+ \left[\sum_{\alpha=0}^{n-1} \hat{\pi}^{\otimes\alpha} \otimes \Upsilon \otimes \hat{P}^{\otimes(n-\alpha-1)} \right] \left(\sum_{j_1=1}^{m} \cdots \sum_{j_{n-1}=1}^{m(n-1)} (\hat{\pi}^1 (A_{j_{n-1},v^{[n-2]},t_{n-1}} \dots A_{j_1,v^{[0]},t_1} f \circ u) \right) \otimes$$

$$\left[\sum_{\alpha=0}^{n-2} \hat{\pi}^{\otimes\alpha} \otimes \Upsilon \otimes \hat{P}^{\otimes(n-\alpha-2)} \right] ((\Upsilon^1 \circ p_{j_{n-1}} \hat{S}_{j_{n-2}+1,v^{[n-3]}t_{n-2}} \dots \hat{S}_{j_1+1,v^{[0]}t_1} u^{n-2})$$

$$\otimes \dots \otimes (P_{n-1} \dots P_2 \Upsilon^1 \circ p_{j_1} u))$$

$$+ \left[\sum_{\alpha=0}^{2} \hat{\pi}^{\otimes\alpha} \otimes \Upsilon \otimes \hat{P}^{\otimes(2-\alpha)} \right]^{n-2} \Bigg\{ \sum_{j_1=1}^{m} \sum_{j_2=1}^{m(2)} (\hat{\pi}^1 A_{j_2,v^{[1]},t_2} A_{j_1,v^{[0]},t_1} f \circ u)(\Upsilon^1 \otimes \hat{P}^1 + \hat{\pi}^1 \otimes \Upsilon^1)$$

$$((\Upsilon^1 \circ p_{j_2} \hat{S}_{j_1+1,v^{[0]}t_1} u) \otimes (P_2 \Upsilon^1 \circ p_{j_1} u)) \Bigg\}$$

$$+(\Upsilon\otimes\hat{P}+\hat{\pi}\otimes\Upsilon)^{n-1}\left\{\sum_{j_1=1}^{m}(\hat{\pi}^1 A_{j_1,v^{[0]},t_1}f\circ u)\otimes(\Upsilon^2\circ p_{j_1}u)\right\}\Bigg](y^{[n+1]}). \qquad (7)$$

One can mention that in general the partial difference quotient $(\Upsilon^{n+1}f\circ u)(y^{[n+1]})$ may depend nontrivially on all components of the vector $y^{[n+1]}$ through several terms in Formula (7). Thus Formula (1) of this Lemma is proved by induction.

10. Corollary. *Let $u\in C^n(\mathbf{K}^s,\mathbf{K}^m)$, $u(\mathbf{K}^s)\subset U$ and $f\in C^n(U,Y)$, where U is an open subset in $\mathbf{K}^m$, $s,m\in\mathbf{N}$, Y is a $\mathbf{K}$-linear space. Then for the partial difference quotient $\bar{\Phi}^n(f\circ u)$ of the composition $f\circ u$ of two functions f and u the formula:*

$$\bar{\Phi}^n(f\circ u)(x^{(n)})=\Bigg[\sum_{j_1=1}^{m}\cdots\sum_{j_n=1}^{m(n)}(B_{j_n,v^{(n-1)},t_n}\dots B_{j_1,v^{(0)},t_1}f\circ u)$$

$$(\bar{\Phi}^1\circ p_{j_n}\hat{S}_{j_{n-1}+1,v^{(n-2)}t_{n-1}}\dots\hat{S}_{j_1+1,v^{(0)}t_1}u^{n-1})$$

$$(P_n\bar{\Phi}^1\circ p_{j_{n-1}}\hat{S}_{j_{n-2}+1,v_0^{(n-3)},t_{n-2}}\dots\hat{S}_{j_1+1,v^{(0)}t_1}u^{n-2})$$

$$\dots(P_n\dots P_2\bar{\Phi}^1\circ p_{j_1}u)+\sum_{j_1=1}^{m}\cdots\sum_{j_{n-1}=1}^{m(n-1)}(\hat{\pi}^1(B_{j_{n-1},v^{(n-2)},t_{n-1}}\dots B_{j_1,v^{(0)},t_1}f\circ u)$$

$$\left[\sum_{\alpha=0}^{n-2}\hat{\pi}^{\otimes\alpha}\otimes\bar{\Phi}\otimes\hat{P}^{\otimes(n-\alpha-2)}\right]((\bar{\Phi}^1\circ p_{j_{n-1}}\hat{S}_{j_{n-2}+1,v^{(n-3)}t_{n-2}}\dots\hat{S}_{j_1+1,v^{(0)}t_1}u^{n-2})\otimes$$

$$\dots\otimes(P_{n-1}\dots P_2\bar{\Phi}^1\circ p_{j_1}u))+\left[\sum_{\alpha=0}^{n-2}\hat{\pi}^{\otimes\alpha}\otimes\bar{\Phi}\otimes\hat{P}^{\otimes(n-\alpha-2)}\right]$$

$$\left(\sum_{j_1=1}^{m}\cdots\sum_{j_{n-2}=1}^{m(n-2)}(\hat{\pi}^1(B_{j_{n-2},v^{(n-3)},t_{n-2}}\dots B_{j_1,v^{(0)},t_1}f\circ u)\right)\otimes$$

$$\left[\sum_{\alpha=0}^{n-3}\hat{\pi}^{\otimes\alpha}\otimes\bar{\Phi}\otimes\hat{P}^{\otimes(n-\alpha-3)}\right]((\bar{\Phi}^1\circ p_{j_{n-2}}\hat{S}_{j_{n-3}+1,v^{(n-4)}t_{n-3}}\dots\hat{S}_{j_1+1,v^{(0)}t_1}u^{n-3})$$

$$\otimes\dots\otimes(P_{n-2}\dots P_2\bar{\Phi}^1\circ p_{j_1}u))+\dots$$

$$+\left[\sum_{\alpha=0}^{2}\hat{\pi}^{\otimes\alpha}\otimes\bar{\Phi}\otimes\hat{P}^{\otimes(2-\alpha)}\right]^{n-3}\left\{\sum_{j_1=1}^{m}\sum_{j_2=1}^{m(2)}(\hat{\pi}^1 B_{j_2,v^{(1)},t_2}B_{j_1,v^{(0)},t_1}f\circ u)(\bar{\Phi}^1\otimes\hat{P}^1\right.$$

$$\left.+\hat{\pi}^1\otimes\bar{\Phi}^1)((\bar{\Phi}^1\circ p_{j_2}\hat{S}_{j_1+1,v^{(0)}t_1}u)\otimes(P_2\bar{\Phi}^1\circ p_{j_1}u))\right\}$$

$$+(\bar{\Phi}\otimes\hat{P}+\hat{\pi}\otimes\bar{\Phi})^{n-2}\left\{\sum_{j_1=1}^{m}(\hat{\pi}^1 B_{j_1,v^{(0)},t_1}f\circ u)\otimes(\bar{\Phi}^2\circ p_{j_1}u)\right\}\Bigg](x^{(n)}) \qquad (1)$$

is valid and $f\circ u\in C^0((\mathbf{K}^s)^{(n)},Y)$ (see notation of Lemma 9), where

$$B_{j,v,t}:=(\hat{S}_{j+1,vt}\otimes t\bar{\Phi}^1\circ p_j)^*\bar{\Phi}^1_j,$$

the partial difference quotient operator $\bar{\Phi}^1$ *is taken for variables* (x,v,t) *or corresponding to them after actions of preceding operations as* $\bar{\Phi}^k$,

$$\bar{\Phi}^1_j f(x,v,t) := [f(x+e_j v_j t) - f(x)]/t,$$

$$(B\otimes A)^*\bar{\Phi}^1 f_i \circ u^i(x,v,t) := \bar{\Phi}^1_j f_i(Bu^i, v, Au^i),$$

$$B : \mathbf{K}^{m(i)} \to \mathbf{K}^{m(i)},\ A : \mathbf{K}^{m(i)} \to \mathbf{K},\ m(i) = m+i-1,\ j_i = 1,\dots,m(i),$$

$$u^1 = u,\quad u^2 := (u^1, t_1\bar{\Phi}^1 \circ p_{j_1}u^1),$$

$$u^n := (u^{n-1}, t_{n-1}\bar{\Phi}^1 \circ p_{j_{n-1}}u^{n-1}),$$

$$\hat{S}_*\bar{\Phi}^1 f(x) := \bar{\Phi}^1 f(\hat{S}_{*}x).$$

Proof. The restriction of operators of Lemma 9 on the domain $W^{(n)}$ from Note 5 gives Formula (1) of this corollary, where $v^{(k)} \in (\mathbf{K}^s)^k \times \mathbf{K}^k$.

11. Lemma. *If* $a \neq 0$, $a \in \mathbf{K}$, *U is an open subset in X, where X and Y are topological vector spaces over* $\mathbf{K}$, $f \in C^1(U,Y)$, $T \in \mathbf{K}$, $T \neq 0$, *then*

(1) $\Upsilon^1 f(x, av, t/a) = a\Upsilon^1 f(x,v,t)$ *and*

(2) $\Upsilon^1 f(x,v,at) = a^{-1}\Upsilon^1 f(x,av,t)$ *and*

(3) $\Upsilon^1 f(x/T,v,t) = T^{-1}\Upsilon^1 f(x/T,v,t/T)$ *for each* $(x,v,t) \in U^{[1]}$ *and* $(x,v,at) \in U^{[1]}$ *and* $(x/T,v,t) \in U^{[1]}$ *respectively.*

Proof. Certainly, we have the identities:

$$\Upsilon^1 f(x,av,t/a) = [f(x+vta/a) - f(x)]/(t/a) = a[f(x+vt) - f(x)]/t = a\Upsilon^1 f(x,v,t),$$

$$\Upsilon^1 f(x,v,at) = [f(x+vta) - f(x)]/(at) = a^{-1}\Upsilon^1 f(x,av,t).$$

Moreover, for the function $g(x) := f(x/T)$ the equality

$$\Upsilon^1 g(x,v,t) = [g(x+vt) - g(x)]/t = [f((x+vt)/T) - f(x/T)]/t =$$

$$= T^{-1}[f(x/T + vt/T) - f(x/T)]/(t/T) = T^{-1}\Upsilon^1 f(x/T,v,t/T)$$

is valid.

12. Lemma. *Let* $u : \mathbf{K} \to \mathbf{K}^b$ *be a polynomial function:*

$$u = \sum_{n=0}^{m} a_n x^n, \tag{1}$$

where $a_n \in \mathbf{K}^b$ *are expansion coefficients,* $x \in \mathbf{K}$, $m \in \mathbf{N}$ *is a natural number. Then the partial difference quotient* $\Upsilon^q u$ *of the function u takes the form:*

$$\Upsilon^q u(x^{[q]}) = \sum_{n=1}^{m} a_n \sum_{k_1=1}^{n} \binom{n}{k_1} \Big\{ \Big[\sum_{k_2=1}^{n-k_1} \binom{n-k_1}{k_2} \dots$$

$$\sum_{k_q=1}^{n-k_1-\cdots-k_{q-1}} \binom{n-k_1-\cdots-k_{q-1}}{k_q} x^{n-k_1-\cdots-k_q}$$

$$\left.({}_1v_1^{[q-1]})^{k_q} t_q^{k_q-1} S_{v^{[q-1]},t_q} ({}_1v_1^{[q-2]})^{k_{q-1}} (t_{q-1})^{k_{q-1}-1} \dots S_{v^{[1]},t_2} (v^{[0]})^{k_1} (t_1)^{k_1-1}\right]$$

$$+\left[x^{n-k_1-\cdots-k_{q-1}} \sum_{k_2=1}^{k_1} \binom{n-k_1}{k_2} \cdots \sum_{k_{q-1}=1}^{n-k_1-\cdots-k_{q-2}} \binom{n-k_1-\cdots-k_{q-2}}{k_{q-1}} \sum_{k_q=1}^{k_{q-1}} \binom{k_{q-1}}{k_q}\right.$$

$$({}_1v_1^{[q-2]})^{k_{q-1}-k_q} ({}_1v_2^{[q-1]})^{k_q} t_q^{k_q-1} S_{v^{[q-1]},t_q} (t_{q-1})^{k_{q-1}-1}$$

$$\left.S_{v^{[q-2]},t_{q-1}} ({}_1v_1^{[q-3]})^{k_{q-2}} (t_{q-2})^{k_{q-2}-1} \dots S_{v^{[1]},t_2} (v^{[0]})^{k_1} (t_1)^{k_1-1}\right] + \dots$$

$$+\left[x^{n-k_1} (v^{[0]})^{k_1} \sum_{k_2=1}^{k_1-1} \binom{k_1-1}{k_2} \cdots \sum_{k_q=1}^{k_{q-1}-1} \binom{k_{q-1}-1}{k_q} t_q^{k_q-1}\right.$$

$$\left.\left.(v_3^{[q-1]})^{k_q} t_{q-1}^{k_{q-1}-k_q-1} \dots (v_3^{[2]})^{k_3} t_2^{k_2-k_3-1} (v_3^{[1]})^{k_2} t_1^{k_1-k_2-1}\right]\right\}, \tag{2}$$

where

$$S_{v^{[q-1]},t_q}\ {}_jx^{[q-1]} := {}_jx^{[q-1]} + {}_jv^{[q-1]} t_q$$

for each j, *while* $x^{[q]} = ({}_1x^{[q]}, {}_2x^{[q]}, \dots)$ *and this shift* $S_{v^{[q-1]},t_q}$ *operator acts on all terms on the right of it in a product.*

Proof. In view of Lemma 3 we have the identity:

$$\Upsilon^1 u(x^{[1]}) = \sum_{n=0}^{m} a_n((x+v^{[0]}t_1)^n - x^n)/t_1 = \sum_{n=1}^{m} a_n \sum_{k_1=1}^{n} \binom{n}{k_1} x^{n-k_1} (v^{[0]})^{k_1} t_1^{k_1-1}, \tag{3}$$

where $\binom{n}{k}$ are binomial coefficients,

$$\Upsilon^2 u(x^{[2]}) = \sum_{n=1}^{m} a_n \sum_{k_1=1}^{n} \binom{n}{k_1} ((x+v_1^{[1]}t_2)^{n-k_1} (v^{[0]}+v_2^{[1]}t_2)^{k_1} (t_1+v_3^{[1]}t_2)^{k_1-1}$$

$$-x^{n-k_1} (v^{[0]})^{k_1} t_1^{k_1-1})/t_2$$

in accordance with the notation of the proof of Lemma 4. Then one gets the formula

$$\Upsilon^2 u(x^{[2]}) = \sum_{n=1}^{m} a_n \sum_{k_1=1}^{n} \binom{n}{k_1} \left\{\left[\sum_{k_2=1}^{n-k_1} \binom{n-k_1}{k_2}\right.\right.$$

$$\left.x^{n-k_1-k_2} (v_1^{[1]})^{k_2} t_2^{k_2-1} (v^{[0]}+v_2^{[1]}t_2)^{k_1} (t_1+v_3^{[1]}t_2)^{k_1-1}\right]$$

$$+\left[x^{n-k_1} \sum_{k_2=1}^{k_1} \binom{k_1}{k_2} (v^{[0]})^{k_1-k_2} (v_2^{[1]})^{k_2} t_2^{k_2-1} (t_1+v_3^{[1]}t_2)^{k_1-1}\right]$$

$$+\left[x^{n-k_1}(v^{[0]})^{k_1}\sum_{k_2=1}^{k_1-1}\binom{k_1-1}{k_2}t_1^{k_1-k_2-1}(v_3^{[1]})^{k_2}t_2^{k_2-1}\right]\Big\}. \tag{4}$$

Therefore, Formulas $(3,4)$ imply Formula (2) for $n=1$ and $n=2$. Let formula (2) be true for $n=1,\dots,q$, we prove it for $n=q+1$. Applying to both sides of Equation (2) the partial difference quotient operator Υ^1 with the help of Formula $7(2)$ or $7(1)$ we get Formula (2) for $n=q+1$ also.

13. Corollary. *Let suppositions of Lemma 12 be satisfied, then the partial difference quotient $\Upsilon^q u$ of the function u satisfies the estimate*

$$|\Upsilon^q u(x^{[q]})| \le \max_{n=0}^{m} |a_n|$$

for each $x^{[q]} \in \mathbf{K}^{[q]}$ with $|x^{[q]}| \le 1$.

Proof. The absolute value of each term on the right side of Formula $12(2)$ in the curled brackets is not greater than one, since binomial coefficients are integer numbers and their non-archimedean absolute value is not greater than one and each component of the vector $x_j^{[q]} \in \mathbf{K}$ has an absolute value not greater than one. Applying the non-archimedean inequality

$$|y+z| \le \max(|y|,|z|)$$

for arbitrary numbers $y,z \in \mathbf{K}$ we get the statement of this corollary.

14. Corollary. *Let u be a polynomial as in Lemma* 13, *then the partial difference quotient $\bar{\Phi}^q u$ of the function u is the following:*

$$\bar{\Phi}^q u(x^{(q)}) = \sum_{n=q}^{m} a_n \sum_{k_1=1}^{n}\sum_{k_2=1}^{n-k_1}\cdots\sum_{k_q=1}^{n-k_1-\dots-k_{q-1}}\binom{n}{k_1}\binom{n-k_1}{k_2}$$

$$\cdots\binom{n-k_1-\dots-k_{q-1}}{k_q}v_1^{k_1}\dots v_q^{k_q}t_1^{k_1-1}\dots t_q^{k_q-1}x^{n-k_1-\dots-k_q}. \tag{1}$$

15. Lemma. *Let $V_j \in \mathbf{R}$ be a positive number $V_j > 0$ for each $j \in \mathbf{N}$ and the limit*

$$\lim_{j\to\infty} V_j = 0$$

is zero. Suppose also that $g \in C^\infty(\mathbf{K}^l,\mathbf{K})$, there exists $R>0$ such that $g(x)=0$ for each $|x|>R$, moreover,

$$|\Upsilon^j g(x^{[j]})| \le C^{j+1}V_j^{-j} \tag{1}$$

for each j and $|x^{[j]}| \le R$, where $C>0$ is a constant. Let also a function u be given by the formula

$$u(x) = (a + \sum_{k_1,k_2=0}^{m}\sum_{i_1,i_2=1}^{l} {}_{i_1,i_2}b_{k_1,k_2}x_{i_1}^{k_1}x_{i_2}^{k_2})g(x/T)$$

for each $x \in \mathbf{K}^l$, where $T \in \mathbf{K}$, $0<|T|\le 1$, a, ${}_{i_1,i_2}b_{k_1,k_2} \in \mathbf{K}$. Then there exists a constant $C_1>0$ independent of a, ${}_{i_1,i_2}b_{k_1,k_2}, j, x$ and T such that

$$\|\Upsilon^j u(x^{[j]})\|_{C^0(B(\mathbf{K}^{[j]},0,R),\mathbf{K})}$$

$$\leq (\max_{i_1,i_2,k_1,k_2}(|a|,|\ _{i_1,i_2}b_{k_1,k_2}|))\max(1,R^m)|T|^{-j}C_1^{j+1}V_j^{-j}. \tag{2}$$

Proof. We apply Lemmas 4 and 11 calculating by induction the partial difference quotients

$$\Upsilon^1(a+bx)(x,v,t) = bv,$$

$$\Upsilon^2(a+bx)(x^{[2]}) = bv_2^{[1]},\dots,\Upsilon^j(a+bx)(x^{[j]}) = bv_2^{[j-1]}$$

for each $j \geq 3$. Therefore, the estimate

$$\|\Upsilon^j(a+bx)\|_{C^0(B(\mathbf{K}^{[j]},0,R),\mathbf{K})} \leq \max(|a|,|b|)R$$

is valid for each $j \geq 0$. In general one can apply Formula 12(2) and Corollary 13. Then by induction from Formula 11(3) it follows, the relation

$$\|\Upsilon^j g(x/T)\|_{C^0(B(\mathbf{K}^{[j]},0,R),\mathbf{K})} = |T|^{-j}\|\Upsilon^j g(x)\|_{C^0(B(\mathbf{K}^{[j]},0,R),\mathbf{K})}$$

for each $j \geq 0$, where $\Upsilon^0 g = g$. Therefore, from Formula 4(1) and the ultra-metric inequality we get the estimate:

$$\|\Upsilon^j u(x^{[j]})\|_{C^0(B(\mathbf{K}^{[j]},0,R),\mathbf{K})}$$

$$\leq (\max_{i_1,i_2,k_1,k_2}(|a|,|\ _{i_1,i_2}b_{k_1,k_2}|))\max(1,R^m)\max_{k=0}^{j}|T|^{-k}\|\Upsilon^k g(x)\|_{C^0(B(\mathbf{K}^{[k]},0,R),\mathbf{K})}$$

$$\leq (\max_{i_1,i_2,k_1,k_2}(|a|,|\ _{i_1,i_2}b_{k_1,k_2}|))\max(1,R^m)\max_{k=0}^{j}|T|^{-k}C^{k+1}V_k^{-k},$$

since $g(x) = 0$ for $|x| > R$ and choosing $C_1 > 0$ such that

$$\infty > C_1 \geq \sup_{j=0}^{\infty}[\sup_{k=0}^{j}(C^{k+1}V_k^{-k}V_j^j|T|^{j-k})^{1/(j+1)}]$$

we get the statement of this Lemma.

16. Lemma. *If U is an open subset in the linear space $\mathbf{K}^b$, $f : U \to \mathbf{K}$ is a marked function, then a space Y_n of functions*

$$\{\Upsilon^n f(x^{[n]}) : v^{[0]},\ _lv_j^{[k]} \in \{0,1\}; j = 1,2; l = 1,2,\dots; k = 1,\dots,n-1\}$$

is finite dimensional over the field $\mathbf{K}$ *whenever it exists such that*

$$dim_{\mathbf{K}}Y_n \leq (2^{m(n-1)}-1)dim_{\mathbf{K}}Y_{n-1},$$

$n \in \mathbf{N}$, $m(n) = 2m(n-1)+1$ *for* $n \in \mathbf{N}$, $m(0) = b$.

Proof. We have the recurrence relation for a number of variables belonging to the field $\mathbf{K}$, $m(n) = 2m(n-1)+1$ for each natural number $n \in \mathbf{N}$ corresponds to the partial difference quotient $\Upsilon^n f(x^{[n]})$, $m(0) = b$ corresponds to $f(x)$. For $n = 1$ we have the identities:

$$\Upsilon^1 f(x,v,t) = (f(x+vt)-f(x))/t = [f(x+vt)-f(x+(v-\ _bve_b)t)]/t$$

$$+[f(x+(v-{}_bve_b)t)-f(x+(v-{}_be_b-{}_{b-1}e_{b-1})t)]/t+\dots+[f(x+{}_1ve_1t)-f(x)]/t,$$

where $v=({}_1v,\dots,{}_bv)$, ${}_lv\in\mathbf{K}$ for each $l=1,\dots,b$. We have that ${}_lv\in\{0,1\}$ may take only two values and the amount of such nonzero vectors v is equal to 2^b-1. Thus the family

$$\{\Upsilon^1 f(x+(v-{}_bve_b-\dots-{}_kve_k),{}_kve_k,t):\ {}_lv\in\{0,1\},l=1,\dots,b\}$$

of functions by the variables $(x,t)\in\mathbf{K}^{b+1}$ spans over $\mathbf{K}$ the space $\{\Upsilon^1 f(x,v,t):\ {}_lv\in\{0,1\},l=1,\dots,b\}$. Its dimension over $\mathbf{K}$ for the given function f is not greater, than 2^b-1.

Let the statement of this lemma be true for $n-1\geq 1$. Then one can apply the partial difference quotient operator Υ^1 to $\Upsilon^{n-1}f(x^{[n-1]})$. Replacing in the proof above the function f on the function $f^{[n-1]}$ we get the statement of this lemma for n also, since

$$\Upsilon^n f(x^{[n]})=\Upsilon^1(\Upsilon^{n-1}f(x^{[n-1]}))((x^{[n-1]})^{[1]})$$

and $x^{[n]}=(x^{[n-1]})^{[1]}$ considering $f^{[n]}(x^{[n]})$ by free variables $(x,t_1,\dots,t_n)$.

17. Corollary. *For any natural numbers $n\in\mathbf{N}$ and $b\in\mathbf{N}$ and a marked function $f:U\to\mathbf{K}$, where U is an open subset in $\mathbf{K}^b$ a finite system Λ_n of vectors $0\neq(y,v)$ with y and $v\in\mathbf{K}^{m(n-1)}$ exists such that*

$$\sum_{(y,v)\in\Lambda_n} C_{(y,v)}\Upsilon^n f(x^{[n-1]}+y,v,t_n)=0 \tag{1}$$

is identically equal to zero as the function of $(x,t_1,\dots,t_n)$, where $(x^{[n-1]}+y,v,t_n)\in U^{[n]}$, $x^{[n-1]}\in U^{[n-1]}$, $v^{[0]}$ and ${}_lv_j^{[k]}\in\{0,1\}$ for each j,k,l, y may depend on the parameters $t_1,\dots,t_n$ polynomially, $0\neq C_{(y,v)}\in\mathbf{K}$ are constants for each (y,v).

Proof. We take the cardinal $card(\Lambda_n)>dim_{\mathbf{K}}Y_n$ and ${}_lv_j^{[k]}\in\{0,1\}$ for each $l=1,\dots,m(k-1)$, $j=1,2$ and $k=0,\dots,n-1$ such that $(x,v,t_1)\in U^{[1]}$. Then one gets

$$\Upsilon^1 f(x,v,t_1)=\Upsilon^1 f(x+(v-{}_bve_b)t_1,{}_bve_b,t_1)$$

$$+\Upsilon^1 f(x+(v-{}_bve_b-{}_{b-1}ve_{b-1})t,{}_{b-1}ve_{b-1},t_1)+\dots+\Upsilon^1 f(x,{}_1ve_1,t_1),$$

hence $\Upsilon^1 f$ on vectors $\{(x,v,t_1);(x+(v-{}_bve_b)t_1,{}_bve_b,t_1);(x+(v-{}_bve_b-{}_{b-1}ve_{b-1})t_1,{}_{b-1}ve_{b-1},t_1);\dots;(x,{}_1ve_1,t_1)\}$ is the $\mathbf{K}$-linearly dependent system of functions by (x,t_1), where ${}_lv\in\{0,1\}$, $l=1,\dots,b$, $C_{(y,v)}\neq 0$.

Let the statement be proved for $n-1$, then we prove it for n. Then we apply to both sides of the equation

$$\sum_{(y^{n-1},v^{n-1})\in\Lambda_{n-1}} C_{(y^{n-1},v^{n-1})}\Upsilon^{n-1}f(x^{[n-2]}+y^{n-1},v^{n-1},t_{n-1})=0$$

the partial difference quotient operator Υ^1, which is $\mathbf{K}$-linear, consequently,

$$\sum_{(y^1,v^1)\in\Lambda_1} C_{(y^1,v^1)}\Upsilon^1(\sum_{(y^{n-1},v^{n-1})\in\Lambda_{n-1}}\Upsilon^{n-1}f)((x^{[n-2]}+y^{n-1},v^{n-1},t_{n-1})+y^1,v^1,t_n)=0,$$

where Λ_1, y^1 and v^1 already correspond to $\Upsilon^{n-1}f(x^{[n-1]})$ instead of $f(x)$. Thus, we get Formula (1) with $C_{(y^n,v^n)} = C_{(y^1,v^1)}C_{(y^{n-1},v^{n-1})} \neq 0$ and with

$$\Upsilon^n f(x^{[n-2]} + y^{n-1}, v^{n-1}, t_{n-1}) + y^1, v^1, t_n) = \Upsilon^n f(x^{[n-1]} + y^n, v^n, t_n).$$

18. Corollary. *If U is an open subset in the topological vector space $\mathbf{K}^b$, $f : U \to \mathbf{K}$ is a marked function, then a space X_n of functions $\{\bar{\Phi}^n f(x^{(n)}) : \ {}_lv_j \in \{0,1\}, l = 1,\dots,b; j = 1,\dots,n\}$ is finite dimensional over the field $\mathbf{K}$ whenever it exists such that $dim_{\mathbf{K}}X_n \leq (2^b - 1)^n$, $n \in \mathbf{N}$. Moreover, there exists a finite system Λ_n of vectors $0 \neq (y,v)$, $y \in \mathbf{K}^b$, $v \in (\mathbf{K}^b)^n$ such that*

$$\sum_{(y,v)\in\Lambda_n} C_{(y,v)}\bar{\Phi}^n f(x+y, v, t_1,\dots,t_n) = 0 \tag{1}$$

is identically equal to zero as the function of $(x, t_1,\dots,t_n)$, where $(x+y, v, t_1,\dots,t_n) \in U^{(n)}$, $x^{(n-1)} \in U^{(n-1)}$, $v^{(0)}$ and ${}_lv_j \in \{0,1\}$ for each $l = 1,\dots,b$, $j = 1,\dots,n$, the variable y may depend on the parameters $t_1,\dots,t_n$ linearly, $0 \neq C_{(y,v)} \in \mathbf{K}$ are constants for each (y,v).

Proof. Restricting in the preceding formulas $\Upsilon^n f(x^{[n]})$ on the domain $W^{(n)}$ and using Lemma 25 and Corollary 26 we get the statement of this corollary.

19. Lemma. *Let U be an open subset in the topological vector space $\mathbf{K}^m$, Y be a $\mathbf{K}$-linear space. If $char(\mathbf{K}) = 0$, then either $f \in C^{[n]}(U,Y) \cap C^{n+1}(U,Y)$ or $f \in C_b^{[n]}(U,Y) \cap C_b^{n+1}(U,Y)$ if and only if either $f \in C^{[n+1]}(U,Y)$ or $f \in C_b^{[n+1]}(U,Y)$. If $char(\mathbf{K}) > 0$, then $C^{[n]}(U,Y) \subset C^n(U,Y)$ and $C_b^{[n]}(U,Y) \subset C_b^n(U,Y)$.*

Proof. If $f \in C^{[n+1]}(U,Y)$ or $f \in C_b^{[n+1]}(U,Y)$, then the restriction $\Upsilon^{n+1}f|_{W^{(n+1)}} = \bar{\Phi}^{n+1}f$ is continuous or uniformly continuous on the domain $W^{(n+1)}$ correspondingly, consequently, $f \in C^{n+1}(U,Y)$ or $f \in C_b^{n+1}(U,Y)$ respectively. Since $C^{[n]}(U,Y) \subset C^{[n+1]}(U,Y)$ or $C_b^{[n]}(U,Y) \subset C_b^{[n+1]}(U,Y)$, the function f belongs to the space $C^{[n]}(U,Y) \cap C^n(U,Y)$ or $f \in C_b^{[n]}(U,Y) \cap C_b^n(U,Y)$ correspondingly (see also Note 5).

Let now $f \in C^{[n]}(U,Y) \cap C^{n+1}(U,Y)$ or $f \in C_b^{[n]}(U,Y) \cap C_b^{n+1}(U,Y)$. For $n = 0$ the statement of this lemma follows from Lemma 2. Suppose that the statement of this lemma is true for $k = 1,\dots,n$, then we prove it for $k = n+1$. In view of Lemma 9 we have, that $f^{[n+1]}(x^{[n]})$ has the expression through the finite sum of terms $(\hat{\Phi}^{n+1} f \circ u^{n+1})h_\beta$ up to minor terms $(\Upsilon^i f)h_\beta$ with $i \leq n$, where $u^{n+1} \in C_b^\infty$ and $h_\beta \in C_b^\infty$ are functions associated with $\mathbf{K}$-linear and polynomial shift operators and their compositions independent of f. We can write this in more details by induction. Now we consider the composite function:

$$g(x^{[q]}) := (\bar{\Phi}^q f)(u(x;\alpha); e_{j_1},\dots,e_{j_q}; a_1\bar{\Phi}^{n(1)}u_{k_1}(x;\ {}_1e; b_1),\dots,$$

$$a_s\bar{\Phi}^{n(s)}u_{k_s}(x;\ {}_se; b_s))\bar{\Phi}^{m(1)}w_1(x;\ {}_1\xi; c_1)\dots\bar{\Phi}^{m(r)}w_r(x;\ {}_r\xi; c_r)$$

appearing from the decomposition of $f^{[q]}$, where $a_1,\dots,a_s$ are polynomials of the variables $t_1,\dots,t_q$ and ${}_lv_j^{[k]}$, $k = 0,\dots,q$, $l = 1,2,\dots$, $j = 1,2,3$ for $k > 0$, $b_l \subset \{t_1,\dots,t_q\}$, ${}_r\xi = ({}_re_{i_1},\dots, {}_re_{i_{m(r)}})$, ${}_se = ({}_se_{j_1},\dots, {}_se_{j_{n(s)}})$; $k_j, s, r, n(s), m(r) \in \mathbf{N}$; α is a parameter, $w_1,\dots,w_r$

are polynomials of the variables $x,t_1,\dots,t_q$, ${}_lv_j^{[k]}$. Here $\alpha,u,a_1,\dots,a_s,w_1,\dots,w_r,e_i$, ${}_j\xi,b_j$ are independent of f. The set of variables

$$(x;t_1,\dots,t_q;\ {}_lv_j^{[k]}\ :\ l=1,2,\dots;k=0,\dots,q;j=1,2,3\}$$

is in the bijective correspondence with the vector $x^{[q]}$. Then acting on this function g by the partial difference quotient operator Υ^1 at the vector $x^{[q+1]}=(x^{[q]},v^{[q]},t_{q+1})$ one gets that

$$\Upsilon^1 g(x^{[q+1]})=[g(x^{[q]}+v^{[q]}t_{q+1})-g(x^{[q]})]/t_{q+1}.$$

For the calculation of $\Upsilon^1 g$ one can apply Formulas 7(2) and 9(2) to the functions g and $(\bar{\Phi}^q f)$ by all variables of functions in this composition and product. These expressions are non-linear by $\bar{\Phi}^q f$. As the result $\Upsilon^1 g$ is the $\mathbf{K}$-linear combination of functions of the same type (1) with $q+1$ instead of q and in general new functions in the composition and product after actions on them operators Υ^1, P_k, $\hat{\pi}$ and S. Shift operators $\hat{S}$ over $\mathbf{K}$ are infinite differentiable and invertible for the field $\mathbf{K}$ of zero characteristic such that we have

$$\sum_{i=1}^{k}\hat{S}=k\hat{S}\neq 0 \text{ and } \sum_{i=1}^{k}t\bar{\Phi}^1 id(x;v;t)=kt\bar{\Phi}^1 id(x;v;t)\neq 0$$

for any natural number $k\in\mathbf{N}=\{1,2,3,\dots\}$ and each $t\neq 0$ and $v\neq 0$.

20. Corollary. *If $char(\mathbf{K})=0$, then a function f belongs to $C^{[n+1]}(U,Y)$ or $C_b^{[n+1]}(U,Y)$ if and only if f belongs to $C^{n+1}(U,Y)$ or $C_b^{n+1}(U,Y)$ respectively, moreover, there exists a constant $0<C_1<\infty$ independent of f such that*

$$\|f\|_n\leq\|f\|_{[n]}\leq C_1\|f\|_n$$

for each $f\in C_b^{n+1}(U,Y)$, where

$$V^{[k]}:=\{x^{[k]}\in U^{[k]}\ :\ |v_1^{[q]}|=1;|\ {}_lv_2^{[q]}t_{q+1}|\leq 1,|v_q^{[3]}|\leq 1\quad\forall l,q\}$$

or

$$V^{(k)}:=\{x^{(k)}\in U^{(k)}:|v_j|=1\quad\forall j\}$$

with norms either

$$\|f\|_{[n]}:=\sup_{k=0,\dots,n;x^{[k]}\in V^{[k]}}|f^{[k]}(x^{[k]})|\ or$$

$$\|f\|_n:=\sup_{k=0,\dots,n;x^{(k)}\in V^{(k)}}|\bar{\Phi}^k f(x^{(k)})|.$$

Proof. We apply Lemma 19 by induction for $k=1,\dots,n$ and use Lemma 2. If g is a bounded continuous function $g:\mathbf{K}^m\to\mathbf{K}$, then $[g(x+vt)-g(x)]/t$ is a bounded continuous function by $(x,v,t)\in\mathbf{K}^m\times\mathbf{K}^m\times(\mathbf{K}\setminus B(\mathbf{K},0,\delta))$, where $\delta>0$ is a constant. If $L(X,Y)$ is the space of all bounded $\mathbf{K}$ linear operators $T:X\to Y$ from a normed space X into a normed space Y over the field $\mathbf{K}$, then operator norms

$$\|T\|_1:=\sup_{0\neq x\in X}\|Tx\|_Y/\|x\|_X,$$

$$\|T\|_2 := \sup_{0<|x|\le 1, x\in X} \|Tx\|_Y/\|x\|_X \text{ and } \|T\|_3 := \sup_{|x|=1, x\in X} \|Tx\|_Y/\|x\|_X$$

are equivalent [99]. In view of Lemma 2 each operator $\bar{\Phi}^j f(x;v_1,\dots,v_j;0,\dots,0)$ is j multilinear over the field $\mathbf{K}$ by vectors $v_1,\dots,v_j \in X$. Therefore, the definition of the C^n norm given above is worthwhile. If $x^{[k]} \in V^{[k]}$, then $|v_1^{[q]}| = 1$ and

$$|\ _lv_1^{[q-1]} + {_lv_2^{[q]}} t_{q+1}| \le 1$$

for each l,q. The inequality $\|f\|_n \le \|f\|_{[n]}$ follows from the restriction $\bar{\Phi}^n f = f^{[n]}|_{W^{(n)}}$. The second inequality $\|f\|_{[n]} \le C_1\|f\|_n$ follows from the decomposition of the partial difference quotient $f^{[q]}$ as a finite $\mathbf{K}$-linear combination of terms having the form 19(1) for each $q = 1,\dots,n$ and since norms of all terms are bounded and expansion coefficients are independent of f, where $f^{[0]} = f$.

When the field $\mathbf{K}$ is of zero characteristic any expansion coefficient b in the mentioned above formulas relating Υ^n and $\bar{\Phi}^j$ which is an integer multiple $b = kp^m$ with a non-zero integer k and a natural number m and a prime number p is non-zero, $b \in \mathbf{K}\setminus\{0\}$. This is the important difference with the case of a field of the positive characteristic p.

21. Lemma. *Let U be an open subset in the topological vector space $\mathbf{K}^b$, $b \in \mathbf{N}$, let also $f : U \to Y$ be a function with values in a topological vector space Y over $\mathbf{K}$. Then $f \in C^{[n]}(U,Y)$ or $f \in C_b^{[n]}(U,Y)$ or $f \in C^n(U,Y)$ or $f \in C_b^n(U,Y)$ if and only if $\Upsilon^k f(x^{[k]}) \in C^0(U^{[k]}_{j(0),j(1),\dots,j(k)},Y)$ or $\Upsilon^k f(x^{[k]}) \in C_b^0(V^{[k]}_{j(0),j(1),\dots,j(k)},Y)$ or $\bar{\Phi}^k f(x;e_{j(1)},\dots,e_{j(k)};t_1,\dots,t_k) \in C^0(U^{(k)}_{j(1),\dots,j(k)},Y)$ or $\bar{\Phi}^k f(x;e_{j(1)},\dots,e_{j(k)};t_1,\dots,t_k) \in C_b^0(V^{(k)}_{j(1),\dots,j(k)},Y)$ for each $k = 0,1,\dots,n$ and for each $j(i) \in \{1,\dots,m_i\}$, $v^{[i]} = e_{j(i)} \in (\mathbf{K}^b)^{[i]}$, $m_i = dim_{\mathbf{K}}(\mathbf{K}^b)^{[i]}$, $i = 0,1,\dots,k$, $x^{[i+1]} = (x^{[i]},v^{[i]},t_{i+1})$ or respectively each $j(1),\dots,j(k) \in \{1,2,\dots,b\}$ with $e_j = (0,\dots,0,1,0,\dots,0) \in \mathbf{K}^b$ is the vector with 1 on the j-th place, where*

$$U^{[k]}_{j(0),\dots,j(l)} := \{x^{[k]} \in U^{[k]} : v^{[i]} = e_{j(i)}, i = 0,\dots,l\},$$

$$V^{[k]}_{j(0),\dots,j(l)} = V^{[k]} \cap U^{[k]}_{j(0),\dots,j(l)},$$

$$U^{(k)}_{j(1),\dots,j(l)} := \{x^{(k)} \in U^{(k)} : v_1 = e_{j(1)},\dots,v_l = e_{j(l)}\},$$

$$V^{(k)}_{j(1),\dots,j(l)} = V^{(k)} \cap U^{(k)}_{j(1),\dots,j(l)}.$$

Moreover, if each partial difference quotient $\Upsilon^k f(z^{[k]})|_{V^{[k]}_{j(0),\dots,j(k)}}$ or $\bar{\Phi}^k f(z;e_{j(1)},\dots,e_{j(n)};t_1,\dots,t_n)$ is locally bounded, then $\Upsilon^k f(z^{[k]})$ or $\bar{\Phi}^k f(z^{(k)})$ is locally bounded respectively.

Proof. If $n = 1$, then

$$\bar{\Phi}^1 f(x;v_1;t_1) = \bar{\Phi}^1 f({_1x}, {_2x} + {_2v_1}t_1,\dots, {_bx} + {_bv_1}t_1; e_1; {_1v_1}t_1)\ {_1v_1}$$
$$+\bar{\Phi}^1 f({_1x}, {_2x}, {_3x} + {_3v_1}t_1,\dots, {_bx} + {_bv_1}t_1; e_2; {_2v_1}t_1)\ {_2v_1} + \dots$$
$$+\bar{\Phi}^1 f({_1x},\dots, {_bx}; e_b; {_bv_1}t_1)\ {_bv_1}, \qquad (1)$$

hence $\bar{\Phi}^1 f(x;v_1;t_1) \in C^0(U^{(1)},Y)$ or $\bar{\Phi}^1 f(x;v_1;t_1) \in C^0_b(V^{(1)},Y)$ if and only if $\bar{\Phi}^1 f(x;e_{j(1)};t_1) \in C^0(U^{(1)}_{j(1)},Y)$ or $\bar{\Phi}^1 f(x;e_{j(1)};t_1) \in C^0_b(V^{(1)}_{j(1)},Y)$ for each $j(1) \in \{1,\dots,b\}$, where $x = ({}_1x,\dots,{}_bx)$, ${}_jx \in \mathbf{K}$ for each j, $\Upsilon^1 f = \bar{\Phi}^1 f$. In accordance with Formula 10(1) or 9(1) we have the expression of the partial difference quotient $\bar{\Phi}^k f(x;v_1,\dots,v_k;t_1,\dots,t_k)$ or $\Upsilon^k f(x^{[k]})$ throughout the sum of terms containing $\bar{\Phi}^k f(x;e_{j(1)},\dots,e_{j(k)};t_1,\dots,t_k)$ or $\Upsilon^k f(z^{[k]})|_{V^{[k]}_{j(0),\dots,j(k)}}$ with multipliers belonging to $C^\infty_b(U,Y)$ or $C^{[\infty]}_b(U,Y)$ putting in Formula 10(1) or 9(1) $u = id : \mathbf{K}^b \to \mathbf{K}^b$, $id(x) = x$ for each x, $s = m = b$. From this the second assertion follows.

Suppose that the first statement of this lemma is proved for all $k = \overline{0,1,\dots,n-1}$. Then one can apply the operator $\bar{\Phi}^1$ to each partial difference quotient $\bar{\Phi}^{n-1} f(x;e_{j(1)},\dots,e_{j(n-1)};t_1,\dots,t_{n-1})$ and in accordance with Formula (1) with $\bar{\Phi}^{n-1} f$ or Υ^1 to each $\Upsilon^{n-1} f(x^{[n-1]})$ with $x^{[n-1]} \in U^{[n-1]}_{j(0),\dots,j(n-1)}$ instead of f we get the same conclusion. Thus $\bar{\Phi}^n f(x;e_{j(1)},\dots,e_{j(n-1)},v_n;t_1,\dots,t_{n-1},t_n)$ belongs to the space C^0 or C^0_b by its variables belonging to $U^{(n)}_{j(1),\dots,j(n-1)}$ or to $V^{(n)}_{j(1),\dots,j(n-1)}$ or $\Upsilon^n f(x^{[n]})$ belongs to the space C^0 or C^0_b by the variables $x^{[n]} \in U^{[n]}_{j(0),\dots,j(n-1)}$ or $x^{[n]} \in V^{[n]}_{j(0),\dots,j(n-1)}$ respectively if and only if $\bar{\Phi}^{n-1} f(x;e_{j(1)},\dots,e_{j(n-1)},e_{j(n)};t_1,\dots,t_{n-1},t_n)$ belongs to the space $bC^0(U^{(n)}_{j(1),\dots,j(n)},Y)$ or $C^0_b(V^{(n)}_{j(1),\dots,j(n)},Y)$ or $\Upsilon^n f(x^{[n]})|_{U^{[n]}_{j(0),\dots,j(n)}} \in C^{[n]}(U^{[n]}_{j(0),\dots,j(n)},Y)$ or $\Upsilon^n f(x^{[n]})|_{V^{[n]}_{j(0),\dots,j(n)}} \in C^{[n]}_b(V^{[n]}_{j(0),\dots,j(n)},Y)$ respectively for each $j(n)$, where $j(0),\dots,j(n)$ are arbitrary. Together with the induction hypothesis this finishes the proof of this lemma.

22. Lemma. *Suppose that U^k is an open subset in the topological vector space $\mathbf{K}^k$ for each $2 \le k \le m$ with a domain $domain(f_k) = U^k$ and from $f_k \circ u \in C^{[n]}(\mathbf{K}^{k-1},Y)$ or $C^{[n]}_b(\mathbf{K}^{k-1},Y)$ or $C^n(\mathbf{K}^{k-1},Y)$ or $C^n_b(\mathbf{K}^{k-1},Y)$ for each $u \in C^{[\infty]}(\mathbf{K}^{k-1},\mathbf{K}^k)$ or $C^{[\infty]}_b(\mathbf{K}^{k-1},\mathbf{K}^k)$ or $C^\infty(\mathbf{K}^{k-1},\mathbf{K}^k)$ or $C^\infty_b(\mathbf{K}^{k-1},\mathbf{K}^k)$ with an image $image(u) \subset U^k$ it follows that $f_k \in C^{[n]}(U^k,Y)$ or $C^{[n]}_b(U^k,Y)$ or $C^n(U^k,Y)$ or $C^n_b(U^k,Y)$ respectively. Then for a domain $domain(f) = U$ open in $\mathbf{K}^m$ from $f \circ u \in C^{[n]}(\mathbf{K},Y)$ or $C^{[n]}_b(\mathbf{K},Y)$ or $C^n(\mathbf{K},Y)$ or $C^n_b(\mathbf{K},Y)$ for each $u \in C^{[\infty]}(\mathbf{K},\mathbf{K}^m)$ or $C^{[\infty]}_b(\mathbf{K},\mathbf{K}^m)$ or $C^\infty(\mathbf{K},\mathbf{K}^m)$ or $C^\infty_b(\mathbf{K},\mathbf{K}^m)$ with the image $image(u) \subset U$ it follows, that $f \in C^{[n]}(\mathbf{K},Y)$ or $C^{[n]}_b(\mathbf{K},Y)$ or $C^n(\mathbf{K},Y)$ or $C^n_b(\mathbf{K},Y)$ respectively.*

Proof. We write $f \circ u$ in the form

$$f \circ u = f \circ u_{m-1} \circ u_{m-2} \circ \dots \circ u_1,$$

where $u_j : \mathbf{K}^j \to \mathbf{K}^{j+1}$ for each j and $u : \mathbf{K} \to \mathbf{K}^m$ of corresponding classes of smoothness. Applying supposition of this lemma for $k = m, m-1, \dots, 2$ we get that the inclusion

$$f \circ u_{m-1} \circ \dots \circ u_j \in C^{[n]}(\mathbf{K}^j,Y)$$

provides $f \in C^{[n]}(U,Y)$ or also for others classes of smoothness correspondingly for each $j = m-1, m-2, \dots, 1$.

23. Lemma. *Let $f : U \to \mathbf{K}^l$, where U is an open subset in the topological vector space $\mathbf{K}^m$. Then a function f belongs to the space $C^{[n]}(U,\mathbf{K}^l)$ if and only if each $\Upsilon^n f(x^{[n]})_{U^{[n]}_{j(0),\dots,j(n)}}$ is continuous for $v_3^{[k]} = 0$ for each $k = 1,\dots,n-1$.*

Proof. In view of Lemma 21 it remains to prove, that a continuity of each partial difference quotient $\Upsilon^n f(x^{[n]})_{U^{[n]}_{j(0),\dots,j(n)}}$ is equivalent to the continuity of this family under the condition $v_3^{[k]} = 0$ for each $k = 1,\dots,n-1$. We prove this by induction. We already have that $v_3^{[k]} \in \{0,1\}$ for each $k = 0,\dots,n-1$. Denote by $\hat{S}_{n,t_n}$ the shift operator

$$\hat{S}_{n,t_n} g(t_{n-1},\beta) := g(t_{n-1}+t_n),$$

where β denotes the family of all other variables of a function g. Then we infer that

$$\Upsilon^n f(x^{[n]}) = [\hat{S}_{n,t_n}\Upsilon^{n-1} f(x^{[n-1]} + w^{[n-1]}t_n) - \Upsilon^{n-1} f(x^{[n-1]})]/t_n$$

$$= [(\hat{S}_{n,t_n} - I)\Upsilon^{n-1} f(x^{[n-1]} + w^{[n-1]}t_n)]/t_n + \Upsilon^n f(x^{[n]})|_{v_3^{[n-1]}=0},$$

where $w^{[n-1]}$ differs from $v^{[n-1]}$ on the vector $v_3^{[n-1]}$ such that in $w^{[n-1]}$ it is zero and in $v^{[n-1]}$ it is one while all others their components coincide such that

$$x^{[n]} = (x^{[n-1]}, v^{[n-1]}, t_n) \text{ and } x^{[n]}|_{v_3^{[n-1]}=0} = (x^{[n-1]}, w^{[n-1]}, t_n).$$

Since

$$[(\hat{S}_{n,t_n} - I)\Upsilon^{n-1} f(x^{[n-1]} + w^{[n-1]}t_n)]/t_n = [\Upsilon^{n-1} f((x^{[n-2]}, v^{[n-2]}, t_{n-1}+t_n) + w^{[n-1]}t_n)$$

$$-\Upsilon^{n-1} f(x^{n-1} + w^{[n-1]}t_n)]/t_n,$$

where $x^{[0]} = x$, $x^{[n-1]} = (x^{[n-2]}, v^{[n-2]}, t_{n-1})$, $n \ge 2$ and $k \ge 1$, the mapping

$$[(\hat{S}_{n,t_n} - I)\Upsilon^{n-1} f(x^{[n-1]} + w^{[n-1]}t_n)]/t_n$$

$$= \Upsilon^1_{s(n-1)}\Upsilon^{n-1} f(x^{[n-1]} + v^{[n-1]}t_n)(t_{n-1}+t_n-t_{n-1})/t_n$$

is continuous, where $s(n-1)$ corresponds to the partial difference quotients by the variable t_{n-1}. Then by induction get that

$$(\hat{S}_{k,t_k} - I)/t_k = \Upsilon^1_{s(k)}$$

for each $k = n-1,\dots,1$ which leads to the assertion of this lemma.

24. Lemma. *Suppose that $f \in C^n(U,Y)$ <u>or</u> $f \in C^{[n]}(U,Y)$, where U is an open subset in the topological vector space $\mathbf{K}^m$, then each $\Phi^n f(x^{(n)})$ has the symmetry property relative to transpositions of pairs (v_j,t_j) characterized by Young's table consisting of one row of length n, each $\Upsilon^n f(x^{[n]})|_{\{U^{[n]}:v_3^{[k]}=0,k=1,\dots,n\}}$ is characterized by Young's table consisting of 2^{n-1} rows, where the first row of length n contains numbers $1,\dots,n$, the second row of length $n-1$ contains numbers $2,\dots,n$, the third and the fourth rows have lengths $n-2$ and contain numbers $3,\dots,n$ and so on, where the number of rows of equal lengths $n-k$ is 2^{k-1}*

for $1 \le k < n-1$. *Moreover, if* $t_{i_1} = 0,\dots,t_{i_l} = 0$ *as arguments of* $\Upsilon^n f$, *then its symmetry becomes higher with the amount of rows* 2^{n-l} *instead of* 2^{n-1}.

Proof. The function $\bar{\Phi}^n f(x; v_1,\dots,v_n; t_1,\dots,t_n)$ is symmetric relative to transpositions $(v_i,t_i) \mapsto (v_j,t_j)$, since

$$[(f(x+v_it_i+v_jt_j)-f(x+v_jt_j))/t_i-(f(x+v_it_i)-f(x))/t_i]/t_j$$

$$=[(f(x+v_it_i+v_jt_j)-f(x+v_it_i))/t_j-(f(x+v_jt_j)-f(x))/t_j]/t_i$$

for each $i \ne j$ and so on by induction.

When $v_3^{[k]} = 0$ for $1 \le k \le n-1$ and $n \ge 2$ we have

$$\Upsilon^{k+1} f(x^{[k+1]}) = \{[\Upsilon^{k-1} f(x^{[k-1]} + (v^{[k-1]} + v_2^{[k]} t_{k+1})t_k + v_1^{[k]} t_{k+1})$$

$$-\Upsilon^{k-1} f(x^{[k-1]} + v_1^{[k]} t_{k+1})]/t_k - [\Upsilon^{k-1} f(x^{[k-1]} + v^{[k-1]} t_k) - \Upsilon^{k-1} f(x^{[k-1]})]/t_k\}/t_{k+1}$$

$$= \{[\Upsilon^{k-1} f(x^{[k-1]} + (v^{[k-1]} + v_2^{[k]} t_{k+1})t_k + v_1^{[k]} t_{k+1})$$

$$-\Upsilon^{k-1} f(x^{[k-1]} + v^{[k-1]} t_k + v_1^{[k]} t_{k+1})]/t_k$$

$$+[\Upsilon^{k-1} f(x^{[k-1]} + v^{[k-1]} t_k + v_1^{[k]} t_{k+1}) - \Upsilon^{k-1} f(x^{[k-1]} + v_1^{[k]} t_{k+1})]/t_k$$

$$-[\Upsilon^{k-1} f(x^{[k-1]} + v^{[k-1]} t_k) - \Upsilon^{k-1} f(x^{[k-1]})]/t_k\}/t_{k+1} \qquad (1)$$

and this expression is symmetric relative to transpositions $(v^{[k-1]}, t_k) \mapsto (v_1^{[k]}, t_{k+1})$. Therefore, it is possible to exclude $v_3^{[k]} = 0$ from the consideration such that $v^{[0]} := v^{[0],1}$, $v^{[1]} = (v^{[1],1}, v^{[1],2}, 0)$, where $v^{[0],1}, v^{[1],1}, v^{[1],2} \in \mathbf{K}^m$. Then by induction vectors $v^{[k],i} \in \mathbf{K}^m$ are defined such that

$$x^{[k]} + v^{[k]} t_{k+1} = (x^{[k-1]} + v_1^{[k]} t_{k+1}, v^{[k-1]} + v_2^{[k]} t_{k+1}, t_k + v_3^{[k]} t_{k+1})$$

with $v_3^{[k]} = 0$ and to this the vector $v^{[k-1],i} + v^{[k],i+2^{k-1}} t_{k+1}$ corresponds such that $v^{[k]}$ is completely characterized by $(v^{[k],i} : i = 1,\dots,2^k)$, where $k \ge 1$. Therefore, by induction one gets that the partial difference quotient $\Upsilon^n f$ is symmetric relative to transpositions $(v^{[k-1],i}, t_k) \mapsto (v^{[k],i}, t_{k+1})$ for each $1 \le i \le 2^{k-1}$, $1 \le k \le n-1$. To $v^{[k],1}$ we pose the first row of length n with numbers $1,\dots,n$ in boxes from left to right, $k = 0,1,\dots,n-1$. To vectors $v^{[k],i}$ with $i = 2^{k-1}+1,\dots,2^k$ and $k \ge 1$ we pose rows in Young's table with such numbers in squares from left to right beginning with $k+1$ and ending with n in each such i-th row.

If $t_{i_1} = 0,\dots,t_{i_l} = 0$ as arguments of $\Upsilon^n f$, then the symmetry of $\Upsilon^n f$ up to notation corresponds to $v^{[i_s-2]} + v_2^{[i_s-1]} t_{i_s} = v^{[i_s-2]}$ and $\Upsilon^n f$ is characterized by less amount of vectors $v^{[k],i}$, since

$$\Upsilon^l_{t_{i_1},\dots,t_{i_l}} f = \bar{\Phi}^l_{t_{i_1},\dots,t_{i_l}} f$$

such that instead of $(v^{[k-1],j} : j = 1,\dots,2^{k-1})$ it is sufficient to take $j = 1,\dots,2^{k-2}$ for $k = i_2$ for $k \ge 2$ and so on excluding excessive vectors by induction on $s = 3,\dots,l$.

25. Lemma. *Suppose that $f \in C^{n-1}(U,Y)$ or $f \in C^{[n-1]}(U,Y)$, where U is an open subset in the topological vector space $\mathbf{K}^m$. Then*

$$f \in C^n(U,Y) \quad or \quad f \in C^{[n]}(U,Y) \tag{1}$$

if and only if $\bar{\Phi}^n f(x;w,\dots,w;t_1,\dots,t_n)$ or $\Upsilon^n f(x^{[n]})|_{\{U^{[n]}:v^{[k],i}=w_s \ \forall 2^{s-1}<i\le 2^s, 0\le s\le k<n\}}$ is continuous for each marked $w \in \mathbf{K}^m$ or $w_0,\dots,w_{n-1} \in \mathbf{K}^m$ respectively;

$$\bar{\Phi}^n f \quad or \quad \Upsilon^n f \tag{2}$$

is not locally bounded if and only if there exists a marked vector $w \in \mathbf{K}^m$ or vectors $w_0,\dots,w_{n-1} \in \mathbf{K}^m$ such that

$$\bar{\Phi}^n f(x;w,\dots,w;t_1,\dots,t_n) \quad or \quad \Upsilon^n f(x^{[n]})|_{\{U^{[n]}:v^{[k],i}=w_s \ \forall 2^{s-1}<i\le 2^s, 0\le s\le k<n\}}$$

is not locally bounded.

Proof. In view of Lemma 11 and Formula 9(2) applied by induction we have

$$\bar{\Phi}^n f(x;w,\dots,w;t_1,\dots,t_n) = \sum_{i_1,\dots,i_n=1}^{m} a_{i_1}\dots a_{i_n} \bar{\Phi}^n f(x + t_1 \sum_{l_1=i_1+1}^{m} a_{l_1} e_{i_1} + \dots$$

$$+ t_n \sum_{l_n=i_n+1}^{m} a_{l_n} e_{l_n}; e_{i_1},\dots,e_{i_n}; a_{i_1} t_1,\dots,a_{i_n} t_n)$$

for each $w = \sum_{i=1}^m a_i e_i$ if at least one $t_i \ne 0$, where $a_i \in \mathbf{K}$, for convenience of notation $\sum_{i=m+1}^m a_i e_i = 0$. Then we consider all numbers $t_1,\dots,t_n \in \mathbf{K}$ such that $0 \ne t_i \to 0$. Due to Lemma 24 and since a_i are arbitrary and can be taken nonzero, each partial difference quotient $\bar{\Phi}^n f(x;e_{i_1},\dots,e_{i_n};a_{i_1}t_1,\dots,a_{i_n}t_n)$ is continuous or locally bounded if and only if $\bar{\Phi}^n f(x;w,\dots,w;t_1,\dots,t_n)$ is continuous or locally bounded for each marked vector $w \in \mathbf{K}^m$. In view of Lemma 21 this provides assertions $(1,2)$ for $\bar{\Phi}^n f$.

On the other hand, we have

$$\hat{P}^n(x^{[n]})|_{(U^{[n]}:v^{[k],i}=w_s \forall 2^{s-1}<i\le 2^s, 0\le s\le k<n)} = x + \sum_{k=0}^{n-1} \phi_{k+1}(t) w_k,$$

where

$$\phi_l(t) = \sum_{1\le i_1<\dots<i_l\le n} t_{i_1}\dots t_{i_l}$$

are linearly independent symmetric polynomials, $l = 1,\dots,n$, $t = (t_1,\dots,t_n)$, in particular, $\phi_1(t) = t_1 + \dots + t_n$. We put $\alpha_{j,l} := a_{j,s}$ for each $j = 1,\dots,m$, $2^{s-1} < l \le 2^s$, $s = 0,\dots,n-1$, where

$$w_s = \sum_{i=1}^m a_{i,s} e_i$$

with $a_{i,s} \in \mathbf{K}$ for each $s = 0,\dots,n-1$.

Applying Formula 11(2) by induction we infer that

$$\Upsilon^n f(x^{[n]})|_{\{U^{[n]}:v^{[k],i}=w_s \quad \forall 2^{s-1}<i\le 2^s, 0\le s\le k<n\}} = \sum_{i_0,\dots,i_{n-1}=1}^{m} \ \sum_{1\le q_k\le 2^k, k=0,\dots,n-1}$$

$$\Big(\prod_{k=0}^{n-1}\alpha_{i_k,q_k}\Big)\Upsilon^n f(x_J^{[n]})|_{\{U^{[n]}:v^{[s],l}=\delta_{l,q_s}e_{i_{s+1}},\tau_{s+1}=\alpha_{i_s,q_s}t_{s+1}\ \forall s=0,\dots,n-1,1\le l\le 2^s\}}$$

for each marked vector w_s if at least one $t_i\neq 0$, where $\delta_{i,j}=1$ for $i=j$ and $\delta_{i,j}=0$ for each $i\neq j$, J is the set $J=\{(i_k,q_k):k=0,\dots,n-1\}$;

$$\hat{\pi}^n(x_J^{[n]})=\hat{P}^n(y),$$

where τ_{k+1} corresponds to $x_J^{[n]}$ instead of t_{k+1} for $x^{[n]}$, a vector $y\in(\mathbf{K}^m)^{[n]}$ corresponds to the set

$$\Big(x;v^{[k],l}=\sum_{j_k\ge i_k+\delta_{l,q_k}}\alpha_{j_k,l}e_{j_k},\ k=0,\dots,n-1,\ 0\le s\le k<n,\ 2^{s-1}<l\le 2^s;t_1,\dots,t_n\Big)$$

in the notation introduced above. Then one considers all numbers $t_1,\dots,t_n\in\mathbf{K}$ such that $0\neq t_i\to 0$. Since $a_{i,s}\in\mathbf{K}$ are arbitrary constants which can be taken nonzero, from Lemmas 21 and 24 the statement of this lemma for $\Upsilon^n f$ as well follows. Indeed,

$$\Upsilon^1(f^{[n-1]}(x^{[n-1]}))(x^{[n]})=[f^{[n-1]}(x^{[n-1]}+v^{[n-1]}t_n)-f^{[n-1]}(x^{[n-1]})]/t_n$$

and $x^{[n]}=(x^{[n-1]},v^{[n-1]},t_n)$ and due to repeated application of Formula 24(1). At the same time the property $g(h(z)e_i,y)\in C^0$ by the variables $(z,y)\in U_1\times U_2$ is equivalent to the inclusion $g(ue_i,y)\in C^0$ by the variables $(u,y)\in h(U_1)\times U_2$ for a continuous function $h(z)$ by the variables $z=(z_1,\dots,z_a)\in U_1$, where U_1 and U_2 are domains in $\mathbf{K}^a$ and $\mathbf{K}^c$, $g(ue_i,y)\in Y$, $h(U_1)\subset\mathbf{K}$.

26. Lemma. *If either $f\in C^{n-1}(U,Y)$ or $f\in C^{[n-1]}(U,Y)$, where U is an open subset in the topological vector space $\mathbf{K}^m$, then respectively either*

$$\bar{\Phi}^n f(x^{(n)})|_{\{U^{(n)}:\exists i\ |t_i|\ge\delta\}}\ \ or\ \ \Upsilon^n f(x^{[n]})|_{\{U^{[n]}):\exists i\ |t_i|\ge\delta,v_3^{[k]}=0\forall k\}}$$

is continuous, where $\delta>0$.

Proof. Since the partial difference quotients fulfill the relations

$$\bar{\Phi}^n f(x^{(n)})=\bar{\Phi}^1(\bar{\Phi}^{n-1}f(x^{(n-1)}))(x^{(n)})$$

and

$$\Upsilon^n f(x^{[n]})=\Upsilon^1(\Upsilon^{n-1}f(x^{[n-1]}))(x^{[n]})$$

whenever they exist and

$$\Upsilon^1 f(x^{[1]})=\bar{\Phi}^1 f(x^{(1)})=[f(x+vt)-f(x)]/t,$$

in view of Lemmas 21 and 24 we get the statement of this lemma. Indeed, the partial difference quotient either $\bar{\Phi}^{n-1}f(x^{(n-1)})$ or $\Upsilon^{n-1}f(x^{[n-1]})$ is respectively continuous and the domain with $|t_i|\ge\delta$ and $t_i+v_3^{[i]}t_{i+1}=t_i$ can be taken, where $v_3^{[i]}=0$.

27. Lemma. *Let $f:\mathbf{K}^b\to\mathbf{K}$ be a function such that the inclusion either $f\circ u\in C^{[n]}(\mathbf{K},\mathbf{K})$ or $f\circ u\in C^n(\mathbf{K},\mathbf{K})$ is satisfied for $n\ge 0$ and either $f\in C^{[n-1]}(\mathbf{K}^b,\mathbf{K})$ or $f\in$*

$C^{n-1}(\mathbf{K}^b,\mathbf{K})$ for $n\ge 1$ for each either $u\in C^{[\infty]}(\mathbf{K},\mathbf{K}^b)$ or $u\in C^{\infty}(\mathbf{K},\mathbf{K}^b)$, where $\mathbf{K}$ is a field with a non-archimedean multiplicative norm and $2\le b\in\mathbf{N}$. Then either $\Upsilon^n f(x^{[n]})$ or $\bar{\Phi}^n f(x^{(n)})$ respectively is the locally bounded function on the topological vector space either $(\mathbf{K}^b)^{[n]}$ or $(\mathbf{K}^b)^{(n)}$ and the function f is continuous.

Proof. At first we prove, that the function f is continuous, when $n=0$, since for $n\ge 1$ we have $C^0\subset C^{n-1}$. Suppose the contrary, that there exists a sequence ${}_jz$ such that

$$\lim_{j\to\infty} {}_jz = z_0$$

and a limit of the sequence $\{f({}_jz):j\}$ either does not exist or is not equal to $f(z_0)$. Take c_j and r_j and $u(x)$ as above, then the equalities

$$\lim_{j\to\infty}(f\circ u)({}_jx)=\lim_{j\to\infty}f({}_jz)\neq f(z_0)=(f\circ u)(y_0)$$

are valid. Hence the composition $f\circ u$ is not continuous at the point y_0 contradicting the assumption of this lemma.

Now we suppose the contrary, that there exists a point either $z_0^{[n]}\in(\mathbf{K}^b)^{[n]}$ or $z_0^{(n)}\in(\mathbf{K}^b)^{(n)}$ such that $\Upsilon^n f$ or $\bar{\Phi}^n f$ is unbounded in a neighborhood of either $z_0^{[n]}$ or $z_0^{(n)}$ correspondingly. As a neighborhood we take the ball $B((\mathbf{K}^b)^{[n]},z_0^{[n]},\varepsilon)$ in the topological vector space $(\mathbf{K}^b)^{[n]}$ containing the point $z_0^{[n]}$ and of radius $\varepsilon>0$ or $B((\mathbf{K}^b)^{(n)},z_0^{(n)},\varepsilon)$. Without loss of generality we can suppose, that $z_0:=z_0^{[0]}=0\in\mathbf{K}^b$ making the shift $\phi(x):=f(x-z_0)$ when $z_0\neq 0$, where z_0 denotes the projection of $z_0^{[n]}$ in $\mathbf{K}^b$. Then a sequence either ${}_kz^{[n]}$ or ${}_kz^{(n)}$ tending to either $z_0^{[n]}$ or $z_0^{(n)}$ when k tends to the infinity exists such that either

$$\lim_{k\to\infty}|\Upsilon^n f({}_kz^{[n]})|=\infty \text{ or } \lim_{k\to\infty}|\bar{\Phi}^n f({}_kz^{(n)})|=\infty$$

respectively, where $|x|=|x|_{\mathbf{K}}$ is the multiplicative norm in the field $\mathbf{K}$. So we choose the sequence $\{{}_kz_0^{[n]}:k=1,2,\dots\}$ such that $|{}_kv^{[n-1]}|\le 1$ and $|{}_kt_j|\le 1$ for each $k\in\mathbf{N}$ and $j=1,\dots,n$. In view of Lemma 25 without loss of generality a marked point either $w\in\mathbf{K}^m$ or points $w_0,\dots,w_{n-1}\in\mathbf{K}^m$ exist such that either $\bar{\Phi}^n f(x;w,\dots,w;t_1,\dots,t_n)$ or $\Upsilon^n f(x^{[n]})|_{\{U^{[n]}:v^{[k],i}=w_s\ \forall 2^{s-1}<i\le 2^s,0\le s\le k<n\}}$ is not locally bounded in a neighborhood of the point either $z_0^{(n)}$ or $z_0^{[n]}$ with the sequence either $\{{}_jz^{(n)}:j\in\mathbf{N}\}$ or $\{{}_jz^{[n]}:j\in\mathbf{N}\}$ such that either $\{{}_jz^{[n]}:j\in\mathbf{N};\ {}_jv_i=w,i=1,\dots,n\}$ or $\{{}_jz^{[n]}:j\in\mathbf{N};\ {}_jv^{[k],i}=w_s\quad\forall 2^{s-1}<i\le 2^s,0\le s\le k<n\}$ respectively. At the same time due to Lemma 26 we can consider, that the limit

$$\lim_{j\to\infty}\max_{i=1}^{n}|{}_jt_i|=0$$

is zero. From Formula 9(1) or 10(1) and Lemma 21 applied to $u=id$ and the conditions of this lemma it follows, that all terms with orders $k<n$ of $B^k_{*}f$ or A^k_{*} are continuous. Therefore, an ordered set $\{j_n,\dots,j_1\}$ exists such that the sequence either

$$\{(B_{j_n,v^{(n-1)},t_n}\dots B_{j_1,v^{(0)},t_1}f\circ u)(\bar{\Phi}^1\circ p_{j_n}\hat{S}_{j_{n-1}+1,v^{(n-2)}t_{n-1}}$$

$$\dots\hat{S}_{j_1+1,v^{(0)}t_1}u^{n-1})(P_n\bar{\Phi}^1\circ p_{j_{n-1}}\hat{S}_{j_{n-2}+1,v^{(n-3)},t_{n-2}}$$

$$\ldots \hat{S}_{j_1+1,v^{(0)}t_1}u^{n-2})\ldots(P_n\ldots P_2\bar{\Phi}^1\circ p_{j_1}u)({}_jz_0^{(n)}):j\in\mathbf{N}\} \quad (1)$$

or

$$\{(A_{j_n,v^{[n-1]},t_n}\ldots A_{j_1,v^{[0]},t_1}f\circ u)(\Upsilon^1\circ p_{j_n}\hat{S}_{j_{n-1}+1,v^{[n-2]}t_{n-1}}$$
$$\ldots\hat{S}_{j_1+1,v^{[0]}t_1}u^{n-1})(P_n\Upsilon^1\circ p_{j_{n-1}}\hat{S}_{j_{n-2}+1,v^{[n-3]}t_{n-2}}$$
$$\ldots\hat{S}_{j_1+1,v^{[0]}t_1}u^{n-2})\ldots(P_n\ldots P_2\Upsilon^1\circ p_{j_1}u)({}_jz_0^{[n]}):j\in\mathbf{N}\} \quad (2)$$

is unbounded for the identity mapping $f = id$.

Now we consider the same Formulas $10(1)$ or $9(1)$ for an arbitrary function u satisfying conditions of this lemma. Again all terms with orders $k < n$ of $B^k_* f\circ u$ or $A^k_* f\circ u$ are continuous and hence bounded in a neighborhood of the point either $z_0^{(n)}$ or $z_0^{[n]}$ respectively. We construct a curve u in several steps leading to the contradiction with the supposition of this lemma.

At first one can mention that $\Upsilon^1 id(y,v^{[0]},t_1) = v^{[0]}$, where $y,v^{[0]}\in\mathbf{K}^b$ and $t_1\in\mathbf{K}$. Then we get the equalities

$$\Upsilon^2 id(y^{[2]}) = (v^{[0]} + v_2^{[1]}t_2 - v^{[0]})/t_2 = v_2^{[1]}$$

and

$$\Upsilon^3 id_j(y^{[3]}) = ({}_jv_2^{[1]} + {}_{j+b}v_2^{[2]}t_3 - v_2^{[1]})/t_3 = {}_{j+b}v_2^{[2]},$$

where $j = 1,\ldots,b$,

$$v^{[k]} = ({}_1v_1^{[k]},\ldots,{}_cv_1^{[k]},\ {}_1v_2^{[k]},\ldots,{}_cv_2^{[k]},v_3^{[k]}),$$

$c = c(k) = 2^{k-1} - k + b(2^k - 1)$, ${}_jv_l^{[k]}\in\mathbf{K}$ for each j,k,l, $id(y) = (id_1(y),\ldots,id_b(y)) = (y_1,\ldots,y_b)$. Therefore, we get by induction the equality:

$$\Upsilon^m id_j(y^{[m]}) = {}_{j(m)}v_2^{[m-1]}$$

for each $m\geq 2$, where $j(1) = j$, $j(2) = j$, $j(3) = j + b$, $j(m) = j + 2^{m-2} - (m-1) + b(2^{m-1} - 1)$ for each $m\geq 4$, since

$$j(m) = j + b + (2b+1) + (2(2b+1)+1) + (2(2(2b+1)+1)+1)$$
$$+\cdots+(2(2(\ldots(2b+1)+1)+1)$$

with 2 in power $m-3$ in the latter term.

We consider equations

$$\bar{\Phi}^k u(x^{(n)}) = \alpha_k\bar{\Phi}^k id(z^{(n)}) \quad (3)$$

or

$$\Upsilon^k u(x^{[n]}) = \alpha_k\Upsilon^k id(z^{[n]}) \quad (4)$$

for $k = 0,1,\ldots,n$ in neighborhoods of the points either $x_0^{(n)}$ and $z_0^{(n)}$ or $x_0^{[n]}$ and $z_0^{[n]}$ with prescribed marked vectors η or $\eta_0,\ldots,\eta_{n-1}$ and w or $w_0,\ldots,w_{n-1}$ respectively, where η or $\eta_0,\ldots,\eta_{n-1}$ are determined from the equations. Here $0\neq\alpha_k\in\mathbf{K}$ are constants specified below for a sequence such that to have

$$\lim_{j\to\infty} g_j = 0,$$

where $0 < q_j := \min_{k=1}^n |\alpha_{j,k}| \le g_j := \max_{k=1}^n |\alpha_{j,k}| < 1$.

If $t_s = 0$, then equations for $\Upsilon^k u$ simplify due to the term D_{t_s} instead of $\Upsilon^1_{t_s}$ for which w_s does not play a role and we can consider $\tau_s = 0$, where τ_s play the same role for $x^{(n)}$ and $x^{[n]}$ as t_s for $z^{(n)}$ and $z^{[n]}$, $s = 1,\dots,n$. If $t_s \ne 0$, then we can take $\tau_s \ne 0$. In view of Lemma 22 one can consider the data $(b-1,b)$ instead of $(1,b)$. Since w or $w_0,\dots,w_{n-1}$ are the fixed vectors independent of j, these equations for marked nonzero vectors either $\eta \in \mathbf{K}^{b-1}$ or $\eta_0,\dots,\eta_{n-1} \in \mathbf{K}^{b-1}$ corresponding to either ${}_jx^{(n)}$ or ${}_jx^{[n]}$ can be resolved. Variables will be ${}_jx \in \mathbf{K}^{b-1}$ and $\tau_1,\dots,\tau_n$ for u instead of ${}_jz \in \mathbf{K}^b$ and $t_1,\dots,t_n$ for f, such that the limit is zero

$$\lim_{j\to\infty} \max_{i=1}^{n} |\tau_i| = 0.$$

In view of Formulas 12(2) and 14(1) it is sufficient to consider a quadratic function having the form:

$$u(h) = z + c \sum_{k_1,k_2=0}^{2} \sum_{i_1,i_2=1}^{b-1} {}_{i_1,i_2}a_{k_1,k_2} h_{i_1}^{k_1} h_{i_2}^{k_2}, \tag{5}$$

where ${}_{i_1,i_2}a_{k_1,k_2} \in \mathbf{K}^b$, $c \in \mathbf{K}$, $|{}_{i_1,i_2}a_{k_1,k_2}| \le 1$ for each i_1,i_2,k_1,k_2, $h = (h_1,\dots,h_{b-1}) \in \mathbf{K}^{b-1}$. Thus we get the expressions:

$$\begin{gathered}
|(B_{j_n,\eta^{\otimes n},\tau_n} \dots B_{j_1,\eta,\tau_1} f\circ u)(\bar{\Phi}^1 \circ p_{j_n} \hat{S}_{j_{n-1}+1,\eta^{\otimes(n-1)}\tau_{n-1}} \\
\dots \hat{S}_{j_1+1,\eta\tau_1} u^{n-1})(P_n\bar{\Phi}^1 \circ p_{j_{n-1}} \hat{S}_{j_{n-2}+1,\eta^{\otimes(n-2)},\tau_{n-2}} \dots \hat{S}_{j_1+1,\eta\tau_1} u^{n-2}) \\
\dots (P_n \dots P_2\bar{\Phi}^1 \circ p_{j_1} u)({}_jx_0^{(n)})| \ge |q_j|^n |\pi|^{l_0+s_0} |(B_{j_n,w^{\otimes n},t_n} \dots B_{j_1,w,t_1} f\circ id) \\
(\bar{\Phi}^1 \circ p_{j_n} \hat{S}_{j_{n-1}+1,w^{\otimes(n-1)}t_{n-1}} \\
\dots \hat{S}_{j_1+1,wt_1} id^{n-1})(P_n\bar{\Phi}^1 \circ p_{j_{n-1}} \hat{S}_{j_{n-2}+1,w^{\otimes(n-2)},t_{n-2}} \dots \hat{S}_{j_1+1,wt_1} id^{n-2}) \\
\dots (P_n \dots P_2\bar{\Phi}^1 \circ p_{j_1} id)({}_jz_0^{(n)})|
\end{gathered} \tag{6}$$

or

$$\begin{gathered}
|(A_{j_n,\eta^{[n-1]},\tau_n} \dots A_{j_1,\eta^{[0]},\tau_1} f\circ u)(\Upsilon^1 \circ p_{j_n} \hat{S}_{j_{n-1}+1,\eta^{[n-2]}\tau_{n-1}} \\
\dots \hat{S}_{j_1+1,\eta^{[0]}\tau_1} u^{n-1})(P_n\Upsilon^1 \circ p_{j_{n-1}} \hat{S}_{j_{n-2}+1,\eta^{[n-3]}\tau_{n-2}} \dots \hat{S}_{j_1+1,\eta^{[0]}\tau_1} u^{n-2}) \\
\dots (P_n \dots P_2\Upsilon^1 \circ p_{j_1} u)({}_jx_0^{[n]})| \ge |q_j|^n |\pi|^{l_0+s_0} |(A_{j_n,w^{[n-1]},t_n} \dots A_{j_1,w^{[0]},t_1} f\circ id) \\
(\Upsilon^1 \circ p_{j_n} \hat{S}_{j_{n-1}+1,w^{[n-2]}t_{n-1}} \\
\dots \hat{S}_{j_1+1,w^{[0]}t_1} id^{n-1})(P_n\Upsilon^1 \circ p_{j_{n-1}} \hat{S}_{j_{n-2}+1,w^{[n-3]}t_{n-2}} \dots \hat{S}_{j_1+1,w^{[0]}t_1} id^{n-2}) \\
\dots (P_n \dots P_2\Upsilon^1 \circ p_{j_1} id)({}_jz_0^{[n]})|
\end{gathered} \tag{7}$$

for each natural number $j \in \mathbf{N}$, where $l_0 \in \mathbf{N}$ is a marked natural number, $s_0 = s_0(j) \in \mathbf{N}$, each vector $w^{[k]}$ corresponds to the marked vectors $w_0,\dots,w_{n-1}$, while $\eta,\eta_0,\dots,\eta_{n-1} \in \mathbf{K}^{b-1}$ are some marked vectors for u, where $w^{\otimes k} := (w,\dots,w) \in X^{\otimes k}$ for $w \in X$ and $k \in \mathbf{N}$.

Now we take a function $\psi \in C^\infty(\mathbf{K},\mathbf{K})$ such that $\psi(x) = 1$ for $|x| \le |\pi|$ and $\psi(x) = 0$, when $|x| > |\pi|$, for example, a locally constant function, where $\pi \in \mathbf{K}$, $0 < |\pi| < 1$. In particular, the characteristic function of the clopen ball $B(\mathbf{K},0,|\pi|)$ is locally analytic, since the normed field $\mathbf{K}$ is totally disconnected with the base of its topology consisting of clopen (closed and open simultaneously) balls, where $B(X,x,R) := \{y \in X, \rho(x,y) \leq R\}$ for a topological space X metrizable by a metric ρ. It is proved further that such ψ after definite scalings suits the construction described below. We next define the functions $u_j(h) := (\xi_j\psi)((h - {}_jx)/T_j)$, where

$$\xi_j(h) := \left[{}_{r_j}z_0 + c_j \sum_{k_1,k_2=0}^{2} \sum_{i_1,i_2=1}^{b-1} {}_{i_1,i_2}a_{k_1,k_2} h_{i_1}^{k_1} h_{i_2}^{k_2} \right]$$

such that $\xi_j(0) = {}_{r_j}z_0$ and put

$$u(x) := \sum_{j=1}^{\infty} u_j(x),$$

where $x = (x_1, \dots, x_{b-1}) \in \mathbf{K}^{b-1}$, each vector ${}_{i_1,i_2}a_{k_1,k_2} \in \mathbf{K}^b$ is marked, every $c_j \in \mathbf{K} \setminus \{0\}$ is a non-zero number. Another data $r_j \in \mathbf{N}$, ${}_jx_i, T_j \in \mathbf{K}$ are chosen below in this section. All functions u_j have disjoint supports, hence the series is convergent, while the function u is of class $C^{[\infty]} := \bigcap_{n=1}^{\infty} C^{[n]}$ in $\mathbf{K} \setminus \{z_0\}$.

We consider the sets

$$\lambda_i := \{j \in \mathbf{N} : \ {}_jt_i = 0\},$$

then either $card(\mathbf{N} \setminus \lambda_i) = \aleph_0$ or $card(\lambda_i) = \aleph_0$ or both cardinalities are $\aleph_0$. There are the intersections $A_1 \cap \dots \cap A_n$, where $A_i = \lambda_i$ or $A_i = \mathbf{N} \setminus \lambda_i$. The union of all such finite intersections is the set $\mathbf{N}$ of all natural numbers. Therefore, one of these intersections is of the cardinality $\aleph_0 = card(\mathbf{N})$. Thus a subsequence $\{j(l) : l \in \mathbf{N}\}$ exists such that ${}_st_{j(l)} = 0$ for each l and every $s \in \{i_1, \dots, i_r\}$ and ${}_st_{j(l)} \neq 0$ for each $s \in \{1, \dots, n\} \setminus \{i_1, \dots, i_r\}$, where $0 \leq r \leq n$. After the enumeration we can consider a sequence with such property. For such sequence we can choose a subsequence which after the enumeration has the property:

$$|\ {}_{j+1}t_i| \leq |\pi|^{s(j)} |\ {}_jt_i| \tag{8}$$

and

$$|\pi|^{r(j)} b_{j+1} \geq b_j \quad \text{for each} \quad j \in \mathbf{N} \quad \text{and} \quad \mathbf{i} \in \{\mathbf{1}, \dots, \mathbf{n}\}, \tag{9}$$

where $s(j), r(j) \in \mathbf{N}$ are sequences specified below;

$$b_j := |(B_{j_n,w^{\otimes n},t_n} \dots B_{j_1,w,t_1} f \circ id)(\bar{\Phi}^1 \circ p_{j_n} \hat{S}_{j_{n-1}+1,w^{\otimes(n-1)}t_{n-1}}$$

$$\dots \hat{S}_{j_1+1,w\tau_1} id^{n-1})(P_n \bar{\Phi}^1 \circ p_{j_{n-1}} \hat{S}_{j_{n-2}+1,w^{\otimes(n-2)},t_{n-2}} \dots \hat{S}_{j_1+1,wt_1} id^{n-2})$$

$$\dots (P_n \dots P_2 \bar{\Phi}^1 \circ p_{j_1} id)(\ {}_jz_0^{(n)})|$$

or with analogous Properties $(8,9)$ for $A^n_* f$ instead of $B^n_* f$.

Now we choose r_j and c_j such that

$$\lim_{j \to \infty} c_j T_j^{-q} = 0$$

for each $q \in \mathbf{N}$, for example, $c_j = T_j^j$, where

$$\lim_{j \to \infty} T_j = 0,$$

$|T_j| > |T_{j+1}|$ for each j, $T_j \neq 0$ for each j. Then we choose natural numbers $r_j \in \mathbf{N}$ such that

$$\max_{l=1}^{n}(|\,_{r_j}t_l|) \leq |c_j|$$

and

$$|_{r_j}z_0|) \leq |c_j|$$

for each j and

$$\lim_{j\to\infty} |c_j^n \Upsilon^n f(\,_{r_j}z_0^{[n]})| = \infty.$$

It is possible to take $_jx \in \mathbf{K}^{b-1}$ such that

$$_jx_i = (\pi^{-1}\sum_{k=1}^{j-1} T_k) + T_j,$$

where $\pi \in \mathbf{K}$, $0 < |\pi| < 1$, $i = 1,\dots,b-1$. Since $|T_j| > |T_{j+1}| > 0$ for each $j \in \mathbf{N}$, the estimates

$$|\,_jx - \,_{j+1}x| = |T_{j+1} + T_j(\pi^{-1} - 1)| = |\pi^{-1}T_j| > |T_j|$$

and

$$|\,_kx - \,_{k+1}x| \geq \min(|\,_kx - \,_{k+1}x|, |\,_{k+1}x - \,_{k+2}x|, \dots, |\,_jx - \,_{j+1}x|) \geq \min(|T_k|, \dots, |T_j|)$$

are satisfied for each $k \leq j$, consequently,

$$B(\mathbf{K}^{b-1},\,_kx, |T_k|) \cap B(\mathbf{K}^{b-1},\,_{j+1}x, |T_{j+1}|) = \emptyset$$

for each $k \leq j$, hence the intersection of supports

$$\mathrm{supp}(u_j) \cap \mathrm{supp}(u_k) = \emptyset$$

is void for each $k < j$. We take numbers $s_0(j+1) \geq s_0(j) + j + 1$ and

$$|\pi|^{s_0(j+1)} < |q_j| \leq g_j \leq |\pi|^{s_0(j)}$$

and $r(j) \geq s_0(j)2n$ and $s(j) \geq s_0(j)$ for each j, where l_0 is such that $0 < |\pi|^{l_0} < 1/2$ (see also $(6-9)$).

Denote the limit

$$y_0 := \lim_{j\to\infty} \,_jx$$

by y_0. Then u is of class $C_b^{[\infty]}$ or C_b^∞ in a neighborhood of the point z_0. To prove this we show, that $\Upsilon^q u(z^{[q]})$ or $\bar{\Phi}^q u(z^{(q)})$ tends to zero as z tends to $z_0 = 0$, where $|z^{[q]}| < \varepsilon$ or $|z^{(q)}| < \varepsilon$, since then for $|t_1| \geq \varepsilon$, or $\dots, |t_n| \geq \varepsilon$, $|v^{[q-1]}| \leq 1$ the continuity will be evident. For this Lemma 15 is useful. One mentions, that

$$\|\Upsilon^q \psi(x)\|_{C^0(B(\mathbf{K}^{[q]},0,R),\mathbf{K})} < \infty$$

for each q and each $R > 0$. Indeed, $\max_{x\in\mathbf{K}} |\psi(x)| = 1$ so that $\Upsilon^0\psi$ is bounded for $q = 0$. For $q = 1$ we have

$$\Upsilon^1\psi(x, v, t_1) = 0$$

for $\max(|x|,|x+vt_1|) \le |\pi|$ or $\min(|x|,|x+vt_1|) > |\pi|$,

$$\Upsilon^1\psi(x,v,t_1) = 1/t_1$$

for either $|x| \le |\pi|$ and $|x+vt_1| > |\pi|$ or $|x| > |\pi|$ and $|x+vt_1| \le |\pi|$. Since we consider the domain $|x^{[1]}| \le R$, the inequality $|v| \le R$ is fulfilled. Therefore,

$$\|\Upsilon^1\psi(x)\|_{C^0(B(\mathbf{K}^{[1]},0,R),\mathbf{K})} \le R|\pi|^{-1},$$

since $|t_1|^{-1} \le |\pi|^{-1}R$ in the considered domain, when $\Upsilon^1\psi(x,v,t_1) \ne 0$. The function $\Upsilon^1\psi(x,v,t_1)$ is the product of the locally constant function by variables (x,v) and the function $1/t_1$ with $|\pi/v| \le |t_1| \le R$, when this function is non-zero and $v \ne 0$, hence

$$|\pi|/R \le |v| \le R,$$

that is $|\pi|/R \le |t_1| \le R$, where $\Upsilon^1\psi(x,0,t_1) = 0$ for each x and t_1. Evidently, by induction the partial difference quotient $\Upsilon^q\psi(x^{[q]})$ belongs to the space $C^0(B(\mathbf{K}^{[q]},0,R),\mathbf{K})$ with the finite norm

$$\|\Upsilon^q\psi(x^{[q]})\|_{C^0(B(\mathbf{K}^{[q]},0,R),\mathbf{K})} \le C^{q+1}V_q^{-q} \le (q+1)(R/|\pi|)^q$$

with $V_q = 1$ and

$$C := \lim_{q\to\infty}[(q+1)(R/|\pi|)^q]^{1/(q+1)}$$

for the non-scaled function ψ for each $q \in \mathbf{N}$ and each $R \ge 1$. In general for the scaled function ψ it is useful to put

$$V_q := \min_{j=1}^{q}|T_j| > 0.$$

At the same time for each x satisfying the conditions

$$|x - {}_jx| \le |T_j| \quad \text{and} \quad |v^{[k]}| \le R \quad \text{and} \quad |t_{k+1}| \le R$$

for each $k = 0,\dots,n-1$ in accordance with Lemmas 4, 12, 15 and Corollary 13 the inequality

$$|\Upsilon^q u(x^{[q]})| \le (\max(1,R^2))|c_j||T_j|^{-q}C_1^{q+1}V_q^{-q} \tag{10}$$

is fulfilled, while its right side tends to zero as j tends to the infinity, since $C_1^{q+1} \le (q+1)(R/|\pi|)^q$, $0 < |T_{j+1}| < |T_j|$ for each j and

$$\lim_{j\to\infty} c_jT_j^{-\beta} = 0$$

for every natural number $\beta \in \mathbf{N}$, where $R \ge 1$.

If each term in Formula 9(1) or 10(1) would be locally bounded, then the partial difference quotient either $\bar{\Phi}^n(f\circ u)({}_{r_j}x^{(n)})$ or $\Upsilon^n(f\circ u)({}_{r_j}x^{[n]})$ would be locally bounded. Since each partial difference quotient either $\bar{\Phi}^k f$ or $\Upsilon^k f$ is locally bounded for $k < n$ by our supposition above, from Formula 9(1) or 10(1) and the condition

$$\lim_{j\to\infty}|c_j^n\Upsilon^n f({}_{r_j}z_0^{[n]})| = \infty \quad \text{or} \quad \lim_{j\to\infty}|c_j^n\bar{\Phi}^n f({}_{r_j}z_0^{(n)})| = \infty$$

it follows, that there exists a term or a finite sum of terms of the type

$$(A_{j_n,v^{[n-1]},t_n}\dots A_{j_1,v^{[0]},t_1}f\circ u)(\Upsilon^1\circ p_{j_n}S_{j_{n-1}+1,v^{[n-2]}t_{n-1}}$$
$$\dots S_{j_1+1,v^{[0]}t_1}u^{n-1})(P_n\Upsilon^1\circ p_{j_{n-1}}S_{j_{n-2}+1,v^{[n-3]}t_{n-2}}\dots S_{j_1+1,v^{[0]}t_1}u^{n-2})$$
$$\dots(P_n\dots P_2\Upsilon^1\circ p_{j_1}u)\ {}_{r_l}x^{[n]})$$

the absolute value of which tends to the infinity for a particular set ω of indices $(j_1,\dots,j_n)$ and a subsequence

$$\{\ {}_{r_l}x^{[n]}:\ j\in\mathbf{N}\}$$

or analogously for $B^n_*f\circ u$ instead of $A^n_*f\circ u$. But this contradicts supposition of this lemma in view of Lemmas 9, 21 and Corollary 10. Therefore, the partial difference quotient either $\Upsilon^n f$ or $\bar{\Phi}^n f$ respectively is locally bounded.

28. Remark. Certainly, the inclusion $u\in C^\infty(\mathbf{K},\mathbf{K}^b)$ does not imply that a function u is locally analytic in general, since the sequence $\{x_j:\ j\}$ converges to the point $y_0\in\mathbf{K}$ and the function u need not to have a power series expansion in a neighborhood of the point y_0 with some positive radius of convergence.

29. Definitions. Let $\phi:(0,\infty)\to(0,\infty)$ be a function such that the limit

$$\lim_{q\to 0}\phi(q)=0$$

is zero. By either $\mathbf{K}(\phi)$ or $\mathbf{K}(u,\phi)$ we denote the $\mathbf{K}$-linear space of all functions $f:\mathbf{K}^m\to\mathbf{K}$ such that for each bounded subset U in the topological vector space $\mathbf{K}^m$ a constant $C>0$ exists so that either

$$|f(x+y)-f(x)|\le C\phi(|y|), \tag{1}$$

when $x\in U$ and $x+y\in U$ or

$$|f(x+ut)-f(x)|\le C\phi(|t|), \tag{2}$$

when $x\in U$ and $x+ut\in U$ respectively, where $u\in\mathbf{K}^m$ is a nonzero vector. In the particular case of $\phi(q)=q^w$, where $0<w\le 1$, we also denote $\mathbf{K}(q^w)=:Lip(w)$ and $\mathbf{K}(u,q^w)=:Lip(u,w)$.

Then we denote by $C^{[n],w}(\mathbf{K}^m,\mathbf{K})$ or $C^{[n]}_\phi(\mathbf{K}^m,\mathbf{K})$ or $C^{n,w}(\mathbf{K}^m,\mathbf{K})$ or $C^n_\phi(\mathbf{K}^m,\mathbf{K})$ the $\mathbf{K}$-linear space of all functions $f\in C^{[n]}(\mathbf{K}^m,\mathbf{K})$ in the first and the second cases or in $C^n(\mathbf{K}^m,\mathbf{K})$ in the third and the fourth cases such that $f^{[n]}(x^{[n]})\in Lip(w)$ or $f^{[n]}(x^{[n]})\in\mathbf{K}(\phi)$ or $\bar{\Phi}^n f(x^{(n)})\in Lip(w)$ or $\bar{\Phi}^n f(x^{(n)})\in\mathbf{K}(\phi)$ respectively.

30. Lemma. *Let suppositions of Lemma 27 be satisfied and moreover $\Upsilon^n(f\circ u)\in\mathbf{K}(\phi)$ or $\bar{\Phi}^n(f\circ u)\in\mathbf{K}(\phi)$ for each function $u\in\underline{C}^{[\infty]}(\mathbf{K},\mathbf{K}^b)$ or $C^{[\infty]}_b(\mathbf{K},\mathbf{K}^b)$ or $u\in C^\infty(\mathbf{K},\mathbf{K}^b)$ or $C^\infty_b(\mathbf{K},\mathbf{K}^b)$. Then $\Upsilon^n f(x^{[n]})\in\mathbf{K}(v,\phi)$ or $\bar{\Phi}^n f(x^{(n)})\in\mathbf{K}(v,\phi)$, where v is a marked vector $v\in(\mathbf{K}^b)^{[n]}$ or $v\in(\mathbf{K}^b)^{(n)}$, where $x\in\mathbf{K}^b$, $x^{[n]}\in(\mathbf{K}^b)^{[n]}$ or $x^{(n)}\in(\mathbf{K}^b)^{(n)}$ correspondingly.*

Proof. Without loss of generality it can be assumed, that the function ϕ is sub-additive and increasing taking the function:

$$\phi_1(q):=\inf\{\sum_{k=1}^n\phi(q_k):\quad \sum_{k=1}^n q_k\ge q, q_k\ge 0\}$$

which is the largest increasing and sub-additive minorant of the function ϕ. For the sub-additive and increasing function ϕ the inequality:

$$\phi(q\varepsilon) \le \phi((1+[q])\varepsilon) \le (1+q)\phi(\varepsilon) \tag{1}$$

is satisfied for any positive numbers $\varepsilon > 0$ and $q > 0$, where $[q]$ denotes the integral part of q, i.e the greatest integer satisfying the inequality $[q] \le q$.

If S is a family of vectors such that it spans the vector space $\mathbf{K}^b$ and the function f belongs to $\mathbf{K}(u,\phi)$ for each $u \in S$, then $f \in \mathbf{K}(\phi)$, since $b \in \mathbf{N}$ and

$$|f(x+y)-f(x)| = |f(x+y)-f(x+y_2e_2+\dots+y_be_b)+f(x+y_2e_2+\dots+y_be_b)$$

$$-f(x+y_3e_3+\dots+y_be_b)+\dots+f(x+y_be_b)-f(x)|$$

$$\le \max(|f(x+y)-f(x+y_2e_2+\dots+y_be_b)|, |f(x+y_2e_2+\dots+y_be_b)$$

$$-f(x+y_3e_3+\dots+y_be_b)|,\dots,|f(x+y_be_b)-f(x)|)$$

$$\le C\max(\phi(|y_1|),\dots,\phi(|y_b|)) \le C\phi(|y|)$$

due to the increasing monotonicity of the function ϕ and the fact that $|y| = \max(|y_1|,\dots,|y_b|)$, where $y = y_1e_1+\dots+y_be_b$, $y_1,\dots,y_b \in \mathbf{K}$, $e_j = (0,\dots,0,1,0,\dots) \in \mathbf{K}^b$ with 1 on the j-th place. Up to a $\mathbf{K}$-linear topological automorphism of the topological vector space $\mathbf{K}^b$ onto itself we can choose such basis as belonging to S, $j = 1,\dots,b$, and $C > 0$ is a constant.

Let us assume that for some point the statement of this lemma is not true. We can suppose, that this is at the zero point $x^{[n]} = (0,\dots,0) \in (\mathbf{K}^b)^{[n]}$ or $x^{(n)} = 0 \in (\mathbf{K}^b)^{(n)}$ respectively making a shift in a case of necessity. Then sequences $b_k > 0$, $h_k \in \mathbf{K}$, $h_k \ne 0$, ${}_kz^{[n]} \in (\mathbf{K}^b)^{[n]}$ exist such that

$$\lim_{k\to\infty} b_k = \infty, \quad \lim_{k\to\infty} h_k = 0, \quad \lim_{k\to\infty} {}_kz^{[n]} = 0$$

and

$$|\Upsilon^n f({}_kz^{[n]}+h_kv) - \Upsilon^n f({}_kz^{[n]})| > b_k\phi(|h_k|) \tag{2}$$

or

$$|\bar{\Phi}^n f({}_kz^{(n)}+h_kv) - \bar{\Phi}^n f({}_kz^{(n)})| > b_k\phi(|h_k|) \tag{2'}$$

with

$$\lim_{k\to\infty} {}_kz^{(n)} = 0$$

respectively, where $0 \ne v \in (\mathbf{K}^b)^{[n]}$ or $0 \ne v \in (\mathbf{K}^b)^{(n)}$ correspondingly, $k = 1,2,3,\dots$.

Let the functions u and u_j be as in the proof of Lemma 27. We choose natural numbers $r_j \in \mathbf{N}$ such that $|{}_{r_j}z_0| \le |c_j|$,

$$\lim_{j\to\infty} |c_j|^{n+1}b_{r_j} = \infty, \quad |h_{r_j}| < |\pi c_jT_j|.$$

Thus $u \in C^\infty(\mathbf{K},\mathbf{K}^b)$. Now we prove that at least for large natural numbers $j \in \mathbf{N}$ accomplished the inequality:

$$|\Upsilon^n(f\circ u)({}_jx^{[n]}+{}_j\nu^{[n]}) - \Upsilon^n(f\circ u)({}_jx^{[n]})| > |\pi c_j^{n+1}|b_{r_j}\phi(|{}_j\nu^{[n]}|)|\pi|^{l_0} \tag{3}$$

or

$$|\bar{\Phi}^n(f\circ u)({}_jx^{(n)} + {}_j\nu^{(n)}) - \bar{\Phi}^n(f\circ u)({}_jx^{(n)})| > |\pi c_j^{n+1}|b_{r_j}\phi(|{}_j\nu^{[n]}|)|\pi|^{l_0} \qquad (3')$$

is accomplished, where

$$|{}_j\nu^{[n]}| = |h_{r_j}v/c_j| \text{ or } |{}_j\nu^{(n)}| = |h_{r_j}v/c_j|$$

with $c_j \neq 0$ for each j. One can take without loss of generality $|v| = 1$. Together with the condition

$$\lim_{j\to\infty} |c_j|^{n+1} b_{r_j} = \infty$$

this will complete the proof. If $|h| < |\pi T_j|$, then

$$u_j(h) = {}_{r_j}z_0 + c_j \sum_{k_1,k_2=0}^{2} \sum_{i_1,i_2=1}^{b-1} {}_{i_1,i_2}a_{k_1,k_2} h_{i_1}^{k_1} h_{i_2}^{k_2} \qquad (4)$$

with $u_j(0) = {}_{r_j}z_0$. In Formula 9(1) or 10(1) all terms with an amount of operators $A_{j,v^{[k-1]},t_k}$ or $B_{j,v^{[k-1]},t_k}$ in it less than n are in the space $C_\phi^{[1]}(\mathbf{K},\mathbf{K})$. As in Lemma 27 we reduce the consideration to the partial difference quotients $\Upsilon^n f({}_kz^{[n]})$ or $\bar{\Phi}^n f({}_kz^{(n)})$ with prescribed fixed vectors $w_0, w_1, \dots, w_{n-1}$ with $v^{[0]} = v^{[k],1} = w_0$ and $v^{[k],i} = w_l$ for each $2^{l-1} < i \leq 2^l$, where $l = 0,1,\dots,k$ and $k = 0,1,\dots,n-1$, vectors $v^{[k],i} \in \mathbf{K}^b$ are formed from the vectors $v^{[k]}$ after excluding all zeros arising from $v_3^{[k]} = 0$, or $v = (w_0,\dots,w_0)$ with ${}_kz^{(n)} = ({}_kz_0; v; {}_kt)$ respectively, where

$${}_kt = ({}_kt_1,\dots,{}_kt_n) \in \mathbf{K}^n.$$

In view of Lemma 16 we can consider the case $(b-1,b)$ instead of $(1,b)$. Thus, from Formulas (2,4) and 12(2) it follows, that expansion coefficients ${}_{i_1,i_2}a_{k_1,k_2} \in \mathbf{K}^b$ exist with

$$|{}_{i_1,i_2}a_{k_1,k_2}| \leq 1$$

for each i_1, i_2, k_1, k_2 and there exists $j_0 \in \mathbf{N}$, for which the estimates

$$|\Upsilon^n f\circ u({}_jx^{[n]} + {}_j\nu^{[n]}) - \Upsilon^n f\circ u({}_jx^{[n]})| \geq |\pi|^{l_0+s_0}|q_j|^n$$

$$|\Upsilon^n f({}_{r_j}z_0^{[n]} + h_jv) - \Upsilon^n f({}_{r_j}z_0^{[n]})| \geq b_{r_j}|c_j^n|\phi(|c_j\, {}_j\nu^{[n]}|)|\pi|^{l_0} \qquad (5)$$

or

$$|\bar{\Phi}^n f\circ u({}_jx^{(n)} + {}_j\nu^{(n)}) - \bar{\Phi}^n f\circ u({}_jx^{(n)})| \geq |\pi|^{l_0+s_0}|q_j|^n$$

$$|\bar{\Phi}^n f({}_{r_j}z_0^{(n)} + h_jv) - \bar{\Phi}^n f({}_{r_j}z_0^{(n)})| \geq b_{r_j}|c_j^n|\phi(|c_j\, {}_j\nu^{(n)}|)|\pi|^{l_0} \qquad (5')$$

are fulfilled for each $j \geq j_0$, since $|\nu_j| < |\pi T_j|$, where $l_0 \in \mathbf{N}$ is a marked natural number, ${}_j\tau_i$, $i = 1,\dots,n$ are parameters corresponding to $t_1,\dots,t_n$, but for the curve u instead of f. A natural number $j_0 \in \mathbf{N}$ exists such that

$$|c_j| \leq \min(1, |\pi|^{-1} - 1)$$

for each $j > j_0$, where $\pi \in \mathbf{K}$, $0 < |\pi| < 1$. In view of Formula (1) for each $j > j_0$ we have the inequalities:

$$\phi(|\,{}_j\nu^{[n]}|) \le (1 + |c_j|^{-1})\phi(|c_j\,{}_j\nu^{[n]}|) \le |\pi c_j|^{-1}\phi(|c_j\,{}_j\nu^{[n]}|).$$

Therefore, the latter formula and Formula (5) imply Formula (3).

31. Lemma. *Let f be a function $f : \mathbf{K} \to \mathbf{K}$ such that $f(0) = 0$ and $|f(t)| \le 1$ for each number $t \in \mathbf{K}$ satisfying the condition $|t| \le |q|a$, where q and a are constants such that $q \in \mathbf{K}$, $|q| > 1$, $a > 0$, and assume that*

$$|f(qt) - qf(t)| \le \max(b, C_1|t|^r) \tag{1}$$

for each number $t \in \mathbf{K}$ with the norm $|t| \le a$, where $0 < r \le 1$, $b > 0$ and $C_1 > 0$ are constants. Then there exists a constant $C_2 > 0$ such that the function f satisfies the condition:

$$|f(t)| \le \max(b, C_2|t|^r) \tag{2}$$

for each $t \in \mathbf{K}$ with $|t| \le |q|a$, where $C_2 = \max(a^{-r}, |q|^{-1}a^rC_1|q|^{-r})$.

Proof. If $t \in \mathbf{K}$ is such that $a \le |t| \le |q|a$, then Inequality (2) is satisfied with the constant $C_2 = a^{-r}$, since $|f(t)| \le 1$ for such t. Now suppose that $0 < |u| < a$ and Inequality (2) is satisfied for $t = qu$, then

$$|q||f(t)| \le \max(|f(qt)|, b, C_1|u|^r) \le \max(b, C_2|qu|^r, C_1|u|^r)$$

$$= \max(b, |u|^r \max(C_1, C_2|q|^r)),$$

consequently,

$$|f(t)| \le |q|^{-1}\max(b, a^r\max(C_1, C_2|q|^r)) \le \max(b, C_2|t|^r)$$

for $C_2 = \max(a^{-r}, |q|^{-1}a^rC_1|q|^{-r})$, since $C_2|q|^r \ge C_1$. On the other hand $B(\mathbf{K}, 0, |q|a) \setminus \{0\}$ is the disjoint union of subsets $B(\mathbf{K}, 0, |q|^ja) \setminus B(\mathbf{K}, 0, |q|^{j-1}a)$ for $j = 1, 0, -1, -2, \dots$. Therefore, proceeding by induction by j we get the statement of this lemma, since $f(0) = 0$.

32. Lemma. *Let Ω be a finite set of vectors $v \in \mathbf{K}^m$ which are pairwise $\mathbf{K}$-linearly independent, $card(\Omega) \ge m$ and each subset of Ω consisting not less than m vectors has the $\mathbf{K}$-linear span coinciding with the vector space $\mathbf{K}^m$, where $m \ge 2$ is the integer. Suppose that for each $v \in \Omega$ a function $g_v : \mathbf{K}^m \to \mathbf{K}$ is given so that:*

$$|g_v(x)| \le 1 \text{ for each } x \in \mathbf{K}^m \text{ with } |x| \le R, \tag{1}$$

$$|g_v(x + tv) - g_v(x)| \le |t|^r \text{ for each } x, x + tv \in \mathbf{K}^m \tag{2}$$

with $|x| \le R$ and $|x + tv| \le R$, where $t \in \mathbf{K}$,

$$\left|\sum_{v \in \Omega} (g_v(x) - g_v(y))\right| \le b \tag{3}$$

for each $|x| \leq R$ and $|y| \leq R$, where R and b are positive constants, $0 < r \leq 1$. Then a positive constant $C > 0$ exists, which may depend on (r, Ω) such that the inequality

$$|g_v(x) - g_v(y)| \leq C \max(b, |x-y|^r) \tag{4}$$

is accomplished for each $x, y \in \mathbf{K}^m$ such that $|x| \leq R$, $|y| \leq R$ and each $v \in \Omega$.

Proof. We prove this lemma by induction on a number n of elements in Ω. For $n = m = 1$ Inequality (4) is the consequence of Inequality (2). For $n = m \geq 2$ vectors $v_1, \dots, v_m$ by the supposition of lemma are $\mathbf{K}$-linearly independent. Then for each $x, y \in \mathbf{K}^m$ numbers $t_1, \dots, t_m \in \mathbf{K}$ exist such that $x = y + t_1 v_1 + \dots + t_m v_m$. If $|x| \leq R$ and $|y| \leq R$, then

$$B(\mathbf{K}^m, 0, R) = B(\mathbf{K}^m, x, R) = B(\mathbf{K}^m, y, R)$$

due to the ultra-metric inequality. On the other hand, $B(\mathbf{K}^m, 0, R)$ is the additive group, hence $y - x \in B(\mathbf{K}^m, 0, R)$. Vectors v_j have coordinates $v_j = (v_j^1, \dots, v_j^m)$, where $v_j^k \in \mathbf{K}$, consequently,

$$|v_j| = \max(|v_j^1|, \dots, |v_j^m|).$$

Thus,

$$|x-y| = \max(|t_1 v_1^1 + \dots + t_m v_m^1|, \dots, |t_1 v_1^m + \dots + t_m v_m^m|) \leq \max(|t_1 v_1|, \dots, |t_m v_m|).$$

One can choose numbers $t_1, \dots, t_m$ such that

$$|y + t_1 v_1 + \dots + t_k v_k| \leq R$$

also for each $k = 1, \dots, m$. Therefore, the estimate

$$|g_{v_j}(x) - g_{v_j}(y)| = |g_{v_j}(x) - g_{v_j}(y + t_1 v_1 + \dots + t_{m-1} v_{m-1})$$

$$+ g_{v_j}(y + t_1 v_1 + \dots + t_{m-1} v_{m-1}) - \dots - g_{v_j}(y + t_1 v_1) + g_{v_j}(y + t_1 v_1) - g_{v_j}(y)|$$

$$\leq \max_{k=1}^{m} |g_{v_j}(y + t_1 v_1 + \dots + t_k v_k) - g_{v_j}(y + t_1 v_1 + \dots + t_{k-1} v_{k-1})|$$

$$\leq \max(b, |t_1|^r, \dots, |t_m|^r) \leq \max(b, |x-y|^r)$$

is satisfied for each $|x| \leq R$ and $|y| \leq R$ as the consequence of Inequalities (2) and (4) and the ultra-metric inequality for each $x, y \in \mathbf{K}^m$ with $|x| \leq R$ and $|y| \leq R$, since $j = 1, \dots, m$.

Further one can proceed by induction on n. From the preceding prove it follows, that the statement of this lemma is true for $n = m$. It is useful to put $\Omega = \Omega_0 \cup \{w\}$, where $w \notin \Omega_0$ and all elements of Ω are pairwise linearly independent over the field $\mathbf{K}$. Assume that the assertion of this lemma is true for Ω_0 and prove it for Ω. For $v \in \Omega_0$ let us denote $h_v(x, u) = h_v(x) = g_v(x + uw) - g_v(x)$, where $|x| \leq R$, and $|x + uv| \leq R$. For these values of x and $x + uv$ the function h_v satisfies Conditions $(1, 2)$. On the other hand, from (2) for $v = w$ and the inequality (3)

$$\left| \sum_{v \in \Omega_0} (h_v(x) - h_v(y)) \right| \leq \max \left(\left| \sum_{v \in \Omega} (g_v(x + uw) - g_v(x)) \right|, \right.$$

$$\left|\sum_{v\in\Omega}(g_v(y+uw)-g_v(y))\right|,|g_w((x+uw)-g_w(x)|,|g_w(y+uw)-g_w(y)|\Big)$$

$$\leq \max(b,|u|^r)$$

for each $|x|\leq R$, $|y|\leq R$ and $|uw|\leq R$, it follows that $\{h_v : v\in\Omega_0\}$ satisfies Condition (3) with $\max(b,|u|^r)$ instead of b. By the induction hypothesis a positive constant $C_1 = const > 0$ exists, which may depend only on Ω_0, r and R such that

$$|h_v(x)-h_v(y)|\leq C_1\max(b,|u|^r,|x-y|^r) \tag{5}$$

for each $|x|\leq R$, $|y|\leq R$ and $|uw|\leq R$ and $v\in\Omega_0$. Taking $y-x=(q-1)uw$ with $q\in\mathbf{K}$ and $|q|>1$ implies $|q-1|>1$ and Inequality (5) will take the form:

$$|g_v(x+quw)-g_v(x+(q-1)uw)-g_v(x+uw)+g_v(x)|$$

$$\leq C_1\max(b,|(q-1)u|^r,|(q-1)uw|^r)\leq C_2\max(b,|u|^r), \tag{6}$$

when $|x|\leq R$, $|quw|\leq R$ and $v\in\Omega_0$, where $C_2\geq C_1|q-1|^r\max(1,|w|^r)$. Now setting $s(u):=g_v(x+uw)-g_v(x)$ for $v\in\Omega_0$ and Formulas (2) and (6) imply the ultra-metric inequality:

$$|s(qu)-qs(u)|\leq C_2\max(b,|u|^r),$$

when $|quw|\leq R$. In view of Lemma 19 the estimate

$$|s(u)|=|g_v(x+uw)-g_v(x)|\leq C_3\max(b,|u|^r) \tag{7}$$

is valid for each $|uw|\leq R$, $|x|\leq R$ and $v\in\Omega_0$, where

$$C_3=\max(a^{-r},|q|^{-1}a^rC_2|q|^{-r}).$$

Interchanging roles of w and one of $v\in\Omega_0$ we obtain (7) with w in place of v, that is, (4) is proved for each $v\in\Omega$.

33. Corollary. *Let $v_1,\dots,v_n$ be pairwise $\mathbf{K}$-linearly independent vectors in the vector space $\mathbf{K}^m$ and each subset consisting not less than m of these vectors has the $\mathbf{K}$-linear span coinciding with $\mathbf{K}^m$ and let g_k be locally bounded functions from $\mathbf{K}^m$ into the field $\mathbf{K}$, $0<r\leq 1$. If $g_k\in Lip(v_k,r)$ for each k and*

$$\sum_{k=1}^n g_k(x)=0$$

identically by the variable $x\in\mathbf{K}^m$, then $g_k\in Lip(r)$ for each k.

Proof. If a non-zero number $c\in\mathbf{K}$, $c\neq 0$, is small enough by its norm $|c|$, then the functions cg_k satisfy assumptions of Lemma 32 with $b=0$.

34. Remark. We can mention, that apart from the classical case over the real field $\mathbf{R}$ this lemma is true for $r=1$ as well due to the ultra-metric inequality, which is stronger than the usual triangle inequality.

35. Definition. Let a non-zero vector $v\in\mathbf{K}^b$ be given, $v\neq 0$. We say that a function $f:\mathbf{K}^b\to\mathbf{K}$ is continuous in the direction v if $f(x+tv)$ converges to $f(x)$ uniformly by x

on bounded closed sets in the finite-dimensional over the field $\mathbf{K}$ topological vector space $\mathbf{K}^b$ as $t \in \mathbf{K}$ tends to zero.

Mention that in a particular case of a locally compact field $\mathbf{K}$ a bounded closed subset in $\mathbf{K}^b$ is compact.

36. Lemma. *Suppose that a continuous function $f \in C^0(\mathbf{K}^b, \mathbf{K})$ is given and its partial difference quotient $\Upsilon^1 f(x,w,t)$ is continuous or uniformly continuous on $V^{[1]}$ in the direction $v^{[1]}$ with $v_2^{[1]} \neq 0$ and $v_3^{[1]} \neq 0$, where $V^{[1]} := \{(x,v,t) \in U^{[1]} : |v| = 1\}$, U is an open subset in the topological vector space $\mathbf{K}^b$. Then $\Upsilon^1 f(x, v_2^{[1]}, t)$ is continuous or uniformly continuous by (x,t), $(x,v,t) \in U^{[1]}$ or $(x,v,t) \in V^{[1]}$ respectively.*

Proof. Assume the contrary, that the partial difference quotient $\Upsilon^1 f(x, v_2^{[1]}, t)$ is not continuous by (x,t). Making a shift in a case of necessity we can suppose that $\Upsilon^1 f(x, v_2^{[1]}, t)$ is not continuous by (x,t) at 0 or is not uniformly continuous on $V^{[1]}$. Therefore, a sequence $\{x_n^{[1]} \in (\mathbf{K}^b)^{[1]} : n \in \mathbf{N}\}$ exists such that $|\Upsilon^1 f(x_n^{[1]}) - \Upsilon^1 f(0)| > \varepsilon$ for each n or with $x_0^{[1]} \in V^{[1]}$ instead of 0 and a family of sequences parametrized by $x_0^{[1]}$ and

$$\sup_{x_0^{[1]} \in V^{[1]}} |\Upsilon^1 f(x_n^{[1]}) - \Upsilon^1 f(x_0^{[1]})| > \varepsilon$$

correspondingly, where $\varepsilon > 0$ is a positive constant, $x_n^{[1]} = (x_n, v_2^{[1]}, t_n)$, while

$$\lim_{n \to \infty} (x_n, t_n) = x_0^{[1]}.$$

But in accordance with Definition 21 a positive constant $\delta > 0$ independent of n exists such that

$$|\Upsilon^1 f(x_n^{[1]} + v^{[1]}\tau) - \Upsilon^1 f(v^{[1]}\tau)| > \varepsilon|\pi|$$

or

$$\sup_{x_0^{[1]} \in V^{[1]}} |\Upsilon^1 f(x_n^{[1]} + v^{[1]}\tau) - \Upsilon^1 f(v^{[1]}\tau)| > \varepsilon|\pi|$$

for each n and each $\tau \in \mathbf{K}$ with $|\tau| \leq \delta$. On the other hand, the equality

$$\Upsilon^1 f(x_n + v_1^{[1]}\tau, w_n + v_2^{[1]}\tau, t_n + v_3^{[1]}\tau) - \Upsilon^1 f(v^{[1]}\tau)$$

$$= [f(x_n + v_1^{[1]}\tau + (w_n + v_2^{[1]}\tau)(t_n + v_3^{[1]}\tau)) - f(x_n + v_1^{[1]}\tau)]/(t_n + v_3^{[1]}\tau)$$
$$-[f(v_1^{[1]}\tau + v_2^{[1]}\tau v_3^{[1]}\tau) - f(v_1^{[1]}\tau)]/(v_3^{[1]}\tau)$$

is satisfied, where $w_n = v_2^{[1]}$. But

$$\lim_{n \to \infty} [f(x_n + v_1^{[1]}\tau + (w_n + v_2^{[1]}\tau)(t_n + v_3^{[1]}\tau))/(t_n + v_3^{[1]}\tau) - f(v_1^{[1]}\tau)/(v_3^{[1]}\tau)] = 0$$

and

$$\lim_{n \to \infty} [f(x_n + v_1^{[1]}\tau)/(t_n + v_3^{[1]}\tau) - f(v_1^{[1]}\tau)]/(v_3^{[1]}\tau) = 0$$

for $v_3^{[1]}\tau \neq 0$ point-wise or uniformly respectively. If $v_3^{[1]}\tau = 0$, then

$$\Upsilon^1 f(x_n + v_1^{[1]}\tau, w_n + v_2^{[1]}\tau, t_n + v_3^{[1]}\tau) - \Upsilon^1 f(v^{[1]}\tau)$$

$$= \Upsilon^1 f(x_n + v_1^{[1]}\tau, w_n + v_2^{[1]}\tau, 0) - \Upsilon^1 f(v_1^{[1]}\tau, w_n + v_2^{[1]}\tau, 0),$$

but the latter difference tends to zero as τ tends to zero uniformly by n or also uniformly by the family of sequences parametrized by $x_0^{[1]}$ respectively in accordance with the supposition of this lemma. Thus we get the contradiction with our supposition, hence $\Upsilon^1 f(x, v_2^{[1]}, t)$ is continuous or uniformly continuous by (x,t) correspondingly.

37. Definition. Denote by either $C^{[n]}_{\phi,b}(U,Y)$ or $C^n_{\phi,b}(U,Y)$ spaces of all functions $f \in C^{[n]}_\phi(U,Y)$ or $f \in C^n_\phi(U,Y)$ such that $f^{[k]}(x^{[k]})$ or $\bar{\Phi}^k f(x^{(k)})$ is uniformly continuous on a subset either

$$V^{[k]} := \{x^{[k]} \in U^{[k]} : |v_1^{[q]}| = 1; |\, {}_l v_2^{[q]} t_{q+1}| \leq 1, |v_3^{[q]}| \leq 1 \quad \forall l,q\}$$

or

$$V^{(k)} := \{x^{(k)} \in U^{(k)} : |v_j| = 1 \quad \forall j\}$$

for each $k = 0,1,\dots,n$ with finite norms either

$$\|f\|_{[n]} := \|f\|_{[n],\phi} := \max(C, \sup_{k=0,\dots,n; x^{[k]} \in V^{[k]}} |f^{[k]}(x^{[k]})|)$$

or

$$\|f\|_n := \|f\|_{n,\phi} := \max(C, \sup_{k=0,\dots,n; x^{(k)} \in V^{(k)}} |\bar{\Phi}^k f(x^{(k)})|),$$

where $0 \leq C < \infty$ is the least constant satisfying 29(1) or 29(2) for $\Upsilon^n f(x^{[n]})$ or $\bar{\Phi}^n f(x^{(n)})$ respectively instead of f. For $\phi(q) = q^r$ we denote $C^{[n]}_\phi(U,Y)$ by $C^{[n],r}(U,Y)$ and $C^n_{\phi,b}(U,Y)$ by $C^{n,r}_b(U,Y)$, $0 \leq r \leq 1$.

In particular, for $r = 0$ we put $C^{[n],0} = C^{[n]}$, $C^{[n],0}_b = C^{[n]}_b$ and $C^{n,0} = C^n$, $C^{n,0}_b = C^n_b$, with $C = 0$ in the definition of the norm. As usually

$$C^{[\infty]}(U,Y) := \bigcap_{k=0}^{\infty} C^{[k]}(U,Y) \text{ and } C^\infty(U,Y) := \bigcap_{k=0}^{\infty} C^k(U,Y)$$

and

$$C^{[\infty]}_b(U,Y) := \bigcap_{k=0}^{\infty} C^{[k]}_b(U,Y) \text{ and } C^\infty(U,Y) := \bigcap_{k=0}^{\infty} C^k_b(U,Y),$$

where the topology of the latter two spaces is given by the family of the corresponding norms.

In the case of a locally compact field $\mathbf{K}$ and a compact clopen (closed and open at the same time) domain U we have the equalities $C^{[k]}_b(U,Y) = C^{[k]}(U,Y)$ and $C^k_b(U,Y) = C^k(U,Y)$, though for non locally compact $\mathbf{K}$ these $\mathbf{K}$-linear spaces are different.

38. Theorem. *Suppose that* $f : \mathbf{K}^m \to \mathbf{K}$, $m \in \mathbf{N}$ *and* $f \circ u \in C^s_\phi(\mathbf{K},\mathbf{K})$ *or* $f \circ u \in C^s_{\phi,b}(\mathbf{K},\mathbf{K})$ *or* $f \circ u \in C^{[s]}_\phi(\mathbf{K},\mathbf{K})$ *or* $f \circ u \in C^{[s]}_{\phi,b}(\mathbf{K},\mathbf{K})$ *for each* $u \in C^\infty(\mathbf{K},\mathbf{K}^m)$ *or* $u \in$

$C_b^\infty(\mathbf{K},\mathbf{K}^m)$ *or* $u \in C^{[\infty]}(\mathbf{K},\mathbf{K}^m)$ *or* $u \in C_b^{[\infty]}(\mathbf{K},\mathbf{K}^m)$, *where* s *is a non-negative integer, a function* $\phi : (0,\infty) \to (0,\infty)$ *is such that the limit*

$$\lim_{y\to 0} \phi(y) = 0$$

is zero. Then $f \in C^s(\mathbf{K}^m,\mathbf{K})$ *or* $f \in C_b^s(\mathbf{K}^m,\mathbf{K})$ *or* $f \in C^{[s]}(\mathbf{K}^m,\mathbf{K})$ *or* $f \in C_b^{[s]}(\mathbf{K}^m,\mathbf{K})$ *respectively.*

Proof. In view of Lemma 21 it is sufficient to prove that the partial difference quotient $\bar{\Phi}^n f(x;e_{j(1)},\dots,e_{j(n)};t_1,\dots,t_n)$ is in $C^0(U^{(n)}_{j(1),\dots,j(n)},Y)$ or $C^0_b(V^{(n)}_{j(1),\dots,j(n)},Y)$ or $\Upsilon^n f(x^{[n]})$ is in $C^0(U^{[n]}_{j(0),\dots,j(n)},Y)$ or $C^0_b(V^{[n]}_{j(0),\dots,j(n)},Y)$ respectively for each $n = 1,2,\dots,s$ and each $j(1),\dots,j(n) \in \{1,\dots,m\}$ or $j(i) \in \{1,\dots,m(i)\}$, $i = 0,1,\dots,n$. If $\Upsilon^{m+1}f$ or $\bar{\Phi}^{m+1}f$ is locally bounded, then $\Upsilon^m f$ or $\Phi^m f$ is continuous respectively.

Applying Lemma 27 by induction we get that the partial difference quotient $\Upsilon^n f(x^{[n]})$ is the locally bounded function on the topological vector space $(\mathbf{K}^m)^{[n]}$ and the partial difference quotient $\bar{\Phi}^n f(x^{(n)})$ is the locally bounded function on $(\mathbf{K}^m)^{(n)}$. In view of Lemma 30 each $\Upsilon^k f(x^{[k]})$ or $\bar{\Phi}^k f(x^{(k)})$ is continuous in each direction v for each $k = 1,\dots,s$, where $v \in (\mathbf{K}^m)^{[k]}$ or $v \in (\mathbf{K}^m)^{(k)}$ correspondingly. On the other hand, by induction on k we have that in accordance with Lemma 36 $\Upsilon^k f(x^{[k]})$ or $\bar{\Phi}^k f(x;e_{j(1)},\dots,e_{j(k)};t_1,\dots,t_k)$ is continuous on $U^{[k]}_{j(0),\dots,j(k)}$ or $U^{(k)}_{j(1),\dots,j(k)}$ or bounded uniformly continuous respectively on the domain either $V^{[k]}_{j(0),\dots,j(k)}$ or $V^{(k)}_{j(1),\dots,j(k)}$ for bounded U for each $j(1),\dots,j(k) \in \{1,\dots,m\}$ or $j(i) \in \{1,\dots,m(i)\}$ for all $i = 0,1,\dots,k$.

39. Theorem. *Let* $f : \mathbf{K}^m \to \mathbf{K}^n$, $m,n \in \mathbf{N}$. *Let also* $f \circ u \in C^{s,r}(\mathbf{K},\mathbf{K}^n)$ *or* $f \circ u \in C_b^{s,r}(\mathbf{K},\mathbf{K}^n)$ *or* $C^{[s],r}(\mathbf{K},\mathbf{K}^n)$ *or* $C_b^{[s],r}(\mathbf{K},\mathbf{K}^n)$ *for each function* $u \in C^\infty(\mathbf{K},\mathbf{K}^m)$ *or* $u \in C_b^\infty(\mathbf{K},\mathbf{K}^m)$ *or* $C^{[\infty]}(\mathbf{K},\mathbf{K}^m)$ *or* $C_b^{[\infty]}(\mathbf{K},\mathbf{K}^m)$ *correspondingly, where* s *is a nonnegative integer,* $0 \le r \le 1$. *Then* $f \in C^{s,r}(\mathbf{K}^m,\mathbf{K}^n)$ *or* $f \in C_b^{s,r}(\mathbf{K}^m,\mathbf{K}^n)$ *or* $C^{[s],r}(\mathbf{K}^m,\mathbf{K}^n)$ *or* $C_b^{[s],r}(\mathbf{K}^m,\mathbf{K}^n)$ *respectively.*

Proof. If $s = 0$ and $0 \le r \le 1$, then the assertion of this theorem follows from Lemmas 27 and 30. For $r > 0$ by Theorem 38 we have that $f \in C^s(\mathbf{K}^m,\mathbf{K}^n)$ or $f \in C_b^s(\mathbf{K}^m,\mathbf{K}^n)$ or $C^{[s]}(\mathbf{K}^m,\mathbf{K}^n)$ or $C_b^{[s]}(\mathbf{K}^m,\mathbf{K}^n)$ respectively. From Lemma 21 we infer, that it is sufficient to prove that $\bar{\Phi}^n f(x;e_{j(1)},\dots,e_{j(n)};t_1,\dots,t_n)$ is in the space either $C^{0,r}(U^{(n)}_{j(1),\dots,j(n)},Y)$ or $C^{0,r}_b(V^{(n)}_{j(1),\dots,j(n)},Y)$ or $\Upsilon^n f(x^{[n]}) \in C^{0,r}(U^{[n]}_{j(0),\dots,j(n)},Y)$ or $C^{0,r}_b(V^{[n]}_{j(0),\dots,j(n)},Y)$ respectively for each $n = 1,2,\dots,s$ and each $j(1),\dots,j(n) \in \{1,\dots,m\}$, $j(i) \in \{1,\dots,m(i)\}$, $i = 0,1,\dots,n$. One can prove this by induction on n. For $n = 0$ it was proved above.

Let it be true for $n = 0,\dots,k$ and we prove it for $n = k+1 \le s$. For this we consider Formula 10(1) or 9(1). On the right hand side of it all terms having a total degree of f by operators B or A less than $k+1$ are in the space either $C^{0,r}(U^{(n)},Y)$ or $C^{0,r}_b(V^{(n)},Y)$ or $C^{0,r}(U^{[n]},Y)$ or $C^{0,r}_b(V^{(n)},Y)$ respectively by the induction hypothesis, since $u \in C^\infty(\mathbf{K},\mathbf{K}^m)$ or $u \in C_b^\infty(\mathbf{K},\mathbf{K}^m)$ or $C^{[\infty]}(\mathbf{K},\mathbf{K}^m)$ or $C_b^{[\infty]}(\mathbf{K},\mathbf{K}^m)$ correspondingly. Therefore, it remains to prove, that the sum

$$(i)\quad \Big[\sum_{j_1,\dots,j_n} (B_{j_n,v^{(n-1)},t_n} \dots B_{j_1,v^{(0)},t_1} f \circ u)(\bar{\Phi}^1 \circ p_{j_n} \hat{S}_{j_{n-1}+1,v^{(n-2)}t_{n-1}}$$

$$\dots \hat{S}_{j_1+1,v^{(0)}t_1}u^{n-1})(P_n\bar{\Phi}^1 \circ p_{j_{n-1}}\hat{S}_{j_{n-2}+1,v^{(n-3)},t_{n-2}}\dots\hat{S}_{j_1+1,v^{(0)}t_1}u^{n-2})$$
$$\dots(P_n\dots P_2\bar{\Phi}^1\circ p_{j_1}u)\Big]$$

is in the space $C^{0,r}(U^{(n)},Y)$ or $C^{0,r}_b(V^{(n)},Y)$ or corresponding sum by compositions of $A_{j_k,v^{[k-1]},t_k}$ is in the space either $C^{0,r}(U^{[n]},Y)$ or $C^{0,r}_b(V^{[n]},Y)$ respectively. In accordance with the proof above it is sufficient to demonstrate this for the vector $v^{(n-1)}=(e_{j(1)},\dots,e_{j(n)})$ for each $j(1),\dots,j(n)\in\{1,\dots,m\}$ or $v^{[i]}=e_{j(i)}$ with $j(i)\in\{1,\dots,m(i)\}$ and $i=0,1,\dots,n$, where $v_0^{(l)}=v_{l+1}=e_{j(l+1)}$, $l=0,\dots,n-1$. By the induction hypothesis the partial difference quotient $\bar{\Phi}^l f(x;e_{j(1)},\dots,e_{j(l)};t_1,\dots,t_l)$ is in the space either $C^{0,r}(U^{(l)}_{j(1),\dots,j(l)},Y)$ or $C^{0,r}_b(V^{(l)}_{j(1),\dots,j(l)},Y)$ or $\Upsilon^l f(x^{[l]})\in C^{0,r}(U^{[l]}_{j(0),\dots,j(l)},Y)$ or $C^{0,r}_b(V^{[l]}_{j(0),\dots,j(l)},Y)$ respectively for each $l=1,2,\dots,k$ and each $j(1),\dots,j(l)\in\{1,\dots,m\}$, $j(i)\in\{1,\dots,m(i)\}$, $i=0,\dots,n$. In view of Corollary 18 and Lemma 32 functions $\bar{\Phi}^n f(x;e_{j(1)},\dots,e_{j(n)};t_1,\dots,t_n)$ or $\Upsilon^n f(x^{[n]})$ belong to the space $Lip(v,r)$ by the variables $(x,t_1,\dots,t_n)$ or $x^{[n]}\in U^{[n]}_{j(0),\dots,j(n)}$, where $v=(e_{j(n)};l_k)\in\mathbf{K}^{m+n}$, $e_j\in\mathbf{K}^m$, $l_k=(0,\dots,0,1,0,\dots,0)\in\mathbf{K}^n$ with 1 on the k-th place, or $v=(e_{j(0)},\dots,e_{j(n)};l_k)$ with $e_{j(i)}\in\mathbf{K}^{m(i)}$ respectively. By Corollary 18 each partial difference quotient either $\bar{\Phi}^n f(x;e_{j(1)},\dots,e_{j(n)};t_1,\dots,t_n)$ or $\Upsilon^n f(x^{[n]})|_{U^{[n]}_{j(0),\dots,j(n)}}$ belongs to the space $Lip(r)$ by $(x,t_1,\dots,t_n)$ or in addition is bounded uniformly lipschitzian on the domain either $V^{(n)}_{j(1),\dots,j(n)}$ or $V^{[n]}_{j(0),\dots,j(n)}$ respectively. In accordance with Lemma 21 this proves the theorem.

40. Theorem. *Let $f:\mathbf{K}^m\to\mathbf{K}^l$, $m\in\mathbf{N}$. Suppose also that the composition is such that either $f\circ u\in C^\infty(\mathbf{K},\mathbf{K})$ or $f\circ u\in C^\infty_b(\mathbf{K},\mathbf{K})$ or $C^{[\infty]}(\mathbf{K},\mathbf{K})$ or $C^{[\infty]}_b(\mathbf{K},\mathbf{K})$ for each function $u\in C^\infty(\mathbf{K},\mathbf{K}^m)$ or $u\in C^\infty_b(\mathbf{K},\mathbf{K}^m)$ or $C^{[\infty]}(\mathbf{K},\mathbf{K}^m)$ or $C^{[\infty]}_b(\mathbf{K},\mathbf{K}^m)$, then $f\in C^\infty(\mathbf{K}^m,\mathbf{K}^l)$ or $f\in C^\infty_b(\mathbf{K}^m,\mathbf{K}^l)$ or $C^{[\infty]}(\mathbf{K}^m,\mathbf{K}^l)$ or $C^{[\infty]}_b(\mathbf{K}^m,\mathbf{K}^l)$ respectively.*

Proof. This follows after the application of either Theorem 39 for each natural number $s\in\mathbf{N}$ and $r=0$ or Theorem 38 for each natural number $s\in\mathbf{N}$ and the function $\phi(q)=q^r$ with the parameter $0<r<1$, since $C^{s,r}(U,Y)\subset C^{s+1}(U,Y)$ and $C^{s,r}_b(U,Y)\subset C^{s+1}_b(U,Y)$ and $C^\infty(U,Y):=\bigcap_{n=0}^\infty C^n(U,Y)$ and $C^\infty_b(U,Y):=\bigcap_{n=0}^\infty C^n_b(U,Y)$ and $C^{[\infty]}(U,Y):=\bigcap_{n=0}^\infty C^{[n]}(U,Y)$ and $C^{[\infty]}_b(U,Y):=\bigcap_{n=0}^\infty C^{[n]}_b(U,Y)$.

41. Theorem. *Let $h_j(y)$ be $C^\infty(\mathbf{K},\mathbf{K})$ functions such that*

$$h_j(0)=0 \textit{ for each } j=0,1,\dots,m \tag{1}$$

$$\lim_{0\neq y\to 0} h_{j-1}(y)/h_j(y)^n=0 \tag{2}$$

for each $n\in\mathbf{N}$ and $j=1,\dots,m$,

$$\lim_{0\neq y\to 0} h_m(y)/y^n=0 \tag{3}$$

for every $n\in\mathbf{N}$. Put $h(y)=(h_1(y),\dots,h_m(y))$ and suppose that $g\in C^\infty(\mathbf{K}^b,\mathbf{K})$ is not identically zero and $g(x)=0$ for each $|x|>1$. Define a function $f:\mathbf{K}^{m+1}\to\mathbf{K}$ by the formula:

$$f(x,y)=g((x-h(y))/h_0(y)) \tag{4}$$

for each $y \neq 0$ *and* $x \in \mathbf{K}^m$, $f(x,0) = 0$ *for each* x. *Then* $f \circ u \in C^\infty(\mathbf{K}^m, \mathbf{K})$ *for each locally analytic function* $u : \mathbf{K}^m \to \mathbf{K}^{m+1}$, *but* f *is discontinuous.*

Proof. First demonstrate that f is not continuous at zero $(0,0)$. We have $f(x,0) = 0$. But take a sequence (x_n, y_n) such that the limit

$$\lim_{n \to \infty} (x_n, y_n) = 0$$

is zero with

$$\varepsilon \leq |(x_n - h(y_n))/h_0(y_n)| \leq 1$$

and $|g(z_n)| \geq \delta$ with $z_n := (x_n - h(y_n))/h_0(y_n)$, which is possible since

$$\lim_{0 \neq y \to 0} |h(y)/h_0(y)| = \infty$$

and the function g is continuous and non zero, where $\varepsilon > 0$ and $\delta > 0$ are positive constants. For this sequence the inequality $|f(x_n, y_n)| \geq \delta$ is fulfilled for each natural number n. But for the sequence (x_n, y_n) such that

$$|(x_n - h(y_n))/h_0(y_n)| > 1$$

we have $f(x_n, y_n) = 0$, since $g(z_n) = 0$ for $|z_n| > 1$. Thus f is discontinuous at $(0,0)$.

Now we take a locally analytic function $u : \mathbf{K}^m \to \mathbf{K}^{m+1}$ and consider the composition $f \circ u$. Take a non-trivial analytic function $w(x,y)$ which maps from a neighborhood of zero in $\mathbf{K}^{m+1}$ into $\mathbf{K}$ such that $w \circ u(y) = 0$ in a neighborhood of $y_0 \in \mathbf{K}^m$, where $u(y_0) = 0$. Prove that for functions $h_j(y)$ satisfying Conditions $(1-3)$ positive constants $C > 0$ and $\delta > 0$ exist such that

$$|w(x,y)| \geq C|h_0(y)|^n, \tag{5}$$

when $|x - h(y)| \leq |h_0(y)|$, $0 < |y| < \delta$. If prove (5), then from $(x,y) = u(t)$ for t in a neighborhood of y_0 it follows, that $w(x,y) = 0$ and by (5) we have that $|x - h(y)| > |h_0(y)|$, hence $f(x,y) = 0$.

Consider an analytic function q which maps from a neighborhood of zero in the topological vector space $\mathbf{K}^l$ into the field $\mathbf{K}$. Then we can write it in the form:

$$q(x) = x_1^k s(x_2, \dots, x_l) + x_1^{k+1} r(x), \tag{6}$$

where s and r are analytic functions and s is not identically zero, $1 \leq l \leq m$, $x_1, \dots, x_l \in \mathbf{K}$. Then the estimate

$$|q(h(y))| \geq |h_1(y)|^k (|s(h_2(y), \dots, h_l(y))| - C|h_1(y)|)$$
$$\geq |h_1(y)|^k (C|h_2(y)|^n - C|h_1(y)|) \geq C|\pi h_1(y)|^{k+n}$$

is valid for each y with $0 < |y| < \delta$ with suitable $\delta > 0$ and the natural number $n \in \mathbf{N}$, where $\pi \in \mathbf{K}$, $0 < |\pi| < 1$. The induction by l gives from (6) that for an arbitrary non-trivial analytic function q which maps from a neighborhood of zero in the topological vector space $\mathbf{K}^m$ into the field $\mathbf{K}$ positive constants $C > 0$ and $\delta > 0$ and a natural number $n \in \mathbf{N}$ exist such that the estimate

$$|q(h(y))| \geq C|h_1(y)|^n \tag{7}$$

is fulfilled for each $0 < |y| < \delta$. In particular, from (7) it follows, positive constants $C > 0$, $\delta > 0$ and a natural number $n \in \mathbf{N}$ exist such that

$$|w(h(y),y)| \geq C|h_1(y)|^n \text{ for each } 0 < |y| < \delta. \quad (8)$$

It remains to show that (8) implies (5). Take any positive constant $C > 0$ so large that the inequality

$$|grad_x w(x,y)| \leq C$$

is satisfied in some neighborhood of zero and assume that $|x - h(y)| \leq |h_0(y)|$. Then a positive constant $\delta > 0$ exist such that the inequalities

$$|w(x,y)| \geq |w(h(y),y)| - C|x - h(y)| \geq C|h_1(y)|^n - C|h_0(y)| \geq C|\pi h_1(y)|^n$$

are valid for each $0 < |y| < \delta$. Thus $f \circ u = 0$ in a neighborhood of each point $y_0 \in \mathbf{K}^m$ such that $u(y_0) = 0$.

For example, we can take either the functions

$$h_j(y) = \sum_n a_n \pi^{n^2(m-j+1)+n}$$

for each $y = \sum_n a_n \pi^n \in \mathbf{K}$, where constants $a_n \in \mathbf{K}$ belong to the finite set of representatives of distinct classes in the finite residue class field $B(\mathbf{K},0,1)/B(\mathbf{K},0,|\pi|)$, $\pi \in \mathbf{K}$, $0 < |\pi| < 1$, $\mathbf{K}$ is a locally compact field of zero characteristic and $|\pi|$ is the largest generator among those less than one of the normalization group $\Gamma_{\mathbf{K}}$ of $\mathbf{K}$. Or we take the functions

$$h_j(y) = \sum_n a_n \theta^{n^2(m-j+1)+n}$$

for each

$$y = \sum_n a_n \theta^n \in \mathbf{F}_{p^k}(\theta),$$

where $a_n \in \mathbf{F}_{p^k}$ is a constant for each n, p is a prime number, $k \in \mathbf{N}$ is a natural number, $\mathbf{F}_{p^k}(\theta)$ is a locally compact field of the positive characteristic $char(\mathbf{F}_{p^k}(\theta)) = p > 0$, $\mathbf{F}_{p^k}$ denotes the finite field consisting of p^k elements.

42. Theorem. *A discontinuous function $f : \mathbf{K}^m \to \mathbf{K}$ exists such that $f \circ u \in C^\infty(\mathbf{K}^{m-1}, \mathbf{K})$ for each locally analytic function $u : \mathbf{K}^{m-1} \to \mathbf{K}^b$, where $m \geq 2$.*

Proof. This theorem follows from Theorem 41. Another its proof is the following. Let $f \in C^\infty(\mathbf{K}^2 \setminus \{0\}, \mathbf{K})$ and let the function f be non constant with $f(x_1,x_2) = 0$, when $x_1 x_2 = 0$. For simplicity let $\mathbf{K}$ be a locally compact field of zero characteristic. Take an analytic function $g : \mathbf{K} \to \mathbf{K}$ such that

$$\lim_{|y| \to \infty} g(y) = 0.$$

Such functions exist due to Example 43.1 of Section 43 in [103]. Moreover, they can be chosen such that

$$|g(y)| \leq \varepsilon_j \quad \text{for} \quad |y| = |\pi|^{-j}$$

for each $j = 0, 1, 2, \ldots$ and a sequence $\{\varepsilon_j > 0 : j\}$, which in particular may also tend to zero. Then consider the function $g(1/x_2)$ and put $h(x_1, x_2) := f(x_1, g(1/x_2))$, where the function f is taken homogeneous of degree zero.

Since $f \in C^\infty(\mathbf{K}^2 \setminus \{0\}, \mathbf{K})$, it remains to show that $f \circ u \in C^\infty$ in a neighborhood of zero $y = 0$, if $u(0) = (0,0)$. If u_1 coincides with zero, then h is identically zero. If $u_1(0) = 0$ and u_1 is not identically zero, then due to analyticity a natural number $k \in \{1, 2, \ldots\}$ exists such that $u_1(t) = t^k v_1(t)$ and the function v_1 is locally analytic and $v_1(0) \neq 0$. From $u_2(0) = 0$ it follows that $g(1/u_2(t)) = t^k v_2(t)$, where the function v_2 is locally analytic and $v_2(0) = 0$. We can take, for example, $\varepsilon_j = |\pi|^{j^2}$. Since $f(x_1, 0) = 0$ and f is homogeneous of degree zero, the identities

$$h(u_1(t), u_2(t)) = f(t^k v_1(t), t^k v_2(t)) = f(v_1(t), v_2(t))$$

are accomplished for each number $t \in \mathbf{K}$. Since $v_1(0) \neq 0$, the composition $f \circ u \in C^\infty$ is infinite differentiable in a neighborhood of zero.

43. Remark. Over the non-archimedean field the exponential function has the finite radius of convergence on $\mathbf{K}$. Therefore, in the proof of Theorem 40 the specific feature of the non-archimedean analysis was used. It was in the utilization of the analytic functions for which an analog of the Louiville theorem is not true (see also [103]).

Using the particular variant of Theorem 38 with $s = r = 0$ it is easy to prove the following theorem.

Theorem. *Let $f : \mathbf{K}^m \to \mathbf{K}^l$, $f \circ u \in C^n(\mathbf{K}^2, \mathbf{K}^l)$ or $f \circ u \in C^n_b(\mathbf{K}^2, \mathbf{K}^l)$ or $C^{[n]}(\mathbf{K}^2, \mathbf{K}^l)$ or $C^{[n]}_b(\mathbf{K}^2, \mathbf{K}^l)$ for each $u \in C^\infty(\mathbf{K}^2, \mathbf{K}^m)$ or $u \in C^\infty_b(\mathbf{K}^2, \mathbf{K}^m)$ or $C^{[\infty]}(\mathbf{K}^2, \mathbf{K}^m)$ or $C^{[\infty]}_b(\mathbf{K}^2, \mathbf{K}^m)$, where $m \geq 2$ and $n \geq 1$. Then $f \in C^n(\mathbf{K}^m, \mathbf{K}^l)$ or $f \in C^n_b(\mathbf{K}^m, \mathbf{K}^l)$ or $C^{[n]}(\mathbf{K}^m, \mathbf{K}^l)$ or $C^{[n]}_b(\mathbf{K}^m, \mathbf{K}^l)$ correspondingly.*

Proof. To demonstrate this theorem we pu

$$u(y) = \sum_{j=1}^{m} y_1^j e_j + w(y_2),$$

where $y = (y_1, y_2) \in \mathbf{K}^2$, $e_j \in \mathbf{K}^m$, $w \in C^\infty(\mathbf{K}, \mathbf{K}^m)$ or $C^\infty_b(\mathbf{K}, \mathbf{K}^m)$ or $C^{[\infty]}(\mathbf{K}, \mathbf{K}^m)$ or $C^{[\infty]}_b(\mathbf{K}, \mathbf{K}^m)$. Therefore, $u \in C^\infty(\mathbf{K}^2, \mathbf{K}^m)$ or $C^\infty_b(\mathbf{K}^2, \mathbf{K}^m)$ or $C^{[\infty]}(\mathbf{K}^2, \mathbf{K}^m)$ or $C^{[\infty]}_b(\mathbf{K}^2, \mathbf{K}^m)$ correspondingly. In view of Formula 10(1) or 9(1)̲ and Lemmas 11, 12 or Corollary 14 for the partial difference quotient either $\Phi^k f \circ u(y^{(k)})$ or $\Upsilon^n f \circ u$ by induction we get that each partial difference quotient either $\bar{\Phi}^k f(w(y_2), e_{j(1)}, \ldots, e_{j(k)}; t_1, \ldots, t_k)$ or $\Upsilon^n f(x^{[n]})|_{V^{[n]}_{j(0),\ldots,j(n)}}$ with $x = w(y_2)$ is continuous or uniformly continuous. Therefore, from Theorem 39̲ with the parameters $s = r = 0$ it follows, that each partial difference quotient either $\bar{\Phi}^k f(x; e_{j(1)}, \ldots, e_{j(k)}; t_1, \ldots, t_k)$ or $\Upsilon^n f(x^{[k]})$ is continuous on $U^{(k)}_{j(1),\ldots,j(k)}$ or $U^{[n]}_{j(0),\ldots,j(n)}$ or uniformly continuous on $V^{(k)}_{j(1),\ldots,j(k)}$ or $V^{[k]}_{j(0),\ldots,j(k)}$ respectively for each $k = 1, \ldots, n$. Therefore, by Lemma 21 $f \in C^n(\mathbf{K}^m, \mathbf{K}^l)$ or $f \in C^n_b(\mathbf{K}^m, \mathbf{K}^l)$ or $C^{[n]}(\mathbf{K}^m, \mathbf{K}^l)$ or $C^{[n]}_b(\mathbf{K}^m, \mathbf{K}^l)$ correspondingly.

6.2.3. Approximate Differentiability of Functions

The results of this section can be used for the random functions as well.

1. Definition and Notations. Henceforth, in this section let $\mathbf{K}$ be an infinite locally compact field with a non-trivial non-archimedean multiplicative norm and $\mathcal{B}(\mathbf{K}) = \mathcal{B}$ be a Borel σ-algebra of subsets of $\mathbf{K}$. A σ-additive σ-finite measure $\mu : \mathcal{B} \to [0,\infty]$ is called the Haar measure, if μ is non zero and $\mu(x+A) = \mu(A)$ for each $x \in \mathbf{K}$ and $A \in \mathcal{B}$. For convenience it is useful to put $\mu(B(\mathbf{K},0,1)) = 1$ and choose an equivalent norm $|x| = |x|_{\mathbf{K}} = \operatorname{mod}_{\mathbf{K}}(x)$ in the field $\mathbf{K}$, where $\operatorname{mod}_{\mathbf{K}}(x)$ is the modular function such that $\mu(xB(\mathbf{K},0,R)) = \operatorname{mod}_{\mathbf{K}}(x)\mu(B(\mathbf{K},0,R))$ for each $x \in \mathbf{K}$, where R belongs to the normalization group

$$\Gamma_{\mathbf{K}} := \{|x| : 0 \neq x \in \mathbf{K}\}$$

of the field $\mathbf{K}$, $A_1A_2 := \{a : a = a_1a_2, a_1 \in A_1, a_2 \in A_2\}$ for $A_1, A_2 \subset \mathbf{K}$, $B(\mathbf{K}^m, y, R) := \{z \in \mathbf{K}^m : |z-y| \leq R\}$, $m \in \mathbf{N}$. For the topological vector space $\mathbf{K}^m$ take the measure $\mu^m = \otimes_{j=1}^m \mu_j$, where $\mu_j = \mu =: \mu^1$ for each j, such that $\mu^m(A_1 \times \cdots \times A_m) = \prod_{j=1}^m \mu(A_j)$ for each $A_1, \dots, A_m \in \mathcal{B}(\mathbf{K})$, while $\otimes_{j=1}^m \mathcal{B}(\mathbf{K})$ denotes the minimal σ-algebra generated by subsets of the form $A_1 \times \cdots \times A_m$. Evidently, the Borel σ-algebra $\mathcal{B}(\mathbf{K}^m)$ of $\mathbf{K}^m$ coincides with $\otimes_{j=1}^m \mathcal{B}(\mathbf{K})$.

2. Remark. Suppose that (X, ρ_X) is a metric space with a set X and a metric ρ_X in it. A non negative measure ν on a σ-algebra $\mathcal{C}$ of X is called Borel regular if and only if $\mathcal{B}(X) \subset \mathcal{C}$ and for each $A \in \mathcal{C}$ an element $H \in \mathcal{B}(X)$ exists so that $\nu(H) = \nu(A)$, where $\mathcal{B}(X)$ denotes the Borel σ-algebra of X which is the minimal σ-algebra generated by open subsets of X. Denote by $\mathcal{M}$ the class of all Borel regular non negative measures ν on X such that each bounded subset A in X has a finite measure $0 \leq \nu(A) < \infty$.

Consider the family $\mathcal{N}$ of all subsets A of X for which an element $G = G(A) \in \mathcal{B}(X)$ exists so that $A \subset G$ and $\nu(G) = 0$. The minimal σ-algebra $\mathcal{A}_\nu = \mathcal{A}_\nu(X)$ generated by $\mathcal{B}(X) \cup \mathcal{N}$ is the ν-completion of $\mathcal{B}(X)$ and it consists of all ν-measurable subsets.

A subset of the form $\{(y,A) : y \in A \subset X\}$ is called a covering relation. If Y is a subset of X, then it is useful to put $V(Y) := \{A :$ there exists $y \in Y, (y,A) \in V\}$, where V is a covering relation in X. Then V is called fine at a point y if and only if $\inf_{(y,A)\in V} diam(A) = 0$, where with each $A \subset X$ is associated its diameter

$$diam(A) := \sup_{x,y \in A} \rho_X(x,y).$$

Consider a covering relation V in X satisfying three conditions:

$(V1)$ $V(X)$ is a family of Borel subsets of X,

$(V2)$ V is fine at each point of X,

$(V3)$ if $C \subset V$ and $Y \subset X$ and C is fine at each point of Y, then $C(Y)$ has a countable disjoint subfamily covering ν almost all of Y.

If for a given measure $\nu \in \mathcal{M}$ a covering relation V satisfies Conditions $(V1-V3)$, then it is called a ν Vitali relation.

Suppose that $\nu \in \mathcal{M}$ and and V is a ν Vitali relation. With each $\lambda \in \mathcal{M}$ another measure λ_ν is associated. It is defined by the formula:

$$\lambda_\nu(A) := \inf\{\lambda(S) : S \in \mathcal{B}(X), \nu(A \setminus S) = 0\},$$

whenever $A \subset X$, hence $\lambda_\nu \leq \lambda$. If a covering relation V is fine at a point $x \in X$ and $f : dom(f) \to [-\infty, \infty]$, where $dom(f) \subset \mathcal{B}(X)$ is the domain of f, then

$$(V) \lim_{S \to x} f(S) := \lim_{0<\varepsilon\to 0} \{f(S) : (x,S) \in V, diam(S) < \varepsilon, S \in dom(f)\},$$

analogously the limits

$(V)\lim_{S\to x} \sup f(S) := \lim_{0<\varepsilon\to 0} \sup\{f(S) : (x,S) \in V, diam(S) < \varepsilon, S \in dom(f)\}$ and $(V)\lim_{S\to x} \inf f$ are defined.

For a subset A in X and a point $x \in X$ the limit $(V)\lim_{S\to x} \nu(S\cap A)/\nu(S)$ is called the (ν, V) density of A at x.

If $g : X \to Y$ is a mapping of a metric space (X, ρ_X) into a (Hausdorff) topological space Y, then $y \in Y$ is called an approximate limit of g at x (relative to a measure $\nu \in \mathcal{M}(X)$ and a ν Vitaly relation V) if and only of for each neighborhood W of y in Y the set $X \setminus g^{-1}(W)$ has zero density at x and it is denoted by $y = (\nu, V)ap\lim_{z\to x} g(z)$. If (ν, V) are specified, then they may be omitted for brevity.

A function g is called (ν, V) approximately continuous if and only if $x \in dom(g)$ and $(\nu, V)ap\lim_{z\to x} g(z) = g(x)$.

3. Definitions. Let Y be a topological vector space over the field $\mathbf{K}$, $g : U \to Y$ be a mapping, where U is an open subset in the topological vector space $\mathbf{K}^m$, $m \in \mathbf{N}$ is a natural number. Then the function g is called approximate differentiable at a point x of U if an open neighborhood W of the point x exists, $W \subset U$, such that the partial difference quotient $\bar{\Phi}^1 g(x; v; t)$ is μ^{2m+1} almost everywhere continuous on the domain $W^{(1)}$ and at x a linear mapping $T : \mathbf{K}^m \to Y$ exists such that

$$(\mu^m, V)ap\lim_{z\to x} |g(z) - g(x) - T(z-x)|/|z-x| = 0,$$

where $V = \mathcal{B}(\mathbf{K}^m)$. This T is also denoted by ${}_{ap}Dg(x)$. If this is satisfied for each point $z \in U$, then the function g is called approximate differentiable on the domain U. The family of all such functions we denote by ${}_{ap}C^1(U,Y)$. Then also we define approximate partial derivatives:

$${}_{ap}D_j g(x) := ap\lim_{t\to 0} [g(x_1, \dots, x_{j-1}, x_j + t, x_{j+1}, \dots, x_m) - g(x)]/t.$$

The family of all functions $f \in C^n(U,Y)$ such that $\bar{\Phi}^n f \in {}_{ap}C^1(U^{(n)}, Y)$ we denote by ${}_{ap}C^{n+1}(U,Y)$.

Suppose now that A is a μ^m measurable subset of the domain U and Y is a normed space, $0 < r \leq 1$. Then ${}_{ap}C^{n,r}(U,A,Y)$ denotes the family of all functions $f \in C^n(U,Y)$ with

$$ap\overline{\lim}_{x^{(n)}\to z^{(n)}} \|\bar{\Phi}^k f(x^{(n)}) - \bar{\Phi}^n f(z^{(n)})\|_{C^0(V_{z,R}^{(n)},Y)}/|x^{(n)} - z^{(n)}|^r < \infty$$

for μ^m almost all $z \in A$ and each $0 < R < \infty$, where the domain $V_{z,R}^{(k)}$ corresponds to $U_{z,R} = U \cap B(\mathbf{K}^m, z, R)$.

4. Lemma. *The families* ${}_{ap}C^{n+1}(U,Y)$ *and* ${}_{ap}C^{n,r}(U,A,Y)$ *are the* $\mathbf{K}$*-linear spaces.*

Proof. 1. Since $C^n(U,Y)$ is the $\mathbf{K}$ linear space, it is sufficient to verify, that the set of all partial difference quotients $\bar{\Phi}^n f$ with $f \in {}_{ap}C^{n+1}(U,Y)$ is $\mathbf{K}$ linear. Therefore, the consideration reduces to the space ${}_{ap}C^1(U,Y)$, where the domain $U^{(n)}$ is denoted for brevity also by U. If $f,g \in {}_{ap}C^1(U,Y)$ and $a,b \in \mathbf{K}$, then

$$\bar{\Phi}^1(af+bg)(x;v;t) = a\bar{\Phi}^1 f(x;v;t) + b\bar{\Phi}^1 g(x;v;t).$$

Since for μ^{2m+1}-almost all points $(x;v;t) \in \mathbf{K}^{2m+1}$ the right side terms of the latter equality are continuous, the left side term is such also. If $x \in U$, then the inequality

$$ap\lim_{z\to x}|(af+bg)(z)-(af+bg)(x)-a\;{}_{ap}Df(x).(z-x)-b\;{}_{ap}Dg(x).(z-x)|/|z-x|$$

$$\le ap\lim_{z\to x}|a||f(z)-f(x)-\;{}_{ap}Df(x).(z-x)|/|z-x|$$

$$+ap\lim_{z\to x}|b||g(z)-g(x)-\;{}_{ap}Dg(x).(z-x)|/|z-x| = 0$$

is satisfied, hence there exists ${}_{ap}D(af+bg)(x) = a\;{}_{ap}Df(x) + b\;{}_{ap}Dg(x)$.

For each function $f \in {}_{ap}C^1(U,Y)$ the operator $T = {}_{ap}Df(x)$ is unique, since the difference $H = T_1 - T_2$ of two such $\mathbf{K}$-linear mappings is subordinated to the condition:

$$ap\lim_{v\to 0}|Hv|/|v| = ap\lim_{z\to a}|H(z-x)|/|z-x| = 0$$

due to Definition 3. Therefore, if $0 < \varepsilon < 1$, then there exists $R > 0$, $R \in \Gamma_{\mathbf{K}}$, such that

$$\mu^m(B(\mathbf{K}^m,0,R)\cap\{v\in\mathbf{K}^m : |Hv| > \varepsilon|v|\}) < \varepsilon^m R^m.$$

If $w \in B(\mathbf{K}^m,0,R)$ and $v \in B(\mathbf{K}^m,w,\varepsilon R)$ with $|Hv| \le \varepsilon|v|$, then $|Hw| \le \max(|H(w-v)|,|Hv|) \le \varepsilon R\max(\|H\|,1)$. Therefore, $\|H\| \le \varepsilon\max(1,\|H\|)$, consequently, $\|H\| = 0$, since $\varepsilon > 0$ can be chosen arbitrary small. At the same time the partial difference quotient $\bar{\Phi}^1 f(x;v;t)$ is unique on the domain $U^{(1)}$ up to a set of μ^{2m+1}-measure zero.

2. The second assertion follows from the inequality

$$ap\overline{\lim}_{x^{(n)}\to z^{(n)}}\|\bar{\Phi}^k(af+bg)(x^{(n)})-\bar{\Phi}^n(af+bg)(z^{(n)})\|_{C^0(V^{(n)}_{z,R},Y)}/|x^{(n)}-z^{(n)}|^r$$

$$\le \max(ap\overline{\lim}_{x^{(n)}\to z^{(n)}}|a|\|\bar{\Phi}^k f(x^{(n)})-\bar{\Phi}^n f(z^{(n)})\|_{C^0(V^{(n)}_{z,R},Y)}/|x^{(n)}-z^{(n)}|^r$$

and from

$$|b|\;ap\overline{\lim}_{x^{(n)}\to z^{(n)}}\|\bar{\Phi}^k g(x^{(n)})-\bar{\Phi}^n g(z^{(n)})\|_{C^0(V^{(n)}_{z,R},Y)}/|x^{(n)}-z^{(n)}|^r < \infty$$

for each numbers $a,b \in \mathbf{K}$ and functions $f,g \in {}_{ap}C^{n,r}(U,A,Y)$.

5. Note. For a locally compact field $\mathbf{K}$ each number $x \in \mathbf{K}$ has the decomposition $x = \sum_n a_n\pi^n$ for $char(\mathbf{K}) = 0$ and $x = \sum_n a_n\theta^n$ for $char(\mathbf{K}) = p > 0$ with $\mathbf{K} = \mathbf{F_p}(\theta)$ (see [99, 111]), where $a_n = a_n(x)$ are expansion coefficients. Here $\pi \in \mathbf{K}$ is a number so that

$$|\pi| = \sup\{|x| : x \in \mathbf{K},\ |x| < 1\}.$$

Introduce on the field $\mathbf{K}$ the linear ordering: $x \prec y$ if and only if an integer number $m \in \mathbf{Z}$ exists such that $a_n(x) = a_n(y)$ for each $n < m$ and $a_m(x) < a_m(y)$ with the natural ordering of these expansion coefficients either in the residue class field $B(\mathbf{K},0,1)/[\pi B(\mathbf{K},0,1)]$ or in the finite field $\mathbf{F}_{p^k}$ respectively. In the contrary case we write $x = y$. This linear ordering is compatible with neither the additive nor the multiplicative structure of $\mathbf{K}$, but it is useful and it was introduced by M. van der Put [103]. Mention that $B(\mathbf{K},0,1)/B(\mathbf{K},0,|\pi|) = B(\mathbf{K},0,1)/[\pi B(\mathbf{K},0,1)]$.

In the vector space $\mathbf{K}^m$ we can consider the linear ordering: $x \prec y$ if and only if a natural number $l \in \mathbf{N}$ exists such that ${}_jx = {}_jy$ for each $1 \le j < l$ and ${}_lx \prec {}_ly$, where $x = ({}_1x, \dots, {}_mx) \in \mathbf{K}^m$, ${}_ix \in \mathbf{K}$ for each $i = 1, \dots, m$.

6. Theorem. *Let ν be a measure on the topological vector space $\mathbf{K}^m$, $\nu \in \mathcal{M}(\mathbf{K}^m)$, let also a function $f : \mathbf{K}^m \to \mathbf{K}$ be ν-measurable and numbers $y_1, y_2, y_3, \dots \in \mathbf{K}$ are pairwise distinct and such that for each $y \in \mathbf{K}$ and every $\varepsilon > 0$ natural numbers $n, k \in \mathbf{N}$ exist such that $|y_n - y| < \varepsilon$ and $y_n \preceq y$ and $|y_k - y| < \varepsilon$ and $y \preceq y_k$. Then there exist ν-measurable subsets $A_1, A_2, A_3, \dots$ in $\mathbf{K}^m$ with characteristic functions $g_j = ch_{A_j}$ such that*

$$f(x) = \sum_{n=1}^{\infty} y_n g_n(x) \quad \textit{for each} \quad x \in \mathbf{K}^m.$$

Proof. If $|y_l - y_j| \le \delta$, where $0 < \delta$, then $\max(|y - y_l|, |y - y_j|) \le \delta$ for each $y_l \preceq y \preceq y_j$ due to the definition of Section 5. In the space $L^\infty(\mathbf{K}^m, \nu, \mathbf{K})$ the $\mathbf{K}$-linear span of characteristic functions of clopen subsets is dense. Consider the set $W_n := \{x \in \mathbf{K} : |f(x)| \le n\}$, where $n \in \mathbf{N}$ is a natural number, then $f|_{W_n} \in L^\infty(\mathbf{K}^m, \nu, \mathbf{K})$. Therefore, the restriction of the function f has the decomposition

$$f(x)|_{W_n \setminus W_{n-1}} = \sum_{k=1}^{\infty} s_{n,k} g_{n,k}(x),$$

where each $g_{n,k}$ is the characteristic function of a ν-measurable subset in the topological vector space $\mathbf{K}^m$, $s_{n,k} \in \mathbf{K}$, $W_0 := \emptyset$. Then the series

$$f(x) = \sum_{k,n=1}^{\infty} s_{n,k} g_{n,k}(x)$$

converges point-wise for each $x \in \mathbf{K}^m$.

Therefore, it is enough to construct the decomposition of the restriction $f|_{W_q \setminus W_{q-1}}$ for arbitrary q. Thus consider the subset of y_n with $|y_n| \le q$. For each $\varepsilon > 0$ there exists a finite ε-net $\{y_{l(s)} : s = 1, \dots, b\}$, $b = b(\varepsilon)$, that is for each number $y \in \mathbf{K}$ there exists s, $1 \le s \le b$, such that $|y - y_{l(s)}| < \varepsilon$.

Take the sequences $\varepsilon_j = |\pi|^j$ and $b_j = b(\varepsilon_j)$. If y_n is the last point of the $|\pi|^j$ net, then new $|\pi|^{j+1}$-net begins and put $u = 1$. If it is not so, then take $u = 0$. Suppose that $\|f - \sum_{j=1}^{n-1} y_j g_j\|_{L^\infty(A_n, \nu, \mathbf{K})} > 0$. Otherwise the decomposition is already found. Put by induction

$$A_n = A_{n,n+l} := \left\{x : y_n \preceq f(x) - \sum_{j=1}^{n-1} y_j g_j(x) \prec y_{n+l}\right\}$$

if $y_n \preceq y_{n-k}$, where $l \ge 1$ is the minimal natural number for which $y_n \prec y_{n+l}$ and

$$|\pi|^{u+1} \|f(x) - \sum_{j=1}^{n-1} y_j g_j(x)\|_{L^\infty(A_n, \nu, \mathbf{K})} < |y_n - y_{n+l}|$$

$$\leq |\pi|^u \Big\| f(x) - \sum_{j=1}^{n-1} y_j g_j(x) \Big\|_{L^\infty(A_n,\nu,\mathbf{K})};$$

$$A_n = A_{n,n+l} := \Big\{ x : y_{n+l} \preceq f(x) - \sum_{j=1}^{n-1} y_j g_j(x) \prec y_n \Big\}$$

if $y_{n-k} \preceq y_n$, where $l \geq 1$ is the minimal natural number for which $y_{n+l} \prec y_n$ and

$$|\pi|^{u+1} \| f(x) - \sum_{j=1}^{n-1} y_j g_j(x) \|_{L^\infty(A_n,\nu,\mathbf{K})} < |y_n - y_{n+l}|$$

$$\leq |\pi|^u \| f(x) - \sum_{j=1}^{n-1} y_j g_j(x) \|_{L^\infty(A_n,\nu,\mathbf{K})},$$

where $k = k(n)$ is a gap on the preceding step as l on this step. In accordance with this algorithm some set A_q may be empty, when $n < q < n+l$ for subsequent numbers of the algorithm, so that $g_q = 0$ for such q.

Then consider the sum $(\sum_{j=1}^{n-1} y_j g_j(x)) + y_n g_n$. Since $|y_n - y_{n+l}| \leq |\pi|^u |y_n - y_{n-k}|$ for each n and each $|\pi|^j$-net is finite, the series $\sum_{j=1}^{\infty} y_j g_j$ converges. In view of the inequalities above it converges to $f|_{W_q \setminus W_{q-1}}$ by the norm of $L^\infty(\mathbf{K}^m, \nu, \mathbf{K})$ for each q. Denote A_j for $f|_{W_q \setminus W_{q-1}}$ by ${}_qA_j$. Then the countable union

$$\bigcup_{q=1}^{\infty} {}_qA_j =: A_j$$

is ν-measurable for each j. Since for each $x \in \mathbf{K}^m$ a natural number q exists so that $x \in W_q \setminus W_{q-1}$, the series of functions

$$\sum_{j=1}^{\infty} y_j g_j(x) = f(x)$$

converges point-wise.

7. Lemma. *Suppose that H is a compact non void subset in the topological space $\mathbf{K}^n \times (\mathbf{K} \setminus \{0\})$ and*

$$X_t := \{ y \in \mathbf{K}^n : |y - z| \leq |x|^r t \text{ for every } (z,x) \in H \}$$

for some $0 \leq t < \infty$, where $0 < r \leq 1$ is a constant, then

$$c := \inf\{ t : X_t \neq \emptyset \} < \infty \text{ and } X_c = \Big(\bigcap_{t>c} X_t \Big) \neq \emptyset.$$

If $q \in X_c$, then $A_q \neq \emptyset$, where $A_q := \{ z : \text{ there exists } (z,x) \in H \text{ with } |q - z| = |x|^r c \}$.

Proof. Each set X_t is compact, since the subset H is compact. Consider the projection $\pi_{n+1} : \mathbf{K}^{n+1} \to \mathbf{K}$ such that $\pi_{n+1}({}_1x, \dots, {}_{n+1}x) = {}_{n+1}x$, where ${}_jx \in \mathbf{K}$ for each

$j = 1, \dots, n+1$. Thus the subset $\pi_{n+1}(H)$ is compact as the continuous image of the compact set. But the set $\pi_{n+1}(H)$ is contained in $\mathbf{K} \setminus \{0\}$, consequently, $\inf_{(z,x)\in H} |x| > 0$. At the same time the supremum is finite $\sup_{(z,x)\in H} |z| < \infty$, hence

$$0 \le \sup\{|z|/|x|^r : (z,x) \in H\} < \infty.$$

Therefore, $0 \in X_t$ for each $t \ge \sup\{|z|/|x|^r : (z,x) \in H\}$. Then we get the formula $X_c = \bigcap_{c<t<\infty} X_t \ne \emptyset$, since $X_t \subset X_q$ for each $0 < t \le q$ and $X_t \ne \emptyset$ for each $t > c$. We put

$$R := \sup\{|x|^r : (z,x) \in H \text{ for some } z\}.$$

Consider points $y, z \in X_c$, a number $\alpha \in \mathbf{K}$ with $|\alpha| \le 1$, then the inclusion $(a,q) \in H$ implies the inequality

$$|\alpha y + (1-\alpha)z - a| \le \max(|\alpha||y-a|, |1-\alpha||z-a|),$$

but $\max(|\alpha|, |1-\alpha|) \le 1$, consequently,

$$|\alpha y + (1-\alpha)z - a| \le |q|^r c \text{ and } \alpha y + (1-\alpha)z \in X_c.$$

Subjecting $\mathbf{K}^n$ to a translation in a case of necessity we can suppose without loss of generality that $0 \in X_c$. If a vector $z \in \mathbf{K}^n$ is of the unit norm, $|z| = 1$, then $|\pi^s z| > 0$ for each integer number $s \in \mathbf{Z}$. We have $\pi^s z \in X_c$ if and only if $|\pi^s z - a| \le |x|^r c$ for each $(a,x) \in H$, but $|a| \le |x|^r c$, since $0 \in X_c$. Thus $\pi^s z \in X_c$ if and only if $|\pi^s z| \le |x|^r c$ for each $(z,x) \in H$, which is equivalent to $|\pi^s| \le |x|^r c$ for each $(a,x) \in H$ and in its turn this is equivalent to $|\pi^s| \le c(\inf_{(z,x)\in H} |x|)^r$. Then an integer number $s_0 \in \mathbf{Z}$ exists so that the inequality $|\pi^s| \le c(\inf_{(z,x)\in H} |x|)^r$ is accomplished for each $s \ge s_0$, $s \in \mathbf{Z}$. We have $H \cap \{(z,x) : |z| = |x|^r c\} \ne \emptyset$, hence $A_q \ne \emptyset$, where $q = 0$ after a suitable translation.

8. Theorem. *If S is a subset in the topological vector space $\mathbf{K}^m$ and $f : S \to \mathbf{K}^n$ is a lipschitzian function with constants $0 < Lip_1(f) := C < \infty$ and $0 < Lip_2(f) := r \le 1$:*

$$|f(x) - f(y) \le C|x-y|^r \text{ for each } x, y \in S \tag{1}$$

then f has a lipschitzian extension $g : \mathbf{K}^m \to \mathbf{K}$ such that $Lip_1(f) = Lip_1(g)$ and $Lip_2(f) = Lip_2(g)$, where $n, m \in \mathbf{N}$.

Proof. With the help of transformation $f \mapsto cf$, where $0 \ne c \in \mathbf{K}$ we can suppose that $0 < Lip_1(f) =: b \le 1$. Consider a class Ψ of all lipschitzian extensions f_j of the function f on some subset T_j of the topological vector space $\mathbf{K}^m$ having the same constants C, r. Then the class Ψ is partially ordered $(f_1, T_1) \preceq (f_2, T_2)$ if $T_1 \subset T_2$ and $f_1|_{T_1} = f_2|_{T_1}$. Each linearly ordered subset Φ in Ψ has a maximal element (h, T): $h|_{T_j} = f_j$, $(f_j, T_j) \preceq (h, T)$ for each $(f_j, T_j) \in \Phi$, where $T \supset \bigcup_j T_j$, since for each T_j, T_k there is the inequality $j \prec k$ if and only if $T_j \subset T_k$ and $f_j|_{T_j} = f_k|_{T_j}$. In view of the Kuratowski-Zorn lemma [19] a maximal element (g, T) in the well-ordered set Ψ exists, $g : T \to \mathbf{K}^n$, where $T \subset \mathbf{K}^m$. It is sufficient to show, that if there exists a point $z \in \mathbf{K}^m \setminus T$, then there exists a vector $y \in \mathbf{K}^n$ such that the inequality $|y - g(x)| \le b|z-x|^r$ is valid for every point $x \in T$, consequently, $g \cup \{(z,y)\} \in \Psi$ and (g, T) would not be maximal in Ψ. Thus we must prove, that

$$\bigcap_{x\in T} B(\mathbf{K}^n, g(x), b|x-z|^r) \ne \emptyset.$$

These balls are compact, hence it is sufficient to prove that

$$\bigcap_{x\in F} B(\mathbf{K}^n, g(x), b|x-z|^r) \neq \emptyset$$

for each finite subset F in T. Take the set X_c from Lemma 7 and a point $q \in X_c$, then $|q-g(x_i)| = |x_i - x|^r c$ for $i = 1,\dots,k$, $g(x_i) \in A_q$, where $1 \le k \in \mathbf{Z}$, $(q,x) \in H$.

We will show that $q \in X_b$. If $0 < c \le b$, then $q \in X_c \subset X_b$. If $c > b$, then $|q-g(x_i)| > |x_i - x|^r b$ for each x_i, that is impossible by the supposition of this theorem.

If x is a limit point in T, then we take a sequence $\{x_n : n\}$ such that

$$\lim_{n\to\infty} x_n = x \text{ and } q = \lim_{n\to\infty} g(x_n),$$

since g is continuous on T and $\{g(x_n) : n\}$ is the Cauchy net in the field $\mathbf{K}$, but the latter uniform space is complete, because $\mathbf{K}$ is a locally compact field. Then we infer the estimate

$$|q-g(y)| = \lim_{n\to\infty} |g(x_n) - g(y)| \le b \lim_{n\to\infty} |x_n - y|^r = b|x-y|^r$$

for each $y \in T$. Thus T is the closed subset in the topological vector space $\mathbf{K}^m$, hence the subset $\mathbf{K}^m \setminus T$ is open. Suppose that $v \in \mathbf{K}^m \setminus T$, then a positive number $\delta := \inf_{x\in T} |v-x| > 0$ exists. Take $\delta \le R < \infty$, then $T \cap B(\mathbf{K}, x_0, R)$ is compact, where $x_0 \in T$ is a marked point. Therefore, a point $v_0 \in T$ exists such that $|v - v_0| = \delta$.

We have that $g(T)$ is locally compact and closed in the topological vector space $\mathbf{K}^m$ and

$$|g(x) - g(y)| \le b|x-y|^r \quad \text{for each} \quad x, y \in T,$$

consequently, $B(\mathbf{K}^m, g(x), b|x-y|^r) = B(\mathbf{K}^m, g(y), b|x-y|^r)$ for each $x, y \in T$. If $|y-v| > |x-v|$, then $|x-y| = |y-v|$, consequently,

$$B(\mathbf{K}^m, g(y), b|y-v|^r) = B(\mathbf{K}^m, g(y), b|x-y|^r)$$

$$= B(\mathbf{K}^m, g(x), b|x-y|^r) \supset B(\mathbf{K}^m, g(x), b|x-v|^r).$$

Hence

$$\bigcap_{x\in T} B(\mathbf{K}^m, g(x), b|x-v|^r) \supset \bigcap_{x\in T, |x-v|=|v_0-v|} B(\mathbf{K}^m, g(x), b|x-v|^r).$$

On the other hand,

$$|g(v_0) - g(x)| \le b|v_0 - x|^r \le b\max(|x-v|^r, |v_0-v|^r) = b|v_0-v|^r$$

for $|x-v| = |v_0 - v|$. Therefore, the equality is valid $B(\mathbf{K}^m, g(x), b|x-v|^r) = B(\mathbf{K}^m, g(v_0), b|v_0 - v|^r)$ and inevitably the formula is satisfied

$$\bigcap_{x\in T} B(\mathbf{K}^m, g(x), b|x-v|^r) \supset B(\mathbf{K}^m, g(v_0), b|v-v_0|^r) \neq \emptyset,$$

since the norm of the field $\mathbf{K}$ is non trivial and $\lim_{k\to\infty} |\pi|^k = 0$.

9. Theorem. *Let U be an open subset in the field $\mathbf{K}$ and let also $g : U \to \mathbf{K}$ be a locally lipschitzian function such that for each $x_0 \in U$ positive constants $0 < C < \infty$ and $\delta > 0$ and $0 < r \leq 1$ exist with*

$$|g(x) - g(y)| \leq C|x-y|^r \text{ for each } \max(|x-x_0|, |y-x_0|) < \delta, \ x,y \in U. \qquad (1)$$

Then the partial difference quotient $\bar{\Phi}^1 g(x;v;t)$ is continuous for μ^3-almost all points in $U^{(1)}$ and $dg(x)/dx = \bar{\Phi}^1 g(x;1;0)$ exists and is continuous for μ-almost all points of U.

Proof. The function g is locally lipschitzian, hence it is continuous. On the other hand, the Haar measure μ^m on $\mathbf{K}^m$ is Radon and regular. Therefore, it satisfies the following three conditions:

if J is a compact subset of $\mathbf{K}$, then

$$\mu^m(J) < \infty; \qquad (2)$$

if V is open in $\mathbf{K}$, then

$$\mu^m(V) = \sup\{\mu^m(J) : J \text{ is compact }, J \subset V\}; \qquad (3)$$

if A is a μ^m-measurable subset, $A \subset \mathbf{K}$, then

$$\mu^m(A) := \inf\{\mu^m(V) : V \text{ is open }, A \subset V\}. \qquad (4)$$

In view of approximation Theorem 2.2.5 [25] for each μ^m-measurable subset A in $\mathbf{K}^m$ with $\mu^m(A) < \infty$ and $\varepsilon > 0$ a compact subset $J \subset A$ exists so that $\mu^m(A \setminus J) \leq \varepsilon$. If $\mathcal{E}$ is a subset of discontinuity in the domain $U^{(1)}$ of the partial difference quotient $\bar{\Phi}^1 g$ and $\mathcal{D}$ is a subset of discontinuity of the derivative $dg(x)/dx$ in U, then it is sufficient to demonstrate, that $\mu^3(\mathcal{E} \cap U^{(1)}_{R,\varepsilon}) = 0$ and $\mu(\mathcal{D} \cap U_{R,\varepsilon}) = 0$ for each $0 < R < \infty$ and $\varepsilon > 0$, where $U_{R,\varepsilon}$ is a subset in $U \cap B(\mathbf{K}, 0, R)$ such that $\mu(U \setminus U_{R,\varepsilon}) < \varepsilon$, since

$$\mu^3((U^{(1)} \setminus U^{(1)}_{R,\varepsilon}) \cap B(\mathbf{K}^3, 0, R)) < 3R^2\varepsilon + 3R\varepsilon^2 + \varepsilon^3.$$

For each compact subset $U_{R,\varepsilon}$ the covering $B(\mathbf{K}, x_0, \delta)$ with $\delta = \delta(x_0) > 0$ has a finite sub-covering, hence the real numbers exist:

$$C = \sup_{x_0 \in U_{R,\varepsilon}} C(x_0) < \infty \text{ and } 0 < r = \inf_{x_0 \in U_{R,\varepsilon}} r(x_0) \leq 1$$

for which inequality (1) is satisfied for each points $x, y \in U_{R,\varepsilon}$. Consider a restriction $g|_{U_{R,\varepsilon}}$ of the function, then by Theorem 8 it has a lipschitzian extension $g_{R,\varepsilon}$ on the field $\mathbf{K}$ with the same constants $0 < C < \infty$ and $0 < r \leq 1$. Therefore, it is sufficient to prove this theorem for a clopen compact subset U in the field $\mathbf{K}$ which is supposed in the proof below.

Since the mapping $x \mapsto x + vt$ is continuous by $(x, v, t) \in \mathbf{K}^3$ and μ^3 is the Haar measure on the topological vector space $\mathbf{K}^3$ such that μ^3 has not any atoms, the partial difference quotient $\bar{\Phi}^1 g(x;v;t) = [g(x+vt) - g(x)]/t$ is continuous for μ^3-almost all points in $U^{(1)}$ if and only if $[g(x) - g(y)]/[x-y]$ is continuous for μ^2-almost all points of $\mathbf{K}^2$.

Consider the following relation

$$V := \{(x, S) : S \text{ is a compact clopen subset in } \mathbf{K}^m, x \in S\},$$

where $m \in \mathbf{N}$. Verify that V is the μ^m Vitaly relation. Indeed,

(1) V is a covering relation, that is a subset of $\{(x,S) : x \in S \subset \mathbf{K}^m\}$;

(2) $V \subset \mathcal{B}(\mathbf{K}^m)$;

(3) for each $0 < R < \infty$ and each $y \in \mathbf{K}^m$ the ball $B(\mathbf{K}^m, y, R)$ belongs to V, hence $\inf\{diam(S) : (x,S) \in V\} = 0$ for each $x \in \mathbf{K}^m$, consequently, V is fine at each point of $\mathbf{K}^m$;

(4) $\mathbf{K}^m$ is locally compact separable and with a countable base of its topology consisting of clopen balls. Thus if $W \subset V$ and $Z \subset \mathbf{K}^m$ and W is fine at each point $z \in Z$, then $W(Z)$ has a countable disjoint subfamily covering almost all of Z. Indeed, $W(Z)$ gives a base of topology inherited from $\mathbf{K}^m$. This base is countable, hence $Z \subset \bigcup_{j=1}^{\infty} U_j$, where each U_j is a clopen compact subset in the topological vector space $\mathbf{K}^m$. Recall that a measure $\nu \in \mathcal{M}(X)$ is called regular, if for each $A \subset X$ there exists a ν-measurable subset G in X such that $A \subset G$ and $\nu(A) = \nu(G)$. With arbitrary measure ν one associates a regular measure by the formula

$$\lambda(A) := \inf\{\nu(G) : A \subset G \text{ and } G \text{ is } \nu\text{- measurable}\}$$

(see also Section 2.1.5 and Lusin's Theorem 2.3.5 [25]). If put $V_1 = U_1$ and $V_j = U_j \setminus \bigcup_{i<j} U_i$, then $V_i \cap V_j = \emptyset$ for each $i \neq j$. Since μ^m is regular, for each V_j there exists a finite subfamily $W_{k,j} \in W$ such that

$$\mu^m\left(V_j \setminus \bigcup_{k=1}^{n} W_{k,j}\right) < \varepsilon_j,$$

where $n = n(\varepsilon_j, j) \in \mathbf{N}$, $W_{k,j} \subset V_j$, $W_{k,j} \cap W_{l,j} = \emptyset$ for each $k \neq l$. We choose numbers $\varepsilon_j = \varepsilon|\pi|^j$ and allow $\varepsilon > 0$ to tend to zero. Thus $\bigcup_{k,j} W_{k,j}$ covers almost all of Z, since

$$0 \le \mu^1\left(Z \setminus \bigcup_{j,k} W_{k,j}\right) \le \lim_{0<\varepsilon\to 0} \sum_j \mu^m\left(V_j \setminus \bigcup_{k=1}^{n(\varepsilon_j,j)} W_{k,j}\right) = 0.$$

Since the field $\mathbf{K}$ is locally compact, there exists a generator $|\pi|$ of the normalization group $\Gamma_{\mathbf{K}}$ such that $|x| = |\pi|^{-\nu(x)}$ for each $x \in \mathbf{K}$, where $|x| = \mod_{\mathbf{K}}(x)$ is the multiplicative norm in $\mathbf{K}$, while $\nu(x) = \nu_{\mathbf{K}}(x) \in \mathbf{Z}$ is called the valuation function or valuation [99, 103, 111]. To prove the first statement of the theorem it is sufficient to demonstrate, that the μ^2 measure of the set

$$\mathcal{A} := \{(x,y) \in U^2 : \lim_{(x_1,y_1)\to(x,y)} [g(x_1) - g(y_1)]/[x_1 - y_1]$$

either does not exist or is not equal to $[g(x) - g(y)]/[x - y]\}$ is zero, since $\mu^2\{(x,x) : x \in \mathbf{K}\} = 0$. Since g is the lipschitzian function and $[g(x_1) - g(y_1)]/[x_1 - y_1] - [g(x) - g(y)]/[x - y] = [(g(x_1) - g(y_1))((x - y) - (x_1 - y_1)) + ((g(x_1) - g(y_1)) - (g(x) - g(y)))(x_1 - y_1)]/[(x_1 - y_1)(x - y)]$, one deduces

(5) $|[g(x_1) - g(y_1)]/[x_1 - y_1] - [g(x) - g(y)]/[x - y]| \le C\max(|x_1 - y_1|^r \max(|x_1 - x|, |y_1 - y|), |x_1 - y_1|\max(|x_1 - x|^r, |y_1 - y|^r))/[|x_1 - y_1||x - y|]$.

Under the suitable affine mapping $q(x) := a(x - x_0)$ the image of the domain U is contained in the clopen ball $B(\mathbf{K}, 0, |\pi|)$, where $0 \neq a \in \mathbf{K}$, $x_0 \in \mathbf{K}$. Therefore, without restriction of generality suppose that $U \subset B(\mathbf{K}, 0, |\pi|)$, since the partial difference quotient $\bar{\Phi}^1 g$

is almost everywhere continuous on the domain $U^{(1)}$ if and only if $\bar{\Phi}^1 g \circ q^{-1}$ is such on $(q(U))^{(1)}$. Consider the sets

(6) $A_{l,n,k} := \{(x,y) \in U^2 : |x-y| = |\pi|^l$, there exists $(x_1,y_1) \in U^2$ such that $\max(|x_1 - x|, |y_1 - y|) \le |\pi|^k, |[g(x_1) - g(y_1)]/(x_1 - y_1) - [g(x) - g(y)]/(x-y)| \ge |\pi|^n\}$, where $l,n,k \in \mathbf{N}$. We have

$$\mu^2((x,y) \in \mathbf{K}^2 : |x-y| = |\pi|^l, y \in B(\mathbf{K}, y_0, |\pi|^s)) = (|\pi|^l - |\pi|^{l+1})|\pi|^s$$

for each natural numbers $l,s \in \mathbf{N}$. If $|x-y| = |x_1 - y_1| > \max(|x_1 - x|, |y_1 - y|)$, then from (5) it follows, that

$$|[g(x_1) - g(y_1)]/[x_1 - y_1] - [g(x) - g(y)]/[x-y]|$$
$$\le C \max(|x-y|^{(r-2)} \max(|x_1 - x|, |y_1 - y|), |x-y|^{-1} \max(|x_1 - x|^r, |y_1 - y|^r)).$$

Choose a natural number $k \in \mathbf{N}$ sufficiently large, $k \ge m_0$, such that

$$C \max[|\pi|^{k+(r-2)l}, |\pi|^{rk-l}] < |\pi|^n,$$

then $\mu^2(A_{l,n,k}) = 0$, since

$$\mu^2(A_{l,n,k} \cap \{y \in B(\mathbf{K}, y_0, |\pi|^l)\}) = 0 \ \text{ for each } y_0 \in \mathbf{K},$$

where $m_0 = m_0(l) \in \mathbf{N}$. Let $l_0 \in \mathbf{N}$ be a large number, take $m_0 = m_0(l_0)$ such that l_0 tends to the infinity if and only if m_0 tends to the infinity, then

$$\mu^2\Big(\bigcup_{n=1}^{\infty} \bigcup_{k, k \ge m_0} \bigcup_{l=1}^{\infty} A_{l,n,k}\Big) \le \sum_{l=l_0}^{\infty} (|\pi|^{l+1} - |\pi|^{l+2}) = |\pi|^{l_0+1},$$

since $B(\mathbf{K}, 0, |\pi|) \setminus \{0\} = \bigcup_{l=1}^{\infty} \{x \in \mathbf{K} : |x| = |\pi|^l\}$ and $\mu(\{0\}) = 0$. Hence

$$\mu^2\Big(\bigcup_{n=1}^{\infty} \bigcap_{m=1}^{\infty} \bigcup_{k, k \ge m} \bigcup_{l=1}^{\infty} A_{l,n,k}\Big) \le |\pi|^{l_0+1},$$

where l_0 is arbitrary large. Therefore, the measure is zero

$$\mu^2\Big(\bigcup_{n=1}^{\infty} \bigcap_{m=1}^{\infty} \bigcup_{k, k \ge m} \bigcup_{l=1}^{\infty} A_{l,n,k}\Big) = 0,$$

consequently, $\mu^2(\mathcal{A}) = 0$, since the inclusions are satisfied

$$\mathcal{A} \subset \{(x,y) \in U^2 : \text{ there exists a sequence } (x^m, y^m) \in U^2 \text{ such that}$$
$$\lim_{m \to \infty} (x^m, y^m) = (x,y) \text{ and}$$
$$\overline{\lim}_{m \to \infty} |[g(x^m) - g(y^m)]/[x^m - y^m] - [g(x) - g(y)]/[x-y]| \ge |\pi|^n \text{ for some } n \in \mathbf{N}\}$$
$$\subset \Big(\bigcup_{n=1}^{\infty} \bigcap_{m=1}^{\infty} \bigcup_{k, k \ge m} \bigcup_{l=1}^{\infty} A_{l,n,k}\Big).$$

Thus the difference quotient $[g(x)-g(y)]/(x-y)$ is μ^2-almost everywhere continuous on the domain U^2.

Now we prove the second statement. For this we mention that the set $\mathcal{A}$ is symmetric relative to the transposition $(x,y)\mapsto(y,x)$. We have that $[g(x)-g(y)]/(x-y)$ is continuous for μ^2-almost all $(x,y)\in U^2$, hence on some everywhere dense subset in U^2. It is sufficient to show that the μ measure of the set $\mathcal{C} := \{x\in U : \lim_{(x_1,y_1)\to(x,x)}[g(x_1)-g(y_1)]/[x_1-y_1]$ either does not exist or is not equal to $\lim_{y\to x}[g(x)-g(y)]/[x-y]$ or the latter limit does not exist $\}$ is zero. For this we consider the sets

(7) $E_{l,n,k} := \{x\in U :$ there exist $y\in U$ and $(x_1,y_1)\in U^2$ such that $|x-y| = |\pi|^l, \max(|x_1-x|,|y_1-y|)\le|\pi|^k, |[g(x_1)-g(y_1)]/(x_1-y_1)-[g(x)-g(y)]/(x-y)|\ge|\pi|^n\}$, where $l,n,k\in\mathbf{N}$. There exists a natural number $m_0\in\mathbf{N}$ such that for each $k\ge m_0$ the inequality $C\max[|\pi|^{k+(r-2)l},|\pi|^{rk-l}]<|\pi|^n$ is satisfied, consequently, $\mu(E_{l,n,k})=0$ for such natural number k, since $\mu^1(E_{l,n,k}\cap\{y\in B(\mathbf{K},y_0,|\pi|^l)\})=0$ for each number $y_0\in\mathbf{K}$. For a natural number $l_0\in\mathbf{N}$ one can take $m_0=m_0(l_0)$ such that l_0 tends to the infinity if and only if m_0 tends to the infinity, then we deduce the estimate

$$\mu\Big(\bigcup_{n=1}^{\infty}\bigcup_{k,k\ge m}\bigcup_{l,l\ge l_0}E_{l,n,k}\Big)\le\sum_{l=l_0}^{\infty}(|\pi|^l-|\pi|^{l+1})=|\pi|^{l_0},$$

consequently,

$$\mu^1\Big(\bigcup_{n=1}^{\infty}\bigcap_{m=1}^{\infty}\bigcup_{k,k\ge m}\bigcap_{s=1}^{\infty}\bigcup_{l,l\ge s}E_{l,n,k}\Big)\le|\pi|^{l_0},$$

where l_0 may be arbitrary large. Therefore, the following measure is zero

$$\mu\Big(\bigcup_{n=1}^{\infty}\bigcap_{m=1}^{\infty}\bigcup_{k,k\ge m}\bigcap_{s=1}^{\infty}\bigcup_{l,l\ge s}E_{l,n,k}\Big)=0,$$

consequently, $\mu(\mathcal{C})=0$, since in view of (7) the formula is accomplished

$$\mathcal{C}\subset\{x\in U:\ \text{there exists a sequence } (z^m,x^m,y^m)\in U^3 \text{ such that}$$

$$\lim_{m\to\infty}(z^m,x^m,y^m)=(x,x,y)\ \text{and}$$

$$\overline{\lim}_{m\to\infty}|[g(x^m)-g(y^m)]/[x^m-y^m]-[g(x)-g(z^m)]/[x-z^m]|\ge|\pi|^n \text{ for some } n\in\mathbf{N}\}$$

$$\subset\Big(\bigcup_{n=1}^{\infty}\bigcap_{m=1}^{\infty}\bigcup_{k,k\ge m}\bigcap_{s=1}^{\infty}\bigcup_{l,l\ge s}E_{l,n,k}\Big).$$

Therefore, the derivative $dg(x)/dx$ exists and is μ-almost everywhere continuous on the domain U.

10. Lemma. *If S is a $\mu^m\otimes\mu^k$ measurable subset of the topological vector space $\mathbf{K}^m\times\mathbf{K}^k$, $\varepsilon>0$ and $\delta>0$ are positive numbers and $T:=\{x\in\mathbf{K}^m:\ \mu^k(\{z:(x,z)\in S,|z|\le R\})\le\varepsilon R^k\}$ is the subset, whenever $0<R<\delta$, then T is μ^m measurable.*

Proof. For each a positive number $0<R<\infty$ the set $S_R := S\cap\{(x,z):|z|\le R\}$ is $\mu^m\otimes\mu^k$ measurable and by the Fubini theorem $\mu^k(\{z:(x,z)\in S_R\})$ is the μ^m measurable function of the variable x, hence T is μ^m measurable, since

$T = \{x \in \mathbf{K}^m : \mu^k(\{z : (x,z) \in S_R\}) \leq \varepsilon R^k\}.$

11. Lemma. *If a function $\phi : \mathbf{K}^m \times \mathbf{K}^k \to \mathbf{R}$ is $\mu^m \otimes \mu^k$ measurable, then*

$$ap\overline{\lim}_{z\to 0}\phi(x,z) \quad and \quad ap\underline{\lim}_{z\to 0}\phi(x,z)$$

are μ^m measurable functions of the variable x.

Proof. For each real number $c \in \mathbf{R}$ applying Lemma 10 to the sets $\{(x,z) : \phi(x,z) > c\}$ and $\{(x,z) : \phi(x,z) < c\}$ we get the statement of this lemma.

12. Lemma. *If $u : \mathbf{K}^n \to \mathbf{K}^m$ is a $\mathbf{K}$-linear epimorphism and A is a μ^m measurable set, then $u^{-1}(A)$ is μ^n measurable.*

Proof. This follows from the fact that a $\mathbf{K}$-linear topological isomorphism $v : \mathbf{K}^n \to \mathbf{K}^m \times \mathbf{K}^{n-m}$ exists so that

$$v \circ u^{-1}(A) = A \times \mathbf{K}^{n-m}.$$

13. Lemma. *If a function $g : \mathbf{K}^m \to \mathbf{K}$ is μ^m measurable and $1 \leq k \leq m$, then the μ^k approximate limit is*

$$ap\lim_{z\to 0} g({}_1x + {}_1z, \dots, {}_kx + {}_kz, {}_{k+1}x, \dots, {}_mx) = g(x)$$

for μ^m almost all x.

Proof. There exists the linear topological isomorphism $v : \mathbf{K}^m \to \mathbf{K}^k \times \mathbf{K}^{m-k}$. Put $h(x,y,z) = (x+z,y)$, where $h : \mathbf{K}^k \times \mathbf{K}^{m-k} \times \mathbf{K}^k \to \mathbf{K}^k \times \mathbf{K}^{m-k}$ is such function. In view of Lemma 12 the composite function $g \circ h$ is $\mu^k \times \mu^{m-k} \times \mu^k$ measurable. From Lemma 11 it follows, that the set

$$A := \{(x,y) : ap\lim_{z\to 0} g(x+z,y) = g(x,y)\}$$

is $\mu^k \otimes \mu^{m-k}$ measurable.

Theorem 2.9.13 [25] states that if a function f maps ν almost all of X into Y, where (X, ρ_X) is a metric space, (Y, ρ_Y) is a separable metric space, then f is ν measurable if and only if f is approximately (ν, V) continuous at almost all points of X.

In accordance with the latter theorem and the Fubini theorem the function $g(x,y)$ is μ^k measurable by the variable x for μ^{m-k} almost all y and

$$\mu^k(\{x : (x,y) \notin A\}) = 0$$

and the complement of the set A has μ^k measure zero.

14. Corollary. *If A is μ^m measurable set in the topological vector space $\mathbf{K}^m$ and $1 \leq k \leq m$, then for μ^m almost all points $y \in A$ the set $\mathbf{K}^k \cap \{x : ({}_1x, \dots, {}_kx, {}_{k+1}y, \dots, {}_my) \notin A\}$ has zero μ^k density at $({}_1y, \dots, {}_ky)$.*

Proof. This follows from Lemma 13 for the characteristic function $g = Ch_A$, where $Ch_A(y) = 1$ for each $y \in A$, $Ch_A(y) = 0$ for each point $y \in \mathbf{K}^m \setminus A$.

15. Theorem. *If a function $f : \mathbf{K}^m \to \mathbf{K}^n$ is μ^m measurable, then the set $A_i := dom\ {}_{ap}D_i f$ is a μ^m measurable, the set*

$$V_i := dom\ {}_{ap}\bar{\Phi}^1 f(x; e_i; t)$$

is μ^{m+1} measurable so that

$$\mu^{m+1}(\mathbf{K}^{m+1}\setminus V_i)=0.$$

Moreover, ${}_{ap}D_i f$ and ${}_{ap}\bar{\Phi}^1 f(x;e_i;t)$ are $\mu^m|_{A_i}$ and $\mu^{m+1}|_{V_i}$ measurable functions respectively,

$$\begin{gathered}{}_{ap}\bar{\Phi}^1 f(x;v;t)=v_1\ {}_{ap}\bar{\Phi}^1 f(x+e_2v_2+\dots+e_mv_m;e_1;v_1t)\\ +v_2\ {}_{ap}\bar{\Phi}^1 f(x+e_3v_2+\dots+e_mv_m;e_2;v_2t)+\dots+v_m\ {}_{ap}\bar{\Phi}^1 f(x;e_m;v_mt)\end{gathered}\qquad(1)$$

for μ^{m+1} almost all points (x,t) in the set $V:=\bigcap_{i=1}^m V_i$ and each vector $v=v_1e_1+\dots+v_me_m\in\mathbf{K}^m$,

$$ {}_{ap}Df(x).v=\sum_{i=1}^m v_i\ {}_{ap}D_i f(x)) \qquad (2)$$

for μ^m almost all points x in $A:=\bigcap_{i=1}^m A_i$ and each $v\in\mathbf{K}^m$.

Proof. Since $f=(f_1,\dots,f_n)$, where $f_j:\mathbf{K}^m\to\mathbf{K}$, we get

$$dom\ {}_{ap}\bar{\Phi}^1 f(x;e_i;t)=\bigcap_{j=1}^n dom\ {}_{ap}\bar{\Phi}^1 f_j(x;e_i;t)$$

and $dom\ {}_{ap}D_i f=\bigcap_{j=1}^n dom\ {}_{ap}D_i f_j$ for each $i=1,\dots,m$, consequently, it is sufficient to prove this theorem for $n=1$. Thus suppose that $n=1$.

In accordance with Theorem 2.9.13 [25] (see its formulation in Section 13 above) the function $\bar{\Phi}^1 f(x;v;t)$ is approximately continuous on the set $\mathbf{K}^m\times\mathbf{K}^m\times(\mathbf{K}\setminus\{0\})$, since the difference quotient $[f(x+vt)-f(x)]/t$ is μ^{2m+1} measurable on the set $\mathbf{K}^m\times\mathbf{K}^m\times(\mathbf{K}\setminus\{0\})$ and inevitably $\mu^{m+1}(\mathbf{K}^{m+1}\setminus V_i)=0$, since $\mu(\{0\})=0$. Therefore, Formula (1) is satisfied for μ^{m+1} almost all points $(x;t)$ in V, where $\mu^{m+1}(\mathbf{K}^{m+1}\setminus V)=0$. So it remains to spread this for zero point $t=0$, but

$$ {}_{ap}\bar{\Phi}^1 f(x;v;0)=ap\lim_{t\to0}\bar{\Phi}^1 f(x;v;t)=\ {}_{ap}Df(x).v$$

and the proof of this theorem reduces to the proof of its statement relative to ${}_{ap}Df(x)$.

The set A_i is μ^m measurable if and only if its complement $\mathbf{K}^m\setminus A_i$ is μ^m measurable. But the complement set is the following:

$$\mathbf{K}^m\setminus A_i=\bigcup_{s=1}^\infty\bigcap_{l,l\ge s}\bigcup_{u=1}^\infty\bigcap_{k,k\ge u}\bigcup_{q=1}^\infty E_{l,k,q},$$

where $E_{l,k,q}:=\{x\in\mathbf{K}^m$: there exist $t_1,t_2\in\mathbf{K}$ such that $\max(|t_1|,|t_2|)=|\pi|^k,|t_1-t_2|=|\pi|^l,|\ {}_{ap}\bar{\Phi}^1 f(x;e_i;t_1)-\ {}_{ap}\bar{\Phi}^1 f(x;e_i;t_2)|\ge|\pi|^q$ or $|\ {}_{ap}\bar{\Phi}^1 f(x;e_i;t_1)|\ge|\pi|^{-q}\}$. Each set $E_{l,k,q}$ is μ^m measurable, since $\mu^{m+1}(\mathbf{K}^{m+1}\setminus V_i)=0$ and due to Lemma 10, hence A_i is μ^m measurable for each $i=1,\dots,m$.

We consider the sets

$$T_{R,i,j}(x):=\{t\in\mathbf{K}:|t|<R,|f(x+te_i)-f(x)-t\ {}_{ap}D_i f(x)|>|t||\pi|^j\}\quad\text{and}$$

$$B_{i,j,q}:=A_i\cap\{x\in\mathbf{K}^m:\mu(T_{R,i,j}(x))\le R|\pi|^j\forall 0<R<|\pi|^q\},$$

whenever $x \in \mathbf{K}^m$, $0 < R < \infty$, $i, j, q \in \mathbf{N}$. For $i > 1$ we denote

$$Z_{R,i,j,q}(x) := \{z \in \mathbf{K}^{i-1} : |z| < R, \text{ either } x + {}_1ze_1 + \cdots + {}_{i-1}ze_{i-1} \notin B_{i,j,q} \text{ or}$$

$$| {}_{ap}D_if(x + {}_1ze_1 + \cdots + {}_{i-1}ze_{i-1}) - {}_{ap}D_if(x)| > R|\pi|^j\}.$$

We also introduce the sets

$$C_{i,j,q,k} := B_{i,j,q} \cap \{x : \mu^{i-1}(Z_{R,i,j,q}(x)) \le R^{i-1}|\pi|^j \forall 0 < R < |\pi|^k\}$$

for $k \in \mathbf{N}$ and $i > 1$, in particular, $C_{1,j,q,k} := B_{1,j,q}$ for each k. In view of Lemmas 10 and 13 the sets $B_{i,j,q}$ and $C_{i,j,q,k}$ are μ^m measurable and

$$A_i = \bigcup_{q=1}^{\infty} B_{i,j,q}$$

for each pair (i, j), moreover,

$$\mu^m\left(B_{i,j,q} \setminus \bigcup_{k=1}^{\infty} C_{i,j,q,k}\right) = 0$$

for each triple (i, j, q), also $B_{i,j,q} \subset B_{i,j,q+1}$ and $C_{i,j,q,k} \subset C_{i,j,q,k+1}$. For a subset S in the union $\bigcap_{i=1}^m A_i$ with $\mu^m(S) < \infty$ and every $\varepsilon > 0$ a sequences $\{q_j : j \in \mathbf{N}\}$ and $\{k_j : j \in \mathbf{N}\}$ of natural numbers exists such that

$$\mu^m(S \setminus B_{i,j,q_j}) < \varepsilon|\pi|^j \quad \text{and} \quad \mu^m(S \cap B_{i,j,q_j} \setminus C_{i,j,q_j,k_j}) < \varepsilon|\pi|^j$$

for $i = 1, \ldots, m$ and $j \in \mathbf{N}$, consequently,

$$\mu^m(S \setminus G) < 2|\pi|(1 - |\pi|)^{-1}m\varepsilon,$$

where $G := \bigcap_{i=1}^m \bigcap_{j=1}^{\infty} C_{i,j,q_j,k_j}$.

We will demonstrate that the function f is uniformly approximately differentiable at the points of G. Consider any point $x \in G$, $j \in \mathbf{N}$ and $0 < R < \min(|\pi|^{q_j}, |\pi|^{k_j})$,

$$S_i := \{v \in \mathbf{K}^m : |v| \le R \text{ and either } i > 1 \text{ and } (v_1, \ldots, v_{i-1}) \in Z_{R,i,j,q_j}(x)$$

or

$$v_i \in T_{R,i,j}(x + v_1e_1 + \cdots + v_{i-1}e_{i-1})\},$$

where $i = 1, \ldots, m$. Since $\mu^{i-1}(Z_{R,i,j,q_j}(x)) \le R^{i-1}|\pi|^j$ and since $\mu(T_{R,i,j}(x)) \le R|\pi|^j$ for $z = x + v_1e_1 + \cdots + v_{i-1}e_{i-1} \in B_{i,j,q_j}$, we get the estimate:

$$\mu^m(S_i) \le R^{i-1}|\pi|^j R^{m-i+1} + R|\pi|^j R^{m-1} = 2R^m|\pi|^j.$$

If $v \in B(\mathbf{K}^m, 0, R) \setminus S_i$, then the inequality

$$|f(x + v_1e_1 + \cdots + v_ie_i) - f(x + v_1e_1 + \cdots + v_{i-1}e_{i-1}) - v_i\, {}_{ap}D_if(x)|$$

$$\le \max(|v_i||\pi|^j, |v_i\, {}_{ap}D_if(x + v_1e_1 + \cdots + v_{i-1}e_{i-1}) - v_i\, {}_{ap}D_if(x)|) \le |v_i||\pi|^j$$

is valid. Putting $S := \bigcup_{i=1}^m S_i$ we get $\mu^m(S)R^{-m} \le 2m|\pi|^j$. The inclusion $v \in B(\mathbf{K}^m, 0, R) \setminus S$ implies that

$$|f(x+v) - f(x) - \sum_{i=1}^m v_i \ {}_{ap}D_i f(x)| \le |\pi|^j \max_{i=1}^m |v_i| = |\pi|^j |v|.$$

16. Lemma. *Let $S \subset A \subset \mathbf{K}^m$, let also $f : A \to \mathbf{K}^n$ be a function, $0 < R < \infty$, $0 < C < \infty$ and let $z \in S$ imply $B(\mathbf{K}^m, z, R) \subset A$ with $|f(x) - f(z)| \le C|x - z|$ for each $x \in B(\mathbf{K}^m, z, R)$. Suppose also that $y \in S$, $\mathbf{K}^m \setminus S$ has the μ^m density zero at each point $z \in U_y$ for some (open) neighborhood U_y of y in $\mathbf{K}^m$ and f is approximately differentiable at a point y. Then f is differentiable at y such that the partial difference quotient $\bar{\Phi}^1 f(y; v; t)$ is continuous on a domain $(U_y \cap A)^{(1)}$ for some neighborhood U_y of y in $\mathbf{K}^m$.*

Proof. Suppose that $L = {}_{ap}Df(y)$, $0 < \varepsilon < 1$, $0 < \delta \le R$ and put

$$W := S \cap \{z : |f(z) - f(y) - L(z-y)| \le \varepsilon|z - y|\}$$

so that $\mu^m(B(\mathbf{K}^m, y, q) \setminus W) < \varepsilon^m R^m$ for each $0 < q < \delta$, $q \in \Gamma_{\mathbf{K}}$. For $x \in B(\mathbf{K}^m, y, \delta)$ we take $q = |x - y|$ and mention that $B(\mathbf{K}^m, x, q) = B(\mathbf{K}^m, y, q)$, also $W \cap B(\mathbf{K}^m, x, q) \ne \emptyset$. Choose a point $z \in B(\mathbf{K}^m, x, \varepsilon q) \cap W$, consequently, $x \in B(\mathbf{K}^m, z, \varepsilon q) \subset B(\mathbf{K}^m, z, R)$. Therefore,

$$\begin{aligned} |f(x) - f(y) - L(x-y)| &\le \max(|f(z) - f(y) - L(z-y)|, |f(x) - f(z)|, |L(z-x)|) \\ &\le \max(\varepsilon|z-y|, C|x-z|, \|L\||x-z|) \\ &\le \varepsilon \max(|x-y|, Cq, \|L\|q) = \varepsilon|x-y| \max(1, C, \|L\|), \end{aligned} \tag{1}$$

since $|z - y| \le \max(|z - x|, |x - y|) \le |x - y| = q$. From the arbitrariness of $0 < \varepsilon < 1$ it follows, that the derivative $L = Df(y)$ exists. From Inequality (1) we get

$$|\bar{\Phi}^1 f(y; v; t) - Lv| \le \varepsilon \max(1, C, \|L\|) \tag{2}$$

for all $v = x - y$, $x \in B(\mathbf{K}^m, y, \delta)$, consequently, the limit

$$\lim_{t \to 0} \bar{\Phi}^1 f(y; v; t) = Lv \tag{3}$$

converges uniformly by the variable $x \in B(\mathbf{K}^m, y, \delta)$, since $\varepsilon > 0$ is arbitrary small. From the existence of the derivative $Df(y)$ it follows that the function f is continuous at y. The condition for the difference of the sets $\mathbf{K}^m \setminus S$ to have μ^m density zero at each point $z \in U_y$ implies that the set S is everywhere dense in U_y for some neighborhood U_y of y. Then we infer that

$$\begin{aligned} |f(y + xt_1) - f(y + vt_2)| &\le \max(|f(y + xt_1) - f(z)|, |f(z) - f(y + vt_2)|) \\ &\le C \max(|z - y - xt_1|, |z - y - vt_2|) \end{aligned}$$

for each point $z \in S$, where $y + xt_1, y + vt_2 \in A$, $B(\mathbf{K}^m, z, R) \subset A$ and $\max(|z - y - xt_1|, |z - y - vt_2|)$ can be chosen equal to $|xt_1 - vt_2|$, since the normalization group $\Gamma_{\mathbf{K}}$ is discrete in the open subset $(0, \infty)$ of the real field and the set S is dense in U_y. Thus the partial difference quotient $\bar{\Phi}^1 f(y;v;t)$ is continuous on the set $U_y^{(1)} \cap (A^{(1)} \setminus \{(y;v;t) : \mathbf{K}^m \ni v \neq 0, \mathbf{K} \ni t \neq 0\})$. Together with (3) this gives the continuity of the partial difference quotient $\bar{\Phi}^1 f(y;v;t)$ on the set $(U_y \cap A)^{(1)}$.

17. Theorem. *If $f : \mathbf{K}^m \to \mathbf{K}^n$ is a locally lipschitzian function such that for each point $x_0 \in \mathbf{K}^m$ positive constants $0 < C < \infty$ and $\delta > 0$ and $0 < r \leq 1$ exist with*

$$|g(x) - g(y)| \leq C|x-y|^r \tag{1}$$

for each $\max(|x-x_0|, |y-x_0|) < \delta$, $x, y \in \mathbf{K}^m$.

Then the function f is differentiable at μ^m almost all points of the topological vector space $\mathbf{K}^m$ on a subset G and the partial difference quotient $\bar{\Phi}^1 f(x;v;t)$ is continuous at μ^{2m+1} almost all points of $\mathbf{K}^{2m+1}$ on a subset V_0 such that for each $\varepsilon > 0$ a closed subset G_ε in G exists with $\mu^m(G \setminus G_\varepsilon) < \varepsilon$ and the derivative Df is continuous on G_ε and $G_\varepsilon^{(1)} \subset V_0$.

Proof. Let the sets A_i and V_i be the same as in Theorem 15. For a point $x \in \mathbf{K}^m$ we consider the map $f_x(h) := f({}_1x, \dots, {}_{i-1}x, h, {}_{i+1}x, \dots, {}_mx)$ for any number $h \in \mathbf{K}$, so this mapping is locally lipschitzian. In accordance with Theorem 9 the partial difference quotient $\bar{\Phi}^1 f_x(h;w;t)$ is continuous for μ^3 almost every points $(h;w;t) \in \mathbf{K}^3$ and the derivative $df_x(h)/dh$ is μ almost everywhere continuous by the variable h on $\mathbf{K}$ for each marked point $({}_1x, \dots, {}_{i-1}x, {}_{i+1}x, \dots, {}_mx) \in \mathbf{K}^{m-1}$. Since A_i is μ^m measurable and V_i is μ^{2m+1} measurable, the measures $\mu^m(\mathbf{K}^m \setminus A_i) = 0$ and $\mu^{2m+1}(\mathbf{K}^{2m+1} \setminus V_i) = 0$ are zero and inevitably $\mu^m(\mathbf{K}^m \setminus G) = 0$ and $\mu^{2m+1}(\mathbf{K}^{2m+1} \setminus V) = 0$, where

$$G := \bigcap_{i=1}^m A_i \quad \text{and} \quad V = \bigcap_{i=1}^m V_i.$$

Thus the restriction $f|_G$ is differentiable and the partial difference quotient $\bar{\Phi}^1 f(x;v;t)|_V$ is μ^{2m+1} measurable. On the other hand, $(\mathbf{K}^{2m+1} \setminus \mathbf{K}^{2m} \times \{0\}) \subset V$, since f is continuous on the topological vector space $\mathbf{K}^m$ and so the partial difference quotient $\bar{\Phi}^1 f(x;v;t)$ is continuous on the set $(\mathbf{K}^{2m+1} \setminus \mathbf{K}^{2m} \times \{0\})$. Therefore, the partial difference quotient $\bar{\Phi}^1 f(x;v;t)$ is continuous by each triple $({}_jx; {}_jv; t)$ and μ^{2m+1} measurable on $G^{(1)}$, $G^{(1)} \subset V$, $\mu^{2m+1}(\mathbf{K}^{2m+1} \setminus G^{(1)}) = 0$, since $\mu^m(\mathbf{K}^m \setminus G) = 0$.

Lusin's Theorem 2.3.5 [25] asserts: if ϕ is a Borel regular nonnegative measure over a metric space X (or a Radon measure over a locally compact Hausdorff space X), if f is a ϕ measurable function with values in a separable metric space Y, A is a ϕ measurable set for which $\phi(A) < \infty$, and $\varepsilon > 0$, then A contains a closed (compact) set C such that $\phi(A \setminus C) < \varepsilon$ and $f|_C$ is continuous.

In view of Lusin's theorem for each $\varepsilon > 0$ and each $0 < R < \infty$ and a point $\xi \in G$ a compact subset $E \subset (G \cap B(\mathbf{K}^m, \xi, R))$ exists such that $\mu^m((G \cap B(\mathbf{K}^m, \xi, R)) \setminus E) < \varepsilon$ and the restriction $Df|_E$ is continuous, hence the restriction $\bar{\Phi}^1 f(x;v;t)|_{E^{(1)}}$ is continuous. Therefore, the restriction of the function f on E is lipschitzian with $Lip_2(f|_E) = 1$. In view of Theorem 8 an extension g_E on $\mathbf{K}^m$ of $f|_E$ exists with $Lip_1(g_E) = Lip_1(f|_E)$ and $Lip_2(g_E) = 1$.

We take in Lemma 16 $S = E$ and $A = \mathbf{K}^m$. Therefore, the function g_E is differentiable at μ^m almost all points of the topological vector space $\mathbf{K}^m$ and the partial difference quotient $\bar{\Phi}^1 g_E(x;v;t)$ is continuous at μ^{2m+1} almost all points of the topological vector space

$\mathbf{K}^{2m+1}$. Since $\xi \in G$, $0 < R < \infty$ and $\varepsilon > 0$ are arbitrary, we can take a disjoint covering $B(\mathbf{K}^m, \xi_j, R_j)$, $R_j \geq 1$, of the set G and

$$E_{i,j} \subset \left[(G \cap B(\mathbf{K}^m, \xi_j, R_j)) \setminus \bigcup_{l, l<i} E_{l,j}\right]$$

such that

$$\mu^m\left((G \cap B(\mathbf{K}^m, \xi_j, R_j)) \setminus \left(\bigcup_{l=1}^{i} E_{l,j}\right)\right) < |\pi|^{i+j+k},$$

where $k \in \mathbf{N}$ is some large fixed number, $i, j \in \mathbf{N}$, $E_{0,j} := \emptyset$.

Then we consider the set

$$G \setminus \bigcup_{j=1}^{s} \bigcup_{i=1}^{s} E_{i,j}$$

and continue this construction by induction. The family of restrictions $f|_{E_{i,j}}$ generates the function $f|_{G_1}$, where

$$G_1 = \bigcup_{i,j=1}^{\infty} E_{i,j} \quad \text{and} \quad \mu^m(G \setminus G_1) = 0.$$

Since $E_{i,j} \subset [B(\mathbf{K}^m, \xi_j, R_j)) \setminus \bigcup_{l,l<i} E_{l,j}]$, each subset $E_{l,j}$ is compact and each subset $[B(\mathbf{K}^m, \xi_j, R_j)) \setminus \bigcup_{l,l<i} E_{l,j}]$ is open in the topological vector space $\mathbf{K}^m$ and $B(\mathbf{K}^m, \xi_j, R_j) \cap B(\mathbf{K}^m, \xi_q, R_q) = \emptyset$ for each $q \neq j$. Then we take the set

$$G_\varepsilon := \bigcup_{j=1}^{\infty} \bigcup_{i=1}^{n(j)} E_{i,j},$$

where $n(j) \in \mathbf{N}$ is a sequence such that $\sum_{j=1}^{\infty} |\pi|^{n(j)+j+k} < \varepsilon$. Each natural number $n(j)$ is finite and the subset $\bigcup_{i=1}^{n(j)} E_{i,j}$ is closed, hence the subset $\bigcup_{j=1}^{\infty} \bigcup_{i=1}^{n(j)} E_{i,j}$ is closed in the topological vector space $\mathbf{K}^m$ and inevitably the derivative Df is continuous on G_ε such that $G_\varepsilon^{(1)} \subset V_0$.

18. Lemma. *If A is a μ^m-measurable subset in the topological vector space $\mathbf{K}^m$ and a function $f : A \to \mathbf{K}^n$ is locally lipschitzian, then the function f has μ^{2m+1}-everywhere in $A^{(1)}$ an approximate partial difference quotient $\bar{\Phi}^1 f(x; v; t)$ and $\Phi^1 f(x; e_i; t)$ for μ^{m+1} almost all points $(x; e_i; t)$ of $(A \times \{e_i\} \times \mathbf{K}) \cap A^{(1)}$ and μ^m-almost everywhere on A an approximate differential $Df(x)$ and approximate partial differentials $D_j f(x) = \partial f(x)/\partial_j x$.*

Proof. In accordance with Theorem 8 and the proof of Theorem 9 for each $\varepsilon > 0$ and $0 < R < \infty$ the function $f|_{A_{R,\varepsilon}}$ has a lipschitzian extension $g : \mathbf{K}^m \to \mathbf{K}^n$ such that g has the same lipschitzian constants $0 < C < \infty$ and $0 < r \leq 1$, where $A_{R,\varepsilon}$ is a compact subset in $A \cap B(\mathbf{K}^m, 0, R)$ such that $\mu^m(A \cap B(\mathbf{K}^m, 0, R) \setminus A_{R,\varepsilon}) < \varepsilon$.

From Theorem 17 it follows that the function g has the partial difference quotient $\bar{\Phi}^1 g$ and the derivative Dg at μ^{2m+1} and μ^m almost all points of the sets $A^{(1)}$ and A respectively.

We recall that one says that a set B is a ϕ hull of a set A if and only if $A \subset B \subset X$, B is ϕ measurable and $\phi(T \cap A) = \phi(T \cap B)$ for every ϕ measurable subset T, where ϕ is a measure over a measurable space X.

Theorem 2.9.11 [25] states that if $A \subset X$ and

$$P = \{x : (V) \lim_{S \to x} \phi(S \cap A)/\phi(S) = 1\}$$

and

$$Q = \{x : (V) \lim_{S \to x} \phi(S \setminus A)/\phi(S) = 0\},$$

then the sets P and Q are ϕ measurable, $\phi(A \setminus P) = 0$, $A \cup P$ is a ϕ hull of A, $\phi(Q \setminus A) = 0$, $X \setminus (A \cap Q)$ is a ϕ hull of the difference set $X \setminus A$. Moreover, ϕ measurability of the set A is equivalent to each of the two conditions: $\phi(P \setminus A) = 0$, $\phi(A \setminus Q) = 0$.

In view of this theorem the sets $\mathbf{K}^{2m+1} \setminus A^{(1)}$ and $\mathbf{K}^m \setminus A$ have zero densities at μ^{2m+1} and μ^m almost all points of the sets $A^{(1)}$ and A respectively. At points where both conditions hold there are ${}_{ap}\Phi^1 g$ and ${}_{ap}Dg$. By Corollary 14 we have

$$D_i g(x) = {}_{ap}D_i g(x) \quad \text{and} \quad \bar{\Phi}^1 g(x;e_i;t) = {}_{ap}\bar{\Phi}^1 g(x;e_i;t)$$

for μ^m almost all points $x \in A$ and μ^{m+1} almost all points $(x;e_i;t)$ of $(A \times \{e_i\} \times \mathbf{K}) \cap A^{(1)}$ correspondingly. From combinatorial Formulas 15(1,2) we get, that f has μ^{2m+1}-everywhere on the set $A^{(1)}$ an approximate partial difference quotient ${}_{ap}\bar{\Phi}^1 f(x;v;t)$ and μ^m-almost everywhere on the set A an approximate differential ${}_{ap}Df(x)$.

19. Theorem. *If $A \subset \mathbf{K}^m$, $m,n \in \mathbf{N}$, $f : A \to \mathbf{K}^n$ and for each point $y \in A$ there exist $\delta = \delta(y) > 0$ and $0 < r = r(y) \leq 1$ such that*

$$ap\overline{\lim}_{x \to z} |f(x) - f(z)|/|x - z|^r < \infty$$

whenever $z \in A$ and $|z - y| < \delta$, then

$$A = \bigcup_{j \in \Lambda} E_j,$$

where $card(\Lambda) \leq \aleph_0$, such that the restriction of the function f to each subset E_j is lipschitzian; moreover, the function f is approximately differentiable such that there exist ${}_{ap}\bar{\Phi}^1 f(x;v;t)$ and ${}_{ap}Df(x)$ for μ^{2m+1} almost all points $(x;v;t)$ of $A^{(1)}$ and μ^m almost all points x of A correspondingly.

Proof. From the supposition of this theorem there follows that the density of $\mathbf{K}^m \setminus A$ is zero and the function f is approximately continuous at each point of the set A. Therefore, the set A is μ^m measurable and the function f is $\mu^m|_A$ measurable in accordance with Theorems 2.9.11 and 2.9.13 [25]. We denote

$$Q_{R,j}(z) := B(\mathbf{K}^m, z, R) \cap \{x : x \notin A \text{ or } |f(x) - f(z)| > |\pi|^{-j}|x - z|^r\}$$

for $|\pi|^j < R < \delta(y)$ and $z \in A$ with $|y - z| < \delta = \delta(y)$ provided by the conditions of this theorem, $0 < R \in \Gamma_{\mathbf{K}}$, $j \in \mathbf{N}$. Each set

$$E_j := A \cap \{z : \mu^m(Q_{R,j}(z)) < R^m/2 \text{ for } 0 < R < |\pi|^j\}$$

is μ^m measurable in accordance with Lemma 10 and

$$A = \bigcup_{j=1}^{\infty} E_j.$$

If $R = |x-z| < \delta(y)$, then

$$\mu^m(Q_{R,j}(z) \cup Q_{R,j}(x)) < R^m = \mu^m(B(\mathbf{K}^m,z,R) \cap B(\mathbf{K}^m,x,R)),$$

since $B(\mathbf{K}^m,z,R) = B(\mathbf{K}^m,x,R)$. Choosing any point $w \in B(\mathbf{K}^m,z,R) \setminus (Q_{R,j}(x) \cup Q_{R,j}(z))$ we get the inequalities:

$$|f(x)-f(z)| \le \max(|f(x)-f(w)|,|f(w)-f(z)|)$$

$$\le |\pi|^j \max(|x-w|^r,|w-z|^r) \le |\pi|^j |x-z|^r.$$

Therefore, if $x,z \in E_j$ and $|x-z| < |\pi|^j < R$, then

$$|f(x)-f(z)| \le |\pi|^{-j}|x-z|^r,$$

since the set A is everywhere dense in the topological vector space $\mathbf{K}^m$. Then each subset E_j is of diameter less than $|\pi|^j$ for a natural number $j \in \mathbf{N}$. Each restriction $f|_{E_j}$ is lipschitzian and Lemma 18 gives that the restriction $f|_{E_j}$ is approximately differentiable, hence f has μ^{2m+1}-everywhere in the subset $A^{(1)}$ an approximate partial difference quotient $\bar{\Phi}^1 f(x;v;t)$ and $\bar{\Phi}^1 f(x;e_i;t)$ for μ^{m+1} almost all points $(x;e_i;t)$ of the subset $(A \times \{e_i\} \times \mathbf{K}) \cap A^{(1)}$ and μ^m-almost everywhere on the subset A an approximate differential $Df(x)$ and approximate partial differentials $D_jf(x) = \partial f(x)/\partial_j x$.

20. Theorem. *If $A \subset W \subset \mathbf{K}^m$, $m,n \in \mathbf{N}$, a subset W is open, a subset A is μ^m measurable, $f: W \to \mathbf{K}^n$ and for each point $y \in A$ there exist $\delta = \delta(y) > 0$ and $0 < r = r(y) \le 1$ such that*

$$\overline{\lim}_{x\to z}|f(x)-f(z)|/|x-z|^r < \infty$$

whenever $z \in A$ and $|z-y| < \delta$, then the function f is differentiable such that the partial difference quotient $\Phi^1 f(x;v;t)$ and the derivative $Df(x)$ exist for μ^{2m+1} almost all points $(x;v;t)$ of a subset

$$W_A^{(1)} := \{(x;v;t) : x \in A, v \in \mathbf{K}^m, t \in \mathbf{K}, x+vt \in W\}$$

and μ^m almost all points x of A respectively.

Proof. The field $\mathbf{K}$ is locally compact with the non archimedean multiplicative norm, hence the subset A has a countable covering by balls $B(\mathbf{K}^m,y_j,\delta_j)$ contained in W, where $y_j \in A$, $\delta_j := \delta(y_j)$. Therefore, the subset A is contained in the countable union of the subsets

$$E_j := W \cap \{z : |f(x)-f(z)| \le |\pi|^j|x-z|^r \text{ for } x,z \in B(\mathbf{K}^m,y_i,\delta_i) \text{ and}$$

$$|\pi|^j \le \delta_i \text{ for some } i \in \mathbf{N}\}.$$

Suppose that there exists a sequence $\zeta_l \in E_j$ converging to a point $z \in \mathbf{K}^m$ as a natural number l tends to the infinity. Take any point $x \in B(\mathbf{K}^m,z,|\pi|^j) \subset B(\mathbf{K}^m,y_i,\delta_i)$. There exists a natural number $l_0 \in \mathbf{N}$ such that

$$\{z,x\} \subset B(\mathbf{K}^m,\zeta_l,|\pi|^j) \subset W$$

for each $l \geq l_0$, hence

$$|f(x) - f(z)| \leq \max(|f(x) - f(\zeta_l)|, |f(z) - f(\zeta_l)|)$$
$$\leq |\pi|^j \max(|x - \zeta_j|^r, |z - \zeta_j|^r) \leq |\pi|^j |x - z|^r,$$

consequently, $z \in E_j$. Thus each subset E_j is closed in the topological vector space $\mathbf{K}^m$. Each subset E_j is of diameter not greater than δ_i with the corresponding i and the restriction $f|_{E_j}$ is lipschitzian.

In view of Theorems 2.9.11 [25] and 19 above the function $f|_{E_j}$ is approximately differentiable such that there exist ${}_{ap}\bar{\Phi}^1 f(x;v;t)$ and ${}_{ap}Df(x)$ for μ^{2m+1} almost all points $(x;v;t)$ of $W^{(1)}_{E_j}$ and μ^m almost all points x of E_j correspondingly, since for $t \neq 0$ we have ${}_{ap}\bar{\Phi}^1 f(x;v;t) = [f(x+vt) - f(x)]/t, x+vt \in W, x \in E_j \subset A$. Moreover, the subset $\mathbf{K}^m \setminus E_j$ has zero density at each point $y \in A$. Then Theorem 8 and Lemma 16 provide that the function f is differentiable such that the partial differnce quotient $\bar{\Phi}^1 f(x;v;t)$ and the derivative $Df(x)$ exist for μ^{2m+1} almost all points $(x;v;t)$ of the subset $A^{(1)}$ and μ^m almost all points x of the subset A respectively.

21. Lemma. *If $G \subset V \subset \mathbf{K}^m$, $h : V \to \mathbf{K} \setminus \{0\}$ is Lispchitzian with $r = Lip_2(h) = 1$, $\{B(\mathbf{K}^m, y, |h(y)|) : y \in G\}$ is the disjoint family, $b \geq Lip_1(h)$, $0 < \alpha \in \Gamma_{\mathbf{K}}$, $0 < \beta \in \Gamma_{\mathbf{K}}$, $b\alpha < 1$ and $b\beta < 1$ and*

$$G_x := G \cap \{y : B(\mathbf{K}^m, x, \alpha|h(x)|) \cap B(\mathbf{K}^m, y, \beta|h(y)|) \neq \emptyset\}$$

for $x \in V$, then

$$[(1 - b\beta)/(1 + b\alpha)] \leq |h(x)|/|h(y)| \leq [(1 + b\beta)/(1 - b\alpha)] \quad (1)$$

for all $y \in G_x$ and

$$card(G_x) \leq [\max(\alpha, \beta(1 + b\alpha)/(1 - b\beta))]^m [(1 + b\beta)/(1 - b\alpha)]^m. \quad (2)$$

Proof. If $y \in G_x$, then

$$|h(x) - h(y)| \leq b|x - y| \leq b\max(\alpha|h(x)|, \beta|h(y)|) \leq b(\alpha|h(x)| + \beta|h(y)|),$$

consequently, $(1 - b\alpha)|h(x)| \leq (1 + b\beta)|h(y)|$ and $(1 - b\beta)|h(y)| \leq (1 + b\alpha)|h(x)|$, since $|x - z|_{\mathbf{K}^m} \geq ||x|_{\mathbf{K}^m} - |z|_{\mathbf{K}^m}|_{\mathbf{R}}$ for each points $x, z \in \mathbf{K}^m$. If $y \neq g \in G$, then the intersection of balls $B(\mathbf{K}^m, y, |h(y)|) \cap B(\mathbf{K}^m, g, |h(g)|) = \emptyset$ is void if and only if the inequality $|y - g| > \max(|h(y)|, |h(g)|)$ is satisfied. Then we infer that

$$|x - y| \leq \max(\alpha|h(x)|, \beta|h(y)|) \leq |h(x)| \max(\alpha, \beta(1 + b\alpha)/(1 - b\beta)),$$

consequently, $B(\mathbf{K}^m, y, |h(y)|) \subset B(\mathbf{K}^m, x, \gamma|h(x)|)$, where $\gamma := \max(\alpha, \beta(1 + b\alpha)/(1 - b\beta))$. Therefore,

$$card(G_x)[(1 - b\alpha)|h(x)|/(1 + b\beta)]^m \leq \sum_{y \in G_x} |h(y)|^m \leq (\beta(1 + b\alpha)/(1 - b\beta))^m |h(x)|^m,$$

since $G_x \subset G$ and the subset G is discrete, $G_x \subset B(\mathbf{K}^m, x, \gamma|h(x)|)$.

22. Theorem. *Suppose Y is a normed vector space over the field $\mathbf{K}$, a subset A is a closed subset in the topological vector space $\mathbf{K}^m$ and to each point $z \in A$ a polynomial function $P_z : \mathbf{K}^m \to Y$ corresponds with a degree $deg(P_z) \le k$. Let also $S \subset A$ and $\delta > 0$ and*

$$\rho(S,\delta) := \sup_{0<|x-z|\le\delta; x,z\in S; j=0,1,\dots,k} \|\bar{\Phi}^j P_x(z;v;t) - \bar{\Phi}^j P_z(z;v;t)\|_{C^0(V_z^{(j)},Y)} |x-z|^{j-k},$$

where $U_z = B(\mathbf{K}^m, z, |\pi|^\zeta)$, $\zeta \in \mathbf{N}$ is a marked number, $v = (v_1,\dots,v_j)$, $t = (t_1,\dots,t_j)$,

$$U^{(j)} := \{(y; v_1,\dots,v_j; t_1,\dots,t_j) : y \in U, y + v_1 t_1 + \dots + v_j t_j \in U,$$

$$v_i \in \mathbf{K}^m, t_i \in \mathbf{K}\ \forall i = 1,\dots,j\},$$

$$V_z^{(j)} := \{(y; v_1,\dots,v_j; t_1,\dots,t_j) \in U_z^{(j)} : v_i \in \mathbf{K}^m, |v_i| = 1, t_i \in \mathbf{K} \quad \forall i = 1,\dots,j\}.$$

If the limit

$$\lim_{0<\delta\to 0} \rho(S,\delta) = 0$$

is zero for each compact subset S of the topological vector space $\mathbf{K}^m$, then a map $g : \mathbf{K}^m \to Y$ of class C^k exists such that the equality $\bar{\Phi}^j g(z;v;t) = \bar{\Phi}^j P_z(z;v;t)$ is valid on the subset $V_z^{(j)}$ for each $j = 0,1,\dots,k$ and $z \in A$.

Proof. Let $\mathcal{F}$ be a family of open subsets of the topological vector space $\mathbf{K}^m$. Take the number $b = |\pi|^{s_0}$ for a marked natural number s_0. Let h_R be a function on the subset $W := \bigcup_{U\in\mathcal{F}} U$ such that

$$h_R(x) := b \sup\{\inf(1, dist(x, \mathbf{K}^m \setminus T)) : T \in \mathcal{F}\}.$$

Since the normalization group $\Gamma_{\mathbf{K}}$ of the field $\mathbf{K}$ is discrete in the open subset $(0,\infty)$ of the real field and the norm $|*|$ is continuous from the field $\mathbf{K}$ into $\Gamma_{\mathbf{K}} \cup \{0\}$, the inclusion is accomplished $h_R(x) \in \Gamma_{\mathbf{K}} \cup \{0\}$ for each $x \in W$ and the function $h_R(x)$ is continuous, hence for each clopen or closed subset G in $\Gamma_{\mathbf{K}} \cup \{0\}$ its counter image $h_R^{-1}(G)$ is clopen or closed in the subset W respectively.

Therefore, W is the disjoint union of the closed set $h_R^{-1}(0)$ and the clopen subsets $h_R^{-1}(u)$ while $u \in \Gamma_{\mathbf{K}}$. Hence a continuous function $h : W \to \mathbf{K}$ exists such that $|h(x)| = h_R(x)$ for each $x \in W$, since h_R is continuous.

In accordance with Theorem 2.8.4 [25] if (X,ρ_X) is a metric space and F is a family of its closed subsets, δ is a nonnegative bounded function on the family F and $1 < \tau < \infty$, then the family F has a disjointed subfamily G such that for each $T \in F$ there exists $H \in G$ with $T \cap H \ne \emptyset$ and $\delta(T) \le \tau\delta(H)$.

With each $H \in F$ it is possible to associate its δ,τ enlargement:

$$\hat{H} := \bigcup\{T : T \in F, T \cap H \ne \emptyset, \delta(T) \le \tau\delta(H)\}.$$

Corollary 2.8.5 [25] states that

$$\bigcup_{T\in F} T \subset \bigcup_{H\in G} \hat{H}.$$

In the considered here situation we take

$$F := \{B(\mathbf{K}^m, x, h_R(x)) : x \in W\},$$

$\delta = diam$, $\tau = |\pi|^{s_1}$ for a marked integer number $s_1 \leq 0$. We choose $G \subset W$ so that the family of clopen balls $\{B(\mathbf{K}^m, y, h_R(y)) : y \in G\}$ is disjoined, $h_R(y) > 0$, and

$$\bigcup_{y \in G} B(\mathbf{K}^m, y, h_R(y)) = W,$$

where numbers s_0, s_1 are subordinated to the condition $|s_1| + 1 < s_0$. Evidently the subfamily G is countable.

By Lemma 21 we get that the inclusion $x \in W$ implies the inequalities

$$(1 - b\beta)/(1 + b\alpha) \leq h_R(x)/h_R(y) \leq (1 + b\beta)/(1 - b\alpha)$$

for all points $y \in G_x$. Taking $\alpha = \beta = b|\pi|^{s_2}$ we get

$$[(1 - |\pi|^{2s_0+s_2})/(1 + |\pi|^{2s_0+s_2}] < |h(x)|/|h(y)| \leq [(1 + |\pi|^{2s_0+s_2})/(1 - |\pi|^{2s_0+s_2}]$$

for every point $y \in G_x$. Moreover, the estimate of the cardinality

$$card(G_x) \leq |\pi|^{s_0+s_2}[(1 + |\pi|^{2s_0+s_2})/(1 - |\pi|^{2s_0+s_2})]^{2m}$$

is valid, where $s_0 \geq 1$, $s_2 \geq -1$ are integers.

Consider the mapping $w_y(x) := Ch_{B(\mathbf{K}^m,0,1)}((x - y)/(\pi h(y))$ for $y \in G$, $x \in \mathbf{K}^m$, where Ch_P is the characteristic function of a subset P in the topological vector space $\mathbf{K}^m$. Therefore, the support is: $supp w_y = B(\mathbf{K}^m, y, |\pi h(y)|)$. This function w_y is of C^∞ class with finite C^n norms for each $n \in \mathbf{N}$, since each partial difference quotient $\bar{\Phi}^j w_y(x; v; t)$ is bounded on $\mathbf{K}^m \times S(\mathbf{K}^m, 0, 1)^j \times \mathbf{K}^j$ for each $j \in \mathbf{N}$, where

$$S(\mathbf{K}^m, z, R) := \{x \in \mathbf{K}^m : |x - z| = R\}$$

for $R \in \Gamma_{\mathbf{K}}$, $|x| = \max_{j=1}^m |\,{}_jx|$, $x = ({}_1x, \dots, {}_mx)$. Indeed, $\bar{\Phi}^1 ch_B(x; v; t) = 0$ for $x, x + vt \in B$ or for $|x| > 1$ and $|x + vt| > 1$, $\bar{\Phi}^1 Ch_B(x; v; t) = 1/t$ for $|x| \leq 1$ and $|x + vt| > 1$ or $|x| > 1$ and $|x + vt| \leq 1$, where $B = B(\mathbf{K}^m, 0, 1)$ is the unit clopen ball. In the latter two cases $|t| > 1$ for $|v| = 1$.

Continuing by induction we get the inequalities:

$$|\bar{\Phi}^j ch_{B(\mathbf{K}^m,0,1)}(x; v; t)| \leq 1$$

for each $x \in \mathbf{K}^m$, $v \in S(\mathbf{K}^m, 0, 1)^j$ and $t \in \mathbf{K}^j$. Hence

$$|\bar{\Phi}^j w_y(x; v; t)| \leq |\pi h(y)|^{-j}$$

for each $(x; v; t) \in \mathbf{K}^m \times S(\mathbf{K}^m, 0, 1)^j \times \mathbf{K}^j$.

Choose now $G_0 \subset G$ with the additional condition that

$$\bigcup_{y \in G_0} B(\mathbf{K}^m, y, |\pi h(y)|) \supset W$$

and

$$B(\mathbf{K}^m, y, |\pi h(y)|) \cap B(\mathbf{K}^m, g, |\pi h(g)|) = \emptyset$$

for each $y \neq g \in G_0$ and consider the function

$$\phi(x) := \sum_{y \in G_0} w_y(x).$$

Then $\phi \in C^\infty$ and $\phi(x)|_W = 1$ for each $x \in W$. Thus the family of functions $\{w_y(x) : y \in G_0\}$ constitute the partition of unity on the subset W associated with the family $\mathcal{F}$. They are of class C^∞ and their supports form a disjoint clopen refinement $\mathcal{F}$ of covering of the subset W.

Consider the subset $U = \mathbf{K}^m \setminus A$ and put $\mathcal{F} = \{U\}$. Since the subset A is closed, the subset U is open in the topological vector space $\mathbf{K}^m$. Applying Lemma 21 we get $h_R(x)/b = \inf\{1, dist(x, A)\}$ for any $x \in U$. For each $y \in S$ we take $\psi(y) \in A$ with $|y - \psi(y)| = dist(y, A)$. Then we define a function $g : \mathbf{K}^m \to Y$ by the formula $g(x) = P_x(x)$ for each $x \in A$ and

$$g(x) = \sum_{y \in S} w_y(x) P_{\psi(x)}(x)$$

for each $x \in U$. Therefore, $g \in C^\infty(\mathbf{K}^m, \mathbf{K})$, since the partial difference quotients $\bar{\Phi}^j g$ are polynomials of $\bar{\Phi}^i w_y$ and $\bar{\Phi}^l P_{\psi(y)}$ with the corresponding arguments, where $1 \leq i \leq j$, $1 \leq l \leq j$ (see Corollary 2.6).

In the particular case of the topological vector space $X = \mathbf{K}^m$ we have

$$\bar{\Phi}^j f(z; y - z, \dots, y - z; 0, \dots, 0)$$

$$= \sum_{l(1), \dots, l(j) \in \{1, \dots, m\}} \bar{\Phi}^j f(z; e_{l(1)}, \dots, e_{l(j)}; 0, \dots, 0)({}_{l(1)}y - {}_{l(1)}z) \dots ({}_{l(j)}y - {}_{l(j)}z)$$

for $f \in C^n(U, Y)$, $j \leq n$, where U is an open subset in the topological vector space $\mathbf{K}^m$, $y, z \in U$. In view of Theorem C.1 (see Appendix C) we have:

$$\bar{\Phi}^j P_x(y^{(j)}) - \bar{\Phi}^j P_z(y^{(j)}) = \sum_{i=0}^{k} \sum_{\bar{l}_i} (\bar{\Phi}^i_y [\bar{\Phi}^j P_x(y^{(j)}) - \bar{\Phi}^j P_z(y^{(j)})](z^{(i+j)}_{\bar{l}_i})).(y - z)^i$$

$$+ \sum_{\bar{l}_{k-j}} R_{k-j}(\bar{\Phi}^j P_x(z^{(j)}) - \bar{\Phi}^j P_z(z^{(j)}); z^{(k)}_{\bar{l}_{k-j}}).(y - z)^{k-j},$$

where $x, z \in S$, $y \in \mathbf{K}^m$, $j \leq k$, $y^{(j)} = (y; v_1, \dots, v_j; t_1, \dots, t_j)$, $z^{(i+j)}_{\bar{l}_i} = (z; v_1, \dots, v_j, e_{l(1)}, \dots, e_{l(i)}; t_1, \dots, t_j, 0, \dots, 0)$, $v_i \in \mathbf{K}^m$, $t_i \in \mathbf{K}$, $\bar{l}_i = (l(1), \dots, l(i)) \subset \{1, \dots, m\}$ for each $i = 1, \dots, j$; also

$$(\bar{\Phi}^i_y [\bar{\Phi}^j P_x(y^{(j)})](z^{(i+j)}_{\bar{l}_i})).(y - z)^i$$

$$= \bar{\Phi}^{i+j} P_x(z^{(i+j)}_{\bar{l}_i})({}_{l(1)}y - {}_{l(1)}z) \dots ({}_{l(i)}y - {}_{l(i)}z);$$

$R_{k-j}(\bar{\Phi}^j P_x(z^{(j)}) - \bar{\Phi}^j P_z(z^{(j)}); z^{(k)}_{\bar{l}_{k-j}})$ is the continuous residue equal to zero for $x = z$ for each $\bar{l}_{k-j}$, where

$$R_{k-j}(f; z^{(j)}_{\bar{l}_{k-j}}).(y - z)^{k-j} := R_{k-j}(f; z^{(j)}_{\bar{l}_{k-j}})({}_{l(1)}y - {}_{l(1)}z) \dots ({}_{l(k-j)}y - {}_{l(k-j)}z).$$

Therefore, we infer that

$$\|\bar{\Phi}^j P_x(y^{(j)}) - \bar{\Phi}^j P_z(y^{(j)})\| \le \max_{0\le i\le k-j} (|y-z|^i |x-z|^{k-j-i}) \rho(S, |x-z|).$$

If $z \in A$, $G = A \cap B(\mathbf{K}^m, z, |\pi|^{-1})$ and $x \in U \cap B(\mathbf{K}^m, z, |\pi|)$, then we choose a point $y \in G$ with $|x-y| = dist(x,A)$. Hence the inequalities

$$|x-y| \le |x-z| \le |\pi|^{-1} \quad \text{also} \quad |\pi|^{s_0}|h(x)| = |x-y| \le |\pi|$$

$$\text{and} \quad |y-z| \le \max(|x-y|, |x-z|) \le |x-z| < |\pi|^{-1}$$

are satisfied. Take a number $s_2 \ge -1$ such that

$$[(1+|\pi|^{2s_0+s_2})/(1-|\pi|^{2s_0+s_2})] \le |\pi|^{-1},$$

consequently, $q \in G_x$ implies

$$|\pi|^{-s_0}|h(q)| \le |\pi|^{-s_0}[(1+|\pi|^{2s_0+s_2})/(1-|\pi|^{2s_0+s_2})]|h(x)| \le 1$$

and $|\pi|^{-s_0}|h(q)| = |q - \psi(q)|$,

$$|q-x| \le \max(|h(q)|, |h(x)|)|\pi|^{s_0+s_2} \le |\pi|^{s_0+s_2} \le 1 \quad \text{and}$$

$$|\psi(q)-z| \le \max(|\psi(q)-q|, |q-x|, |x-z|) \le \max(|\pi|^{-1}, 1, |\pi|^{-1}) = |\pi|^{-1}$$

and

$$|\psi(q)-y| \le \max(|\psi(q)-q|, |q-x|, |x-y|) \le \max(|\pi|^{-1}, 1, |\pi|^{-1}) = |\pi|^{-1},$$

since $s_0 + s_2 \ge 0$, where $\psi(q) \in G$.

If $f \in C^n(U, \mathbf{K})$ and $g \in C^n(U, Y)$, where a subset U is open in the topological vector space $\mathbf{K}^m$, then

$$(i) \quad \bar{\Phi}^n(fg)(x^{(n)}) = \sum_{0\le a, 0\le b, a+b=n} \sum_{j_1<\dots<j_a; s_1<\dots<s_b; \{j_1,\dots,j_a\}\cup\{s_1,\dots,s_b\}=\{1,\dots,n\}}$$

$$\bar{\Phi}^a f(x; v_{j_1}, \dots, v_{j_a}; t_{j_1}, \dots, t_{j_a}) \bar{\Phi}^b g(x + v_{j_1}t_{j_1} + \dots + v_{j_a}t_{j_a};$$

$$v_{s_1}, \dots, v_{s_b}; t_{s_1}, \dots, t_{s_b})$$

(see also Corollary 2.6).

Take $U_z = U_y = B(\mathbf{K}^m, z, |\pi|^\zeta)$, that is, $y \in B(\mathbf{K}^m, z, |\pi|^\zeta)$ with $\zeta \in \mathbf{N}$, then one deduces the estimates

$$\|\bar{\Phi}^i g(x^{(i)}) - \bar{\Phi}^i P_y(x^{(i)})\|_{C^0(V_y^{(i)}, \mathbf{K})}$$

$$\le \sup_{q\in G_x} \max_{0\le j\le i} \|\bar{\Phi}^{i-j} w_q\| \max_{0\le l\le j} (|x-y|^i |\psi(q)-y|^{j-i}) \rho(G, |\psi(q)-y|)$$

$$\le |\pi|^{-i} \rho(G, |\psi(q)-y|),$$

since

$$\bar{\Phi}^i g(x^{(i)}) - \bar{\Phi}^i P_y(x^{(i)}) = \sum_{q\in G} \bar{\Phi}^i (w_q(P_{\psi(q)} - P_y))(x^{(i)})$$

and due to Formula (i), also one has

$$\max_{0\le j\le i}\|\bar{\Phi}^{i-j}w_q\| \le \max_{0\le j\le i}|\pi h(q)|^{i-j} \le \max_{0\le j\le i}|\pi|^{(s_0+1)(i-j)} \le 1.$$

Thus, the inequalities follow

$$(ii)\quad \|\bar{\Phi}^i g(x^{(i)}) - \bar{\Phi}^i P_z(x^{(i)})\|_{C^0(V_z^{(i)},\mathbf{K})}$$

$$\le \max(\|\bar{\Phi}^i g(x^{(i)}) - \bar{\Phi}^i P_y(x^{(i)})\|_{C^0(V_z^{(i)},\mathbf{K})}, \|\bar{\Phi}^i P_y(x^{(i)}) - \bar{\Phi}^i P_z(x^{(i)})\|_{C^0(V_z^{(i)},\mathbf{K})}$$

$$\le |\pi|^{-i}\max(\rho(G,|\psi(q)-y|),\rho(G,|y-z|))$$

for each $i \le k$. Therefore, by induction relative to i we get

$$\bar{\Phi}^i g(x^{(i)}) = \bar{\Phi}^i P_x(x^{(i)})$$

for each point $x \in A$. From Formula (ii) it follows that the limit exists:

$$\lim_{x\to z}\|\bar{\Phi}^i g(x^{(i)}) - \bar{\Phi}^i P_z(x^{(i)})\|_{C^0(V_z^{(i)},\mathbf{K})}|x-z|^{i-k} = 0$$

for all points $z \in A$ and $1 \le i \le k$, hence $\bar{\Phi}^i g$ is continuous at each point z and for $i < k$ and $1 \le l \le k-i$ there are the identities

$$\bar{\Phi}^l(\bar{\Phi}^i g(z^{(i)}))(z^{i+l}) = \bar{\Phi}^l(\bar{\Phi}^i P_z(z^{(i)}))(z^{i+l}) = \bar{\Phi}^{i+l}P_z(z^{(i+l)}) = \bar{\Phi}^{i+l}g(z^{(i+l)}),$$

where $\bar{\Phi}^i g(z^{(i)})$ and $\bar{\Phi}^i P_z(z^{(i)})$ are defined on the domain $U_z^{(i)}$ such that $z \in U_z$, a subset U_z is open in the topological vector space $\mathbf{K}^m$ and it is sufficient to consider the clopen ball $U_z = B(\mathbf{K}^m, z, |\pi|^{\zeta})$.

23. Theorem. *If $A \subset W \subset \mathbf{K}^m$, $f \in C^k(W,\mathbf{K}^n)$ and*

$$\overline{\lim}_{x^{(k)}\to z^{(k)}}\|\bar{\Phi}^k f(x^{(k)}) - \bar{\Phi}^k f(z^{(k)})\|_{C^0(V_{z,R}^{(k)},\mathbf{K}^n)}/|x^{(k)}-z^{(k)}|^r < \infty$$

for each $z \in A$ and each $0 < R < \infty$, where a subset W is open in the topological vector space $\mathbf{K}^m$ and $0 < r \le 1$, a domain $V_{z,R}^{(k)}$ corresponds to the set $U_{z,R} = W \cap B(\mathbf{K}^m, z, R)$, then for each $\varepsilon > 0$ a mapping $g \in C^{k+1}(\mathbf{K}^m,\mathbf{K}^n)$ exists such that

$$\mu^m(A \setminus \{x : f(x) = g(x)\}) < \varepsilon.$$

Proof. In accordance with Theorem 20 we have

$$\mu^{m+(m+1)(k+1)}(A^{(k+1)} \setminus dom(\bar{\Phi}^{k+1}f)) = 0$$

and due to Theorem 15 the partial difference quotient $\bar{\Phi}^{k+1}f$ is $\mu^{m+(m+1)(k+1)}|_{dom(\bar{\Phi}^{k+1}f)}$ measurable.

From Lousin's Theorem 2.3.5 [25] it follows, that a closed subset E in $A^{(k+1)}$ exists such that the restriction $\bar{\Phi}^{k+1}f(x^{(k+1)})|_E$ is continuous and $\mu^{m+(m+1)(k+1)}$ $(A^{(k+1)} \setminus E) < \varepsilon$, where a set $A^{(k+1)}$ is defined analogously to $U^{(k+1)}$. Practically the partial difference quotient $\bar{\Phi}^{k+1}f(x^{(k+1)})$ is also continuous on the set $W^{(k)} \times \mathbf{K}^m \times \mathbf{K} \setminus \{0\}$, that is, for $t_{k+1} \neq 0$. We put

$$\phi_q(z^{(k)}) := \sup_{0 \le j \le k} \{\|\bar{\Phi}^j f(x^{(j)}) - \bar{\Phi}^j f(z^{(j)})$$

$$-\bar{\Phi}^1(\bar{\Phi}^j f(z^{(j)}))(z^{(j)}; x^{(j)} - z^{(j)}; 1)\| / |x^{(j)} - z^{(j)}|^r : x^{(j)} \in V^{(j)}_{z,|\pi|^q}; x^{(j)} \neq z^{(j)}\},$$

where $z^{(k)} = (z; v_1, \dots, v_k; t_1, \dots, t_k) \in A^{(k)}$, $z^{(j)} = (z; v_1, \dots, v_j; t_1, \dots, t_j)$, $q = 1, 2, \dots$. Then the limit

$$\lim_{q \to \infty} \phi_q(z^{(k)}) = 0$$

exists for each point $z \in E$. Each function ϕ_q is borelian.

Theorem 2.2.2 [25] states: suppose non-negative ϕ is a measure over a metric space X, all open subsets of X are ϕ measurable, and B is a Borel set;

(1) if $\phi(B) < \infty$ and $\varepsilon > 0$, then the set B contains a closed subset C for which $\phi(B \setminus C) < \varepsilon$;

(2) if B is contained in the union of countably many open sets V_i with $\phi(V_i) < \infty$, and if $\varepsilon > 0$, then the set B is contained in an open set W for which $\phi(W \setminus B) < \varepsilon$.

While Egoroff's Theorem 2.3.7 [25] is: suppose $f_1, f_2, \dots$ and g are ϕ measurable functions with values in a separable metric space Y, where nonnegative ϕ is a measure over X; if $\phi(A) < \infty$, $A \subset X$, $f_n(x) \to g(x)$ for almost all x in A, and $\varepsilon > 0$, then a ϕ measurable set B exists such that $\phi(A \setminus B) < \varepsilon$ and $f_n(x) \to g(x)$, uniformly for $x \in B$, as $n \to \infty$.

In view of Egoroff's 2.3.7 and 2.2.2 Theorems [25] for each $\varepsilon > 0$ a closed subset F in $A^{(k)}$ exists such that

$$\mu^{m+(m+1)k}(A^{(k)} \setminus F) < \varepsilon \quad \text{and} \quad \lim_{q \to \infty} \sup_{z \in J} \phi_q(z) = 0$$

for each compact subset J in F.

We take the sets G and G_ε from Theorem 17 and for a point $z \in G_\varepsilon$ consider the Taylor expansion of Theorem C.1 with k here instead of n there and put

$$P_z(y) = f(z) + \sum_{j=1}^{k+1} \bar{\Phi}^j f(z; y - z, \dots, y - z; 0, \dots, 0),$$

hence $P_y(y) = f(y)$. On the other hand $\mu^m(A \setminus G) = 0$ due to Theorems 17 and 19. We show that the suppositions of Theorem 22 are satisfied for the topological vector space $Y = \mathbf{K}^n$. Let H be a compact subset in the set G_ε and $y, z \in H$. In accordance with Theorem C.1 we have:

$$\bar{\Phi}^i P_y(y^{(i)}) - \bar{\Phi}^i P_z(y^{(i)}) = \bar{\Phi}^i f(y^{(i)}) - \bar{\Phi}^i f(z^{(i)})$$

$$-\Big[\sum_{j=i+1}^{k+1} \bar{\Phi}^j f(z; v_1, \dots, v_i, y - z, \dots, y - z; t_1, \dots, t_i, 0, \dots, 0)\Big]$$

for each $i = 0, \dots, k-1$ and $0 < |y - z| < |\pi|^q$, where $y^{(i)} = (y; v_1, \dots, v_i; t_1, \dots, t_i)$ and $z^{(i)} = (z; v_1, \dots, v_i; t_1, \dots, t_i)$. Therefore, we infer that

$$\sup_{0 < |y-z| \le |\pi|^q; y, z \in H; 0 \le i \le k} |\bar{\Phi}^i P_y(y^{(i)}) - \bar{\Phi}^i P_z(y^{(i)})| |y - z|^{i-k-1} \le \phi_q(z^{(k)}) < \infty,$$

where $z^{(k)} = (z; v_1, \dots, v_i, y-z, \dots, y-z; t_1, \dots, t_i, 0, \dots, 0)$. Since $\bar{\Phi}^{k+1} f$ is continuous on G, the difference of the partial difference quotients $\bar{\Phi}^{k+1} P_y(y^{(k+1)}) - \bar{\Phi}^{k+1} P_z(y^{(i)}) = \bar{\Phi}^{k+1} f(y^{(k+1)}) - \bar{\Phi}^{k+1} f(z^{(k+1)})$ is sufficiently small by its norm for small enough $|y^{(k+1)} - z^{(k+1)}|$. Thus from Theorem 22 the statement of this theorem follows.

24. Theorem. 1. *If $A \subset \mathbf{K}^m$, $f: A \to \mathbf{K}^n$ and an approximate limit $ap\overline{\lim}_{x \to z} |f(x) - f(z)|/|x-z|^r$ exists for μ^m-almost all points $z \in A$, where $0 < r \le 1$, then for each $\varepsilon > 0$ a mapping $g \in C^1(\mathbf{K}^m, \mathbf{K}^n)$ exists such that*

$$\mu^m(A \setminus \{x : f(x) = g(x)\}) < \varepsilon.$$

2. *If $A \subset W \subset \mathbf{K}^m$, $f \in C^k(W, \mathbf{K}^n)$ and an approximate limit*

$$ap\overline{\lim}_{x^{(k)} \to z^{(k)}} \|\bar{\Phi}^k f(x^{(k)}) - \bar{\Phi}^k f(z^{(k)})\|_{C^0(V^{(k)}_{z,R}, \mathbf{K}^n)} / |x^{(k)} - z^{(k)}|^r < \infty$$

exists for μ^m almost all points $z \in A$ and each $0 < R < \infty$, where a subset W is open in the topological vector space $\mathbf{K}^m$ and $0 < r \le 1$, the subset A is μ^m measurable, a domain $V^{(k)}_{z,R}$ corresponds to the intersection $U_{z,R} = W \cap B(\mathbf{K}^m, z, R)$, $k \ge 1$, then for each $\varepsilon > 0$ a mapping $g \in C^{k+1}(\mathbf{K}^m, \mathbf{K}^n)$ exists such that

$$\mu^{m+(m+1)k}(W^{(k)}_A \setminus \{x^{(k)} : \bar{\Phi}^k f(x^{(k)}) = \bar{\Phi}^k g(x^{(k)})\}) < \varepsilon,$$

where $W^{(k)}_A := \{(x; v_1, \dots, v_k; t_1, \dots, t_k) : x \in A; v_1, \dots, v_k \in \mathbf{K}^m; t_1, \dots, t_k \in \mathbf{K}; x + v_1 t_1 + \dots + v_j t_j \in W \quad \forall 1 \le j \le k\}$.

Proof. In accordance with Theorems 19 and 20 compact subsets $H_1, H_2, \dots$ exist such that

$$H_i \subset B(\mathbf{K}^{m+(m+1)k}, 0, |\pi|^{-i}) \setminus B(\mathbf{K}^{m+(m+1)k}, 0, |\pi|^{-i+1})$$

for $i > 1$ and $H_1 \subset B(\mathbf{K}^{m+(m+1)k}, 0, 1/|\pi|)$ and the restrictions of the partial difference quotients $\bar{\Phi}^k f|_{H_i}$ are lipschitzian,

$$\mu^{m+(m+1)k}(A^{(k)} \cap [(B(\mathbf{K}^{m+(m+1)k}, 0, |\pi|^{-i}) \setminus B(\mathbf{K}^{m+(m+1)k}, 0, |\pi|^{-i+1})) \setminus H_i]) < \varepsilon 2^{-i}$$

for $i > 1$ and $\mu^{m+(m+1)k}(W^{(k)}_A \cap [B(\mathbf{K}^{m+(m+1)k}, 0, 1/|\pi|) \setminus H_1]) < \varepsilon/2$, where $W^{(0)}_A = A$, $\bar{\Phi}^0 f = f$, $k = 0$ in the first case and $k > 0$ in the second case. Using Theorem 8 by induction we construct a function $h: \mathbf{K}^{m+(m+1)k} \to \mathbf{K}^n$ such that its restriction $h|_{B(\mathbf{K}^{m+(m+1)k}, 0, |\pi|^{-i})}$ is a lipschitzian extension of the restricted function $h|_{B(\mathbf{K}^{m+(m+1)k}, 0, |\pi|^{-i+1})} \cup \bar{\Phi}^k f|_{H_i}$. Therefore, h is locally lipschitzian and

$$\mu^{m+(m+1)k}(W^{(k)}_A \setminus \{x^{(k)} : \bar{\Phi}^k f(x^{(k)}) = h(x^{(k)})\}) < \varepsilon.$$

Applying Theorem 23 with $\mathbf{K}^m, \mathbf{K}^n, h, k$ instead of A, W, f, k gives the assertion of this theorem.

25. Theorem. *Let $f: \mathbf{K}^m \to \mathbf{K}^n$, $m, n \in \mathbf{N}$. Let also $f \circ u \in C^{s,r}(\mathbf{K}, \mathbf{K}^n)$ for each function $u \in C^\infty(\mathbf{K}, \mathbf{K}^m)$, where s is a nonnegative integer, $0 < r \le 1$, a μ^m-measurable subset G in* *the topological vector space* $\mathbf{K}^m$ *exists such that* $f \in C^{s+1}(G, \mathbf{K}^n)$*, where* $\mu^m(\mathbf{K}^m \setminus G) = 0$. *Moreover, for each* $\varepsilon > 0$ *a mapping* $g \in C^{s+1}(\mathbf{K}^m, \mathbf{K}^n)$ *exists such that*

$$\mu^m(\mathbf{K}^m \setminus \{x : f(x) = g(x)\}) < \varepsilon.$$

Proof. In accordance with Theorem 2.33 we have the inclusion $f \in C^{s,r}(\mathbf{K}^m, \mathbf{K}^n)$. Then by Theorem 9 the partial difference quotient $\bar{\Phi}^{s+1} f(x^{(s+1)})$ is continuous for almost all points $x^{(s+1)} \in \mathbf{K}^m \times (\mathbf{K}^m)^{s+1} \times (\mathbf{K})^{s+1}$ on a set $G_{0,s+1}$ and the derivative $D^i \bar{\Phi}^j f(x^{(j)})$ is continuous for almost all points of the topological vector space $\mathbf{K}^m \times (\mathbf{K}^m)^j \times (\mathbf{K})^j$ on a set $G_{i,j}$ for each $0 \le i, j \le s$ with $i + j \le s+1$. Since $G^{(i)}_{i,j} \subset G_{0,i+j}$ for each i, j and $G_{0,k} = \mathbf{K}^m \times (\mathbf{K}^m)^k \times (\mathbf{K})^k$ for each $k \le s$, the inclusion $G \supset G_{0,s+1}$ is fulfilled and $\mu^m(\mathbf{K}^m \setminus G) = 0$.

The second statement follows from Theorem 23.

26. Theorem. 1. *Suppose* $f : \mathbf{K}^m \to \mathbf{K}^n$*,* $m, n \in \mathbf{N}$ *and* $f \circ u \in {}_{ap}C^{s+1}(\mathbf{K}, \mathbf{K}^n)$ *for each function* $u \in C^\infty(\mathbf{K}, \mathbf{K}^m)$*, where* s *is a nonnegative integer, then the inclusion* $f \in {}_{ap}C^{s+1}(\mathbf{K}^m, \mathbf{K}^n)$ *is accomplished.*

2. *If* $A \subset \mathbf{K}$*,* $f \circ u \in {}_{ap}C^{s,r}(\mathbf{K}, A, \mathbf{K}^n)$ *for each* $u \in C^\infty(\mathbf{K}, \mathbf{K}^m)$*, where* $0 < r \le 1$*, a subset* A *is* μ *measurable, then for each* $\varepsilon > 0$ *a mapping* $g \in C^{s+1}(\mathbf{K}^m, \mathbf{K}^n)$ *exists such that*

$$\mu^{m+(m+1)s}((\mathbf{K}^m_{A^m})^{(s)} \setminus \{x^{(s)} : \bar{\Phi}^s f(x^{(s)}) = \bar{\Phi}^s g(x^{(s)})\}) < \varepsilon.$$

Proof. In view of Theorem 2.39 the inclusion $f \in C^s(\mathbf{K}^m, \mathbf{K}^n)$ is valid.

1. It remains to prove, that $\bar{\Phi}^s f \in {}_{ap}C^1((\mathbf{K}^m)^{(s)}, \mathbf{K}^n)$, where the decomposition $(\mathbf{K}^m)^{(s)} = \mathbf{K}^m \times (\mathbf{K}^m)^s \times (\mathbf{K})^s$ is used and $\bar{\Phi}^s f \circ u \in {}_{ap}C^1((\mathbf{K})^{(s)}, \mathbf{K}^n)$ for each infinite differentiable function $u \in C^\infty(\mathbf{K}, \mathbf{K}^m)$.

Therefore, it is sufficient to prove, assertion 1 for $s = 0$ up to a choice of our notation. But $f \circ u \in {}_{ap}C^1(\mathbf{K}, \mathbf{K}^n)$ for each $u \in C^\infty(\mathbf{K}, \mathbf{K}^m)$ means in particular this for $u(t_0) = x$, $u(t_1) - u(t_0) = e_i t$ with $t_0 \ne t_1 \in \mathbf{K}$, that the partial difference quotient $\bar{\Phi}^1 f(x; e_i; t)$ is μ^{m+1}-almost everywhere continuous on the topological vector space $\mathbf{K}^m \times \mathbf{K}$ and a linear mapping $T_i : \mathbf{K} \to \mathbf{K}^n$ exists such that

$$(\mu, V) ap \lim_{t \to 0} |f(x + e_i t) - f(x) - T_i t| / |t| = 0$$

for each $i = 1, \dots, m$, where $V = \mathcal{B}(\mathbf{K})$.

Since i is arbitrary, the partial difference quotient $\bar{\Phi}^1 f(x; v; t)$ is μ^{2m+1} almost everywhere on $\mathbf{K}^{2m+1}$ continuous due to combinatorial Formula 2.9(2) and a linear mapping $T : \mathbf{K}^m \to \mathbf{K}^n$ exists such that $(\mu^m, V) ap \lim_{z \to x} |f(z) - f(x) - T(z - x)| / |z - x| = 0$, where $V = \mathcal{B}(\mathbf{K}^m)$, since $\mu^m = \otimes_{i=1}^m \mu$. Indeed, $\mathcal{B}(\mathbf{K}^m) = \otimes_{i=1}^m \mathcal{B}(\mathbf{K})$ is the minimal σ algebra generated from the family of all subsets of the form $A_1 \times \dots \times A_m$ with $A_i \in \mathcal{B}(\mathbf{K})$ for each $i = 1, \dots, m$, consequently, the inclusion $f \in {}_{ap}C^1(\mathbf{K}^m, \mathbf{K}^n)$ is satisfied, where $Te_i t = T_i t$ for each $t \in \mathbf{K}$ and each $e_i = (0, \underset{i}{\dots}, 0, 1, 0, \dots, 0) \in \mathbf{K}^m$.

2. We will prove, that $\bar{\Phi}^s f \in {}_{ap}C^{0,r}((\mathbf{K}^m)^{(s)}, A^m, \mathbf{K}^n)$, then the second statement will follow from Theorem 24.2. Thus up to the notation it is sufficient to prove the second assertion for $s = 0$. We have the inclusion $f \circ u \in {}_{ap}C^{0,r}(\mathbf{K}, A, \mathbf{K}^n)$ for each infinite differentiable function $u \in C^\infty(\mathbf{K}, \mathbf{K}^m)$, particularly, for $u(t_0) = z$, $u(t_1) - u(t_0) = e_i t$ with $t_0 \ne t_1 \in \mathbf{K}$. Thus the approximate limit

$$(\mu, V) ap \overline{\lim}_{t \to 0} \| f(z + e_i t) - f(z) \|_{C^0(W_{t_0,R}, \mathbf{K}^n)} / |t|^r < \infty$$

exists for μ-almost all points $t_0 \in A$ and each $0 < R < \infty$, where $W_{t_0,R} = B(\mathbf{K}, t_0, R)$ is taken to be the clopen ball, also $V = \mathcal{B}(\mathbf{K})$, consequently,

$$(\mu^m, V)ap\overline{\lim}_{x\to z}\|f(z) - f(x)\|_{C^0(U_{z,R},\mathbf{K}^n)}/|x-z|^r$$

$$\leq \max_{i=1}^{m} ap\overline{\lim}_{\max_i |t_i|\to 0}\|f(z+e_1t_1+\cdots+e_it_i)$$

$$-f(z+e_1t_1+\cdots+e_{i-1}t_{i-1})\|_{C^0(W_{t_{0,i},R},\mathbf{K}^n)}/|t_i|^r < \infty \quad (1)$$

for μ^m almost all points $t_0 = (t_{0,1},\dots,t_{0,m}) \in A^m$ and each $0 < R < \infty$, where $V = \mathcal{B}(\mathbf{K}^m)$. Indeed, $\mu^m = \otimes_{i=1}^m \mu$ and $\mathcal{B}(\mathbf{K}^m) = \otimes_{i=1}^m \mathcal{B}(\mathbf{K})$ is the minimal σ algebra generated from the family of all subsets of the form $A_1 \times \cdots \times A_m$ with $A_i \in \mathcal{B}(\mathbf{K})$ for each $i = 1,\dots,m$, where $U_{z,R} = B(\mathbf{K}^m, z, R)$, $x = z + e_1t_1 + \cdots + e_mt_m$, $t_i = t_{1,i} - t_{0,i}$, ${}_iz = {}_iu(t_{0,i})$, $z = ({}_1z,\dots,{}_mz)$. Thus (1) is satisfied for μ^m almost all points $z \in A^m$ and the application of Theorem 24.2 implies assertion 2.

6.3. Appendix C. Taylor Formula

1. Theorem. *Let $f \in C^{n+1}(U,Y)$, where X and Y are topological vector spaces over a field $\mathbf{K}$ of zero or a positive characteristic, $n \in \mathbf{N}$ is a natural number. Let also either U be clopen in X or X be locally convex. Then for each x and $y \in U$ the formula*

$$f(x) = f(y) + \sum_{j=1}^{n} \bar{\Phi}^j f(y; x-y,\dots,x-y; 0,\dots,0) + R_{n+1}(f;x,y).(x-y)^{\otimes(n+1)} \quad (1)$$

holds, where $R_{n+1}(f;x,y) = R_{n+1}(x,y) : U^2 \to L_{n+1}(X^{\otimes(n+1)}, Y)$ with

$$\lim_{x\to y} R_{n+1}(x,y) = 0,$$

$L_n(X^{\otimes n}, Y)$ denotes the space of n polylinear continuous operators from the n-fold product space $X^{\otimes n}$ into Y.

Proof. If $j \leq n+1$, then $\bar{\Phi}^j f(z; v_1,\dots,v_j; 0,\dots,0)$ is the j polylinear operator by vectors $v_1,\dots,v_j$ as follows from application of Lemma B.2.2. For $n = 0$ we take

$$R_1(x,y).(x-y) := f(x) - f(y) - \bar{\Phi}^1 f(y; x-y; 0) = \bar{\Phi}^1 f(y; x-y; 1) - \bar{\Phi}^1 f(y; x-y; 0).$$

For $n = 1$ from the definition of $\bar{\Phi}^2 f$ we have $f(x) - f(y) = \bar{\Phi}^1 f(y; x-y; 1)$ and $R_2(x,y).(x-y)^{\otimes 2} := \bar{\Phi}^2 f(y; x-y, x-y; 0, 1) - \bar{\Phi}^1 f(y; x-y, x-y; 0, 0)$. Let the statement be true for $n-1$, then from

$$\bar{\Phi}^n(y + t_{n+1}(x-y); x-y,\dots,x-y; t_1,\dots,t_n) = \bar{\Phi}^n(y; x-y,\dots,x-y; t_1,\dots,t_n)$$

$$+(\bar{\Phi}^1(\bar{\Phi}^n f(y; x-y,\dots,x-y; t_1,\dots,t_n))(y; x-y; t_{n+1}))t_{n+1}$$

and the continuity of $\bar{\Phi}^{n+1} f$ it follows that

$$R_{n+1}(x,y).(x-y)^{\otimes(n+1)} = \bar{\Phi}^n f(y; x-y,\dots,x-y; 0,\dots,0,1)$$

$$-\bar{\Phi}^n f(y;x-y,\dots,x-y;0,\dots,0,0),$$

hence $R_{n+1}(f;x,y) = R_{n+1}(x,y) : U^2 \to L_{n+1}(X^{\otimes(n+1)},Y)$. Since $f \in C^{n+1}(U,Y)$, the limit

$$\lim_{x\to y} R_{n+1}(x,y) = 0$$

is zero. In general the entire correction term $R_{n+1}(f;x,y).(x-y)^{\otimes(n+1)}$ need not be poly-linear, because $R_{n+1}(f;x,y)$ may be nonlinear by $x-y$, but it is useful to write it in such form.

Considering given $x,y \in U$ we can associate with it an intersection of $\{tx+(1-t)y : t \in \mathbf{K}\}$. If U is clopen, then f has a C^{n+1} extension on X, so we can consider a $\mathbf{K}$ convex clopen subset U_1 such that $U \subset U_1 \subset X$ instead of an initial one and denote it also by U. Thus, under suppositions of this theorem on U and X we can consider the case $\{tx+(1-t)y : t \in B(\mathbf{K},0,1)\} \subset U$ for each $x,y \in U$ without loss of generality. Therefore,

$$R_{n+1}(y+vt,y).(tv)^{\otimes(n+1)} = t^{n+1}R_{n+1}(y+vt,y).v^{\otimes(n+1)}$$

$$= (\bar{\Phi}^1(\bar{\Phi}^n f(y;vt,\dots,vt;0,\dots,0))(y;v;t))t$$

$$= t^{n+1}(\bar{\Phi}^1(\bar{\Phi}^n f(y;v,\dots,v;0,\dots,0))(y;v;t)),$$

where $x-y=vt$. Thus, the consideration can be reduced to x and y along lines containing x and y. This gives

$$R_{n+1}(y+vt,y).v^{\otimes(n+1)} = (\bar{\Phi}^1(\bar{\Phi}^n f(y;v,\dots,v;0,\dots,0))(y;v;t)),$$

where $\bar{\Phi}^n f(y;v_1,\dots,v_n;0,\dots,0) \in L_n(X^{\otimes n},Y)$ and

$$\bar{\Phi}^1(\bar{\Phi}^n f(y;v_1,\dots,v_n;t_1,\dots,t_n))(y;v_{n+1};t_{n+1}))$$

$$= \bar{\Phi}^{n+1} f(y;v_1,\dots,v_{n+1};t_1,\dots,t_{n+1})$$

is continuous by $(y;v_1,\dots v_{n+1};t_1,\dots,t_n,t_{n+1}) \in U^{(n+1)}$ with

$$\lim_{v_{n+1}\to 0} \bar{\Phi}^1(\bar{\Phi}^n f(y;v_1,\dots,v_n;0,\dots,0))(y;v_{n+1};t)) = 0$$

for each t such that $(y;v_1,\dots,v_{n+1};0,\dots,0,t) \in U^{(n+1)}$, since

$$\bar{\Phi}^{n+1} f(y;v_1,\dots,v_{n+1};0,\dots,0) \in L_{n+1}(X^{\otimes(n+1)},Y)$$

so that

$$\bar{\Phi}^{n+1} f(y;v_1,\dots,v_n,0;0,\dots,0) = 0,$$

while

$$\bar{\Phi}^{n+1} f(y;v_1,\dots v_n,0;t_1,\dots,t_n,t_{n+1}) = 0$$

for $t_{n+1} \neq 0$ due to the definition of $\bar{\Phi}^{n+1} f$ and $\bar{\Phi}^1 f$. This proves the desired limit property of the residue R_{n+1} due to the continuity of $\bar{\Phi}^j f$ for each $j \leq n+1$, since the addition of vectors and multiplication of vectors on scalars are continuous operations in X.

Here the derivative operator D^i is not used, but the partial difference quotients $\bar{\Phi}^i$ are used instead, then multipliers $1/i!$ does not appear. If use derivatives $f^{(j)} = D^j f$ so that

$$\bar{\Phi}^j f(z; v_1, \dots, v_j; 0, \dots, 0) = f^{(j)}.(v_1, \dots, v_j)/j!$$

in the Taylor formula, then the restriction $n+1 < char(\mathbf{K})$ would be necessary, but in Formula (1) they are not used and the restriction on n is not necessary in this theorem. Thus the Taylor formula (1) is true for $char(\mathbf{K}) = p > 0$ with the decomposition up to terms of order not only $n+1 < p$ but also $n+1 \geq p$.

This Formula (1) uses only the partial difference quotients $\bar{\Phi}^i$, so in this respect they are more important, than the partial difference quotients Υ^j of the second type given in the preceding sections A and B.

6.4. Appendix D. Anti-derivation Operators

1. Definitions and Notes. Let X be a Banach space over a local field $\mathbf{K}$. Suppose M is an analytic manifold modelled on a Banach space X over $\mathbf{K}$ with an atlas $At(M)$ consisting of disjoint clopen charts (U_j, ϕ_j), $j \in \Lambda_M$, $\Lambda_M \subset \mathbf{N}$. That is, U_j and $\phi_j(U_j)$ are clopen subsets in M and X respectively, $\phi_j : U_j \to \phi_j(U_j)$ are homeomorphisms, $\phi_j(U_j)$ are bounded in X. Let $X = c_0(\alpha, \mathbf{K})$ be the Banach space (see above) with the standard orthonormal base $(e_i : i \in \alpha)$, where α is an ordinal, $\alpha \geq 1$ [53]. Its cardinality is called a dimension $card(\alpha) =: dim_{\mathbf{K}} c_0(\alpha, \mathbf{K})$ of X over the field $\mathbf{K}$.

Then $C^t(M, Y)$ for the manifold M with a finite atlas $At(M)$, $card(\Lambda_M) < \aleph_0$, denotes a Banach space of functions $f : M \to Y$ supplied with an ultra-norm

$$\|f\|_t = \sup_{j \in \Lambda_M} \|f|_{U_j}\|_{C^t(U_j, Y)} < \infty, \tag{1}$$

where $Y := c_0(\beta, \mathbf{K})$ is the Banach space over the field $\mathbf{K}$, $0 \leq t \in \mathbf{R}$, their restrictions $f|_{U_j}$ are in $C^t(U_j, Y)$ for each j, $\beta \geq 1$.

For $b \in \mathbf{R}$, $0 < b < 1$, we consider the following mapping:

$$j_b(\zeta) := p^{b \times ord_p(\zeta)} \in \blacksquare_{\mathbf{p}} \tag{2}$$

for $\zeta \neq 0$, $j_b(0) := 0$, such that $j_b(*) : \mathbf{K} \to \blacksquare_{\mathbf{p}}$, where $p = char(k)$ is the characteristic of the residue class field of the local field $\mathbf{K}$, $p^{-ord_p(\zeta)} := |\zeta|_{\mathbf{K}}$, $\blacksquare_{\mathbf{p}}$ is a field with a normalization group $\{|x| : 0 \neq x \in \blacksquare_{\mathbf{p}}\} = (0, \infty) \subset \mathbf{R}$ such that $\mathbf{K} \subset \blacksquare_{\mathbf{p}}$ [17,99,103,111]. The multiplicative norm on $\blacksquare_{\mathbf{p}}$ is the extension of that of on $\mathbf{K}$. We suppose that the field $\blacksquare_{\mathbf{p}}$ is complete relative to its norm. Then we denote $j_1(x) := x$ for each $x \in \mathbf{K}$. Let

$$\Phi^b F(x; h; \zeta) := (F(x + \zeta h) - F(x))/j_b(\zeta) \in Y_{\blacksquare_{\mathbf{p}}} \tag{3}$$

be partial difference quotients of order b for $0 < b < 1$, $x + \zeta h \in U$, $\zeta h \neq 0$, $\Phi^0 F := F$, where $Y_{\blacksquare_{\mathbf{p}}}$ is a Banach space obtained from Y by extension of a scalar field from $\mathbf{K}$ to $\blacksquare_{\mathbf{p}}$. By induction we define partial difference quotients of orders $n+1$ and $n+b$:

$$\Phi^{n+1} F(x; h_1, \dots, h_{n+1}; \zeta_1, \dots, \zeta_{n+1}) :$$

$$= \{\Phi^n F(x+\zeta_{n+1}h_{n+1};h_1,\dots,h_n;\zeta_1,\dots,\zeta_n) - \Phi^n F(x;h_1,\dots,h_n;$$
$$\zeta_1,\dots,\zeta_n)\}/\zeta_{n+1} \text{ and } (\Phi^{n+b}F) = \Phi^b(\Phi^n F) \tag{4}$$

and derivatives $F^{(n)} = (F^{(n-1)})'$, $D^bF(x).h = \lim_{\zeta\to 0}\Phi^b F(x;h;\zeta)$, $D^{n+b}F = (n+sign(b))!\lim_{\zeta_1,\dots,\zeta_{n+1}\to 0}\Phi^{n+b}F(x;h_1,\dots,h_{n+1};\zeta_1,\dots,\zeta_{n+1})$, where $sign(\alpha) = 1$ for $\alpha > 0$, $sign(\alpha) = -1$ for $\alpha < 0$, while $sign(0) = 0$. In particular, $\partial^j F(x) := D^j F(x).(e_{j_1},\dots,e_{j_m})$ for $j = e_{j_1}+\dots+e_{j_m}$, $m\in\mathbf{N}$.

2. Let X, Y and M be the same as in §1 for the local field $\mathbf{K}$. We take the standard orthonormal bases $\{e_j : j\}$ in X and $\{q_i : i\}$ in Y. When X or Y are infinite-dimensional over the field $\mathbf{K}$, the Banach space $C^t(M,Y)$ is in general of non-separable type over $\mathbf{K}$ for $0\le t\in\mathbf{R}$.

We denote by $C_0^t(M,Y)$ a completion of a subspace of cylindrical functions restrictions of which on each chart $f|_{U_l}$ are finite $\mathbf{K}$-linear combinations of basic functions $\{\bar{Q}_{\bar{m}}(x_{\bar{m}})q_i|_{U_l} : i\in\beta, m\}$ relative to the following norm:

$$\|f\|_{C_0^t(M,Y)} := \sup_{i,m,l}|a(m,f^i|_{U_l})|J_l(t,m), \tag{1}$$

where multipliers $J_l(t,m)$ are defined as follows:

$$J_l(t,m) := \|\bar{Q}_{\bar{m}}|_{U_l}\|_{C^t(\phi_l(U_l)\cap\mathbf{K^n},\mathbf{K})}, \tag{2}$$

$m\in c_0(\alpha,\mathbf{Q_p})$ with components $m_i\in\mathbf{N_o} := \{0,1,2,\dots,\}$, non-zero components of m are $m_{i_1},\dots,m_{i_n}$ with $n\in\mathbf{N}$, $\bar{m} := (m_{i_1},\dots,m_{i_n})$ for each $m\ne 0$, $x_{\bar{m}} := (x^{i_1},\dots,x^{i_n})\in\mathbf{K^n}\hookrightarrow X$, $Q_0 := 1$. Here each basic function $\bar{Q}_{\bar{m}}(x_{\bar{m}})$ is the product of basic Amice polynomials $Q_{m_j}(x^j)$,

$$\bar{Q}_{\bar{m}}(x_{\bar{m}}) = \prod_{j=1}^n Q_{m_j}(x^j),$$

when $\phi_l(U_l)$ is the ball in X, where $n = n(m)$, x^j are local coordinates in the chart U_l (see also [3]).

3. Lemma. *If $f\in C_0(t,M\to Y)$, then*

$$(f|_{U_j})(x) = \sum_{i,m} a(m,f^i|_{U_j})\bar{Q}_{\bar{m}}(x_{\bar{m}})q_i|_{U_j} \tag{1}$$

for each $j\in\Lambda_M$, where $a(m,f^i|_{U_j})\in\mathbf{K}$ are expansion coefficients such that for each $\varepsilon>0$ a set

$$\{(i,m,j) : |a(m,f^i|_{U_j})|J(t,m) > \varepsilon\} \text{ is finite }. \tag{4}$$

Proof. This follows immediately from the definition, since the decomposition

$$f(x) = \sum_{i\in\beta} f^i(x)q_i \tag{2}$$

has the convergent series, where $f^i(*)\in C_0^t(M,\mathbf{K})$.

In view of Formulas $(1-5)$ the space $C_0^t(M,Y)$ is of separable type over the field $\mathbf{K}$, when $card(\alpha\times\beta\times\Lambda_M)\le\aleph_0$. Evidently, for compact M the spaces $C_0^t(M,Y)$ and $C_0^t(M,Y)$ are isomorphic.

4. Now we define uniform spaces of the corresponding mappings from one manifold into another, which are necessary for the subsequent definitions of wrap semigroups and groups.

Let N be an analytic manifold modelled on Y with an atlas

$$At(N)=\{(V_k,\psi_k):\ k\in\Lambda_N\},\text{ such that }\psi_k:V_k\to\psi_k(V_k)\subset Y \tag{1}$$

are homeomorphisms, $card(\Lambda_N)\le\aleph_0$ and let $\theta:M\to N$ be a $C^{t'}$-mapping, also $card(\Lambda_M)<\aleph_0$, where each subset V_k is clopen in N, $t'\ge\max(1,t)$ is the index of a class of smoothness, that is, for each admissible (i,j) the following two conditions are satisfied:

$$\theta_{i,j}\in C_*(t',U_{i,j}\to Y) \tag{2}$$

with $*$ empty or an index $*$ taking value 0 respectively,

$$\theta_{i,j}:=\psi_i\circ\theta|_{U_{i,j}}, \tag{3}$$

where $U_{i,j}:=[U_j\cap\theta^{-1}(V_i)]$ are non-void clopen subsets. We denote by $C_*^{\theta,\xi}(M,N)$ for $\xi=t$ with $0\le t\le\infty$ a space of all mappings $f:M\to N$ such that

$$f_{i,j}-\theta_{i,j}\in C_*^{\xi}(U_{i,j},Y). \tag{4}$$

In view of Formulas $(1-4)$ we supply it with an ultra-metric

$$\rho_*^{\xi}(f,g)=\sup_{i,j}\|f_{i,j}-g_{i,j}\|_{C_*^{\xi}(U_j,Y)} \tag{5}$$

for each $0\le\xi<\infty$.

5. Let M and N be two analytic manifolds with finite atlases, let also the manifold M be finite dimensional over a local field $\mathbf{K}$, $dim_{\mathbf{K}}M=n\in\mathbf{N}$, $\theta_{i,j}\in C^\infty(U_j,Y)$ for each i,j.

We denote by $C_0^{\theta,(t,s)}(M\to N)$ a completion of a locally $\mathbf{K}$-convex space

$$\{f\in C_0^{\theta,(t+sn)}(M,N):\ \rho_0^{(t,s)}(f,\theta)<\infty \tag{1}$$

and for each $\varepsilon>0$ a set $\{(k,m):\ \sum_{i,j}|a(m,f_{i,j}^k-\theta_{i,j}^k)|J((t,s),m)>\varepsilon\}$ is finite relative to an ultra-metric

$$\rho_0^{(t,s)}(f,g):=\sup_{i,j,m,k}|a(m,f_{i,j}^k-g_{i,j}^k)|J_j((t,s),m), \tag{2}$$

where $s\in\mathbf{N_0}:=\{0,1,2,\dots\}$, $0\le t<\infty$;

$$J_j((t,s),m):=\max_{(v\le[t]+sign(t)+sn)}\|(\bar{\Phi}^v\bar{Q}_m|_{U_j})(x;$$

$$h_1,\dots,h_v;\zeta_1,\dots,\zeta_v)\|_{C_0(0,U_j\to Y)} \tag{3}$$

with

$$h_1 = \cdots = h_\gamma = e_1, \ldots, h_{(n-1)\gamma+1} = \cdots = h_{n\gamma} = e_n \quad (4)$$

for each integer γ such that $1 \le \gamma \le s$ and for each $v \in \{[t] + \gamma n, t + \gamma n\}$.

In view of Formulas $(1-3)$ this space is separable, when N is separable, since M is locally compact.

6. For infinite atlases we use the traditional procedure of inductive limits of spaces. For the manifold M with the infinite atlas let $card(\Lambda_M) = \aleph_0$, and the Banach space Y over the field **K** we denote by $C^\theta_*(\xi, M \to Y)$ for $\xi = t$ with $0 \le t \le \infty$ or for $\xi = (t,s)$ a locally **K**-convex space, which is the strict inductive limit

$$C^\theta_*(\xi, M \to Y) := str - ind\{C^\theta_*(\xi, (U^E \to Y), \pi^F_E, \Sigma\}, \quad (1)$$

where $E \in \Sigma$, Σ is the family of all finite subsets of Λ_M directed by the inclusion $E < F$ if $E \subset F$, $U^E := \bigcup_{j \in E} U_j$ (see also [87]).

For mappings from one manifold into another $f : M \to N$ we therefore get the corresponding uniform spaces. We denote them by $C^{\theta,\xi}_*(M,N)$.

We introduce notations

$$G^\xi(M) := C^{\theta,\xi}_0(M,M) \cap Hom(M), \quad (2)$$

$$Diff^\xi(M) = C^{\theta,\xi}(M,M) \cap Hom(M), \quad (3)$$

that are called groups of diffeomorphisms (and homeomorphisms for $0 \le t < 1$ and $s = 0$), $\theta = id$, $id(x) = x$ for each $x \in M$, where

$$Hom(M) := \{f : \ f \in C(0, M \to M),$$

f is bijective, $f(M) = M$, f and $f^{-1} \in C(0, M \to M)\}$ denotes the usual homeomorphism group. For $s = 0$ we may omit it from the notation, which is always accomplished for M infinite-dimensional over **K**.

7. Notes. Henceforth, ultra-metrizable separable complete manifolds $\bar{M}$ and N are considered. In accordance with Theorem 7.3.3 [19] for each separable metrizable space X a small inductive dimension $ind(X)$, a large inductive dimension $Ind(X)$ and a dimension in the sense of coverings $dim(X)$ coincide. Since a large inductive dimension $Ind(\bar{M}) = 0$ of the manifold $\bar{M}$ is zero due to Theorem 7.3.3 [19], the manifold $\bar{M}$ has not boundaries in the usual topological sense. Therefore, the atlas

$$At(\bar{M}) = \{(\bar{U}_j, \bar{\phi}_j) : \ j \in \Lambda_{\bar{M}}\} \quad (1)$$

has a refinement $At'(\bar{M})$ which is countable and its charts $(\bar{U}'_j, \bar{\phi}'_j)$ are clopen and disjoint and homeomorphic with the corresponding balls $B(X, y_j, \bar{r}_j)$ in the Banach space X, where

$$\bar{\phi}'_j : \ \bar{U}'_j \to B(X, y'_j, \bar{r}_j) \text{ for each } j \in \Lambda'_{\bar{M}} \quad (2)$$

are homeomorphisms. For the manifold $\bar{M}$ we fix such atlas $At'(\bar{M})$.

We define topologies of the groups $G^\xi(\bar{M})$ and locally **K**-convex spaces $C^\xi_*(\bar{M}, Y)$ relative to the atlas $At'(\bar{M})$, where Y is the Banach space over the field **K**. Therefore, we suppose also that $\bar{M}$ and N are clopen subsets of the Banach spaces X and Y respectively. Up to the isomorphism of wrap monoids (see below their definition) we can suppose that $s_0 = 0 \in \bar{M}$ and $y_0 = 0 \in N$.

For $M = \bar{M} \setminus \{0\}$ let the atlas $At(M)$ consist of charts (U_j, ϕ_j), $j \in \Lambda_M$, while the atlas $At'(M)$ consists of charts (U'_j, ϕ'_j), $j \in \Lambda'_M$, where due to Formulas $(1,2)$ we define the charts

$$U_1 = \bar{U_1} \setminus \{0\},\ \phi_1 = \bar{\phi}_1|_{U_1};\ U_j = \bar{U_j} \text{ and } \phi_j = \bar{\phi}_j \text{ for each } j > 1, \tag{3}$$

$$0 \in \bar{U_1},\ \Lambda_M = \Lambda_{\bar{M}},\ U'_1 = \bar{U}'_1 \setminus \{0\},\ \phi'_1 = \bar{\phi}'_1|_{U'_1},\ U'_j = \bar{U}'_j \text{ and } \phi'_j = \bar{\phi}'_j$$

$$\text{for each } j > 1,\ j \in \Lambda'_M = \Lambda'_{\bar{M}},\ \bar{U}'_1 \ni 0.$$

8. Definition. Let a function $f(x)$ be in the space $C^{(t,s-1)}(X', \mathbf{K})$, where $\mathbf{K}$ is a local field. If the field $\mathbf{K}$ is of zero characteristic numbers l and s can be any natural, if the field $\mathbf{K}$ is of a positive characteristic $char(\mathbf{K}) = p > 0$ we suppose that the natural numbers l and s satisfy the restriction $l < p$. Then an anti-derivation $P(l,s)$ is defined by the formula:

$$P(l,s)f(x) := \sum\{(\partial^j f(x_m))(x_{m+\bar{u}} - x_m)^{(j+\bar{u})}/(j+\bar{u})! : m \in \mathbf{N_o^n}, \tag{1}$$

$$j = j' + s'\bar{u},\ s' \in \{0,1,\dots,s-1\},\ |\, j' \,| = 0,\dots,l-1\},$$

where $\partial^j = \partial_1^{j(1)} \dots \partial_n^{j(n)}$, $j = (j(1),\dots,j(n))$, $\bar{u} = (1,\dots,1) \in \mathbf{N^n}$, $x_m = \sigma_m(x)$,

$$\{\sigma_m : m \in \mathbf{No^n}\} \text{ is an approximation of the identity in } X',$$

X' is a clopen subset in $B(\mathbf{K^n}, 0, R)$, $\infty > R > 0$, $1 \le t \in \mathbf{R}$, $l = [t] + 1$, $n \in \mathbf{N}$.

9. Lemma. *Let $f \in C^{(t,s-1)}(X', \mathbf{K})$, $t = l + b - 1$, $0 < b < 1$ and $\partial^{\bar{u}} f(x) \in C^{(t,s-1)}(X', \mathbf{K})$. Suppose in addition that*

$$f(x) = f(y) + \sum_{(1 \le |j'| < l,\ j = j' + s'\bar{u},\ s' \in \{0,1,\dots,s\})} (\partial^j f(y))(x-y)^j / j!$$

$$+ \sum_{(|j'| = l-1,\, j = j' + s\bar{u})} (x-y)^j R(n,j;x,y), \tag{1}$$

where $R(n,j;x,y) \in C^b(X' \times X', \mathbf{K})$ and they are zero on the diagonal $((x,y) \in X' \times X' : x = y)$, l may be any non-negative integer for $char(\mathbf{K}) = 0$, $l < p$ for $char(\mathbf{K}) = p > 0$. Then for each $z \in X'$:

$$(k+qn)! \lim_{x,y \to z;\, |\zeta_1| + \dots + |\zeta_n| \to 0} J_j^{k,q} f(x,y;\zeta_1,\dots,\zeta_{qn}) = \partial^{q\bar{u}} f^{(k)}(z), \tag{2}$$

for each $k = 1,\dots,l$ and

$$(qn)! \lim_{x,y \to z} J_j^{l-1,q} f(x,y;\zeta_1,\dots,\zeta_{qn})(x-y)^{l-1}/j_b(\zeta) = \partial^{q\bar{u}}(\Phi^t f)(z,\dots,z), \tag{3}$$

where

$$(J_j^{k,q} f(x,y;\zeta_1,\dots,\zeta_{qn}))(x-y)^k := (\bar{\Phi}^{k+qn} f)\ (y; x-y,\dots,x-y, h_1,\dots,h_{qn};$$

$$0,\dots,0,1,\dots,1,\zeta_1,\dots,\zeta_{qn}) \tag{4}$$

has j zeros in $(\dots)$, the function $j_b(\zeta)$ was defined in §1 above, $x = y + \zeta e_i$ in Formula (3), $h_1 = h_2 = \dots = h_q = e_1$, $h_{q+1} = \dots = h_{2q} = e_2, \dots h_{q(n-1)+1} = \dots = h_{qn} = e_n$, $\zeta_i \in \mathbf{K}$, $q \in \{0,1,\dots,s\}$, $x,y \in X'$, $y + \zeta_i h_i \in X'$.

Proof. By the Taylor formula (see Appendix C) we have the decomposition:

$$f(x) = f(y) + \sum_{1\le|j|\le l-1} (\partial^j f(y))(x-y)^j/j!$$

$$+(\bar{\Phi}^l f)(y;x-y,\dots,x-y;1,\dots,1,1) - f^{(l)}(y)(x-y)^l/l!. \tag{5}$$

Let n be a natural number, X be a subset in $\mathbf{K}$ dense in itself, $f \in C^{n-1}(X,\mathbf{K})$ and $x \neq y \in X$. We set $\rho_1 f(x,y) := \bar{\Phi}^n f(x;y-x,\dots,y-x;1,\dots,1)$, $\rho_2 f(x,y) := \bar{\Phi}^n f(x;y-x,\dots,y-x;0,1,\dots,1)$, $\rho_n f(x,y) := \bar{\Phi}^n f(x;y-x,\dots,y-x;0,\dots,0,1)$. Let $D_n f(z) := \bar{\Phi}^n(z;1,\dots,1;0,\dots,0)$. Then the functions $\rho_1 f,\dots,\rho_n f$ are continuous on the domain $\nabla^2 X := \{(x,y) \in X^2 : x \neq y\}$ and satisfy the identities:

$$\bar{\Phi}^j D_{n-j} f(x;y-x,\dots,y-x;1,\dots,1) = \sum_{k=j-1}^{n-1} \binom{k}{j-1} \rho_{n-k} f(x,y) \tag{6}$$

for each $j = 1,2,\dots,n-1$. If in addition $f' \in C^{n-1}(X,\mathbf{K})$ and $\lim_{x,y\to z} \rho_1 f(x,y) = D_n f(z)$ for some $z \in X$, then

$$\lim_{x,y\to z} \rho_j f(x,y) = D_n f(z) \tag{7}$$

for each $j = 2,3,\dots,n$. Then using Formula $(6,7)$ by induction by each variable and Formulas $8(1)$ and $9(1,4)$ above we get Formulas $(2,3)$.

10. Lemma. *Let $f \in C^{(t,s-1)}(B,\mathbf{K})$, $B = B(\mathbf{K^n},0,1)$ and $S = B(\mathbf{K},0,r)$ be balls in the normed space $\mathbf{K^n}$ and in the field $\mathbf{K}$ respectively, $t = l$, $l \in \mathbf{N}$. Suppose that $J_j^{k,q} f(x,y;\zeta_1,\dots,\zeta_{qn})(x-y)^k \in S$ for each $x,y \in B$, $\zeta_1,..,\zeta_{qn} \in B(\mathbf{K},0,1)$ and for each $0 < j < n+1$, $k < l+1$. Then $(\bar{\Phi}^{k+qn} f)(y;x_1,\dots,x_k,h_1,\dots,h_{qn};\zeta_1,\dots,\zeta_{k+qn}) \in S$ for each $x_i \in B$ and $|\zeta_i| \le 1$, where h_i are the same as in §9.*

Proof. The clopen balls B and S are complete, since the local field $\mathbf{K}$ is complete. Lemma 81.2 [103] states that is $n \in \mathbf{N}$, $f \in C^{n-1}(X,\mathbf{K})$, B and S are balls in $\mathbf{K}$, $\rho_j f(x,y) \in S$ for all $x \neq y \in B \cap X$, then $\bar{\Phi}^n f(x;v_1,\dots,v_n;1,\dots,1) \in S$ for all $x, x+v_1,\dots,x+v_n \in (X^{n+1} \cap B^{n+1}) \setminus \Delta$, where $\Delta := \{(x,\dots,x) \in X^n : x \in X\}$. Applying this Lemma 81.2 [103] by induction by each variable we get the statement of the lemma, since $f^{(k)}(y) = k!\bar{\Phi}^k(y;x-y,\dots,x-y;0,\dots,0) = k!D_k f(y)$.

11. Lemma. *Let $f \in C^{(t,s-1)}(X',\mathbf{K})$ and $\partial^{\bar{u}} f \in C^{(t,s-1)}(X',\mathbf{K})$ and*

$$f(x) = f(y) + \sum_{(1\le|j'|\le l,\ j=j'+s'\bar{u},\ s'\in\{0,1,\dots,s\})} (\bar{\Phi}^j f(y))(x;x-y,\dots,x-y;0,\dots,0)$$

$$+ \sum_{(|v'|=l,\ v=v'+s\bar{u})} (x-y)^v \times R(n,v;x,y), \tag{1}$$

where $R(n,v;x,y)/j_b(\zeta)$ are continuous functions zero on the diagonal for $x-y=\zeta e_i$ with $\zeta \in \mathbf{K}$, $\bar{u} := (1,\dots,1) \in \mathbf{N}^k$, $X' := \mathbf{K}^k$, $k \in \mathbf{N}$. Then $f \in C^{(t,s)}(X',\mathbf{K})$.

Proof. For $s=1$ by assumption $\partial^{\bar{u}} f \in C^{(t,0)}(X',\mathbf{K})$, hence $f \in C^{(t,1)}(X',\mathbf{K})$. Then by induction applying Lemmas 9 and 10 we get the statement of this lemma for each $s \in \mathbf{N}$.

12. Theorem. *Let $f \in C^{(t,s-1)}(X',\mathbf{K})$, l is any non-negative integer for $char(\mathbf{K})=0$, or $l<p$ for $char(\mathbf{K})=p>0$. Then*

$$P(l,s)f(x) - P(l,s)f(y) = \sum_{(j=j'+s'\bar{u},\ 0\le|j'|<l,\ s'\in\{1,\dots,s\})} (\partial^{j'} f(y))(x-y)^j/j!$$

$$+ \sum_{(v=v'+s\bar{u},\ |v'|=l-1)} (x-y)^v R(n,v;x,y), \quad (1)$$

where $R(n,v;x,y)$ and $R(n,v;x,y)/j_b(\zeta)$ (with $x-y=\zeta e_i$, $\zeta \in \mathbf{K}$ for $i=1,\dots,n$ in the latter case) are continuous functions equal to zero on the diagonal.

Proof. Lemma 78.2 [103] states that if $n \in \mathbf{N}$, $f \in C^n(X,\mathbf{K})$, then $D_{n-j}f \in C^j(X,\mathbf{K})$ and $D_j D_{n-j} f = \binom{n}{j} D_n f$. Applying Lemma 78.2 [103] by each variable and using Lemma 11 we get $\partial^j f \in C^{(t-|j|,s-1)}(X',\mathbf{K})$. In view of Formula 11(1) there are continuous functions $A(j,v;*)$ together with $A(j,v;x,y)/j_b(\zeta)$ (for $x-y=\zeta e_i$, $i=1,\dots,n$ in the latter case), such that

$$\partial^j f(x_m) = \sum_{(|q|=0,\dots,l-|j|-1)} (\partial^{j+q} f(y))(x_m-y)^q/q!$$

$$+ \sum_{(|v|=l-|j|-1)} (x_m-y)^v A(j,v;x_m,y)$$

and

$$P(l,s)f(x) = \sum_{(m\in\mathbf{N_o^n},\ |j'|=0,\dots,l-1,\ j=j'+s'\bar{u},\ s'\in\{0,1,\dots,s-1\})} \Big[\sum_{(|q|=0,\dots,l-|j|-1)} (\partial^{j+q} f(y))$$

$$(x_m-y)^q/q! + \sum_{|v|=l-|j|-1} (x_m-y)^v A(j,v;x_m,y)(x_{m+\bar{u}}-x_m)^{j+\bar{u}}/(j+\bar{u})!\Big]$$

$$= \sum_{(\ |j'|=0,\dots,l-1,\ j=j'+s'\bar{u},\ s'\in\{0,1,\dots,s-1\})} \Big\{ (\partial^j f(y))/(j+\bar{u})![(x-y)^{j+\bar{u}} + (-1)^n (x_0-y)^{j+\bar{u}}]$$

$$+ \sum_{(m\in\mathbf{N_o^n},|v|=l-|j|-1)} (x_m-y)^v (x_{m+\bar{u}}-x_m)^{j+\bar{u}} A(j,v;x_m,y)/(j+\bar{u})! \Big\}, \quad (3)$$

analogously for $P(l,s)f(y)$. From Formulas $(2,3)$ and $11(1)$ we get

$$P(l,s)f(x) - P(l,s)f(y) =: \sum_{(|v'|=l-1,\ v=v'+s\bar{u})} (x-y)^v R(l,v;x,y)$$

$$= \sum_{(m\in\mathbf{No^n},\ |j|=0,\dots,l-1;\ |v'|=l-|j|-1,\ v=v'+s\bar{u})} [(x_m-y)^v$$

$$(x_{m+\bar{u}}-x_m)^{j+\bar{u}} A(j,v;x_m,y) - (y_m-y)^v (y_{m+\bar{u}}-y_m)^{j+\bar{u}} A(j,v;y_m,y)]/(j+\bar{u})!. \quad (4)$$

If $\varepsilon > 0$, there is $s \in \mathbf{N}$ so that $|u-z| < \rho^s$, $|v-z| < \rho^s$ implies $|A(j,v;u,v)| < \varepsilon\rho^n/|n!|$, $j = 0,1,\dots,n-1$. Let now $|x-z| < \rho^s$, $|y-z| < \rho^s$, $x \ne y$, then $x_0 = y_0, \dots, x_t = y_t$, $x_{t+1} \ne y_{t+1}$ for some $t \ge s$. Then $\rho^{t+1} \le |x-y| < \rho^s$. For $m < t$ the terms in (4) vanish. For $m \ge t$ we have $|x_m - y| \le \max(|x_m - x|, |x-y|) \le \max(\rho^m, \rho^t) = \rho^t \le |x-y|/\rho$. Similarly, we deduce $\max(|x_{m+1}-x_m|, |y_m - y|, |y_{m+1}-y_m|) \le |x-y|/\rho$. Therefore, we infer that

$$\Big|\sum_{(|v'|=l-1,\ v=v'+s\bar{u})} (x-y)^v R(l,v;x,y)\Big| \le \max_{(|v'|=l-1,\ v=v'+s\bar{u})} |(x-y)^v R(l,v;x,y)|$$

$$\le \rho^{-l+1+sq}|v!|\Big[\prod_{i=1}^{q}|x^i-y^i|^{v^i}\Big]\varepsilon\rho^{l-1-sq}/|v!| = \varepsilon \max_{(|v'|=l-1,\ v=v'+s\bar{u})}\Big[\prod_{i=1}^{q}|x^i-y^i|^{v^i}\Big],$$

where $q = dim_{\mathbf{K}}X'$, $X' = \mathbf{K}^q$, $x = (x^1,\dots,x^q)$, $x^i \in \mathbf{K}$ for each i, $v = (v^1,\dots,v^q)$, each v^i is an integer non-negative number, $v! := \prod_{i=1}^q v^i!$, $\bar{u} = (1,\dots,1) \in \mathbf{N}^q$.

13. Corollary. *Let $1 \le t \in \mathbf{R}$, where l is any non-negative integer for $char(\mathbf{K}) = 0$, or $l < p$ for $char(\mathbf{K}) = p > 0$. Then each $f \in C^{(t,s-1)}(X',\mathbf{K})$ has a $C^{(t,s)}(X',\mathbf{K})$-antiderivative:*

$$\partial^{\bar{u}}(P(l,s)f)(x) = f(x) \text{ for each } x \in X', \tag{1}$$

moreover, for each $j = (j(1),\dots,j(n))$ with $0 \le j(i) < 2$ for each $i = 1,\dots,n$, and each $x = (x^1,\dots,x^n)$ the following equation is fulfilled:

$$\partial^j P(l,s)f(x)\,|_{\{\text{ there is } x^i = x^i_0\}} = 0. \tag{2}$$

Proof. Theorem 29.12 in [103] states that if X is a subset in $\mathbf{K}$ so that X is dense in itself, $f: X \to \mathbf{K}$, then the following conditions are equivalent:

(α) $f \in C^n(X,\mathbf{K})$ and $D_j f = 0$ for $j = 1,\dots,n$, where $D_n f(z) := \bar{\Phi}^n(z;1,\dots,1;0,\dots,0)$;

(β) $\lim_{(x,y)\to(z,z)}[f(x)-f(y)]/(x-y)^n = 0$ for each $z \in X$.

From this theorem and Formulas 8(1) and 12(1) the statement of this corollary follows.

6.5. Appendix E. Wrap Groups

1. Definitions and Notes. Let the spaces be the same as in §6 of Appendix D (see Formulas $(1-3)$) with the atlas of M defined by Conditions 7(3). Then we consider their subspaces of all mappings preserving marked points:

$$C_0^{\theta,\xi}((M,s_0);(N,y_0)) := \{f \in C_0^{\theta,\xi}(\bar{M},N) : \lim_{|\zeta_1|+\dots+|\zeta_k|\to 0} \bar{\Phi}^v(f-\theta)(s_0;$$

$$h_1,\dots,h_k;\zeta_1,\dots,\zeta_k) = 0 \text{ for each } v \in \{0,1,\dots,[t],t\},\ k = [v] + sign\{v\}\}, \tag{1}$$

where for $s \ge 0$ and $\xi = (t,s)$ in addition Condition 5(4) is satisfied for each $1 \le \gamma \le s$ and for each $v \in \{[t]+n\gamma, t+n\gamma\}$, and the following subgroup:

$$G_0^{\xi}(M) := \{f \in G^{\xi}(\bar{M}) :\ f(s_0) = s_0\} \tag{2}$$

of the diffeomorphism group, where $s \in \mathbf{N_0}$ for $dim_{\mathbf{K}}M < \aleph_0$ and $s = 0$ for $dim_{\mathbf{K}}M = \aleph_0$.

With the help of them we define the following equivalence relations K_ξ: $fK_\xi g$ if and only if there exist sequences

$$\{\psi_n \in G_0^\xi(M) : n \in \mathbf{N}\}, \quad \{f_n \in C_0^{\theta,\xi}(M,N) : n \in \mathbf{N}\} \text{ and}$$

$$\{g_n \in C_0^{\theta,\xi}(M,N) : n \in \mathbf{N}\}$$

such that

$$f_n(x) = g_n(\psi_n(x)) \text{ for each } x \in M \text{ and } \lim_{n\to\infty} f_n = f \text{ and } \lim_{n\to\infty} g_n = g. \tag{3}$$

Due to Condition (3) these equivalence classes are closed, since $(g(\psi(x))' = g'(\psi(x))\psi'(x)$, $\psi(s_0) = s_0$, $g'(s_0) = 0$ for $t + s \geq 1$. We denote such equivalence classes by $< f >_{K,\xi}$. Then for $g \in < f >_{K,\xi}$ we write $gK_\xi f$ also. The quotient space $C_0^{\theta,\xi}((M,s_0);(N,y_0))/K_\xi$ we denote by $\Omega_\xi(M,N)$, where $\theta(M) = \{y_0\}$.

2. Let as usually $A \vee B := A \times \{b_0\} \cup \{a_0\} \times B \subset A \times B$ be the wedge product of pointed spaces (A,a_0) and (B,b_0), where A and B are topological spaces with marked points $a_0 \in A$ and $b_0 \in B$. Then the composition $g \circ f$ of two elements $f,g \in C_0^{\theta,\xi}((M,s_0);(N,y_0))$ is defined on the domain $\bar{M} \vee \bar{M} \setminus \{s_0 \times s_0\} =: M \vee M$.

Let $M = \bar{M} \setminus \{0\}$ be as in 7 of Appendix D. We fix an infinite atlas $\tilde{A}t'(M) := \{(\tilde{U}'_j, \phi'_j) : j \in \mathbf{N}\}$ such that $\phi'_j : \tilde{U}'_j \to B(X, y'_j, r'_j)$ are homeomorphisms,

$$\lim_{k\to\infty} r'_{j(k)} = 0 \text{ and } \lim_{k\to\infty} y'_{j(k)} = 0$$

for an infinite sequence $\{j(k) \in \mathbf{N} : k \in \mathbf{N}\}$ such that $cl_{\bar{M}}[\bigcup_{k=1}^\infty \tilde{U}'_{j(k)}]$ is a clopen neighborhood of 0 in the manifold $\bar{M}$, where $cl_{\bar{M}}A$ denotes the closure of a subset A in the manifold $\bar{M}$. In the wedge product $M \vee M$ we choose the following atlas $\tilde{A}t'(M \vee M) = \{(W_l, \xi_l) : l \in \mathbf{N}\}$ such that $\xi_l : W_l \to B(X, z_l, a_l)$ are homeomorphisms,

$$\lim_{k\to\infty} a_{l(k)} = 0 \text{ and } \lim_{k\to\infty} z_{l(k)} = 0$$

for an infinite sequence $\{l(k) \in \mathbf{N} : k \in \mathbf{N}\}$ such that $cl_{\bar{M}\vee\bar{M}}[\bigcup_{k=1}^\infty W_{l(k)}]$ is a clopen neighborhood of 0×0 in the wedge product space $\bar{M} \vee \bar{M}$ and

$$card(\mathbf{N} \setminus \{l(k) : k \in \mathbf{N}\}) = card(\mathbf{N} \setminus \{j(k) : k \in \mathbf{N}\}).$$

Then we fix a $C(\infty)$-diffeomorphisms $\chi : M \vee M \to M$ such that

$$\chi(W_{l(k)}) = \tilde{U}'_{j(k)} \text{ for each } k \in \mathbf{N} \tag{1}$$

and

$$\chi(W_l) = \tilde{U}'_{\kappa(l)} \text{ for each } l \in (\mathbf{N} \setminus \{l(k) : k \in \mathbf{N}\}), \tag{2}$$

where

$$\kappa : (\mathbf{N} \setminus \{l(k) : k \in \mathbf{N}\}) \to (\mathbf{N} \setminus \{j(k) : k \in \mathbf{N}\}) \tag{3}$$

is a bijective mapping for which

$$p^{-1} \leq a_{l(k)}/r'_{j(k)} \leq p \text{ and } p^{-1} \leq a_l/r'_{\kappa(l)} \leq p. \tag{4}$$

This induces the continuous injective homomorphism

$$\chi^* : C_0^{\theta,\xi}((M \vee M, s_0 \times s_0); (N, y_0)) \to C_0^{\theta,\xi}((M, s_0); (N, y_0)) \tag{5}$$

such that

$$\chi^*(g \vee f)(x) = (g \vee f)(\chi^{-1}(x)) \tag{6}$$

for each point $x \in M$, where $(g \vee f)(y) = f(y)$ for $y \in M_2$ and $(g \vee f)(y) = g(y)$ for $y \in M_1$, $M_1 \vee M_2 = M \vee M$, $M_i = M$ for $i = 1, 2$. Therefore the mapping

$$g \circ f := \chi^*(g \vee f) \tag{7}$$

may be considered as defined on the manifold M also. That is, to $g \circ f$ the unique element in the uniform space $C_0^{\theta,\xi}((M, s_0); (N, y_0))$ corresponds.

3. The composition in the set $\Omega_\xi(M,N)$ is defined due to the following inclusion $g \circ f \in C_0^{\theta,\xi}((M, s_0); (N, y_0))$ (see Formulas $2(1-7)$) and then using the equivalence relations K_ξ (see Condition 1(3)). The set $\Omega_\xi(M,N)$ can be supplied with the quotient space topology which is the finest topology τ on $\Omega_\xi(M,N)$ relative to which the quotient mapping $\psi : C_0^{\theta,\xi}((M, s_0); (N, y_0)) \to C_0^{\theta,\xi}((M, s_0); (N, y_0))/K_\xi = \Omega_\xi(M,N)$ is continuous. That is, the topology τ is the family of all sets V in $\Omega_\xi(M,N)$ so that $\psi^{-1}(V)$ is open in $C_0^{\theta,\xi}((M, s_0); (N, y_0))$ (see also §2.4 in [19]).

It is shown in [56, 62, 82] that $\Omega_\xi(M,N)$ is the monoid, which we call the wrap monoid.

4. Theorem. *The space $\Omega_\xi(M,N)$ from §3 is the complete separable Abelian topological Hausdorff monoid. Moreover, it is non-discrete, topologically perfect and has the cardinality $\mathsf{c} := card(\mathbf{R})$.*

5. Note. For each chart (V_i, ψ_i) of the atlas $At(N)$ (see Equality 4(1) in Appendix D) there are local normal coordinates $y = (y^j : j \in \beta) \in B(Y, a_i, r_i)$, $Y = c_0(\beta, \mathbf{K})$. Moreover, $TV_i = V_i \times Y$, consequently, TN has the disjoint atlas $At(TN) = \{(V_i \times X, \psi_i \times I) : i \in \Lambda_N\}$, where $I_Y : Y \to Y$ is the unit mapping, $\Lambda_N \subset \mathbf{N}$, TN is the tangent vector bundle over the manifold N.

Suppose V is an analytic vector field on N (that is, by definition $V|_{V_i}$ are analytic for each chart and $V \circ \psi_i^{-1}$ has the natural extension from $\psi_i(V_i)$ on the balls $B(X, a_i, r_i)$). Then by analogy with the classical case we can define the following mapping

$$\bar{\exp}_y(zV) = y + zV(y) \text{ for which } \partial^2 \bar{\exp}_y(zV(y))/\partial z^2 = 0$$

(this is the analog of the geodesic), where $\|V(y)\|_Y |z| \leq r_i$ for $y \in V_i$ and $\psi_i(y)$ is also denoted by y, $z \in \mathbf{K}$, $V(y) \in Y$. Moreover, a refinement $At''(N) = \{(V''_i, \psi''_i) : i \in \Lambda''_N\}$ of the atlas $At(N)$ exists. This atlas $At''(N)$ is embedded into $At(N)$ by charts such that it is also disjoint and analytic and $\psi''_i(V''_i)$ are $\mathbf{K}$-convex in Y. The latter means that $\lambda x + (1-\lambda)y \in \psi''_i(V''_i)$

for each $x,y \in \psi"_i(V"_i)$ and each $\lambda \in B(\mathbf{K},0,1)$. Evidently, we can consider $\bar{\exp}_y$ injective on $V"_i$, $y \in V"_i$. The atlas $At"(N)$ can be chosen such that

$$(\bar{\exp}_y|_{V"_i}) : V"_i \times B(Y,0,\tilde{r}_i) \to V"_i$$

to be the analytic homeomorphism for each $i \in \Lambda"_M$, where $\infty > \tilde{r}_i > 0$, $y \in V"_i$,

$$\bar{\exp}_y : (\{y\} \times B(Y,0,\tilde{r}_i)) \to V"_i$$

is the isomorphism. Therefore, $\bar{\exp}$ is the locally analytic mapping, $\bar{\exp} : \tilde{T}N \to N$, where $\tilde{TN}$ is the corresponding neighborhood of N in TN.

Then we get

$$T_f C_*^{\theta,\xi}(M,N) = \{g \in C_*^{(\theta,0),\xi}(M,TN) : \pi_N \circ g = f\}, \tag{1}$$

consequently,

$$C_*^{\theta,\xi}(M,TN) = \bigcup_{f \in C_*^{\theta,\xi}(M,N)} T_f C_*^{\theta,\xi}(M,N) = TC_*^{\theta,\xi}(M,N), \tag{2}$$

where $\pi_N : TN \to N$ is the natural projection, $* = 0$ or $* = \emptyset$ ($\emptyset$ is omitted). Therefore, the following mapping

$$\omega_{\bar{\exp}} : T_f C_*^{\theta,\xi}(M,N) \to C_*^{\theta,\xi}(M,N) \tag{3}$$

is defined by the formula given below

$$\omega_{\bar{\exp}}(g(x)) = \bar{\exp}_{f(x)} \circ g(x), \tag{4}$$

that gives charts in the uniform space $C_*^{\theta,\xi}(M,N)$ induced by charts in $C_*^{\theta,\xi}(M,TN)$.

6. Definition and Note. In view of Equalities $5(1,2)$ the uniform space $C_0^{\theta,\xi}(\bar{M},N)$ is isomorphic with the uniform space $C_0^{\theta,\xi}((M,s_0),(N,y_0)) \times N^{\xi}$, where $y_0 = 0$ is the marked point of N. Here

$$N^{\xi} := N \otimes \Big(\bigotimes_{j=1}^{d} \tilde{L}_{\xi}(X^j \to Y) \Big) \text{ for } t \in \mathbf{N_0} \text{ with } t+s > 0; \tag{1}$$

$$N^{\xi} = N \text{ for } t+s = 0; \tag{2}$$

$$N^{\xi} = N \otimes \Big(\bigotimes_{j=1}^{d} \tilde{L}_{\xi}(X^j \to Y) \Big) \otimes C_0^{0,0}(M^k, Y_{\lambda}) \text{ for } t \in \mathbf{R} \setminus \mathbf{N}, \tag{3}$$

where N^{ξ} is supplied with the product (Tychonoff) topology, $d = [t]$ for $\xi = t$, $d = [t] + n\alpha$ for $\xi = (t,s)$ with $\alpha = dim_{\mathbf{K}} M < \aleph_0$, when $s > 0$, $k = d + sign\{t\}$, $Y_{\lambda} := c_0(\beta,\lambda)$, λ is the least subfield of $\mathbf{C_p}$ such that $\lambda \supset \mathbf{K} \cup j_{\{t\}}(\mathbf{K})$ (see §1 in Appendix D). Then $\tilde{L}_{\xi}(X^j,Y)$ denotes the Banach space of continuous j-linear operators $f_j : X^j \to Y$ with the finite norm:

$$\|f_j\|_{\tilde{L}_{\xi}(X^j,Y)} := \sup_{i,m} \|f_j^i\|_m \tag{4}$$

and

$$\lim_{i+|m|+k\to\infty} \|f_j^i\|_m = 0, \tag{5}$$

where

$$\|f_j^i\|_m := \sup_{0\neq h_l\in \mathbf{K}^k, l=1,\dots,j} \|f_j^i(h_1,\dots,h_j)\|_Y J'(\xi,m)/(\|h_1\|_X \cdots \|h_j\|_X), \tag{6}$$

$\mathbf{K}^k := span_{\mathbf{K}}(e_1,\dots,e_k) \hookrightarrow X$ is a $\mathbf{K}$-linear span of the standard basic vectors, $m = (m_1,\dots,m_k)$, $|m| = m_1+\dots+m_k$, $k\in\mathbf{N}$; $h_1 = \dots = h_{m_1},\dots,h_{m_{k-1}+1} = \dots = h_{m_k}$ for $s=0$; in addition Condition 5(4) in Appendix D is satisfied for each $0<\gamma\leq s$, when $s>0$; $f = (f_0, f_1,\dots,f_j,\dots)\in N^{\xi}$, $\sum_i f_j^i q_i = f_j$, $f_j^i : X^j \to \mathbf{K}$,

$$J'(\xi,m) := |\partial^m \bar{Q}_m(x)|_{x=0}|_{\mathbf{K}}$$

(see §§1 and 2 in Appendix D).

7. Theorem. *Let $G = \Omega_\xi(M,N)$ be the same monoid as in §3.*
If $1\leq t+s$, $0\leq t\in\mathbf{R}$ and $\xi = (t,s)$, $s\in\mathbf{N_o}$, then
(1) G is an analytic manifold and for it the mapping $\tilde{E} : \tilde{T}G \to G$ is defined, where $\tilde{T}G$ is the neighborhood of G in TG such that $\tilde{E}_\eta(V) = \exp_{\eta(x)} \circ V_\eta$ from some neighborhood $\bar{V}_\eta$ of the zero section in $T_\eta G \subset TG$ onto some neighborhood $W_\eta \ni \eta \in G$, $\bar{V}_\eta = \bar{V}_e \circ \eta$, $W_\eta = W_e \circ \eta$, $\eta \in G$ and $\tilde{E}$ belongs to the class $C(\infty)$ by V, $\tilde{E}$ is the uniform isomorphism of uniform spaces $\bar{V}$ and W;
(2) if the atlas $At(M)$ is finite, then there are atlases $\tilde{A}t(TG)$ and $\tilde{A}t(G)$ for which the mapping $\tilde{E}$ is locally analytic. Moreover, G is not locally compact for each $0\leq t$.

8. Note and Definition. For a commutative monoid $\Omega_\xi(M,N)$ with the unity and the cancelation property (see Theorem 7 and Condition 4(5)) there exists a commutative group $L_\xi(M,N)$ equal to the Grothendieck group. This group is the quotient group F/B, where F is a free Abelian group generated by the wrap monoid $\Omega_\xi(M,N)$ and B is a closed subgroup of F generated by all elements of the form $[f+g]-[f]-[g]$, f and $g\in\Omega_\xi(M,N)$, $[f]$ denotes an element of F corresponding to f. In view of §9 [54] and [107] the natural mapping

$$\gamma :\ \Omega_\xi(M,N) \to L_\xi(M,N) \tag{1}$$

is injective. We supply F with a topology inherited from the (Tychonoff) product topology of $\Omega_\xi(M,N)^{\mathbf{Z}}$, where each element z of F can be written as

$$z = \sum_f n_{f,z}[f], \tag{2}$$

$n_{f,z}\in\mathbf{Z}$ for each $f\in\Omega_\xi(M,N)$,

$$\sum_f |n_{f,z}| < \infty. \tag{3}$$

In particular $[nf]-n[f]\in\mathsf{B}$, where $1f = f$, $nf = f\circ(n-1)f$ for each $1<n\in\mathbf{N}$, $f+g := f\circ g$. We call $L_\xi(M,N)$ the wrap group.

9. Theorem. *The topological space $L_\xi(M,N)$ from §8 is the complete separable Abelian Hausdorff topological group; it is non-discrete, dense in itself and has the cardinality* c.

10. Theorem. *Let $G = L_\xi(M,N)$ be the same group as in §8, $\xi = (t,s)$ or $\xi = t$ with $0 \le t \in \mathbf{R}$, $s_0 \in \mathbf{N_0}$.*

(1) *If the atlas $At'(\bar{M})$ has the cardinality $card(\Lambda'_{\bar{M}}) \ge 2$, then G is isomorphic with $G_1 = L_\xi(\tilde{M},N)$, where $\tilde{M} = U'_1 \cup U'_2$ (see §7 in Appendix D). Moreover, $T_\eta G$ is the Banach space for each $\eta \in G$ and G is ultra-metrizable.*

(2) *If $1 \le t+s$, then the group G has a structure of an analytic manifold also and for it the mapping $\tilde{E} : \tilde{T}G \to G$ is defined, where $\tilde{T}G$ is the neighborhood of G in TG such that $\tilde{E}_\eta(V) = \bar{\exp}_{\eta(x)} \circ V_\eta$ from some neighborhood $\bar{V}_\eta$ of the zero section in $T_\eta G \subset TG$ onto some neighborhood $W_\eta \ni \eta \in G$, $\bar{V}_\eta = \bar{V}_e \circ \eta$, $W_\eta = W_e \circ \eta$, $\eta \in G$ and $\tilde{E}$ belongs to the class $C(\infty)$ by V, $\tilde{E}$ is the uniform isomorphism of the uniform spaces $\bar{V}$ and W.*

(3) *There are atlases $\tilde{A}t(TG)$ and $\tilde{A}t(G)$ for which the mapping $\tilde{E}$ is locally analytic. Moreover, the topological group G is not locally compact for each $0 \le t$.*

11. Note. Now let us describe dense wrap submonoids which are necessary for the investigation of random functions with values in the entire monoid. For a finite atlas $At(M)$ and $\xi = (t,s)$ let $C^{\theta,\xi}_{0,\{k\}}(M,Y)$ be a subspace of $C^{\theta,\xi}_0(M,Y)$ consisting of all mappings f for which the following norm is finite:

$$\|f-\theta\|_{C^{\theta,\xi}_{0,\{k\}}(M,Y)} := \sup_{i,m,j} |a(m,f^i|_{U_j})|_{\mathbf{K}} J_j(\xi,m) p^{k(i,m)} < \infty \tag{1}$$

and

$$\lim_{i+|m|+Ord(m)\to\infty} \sup_j |a(m,f^i|_{U_j})|_{\mathbf{K}} J_j(\xi,m) p^{k(i,m)} = 0, \tag{2}$$

where $k(i,m) := c' \times i + c \times (|m| + Ord(m))$, c' and c are non-negative constants, $|m| := \sum_i m_i$,

$$Ord(m) := \max\{i : \ m_i > 0 \text{ and } m_l = 0 \text{ for each } l > i\}$$

(see also §§1 and 2 and Formulas 3(2) and 5(3) in Appendix D).

For a finite-dimensional manifold M over the local field $\mathbf{K}$ this space is linearly topologically isomorphic with the space $C^{\theta,\xi}_{0,\{k'\}}(M,Y)$, where $k'(i,m) = c' \times i + c \times |m|$. For the finite-dimensional Banach space Y over the field $\mathbf{K}$ the space $C^{\theta,\xi}_{0,\{k\}}(M,Y)$ is linearly topologically isomorphic with the space $C^{\theta,\xi}_{0,\{k''\}}(M,Y)$, where $k''(i,m) = c \times (|m| + Ord(m))$. For $c' = c = 0$ this space coincides with the space $C^{\theta,\xi}_0(M,Y)$ and in this case we omit $\{k\}$.

Then as in §§1-3 we define spaces $C^{\theta,\xi}_{0,\{k\}}((M,s_0);(N,0))$, groups

$$G^{\{k\},\xi}(M) := C^{id,\xi}_{0,\{k\}}(M,M) \cap Hom(M), \tag{3}$$

$$G^{\{k\},\xi}_0(M) := \{\psi \in G^{\{k\},\xi}(M) : \ \psi(s_0) = s_0\} \tag{4}$$

and the equivalence relation $K_{\xi,\{k\}}$ in it for each M and N from §§4-7 of Appendix D.

Therefore,

$$G' := \Omega_\xi^{\{k\}}(M,N) := C^{0,\xi}_{0,\{k\}}((M,s_0);(N,0))/K_{\xi,\{k\}} \quad (5)$$

is the dense submonoid in the monoid $\Omega_\xi(M,N)$.

12. Note. Let $\Omega_\xi^{\{k\}}(M,N)$ be the same submonoid as in §11 such that $c > 0$ and $c' > 0$. Then it generates the wrap group $G' := L_\xi^{\{k\}}(M,N)$ as in §8 such that G' is the dense subgroup in $G = L_\xi(M,N)$. The proofs of the formulated above theorems in this appendix are written in the articles [56,62,82].

6.6. Appendix F. Fiber Bundles

Let E, N, F be all either $C^{\alpha'}_\beta$-manifolds or $C^{\alpha'}_\beta$-differentiable spaces over the field $\mathbf{K}$. Certainly a manifold is a particular case of a differentiable space. Let also G be a $C^{\alpha'}_\beta$ Lie group over the field $\mathbf{K}$, $\alpha \le \alpha' \le \infty$. That is, G has a structure of a $C^{\alpha'}_\beta$ manifold over the field $\mathbf{K}$ and the operation $G^2 \ni (h,g) \mapsto hg^{-1} \in G$ is of the same class of smoothness $C^{\alpha'}_\beta$. We suppose that a projection $\pi : E \to N$ is given together with an atlas $\Psi = \{\psi_j\}$ of E so that

$(F1)$ to each chart $\psi_j \in \Psi$ an open subset V_j in the manifold N is counterposed and $(F2)$ the mapping $\psi_j : \pi^{-1}(V_j) \to V_j \times F$ is the $C^{\alpha'}_\beta$ diffeomorphism so that $\psi_j(\pi^{-1}(x)) = \{x\} \times F$: $pr_{V_j} \circ \psi_j = \pi|_{\pi^{-1}(V_j)}$, where $pr_{V_j}(x \times y) = x$ for each $x \in V_j$ and $y \in F$;

$(F3)$ a system of open subsets $\{V_j : j \in J\}$ forms a covering of the manifold N. We get from $(F1-F3)$ that $\pi : E \to N$ is open and surjective. Moreover,

$(F4)$ the mapping $\psi_{j,x} = pr_F \circ \psi_j|_{\pi^{-1}(x)} : \pi^{-1}(x) \to F$ defines the $C^{\alpha'}_\beta$ diffeomorphism of the fiber $F_x := \pi^{-1}(x)$ on the typical fiber F, where $pr_F : (x \times y) = y$ for all $x \in V_j$ and $y \in F$.

Using restrictions of mappings ψ_j we can choose V_j as domains in the manifold N which are charts. Thus charts on $N \times F$ are transferred by ψ_j^{-1} onto charts on $\pi^{-1}(V_j)$.

Let $\psi_j, \psi_l \in \Psi$ and $V_j \cap V_k \ne \emptyset$. In view of $(F4)$ we specify:

$(F5)$ for each point $x \in V_j \cap V_k$ the mapping is defined: $g_{j,k} : V_j \cap V_k \ni x \mapsto g_{j,k}(x) = \psi_{k,x} \circ \psi_{j,x}^{-1} \in Diff^{\alpha'}_\beta(F)$. That is to each point $x \in V_j \cap V_k$ a diffeomorphism of F corresponds. Moreover, these mappings $g_{j,k}$ satisfy the following conditions:

$(F6)$ $g_{j,k}(x) = (g_{k,j}(x))^{-1}$, $g_{j,j}(x) = id_F$, where $id_F(y) = y$ for each $y \in F$, $g_{l,j}(x) = g_{l,k}(x) \circ g_{k,j}(x)$ for each $x \in V_l \cap V_j \cap V_k$;

$(F7)$ a $C^{\alpha'}_\beta$ differentiable transformation group (G,F) is given so that $g_{j,k}(x) \in G$ for each $x \in V_j \cap V_k$, when $V_k \cap V_j \ne \emptyset$, $g_{j,j}(x) = e \in G$, where e denotes the unit element in G. Moreover, we put $l_{g_{j,k}(x)} = \psi_{j,x} \circ \psi_{k,x}^{-1}$ and this mapping $l_g : F \ni f \mapsto l_g(f) = gf \in F$ is the $C^{\alpha'}_\beta$ diffeomorphism of F onto itself so that $l_{gh} = l_g \circ l_h$ for all $g, h \in G$. This means that G acts on F from the left.

If Conditions $(F1-F7)$ are satisfied then $E(N,F,G,\pi,\Psi)$ is called a fiber bundle with a fiber space E, a base space N, a typical fiber F and a structural group G and an atlas Ψ, while the mappings $g_{j,k}$ are called the transition functions.

Local trivializations $\phi_j \circ \pi \circ \Psi_k^{-1} : V_k(E) \to V_j(N)$ induce the $C^{\alpha'}_{\beta}$-uniformity in the family $\mathcal{W}$ of all principal $C^{\alpha'}_{\beta}$-fiber bundles $E(N,F,G,\pi,\Psi)$, where $V_k(E) = \Psi_k(U_k(E)) \subset X(N) \times X(F)$, $V_j(N) = \phi_j(U_j(N)) \subset X(N)$, where $X(F)$, $X(G)$ and $X(N)$ are **K**-vector spaces on which F, G and N are modelled, $(U_k(E),\Psi_k)$ and $(U_j(N),\phi_j)$ are charts of the atlases of E and N, $\Psi_k = \Psi_k^E$, $\phi_j = \phi_j^N$.

If $G = F$ and G acts on itself by left shifts, then a fiber bundle is called the principal fiber bundle and is denoted by $E(N,G,\pi,\Psi)$. In the trivial case $G = F = \{e\}$, the fiber bundle E reduces to the manifold N.

Notation

$Bco(X)$ §2.3.35;
$B(X,x,r)$ §1.2.1;
$Bf(X)$ §4.2.2;
$\mathcal{B}_{(t,s)}M$ §4.3.6;
$c_0(\alpha,\mathbf{K})$ §2.3.1.1;
$C^0(B,X)$ §3.5.1;
$C_0((t,s-1),U\to\mathbf{K})$, ${}_PC_0((t,s-1),U\to\mathbf{K})$ §4.3.3;
$C_b^\alpha(M,N)$ §6.A.2;
C^n, $C^{[n]}$ §6.B.1;
$char(\mathbf{K})$ §1.2.1;
Ch_A §1.2.1;
$\mathbf{C_p}^+$ §3.4.1;
D^b §5.2.7;
$Diff_b^\alpha(M)$ §6.A.2;
exp §4.3.7;
Exp, EXP §3.4.1;
$\mathcal{F}_{(t,s)}M$ §4.3.6;
$\mathbf{F_p}(\theta)$ §1.2.1;
$\|f\|_\phi$, $\|f\|_{\phi,u}$, $\|f\|_{q,\mu,u}$ §§2.2.1, 2.3.16;
Φ^n §6.B.1;
$\Gamma(\phi_j)$ §4.3.6;
$\mathcal{G}_x(a,E)$ §4.3.15;
$\|A\|_\mu$, $\|A\|_{\mu,u}$ §§2.2.1, 2.3.14;
$l^\infty(\alpha,\mathbf{K})$ §2.3.1;
$Lc(X)$ §2.3.1;
$L(\mu,X)$ §2.2.1;
$Lin(X,Y)$ §2.3.1;
$L^2(\xi,\mathbf{K})$ §2.2.17;
$L^2\{g\}$ §2.3.17;
$L^q(T,\mathcal{R},\eta,H)$ §4.2.1;
$L_q^b(H,\mathsf{B_c}(H),\nu,\mathsf{b})$ §2.3.10;
$L_q(E,H)$ §3.5.9;

References

[1] Aigner, M. *Combinatorial theory.* Springer-Verlag, Moscow (1979)

[2] Albeverio, S., Karwoski, W. Diffusion on p-adic numbers. In: K. Ito, T. Hida (Editors). *Gaussian random fields.* Nagoya 1990. World Scientific, River Edge, NJ (1991)

[3] Amice, Y. Interpolation p-adique. *Bull. Soc. Math. France* textbf92, 117–180 (1964)

[4] Aref'eva, I.Ya., Dragovich, B., Volovich, I.V. On the p-adic summability of the anharmonic oscillator. *Phys. Lett.* **B200**, 512–514 (1988)

[5] Bachturin Yu.A. *Basic structures of modern algebra.* Moscow, Nauka (1990).

[6] Belopolskaya, Ya. I., Dalecky, Yu. L. *Stochastic equations and differential geometry.* Dordrecht, Kluwer (1989)

[7] Bertram, W., Glöckner, H., Neeb, K-H. Differential calculus over general base fields and rings. *Expo. Math.* **22**, 213–282 (2004)

[8] Bikulov, A.N., Volovich, I.V. P-adic Brownian motion. *Izv. Akad. Nauk. Ser. Mathem.* **61: 3**, 75–90 (1997)

[9] Bourbaki, N. *Lie groups and algebras. Chapters I-III.* Moscow, Mir (1976)

[10] Bourbaki, N.: *Intégration. Livre VI. Fasc. XIII, XXI, XXIX, XXXV. Ch. 1-9.* Parsi, Hermann (1965, 1967, 1963, 1969)

[11] Bourbaki, N. *Variétés différentielles et analytiques. Fasc. XXXIII.* Paris, Hermann (1967)

[12] Brekke, L., Freund, P.G.O., Olson, M. Non-archimedean string dynamics. *Nuclear Physics.* **B302**, 365–402 (1988)

[13] Castro, C. Fractals, strings as an alternative justification for El Naschie's cantorian spacetime and the fine structurte constants. *Chaos, Solitons and Fractals.* **14**, 1341-1351 (2002)

[14] Christensen, J.P.R. *Topology and Borel structure.* Amsterdam, North-Holland (1974)

[15] Cox, D.R., Miller, H.D. *The theory of stochastic processes.* London, Chapman and Hall (1995)

[16] Dalecky, Yu.L., Fomin, S.V. *Measures and differential equations in infinite-dimensional spaces.* Dordrecht, Kluwer Academic Publishers, (1991)

[17] Diarra, B. Ultraproduits ultrametriques de corps values. *Ann. Sci. Univ. Clermont II. Sér. Math.* **22**, 1–37 (1984)

[18] Djordević, G.S., Dragovich, B. *P*-adic and adelic harmonic oscillator with a time-dependent frequency. *Theor. and Math. Phys.* **124: 2**, 1059-1067 (2000)

[19] Engelking, R. *General topology.* Berlin, Heldermann (1989)

[20] Escassut, A. *Analytic elements in p-adic analysis.* Singapore, World sceintific (1995)

[21] Evans, S.N. Continuity properties of Gaussian stochastic processes indexed by a local field. *Proceedings London Mathematical Society, Series 3.* **56**, 380–416 (1988)

[22] Evans, S.N. Local field Gaussian measures. In: E. Cinlar, et.al. (Editors) *Seminar on Stochastic Processes 1988.* 121-160. Boston, Birkhäuser (1989)

[23] Evans, S.N. Equivalence and perpendicularity of local field Gaussian measures. In: E. Cinlar, et.al. (Editors) *Seminar on Stochastic Processes 1990.* 173–181. Boston, Birkhäuser (1991)

[24] Evans, S.N. Local field Brownian motion. *J. Theoret. Probab.* **6**, 817-850 (1993)

[25] Federer, H. *Geometric measure theory.* Berlin, Springer (1968)

[26] Fell, J.M.G., Doran, R.S. *Representations of* $*$*-Algebras, Locally Compact Groups, and Banach* $*$*-Algebraic Bundles.* **1 , 2** Bosoton, Acad. Press (1988)

[27] Feller, W. *An introduction to probability theory and its applications* **1, 2**. New York, John Wiley and Sons, Inc. (1966)

[28] Fidaleo, F. Continuity of Borel actions of Polish groups on standard measure algebras. *Atti Sem. Mat. Fis. Univ. Modena.* **48**, 79–89 (2000)

[29] Fresnel, J., Put, M. van der. *Géométrie analytique rigide et applications.* Boston, Birkhäuser (1981)

[30] Gelfand, I.M., Vilenkin, N.Ya. *Some applications of harmonic analysis. Generalized functions.* **4** Moscow, Fiz.-Mat. Lit. (1961)

[31] Gihman, I.I., Skorohod, A.V. *Introduction in the theory or random processes.* Moscow, Nauka (1977)

[32] Gihman, I.I., Skorohod, A.V. *Stochastic differential equations and their apllications.* Kiev, Naukova Dumka (1982)

[33] Gihman, I.I., Skorohod, A.V. *Theory of stochastic processes.* **1-3** Moscow, Nauka (1975)

[34] Gruson, L. Théorie de Fredholm p-adique. *Bull. Soc. Math. France.* **94**, 67–95 (1966)

[35] Hennequin, P., Tortrat, A. *Probability theory and some its applications.* Moscow, Nauka (1974)

[36] Hensel, K. *Jahresber. Deutsch. Math. Ver.* **6: 1**, 83–88 (1899)

[37] Hewitt, E., Ross, A. *Abstract harmonic analysis.* Berlin, Springer (1979)

[38] Ikeda, N., Watanabe, S. *Stochastic differential equations and diffusion processes.* Moscow, Nauka (1986)

[39] Itô, K., McKean, H.P. *Diffusion processes and their sample paths.* Berlin, Springer (1996)

[40] Jang, Y. Non-Archimedean quantum mechanics. *Tohoku Mathem. Publications.* **10** Tohoku, Toh. Univ., Math. Inst. (1998)

[41] Khrennikov, A.Yu. Mathematical methods of non-archimedean physics. *Russ. Math. Surv.* **45: 4**, 79–110 (1990)

[42] Khrennikov, A.Yu. Generalized functions and Gaussian path integrals. *Izv. Acad. Nauk. Ser. Mat.* **55**, 780–814 (1991)

[43] Khrennikov, A.Yu. *Interpretations of probability.* Utrecht, VSP (1999)

[44] Khrennikov, A.Yu. Ultrametric Hilbert space representation of quantum mechanics with a finite exactness. *Found. Phys.* **26**, 1033–1054 (1996)

[45] Khrennikov, A.Yu. *Non-Archimedean analysis: quantum paradoxes, dynamical systems and biological models.* Dordrecht, Kluwer (1997)

[46] Khrennikov, A., Kozyrev, S.V. *Ultrametric random field. Infinite Dimensional Analysis, Quantum Probability and Related Topics.* **9: 2**, 199–213 (2006)

[47] Klingenberg, W. *Riemannian geometry.* Berlin, Walter de Gruyter (1982)

[48] Koblitz, N. *P-adic numbers, p-adic analysis and zeta functions.* New York, Springer-Verlag (1977)

[49] Kochubei, A.N. Pseudo-differential equations and stochastics over non-archimedean fields. *Monogr. Textbooks Pure Appl. Math.* **244** New York, Marcel Dekker, Inc. (2001)

[50] Kochubei, A.N. Limit theorems for sums of p-adic random variables. *Expo. Math.* **16**, 425–440 (1998)

[51] Kochubei, A.N. Analysis and probability over infinite extension of a local field, II: a multiplicative theory. In: Ultrametric functional analysis. Seventh International Conference on p-adic functional analysis, June 17-21, 2002, Univ. of Nijmegen, The Netherlands. *Contemporary Mathematics.* **319**, 167–177 (2003)

[52] Kolmogorov, A.N. *Foundations of the theory of probability.* New York, Chelsea Pub. Comp. (1956)

[53] Kunen, K. *Set theory.* Amsterdam, North-Holland Publ. Comp. (1980)

[54] Lang, S. *Algebra.* New York, Addison-Wesley (1965)

[55] Ludkovsky, S.V. Measures on groups of diffeomorphisms of non-archimedean Banach manifolds. *Russ. Math. Surveys.* **51: 2**, 338–340 (1996)

[56] Ludkovsky, S.V. Quasi-invariant measures on non-archimedean semigroups of loops. *Russ. Math. Surveys.* **53: 3**, 633–634 (1998)

[57] Ludkovsky, S.V. Embeddings of non-archimedean Banach manifolds into non-archimedean Banach spaces. *Russ. Math. Surv.* **53**, 1097–1098 (1998)

[58] Ludkovsky, S.V. Irreducible unitary representations of non-archimedean groups of diffeomorphisms. *Southeast Asian Mathem. Bull.* **22: 3**, 301–319 (1998)

[59] Ludkovsky, S.V. Poisson measures for topological groups and their representations. *Southeast Asian Bull. Math.* **25: 4**, 653–680 (2002)

[60] Ludkovsky, S.V. Properties of quasi-invariant measures on topological groups and associated algebras. *Annales Mathématiques Blaise Pascal.* **6: 1**, 33–45 (1999)

[61] Ludkovsky, S.V. Measures on groups of diffeomorphisms of non-archimedean manifolds, representations of groups and their applications. *Theoret. and Math. Phys.* **119: 3**, 698–711 (1999)

[62] Ludkovsky, S.V. Quasi-invariant measures on non-archimedean groups and semigroups of loops and paths, their representations. I, II. *Annales Mathématiques Blaise Pascal.* **7: 2**, 19–53, 55–80 (2000)

[63] Ludkovsky, S.V. Non-Archimedean polyhedral expansions of ultrauniform spaces. *Russ. Math. Surv.* **54: 5**, 163–164 (1999)

[64] Ludkovsky, S.V. Non-Archimedean polyhedral decompositions of ultrauniform spaces. *Fundam. i Prikl. Math.* **6: 2**, 455–475 (2000)

[65] Ludkovsky, S.V. Semidirect products of loops and groups of diemorphisms of real, complex and quaternion manifolds and their representations. In the book: *Focus on Groups Theory Research.* Editor L.M. Ying New York, Nova Science Publishers 59–136 (2006). Parallel publication in: *Intern. J. of Mathem. Game Theory and Algebra.* **16: 4**, 289–357 (2006)

[66] Ludkovsky, S.V. Topological groups of transformations of manifolds over non-archimedean fields, their representations and quasi-invariant measures. I, II. *J. Mathem. Sci.I.* **147: 3**, 6703–6846 (2008)

II. **150: 4**, 2123–2223 (2008) (English translation from: *Sovrem. Mathem. and its Applications.* **39** (2006); *Sovrem. Mathem. Fundam. Napravl.* **18**, 5-100 (2006))

[67] Ludkovsky, S.V. Stochastic processes on groups of diffeomorphisms and loops of real, complex and non-archimedean manifolds. *Fundam. i Prikl. Mathem.* **7: 4**, 1091–1105 (2001)

[68] Ludkovsky, S.V. Representations of topological groups generated by Poisson measures. *Russ. Math. Surv.* **56: 1**, 169-170 (2001)

[69] Ludkovsky, S.V. Quasi-invariant and pseudo-differentiable real-valued measures on a non-archimedean Banach space. *Analysis Math.* **28**, 287-316 (2002)

[70] Ludkovsky, S.V. Differentiability of functions: approximate, global and differentiability along curves over non-archimedean fields. *J. Mathem. Sci.* **157: 2**, 311–366 (2009) (transl. from: *Contemporary Mathem. and its Applic.* **52**, Functional Anal.)

[71] Ludkovsky, S.V. Stochastic processes on non-archimedean Banach spaces. *Int. J. of Math. and Math. Sci.* **2003: 21**, 1341–1363 (2003)

[72] Ludkovsky, S.V. Stochastic antiderivational equations on non-archimedean Banach spaces. *Int. J. of Math. and Math. Sci.* **2003: 41**, 2587–2602 (2003)

[73] Ludkovsky, S.V. Stochastic processes on totally disconnected topological groups. *Int. J. of Math. and Math. Sci.* **2003: 48**, 3067–3089 (2003)

[74] Ludkovsky, S.V. Stochastic processes and antiderivational equations on non-archimedean manifolds. *Int. J. of Math. and Math. Sci.* **31: 1**, 1633–1651 (2004)

[75] Ludkovsky, S.V. Non-Archimedean valued quasi-invariant descending at infinity measures. *Int. J. of Math. and Math. Sci.* **2005: 23**, 3799–3817 (2005)

[76] Ludkovsky S.V. Quasi-invariant and pseudo-differentiable measures on non-archimedean Banach spaces with values in non-archimedean fields. *J. Math. Sci.* **122: 1**, 2949–2983 (2004)

[77] Ludkovsky, S.V. Quasi-invariant and pseudo-differentiable measures on non-archimedean Banach spaces. *Russ. Math. Surv.* **58: 2**, 167–168 (2003)

[78] Ludkovsky, S.V. Infinitely divisible distributions over locally compact non-archimedean fields. *Far East Journal of Theoretical Statistics.* **27: 1**, 1–40 (2009)

[79] Ludkovsky, S.V. Spectral functions of stochastic processes over non-archimedean fields. *Indian J. Math.* **52: 1**, 115-151 (2010)

[80] Ludkovsky, S.V. Stochastic processes in infinite-dimensional spaces over non-archimedean fields and their spectral representations. *Indian. J. Math.* **52: 1**, 153-184 (2010)

[81] Ludkovsky, S.V. Quasi-invariant and Pseudo-differentiable Measures in Banach Spaces. New York, Nova Science Publishers, Inc. (2009)

[82] Ludkovsky, S.V. Groups of diffeomorphisms and wraps of manifolds over non-archimedean fields. In the book: *Lie Groups: New Research.* Editor A.B. Canterra. New York, Nova Science Publishers, Inc., (2009)

[83] Ludkovsky, S.V., Diarra, B. Spectral integration and spectral theory for non-archimedean Banach spaces. *Int. J. of Math. and Math. Sciences.* **31: 7**, 421–442 (2002)

[84] Ludkovsky, S., Khrennikov, A. Stochastic processes on non-archimedean spaces with values in non-archimedean fields. *Markov Processes and Related Fields.* **9: 1**, 131–162 (2003)

[85] Malliavin, P. *Stochastic analysis.* Berlin, Springer (1997)

[86] Mc Kean, H.P. *Stochastic integrals.* Moscow, Mir (1972)

[87] Narici, L., Beckenstein, E. *Topological vector spaces.* New York, Marcel-Dekker (1985)

[88] Neeb, K.-H. On a theorem of S. Banach. *J. of Lie Theory.* **8**, 293–300 (1997)

[89] Øksendal, B. *Stochastic differential equations.* Berlin, Springer (1995)

[90] Ostrowski, A. Untersuchungen zur arithmetischen Theorie der Körper. *Math. Zeit.* **39**, 269–404 (1935)

[91] *P*-Adic Mathematical Physics. 2nd International Conference (Belgrad, 2005). Editors: Khrennikov, A.Yu., Rakić, Z., Volovich, I.V. *AIP Conference Proceedings* **826**, New York (2006)

[92] Partasarathy, K.R., Rao Ranga, R., Varadhan, S.R.S. Probability distributions on locally compact abelian groups. *Illinois J. Math.* **7**, 337–369 (1963)

[93] Partasarathy, K.R. Probability measures on metric spaces. *Probab. Math. Statist.* **3** New York, Academic Press, Inc. (1967)

[94] Petrov, V.V. *Sums of independent random variables.* Moscow, Nauka (1987)

[95] Pietsch, A. *Nukleare lokalkonvexe Räume.* Berlin, Akademie-Verlag (1965)

[96] Put, M. van der. The ring of bounded operators on a non-archimedean normed linear space. *Indag. Math.* **71: 3**, 260–264 (1968)

[97] Rachdi, M., Monsan, V. Spectral density estimation for *p*-adic stationary process. *Ann. Math. B. Pascal.* **5: 1**, 25–41 (1998)

[98] Reed, M., Simon, B. *Methods of Modern Mathematical Physics.* **1**. *Functional Analysis.* New York, Academic Press (1977)

[99] Rooij, A.C.M. van. *Non-Archimedean fucntional analysis.* Ser. Pure and Appl. Math. **51** New York, Marcel Dekker (1978)

[100] Rooij, A.C.M. van, Schikhof, W.H. Non-Archimedean commutative C^*-algebras. *Indag. Math.* **35**, 381–389 (1973)

[101] Sato, T. Wiener measure on certain Banach spaces over non-archimedean local fields. *Compositio Math.* **93**, 81–108 (1994)

[102] Schaefer, H. *Topological vector spaces.* Moscow, Mir (1971)

[103] Schikhof, W.H. *Ultrametric calculus.* Cambridge, Cambidge University Press (1984)

[104] Schikhof, W.H. *Non-Archimedean calculus. Nijmegen: Math. Inst., Cath. Univ., Report* **7812**, (1978)

[105] Shimomura, H. Poisson measures on the configuration space and unitary representations of the group of diffeomorphisms. *J. Math. Kyoto Univ.* **34**, 599–614 (1994)

[106] Shirjaev, A.N. *Probability.* Moscow, Nauka (1989)

[107] Swan, R.C. The Grothendieck ring of a finite group. *Topology* **2**, 85–110 (1963)

[108] Vahanija, N.N., Tarieladze, V.I., Chobanjan, S.A. *Probability distributions in Banach spaces.* Moscow, Nauka (1985)

[109] Vladimirov, V.S., Volovich, I.V. *Comm. Math. Phys.* **123**, 659–676 (1989)

[110] Vladimirov, V.S., Volovich, I.V., Zelenov, E.I. *P-adic analysis and mathematical physics.* Moscow, Fiz.-Mat. Lit. (1994)

[111] Weil, A. *Basic number theory.* Berlin, Springer (1973)

[112] Yasuda, K. Semi-stable processes on local fields. *Tohoku Math. J. V.* **58**, 419–431 (2006)

Index